SPSS for Windows Step by Step

A Simple Guide and Reference
13.0 Update

Sixth Edition

Darren George
Canadian University College

Paul Mallery
La Sierra University

Boston New York San Francisco
Mexico City Montreal Toronto London Madrid Munich Paris
Hong Kong Singapore Tokyo Cape Town Sydney

ISBN 0-205-48071-3

Printed in the United States of America

10 9 8 7 6 5 4 3 09 08 07 06

To Elizabeth

– D.G.

To my parents, Lynn and Janet

– P.M.

SPSS for Windows Step By Step

A Simple Guide and Reference

Table of Contents

BASE SYSTEM MODULE

PREFACE

SPSS is a powerful tool that is capable of conducting just about any type of data analysis used in the social sciences, the natural sciences, or in the business world. While mathematics is generally thought to be the language of science, data analysis is the language of research. Research in many fields is critical for human progress, and as long as there is research, there will be the need to analyze data. The present book is designed to make data analysis more comprehendible and less toxic.

In our teaching, we have frequently encountered students so traumatized by the professor who cheerily says "Analyze these data on SPSS; get the manuals if you don't know how," that they dropped the course rather than continue the struggle. It is in response to this anguish that the present book was conceived. Darren George's background has been teaching high school mathematics, and Paul Mallery worked his way through college training people to use computers and programming computers. Both of us find great pleasure in the challenge of making a process that is intrinsically complex as clear as possible. The ultimate goal in all our efforts with the present book has been to make SPSS procedures, above all else, clear.

As the book started to take shape, a second goal began to emerge. In addition to making SPSS procedures clear to the beginner, we wanted to create a tool that was an effective reference for anyone conducting data analysis. This involved the expansion of the original concept to include essentially all major statistical procedures that SPSS covers in the base module and much of the advanced and regression modules as well. The result of years of effort you now hold in your hands.

The most significant changes in this edition are designed to make the book friendlier. For example, one of the great new features in SPSS version 13 is the splitter control in the data editor, so you can see you ID numbers, names, or demographics along the left column at the same time you see variables far to the right in your file; this is described in Chapter 3. A major visual overhaul has been completed with this edition; nearly all screens and most of the output has been updated to more closely match the SPSS windows and output. In all, 226 screen shots and output tables have been updated.

While the first 16 chapters of the book cover basic topics and would be understandable to many with very limited statistical background, the final 12 chapters involve procedures that progressively require a more secure statistical grounding. Those 12 chapters have provided our greatest challenge. At the beginning of each chapter we spend several pages describing the procedure that follows. But, how can one adequately describe, for instance, factor analysis or discriminant analysis in five or six pages? The answer is simple: The procedures must be described at a common sense, conceptual level that avoids excessive detail and excessive emphasis on computation. However, writing those introductions has not been at all simple. The chapter introductions are the most painstakingly worked sections of the entire book. Although we acknowledge the absence of much detail in our explanation of most procedures, we feel that we have done an adequate job at a project that few would even attempt. How successful have we been at achieving clarity in very limited space? You, the reader, will be our ultimate judge.

SPSS Inc. has produced several manuals to describe everything that their package of programs attempts to accomplish. These volumes, over 3000 pages of fine print documentation, are, if nothing else, comprehensive. Anything that SPSS is able to do is described in the manuals. For the experienced researcher, ownership of the manuals is required. We cannot cover in 380 pages what SPSS does in 3000. However, we are convinced that 95% of analyses that are *ever* conducted in the sciences or business could be accomplished with the information presented in our book. For the additional 5% we frequently refer the reader to the SPSS manuals when the level of specificity required extends beyond the scope of the present volume.

AUTHORS' BIOGRAPHICAL SKETCHES AND PRESENT ADDRESSES

Darren George is currently a professor of Psychology at

Canadian University College
4914 College Avenue
Lacombe, AB, T4L 1Z2
403-782-3381, Ext. 4082
dgeorge@cauc.ca

where he teaches personality psychology, social psychology, and research methods. He completed his M.A. in Experimental Psychology (1982) at California State University, Fullerton; taught high school mathematics for nine years (1980-1989) at Mark Keppel High School (Alhambra, CA) and Mountain View High School (El Monte, CA); then completed a Psychology Ph.D. at UCLA (1992) with emphases in personality psychology, social psychology, and measurement & psychometrics.

Paul Mallery is currently an associate professor of Psychology at

La Sierra University
4500 Riverwalk Parkway
Riverside, CA, 92515
951-785-2528
pmallery@lasierra.edu

where he teaches social psychology and related courses, and experimental methodology (including the application of SPSS). He received his Ph.D. in Social Psychology from UCLA (1994), with emphases in statistics and political psychology. Paul formerly worked as a computer specialist, both programming and teaching computer usage.

ACKNOWLEDGMENTS

As we look over the creative efforts of the past years, we wish to acknowledge several people we have never met. These are individuals who have reviewed our work and offered invaluable insight and suggestions for improvement. Our gratitude is extended to Richard Froman of John Brown University, Michael A. Britt of Marist College, Marc L. Carter of the University of South Florida, and Randolph A. Smith of Ouachita Baptist University. And then there's the standard (but no less appreciated) acknowledgment of our families and friends who endured us while we wrote this. Particular notice goes to Marcus George (son of the first author) who contributed substantially to the design and format of the present volume, to our wives Elizabeth George and Suzanne Mallery for their support and encouragement, and to our children who figure out ways to spend the royalty checks.

An Overview of SPSS for Windows Step by Step

THIS BOOK is designed to give you the step-by-step instructions necessary to do most major types of data analysis using SPSS for Windows. This software was originally created by three Stanford graduate students in the late 1960's, and SPSS (once "Statistical Package for the Social Sciences," then "Statistical Product and Service Solutions," and now just "SPSS"), a Chicago-based firm, has grown to be one of the world's largest statistical software companies.

NECESSARY SKILLS

For this book to be effective when you conduct data analysis with SPSS, you should have certain limited knowledge of statistics and a general acquaintance with the use of a computer. Each issue is addressed in the next two paragraphs.

Statistics: You should have had at least a basic course in statistics or be in the process of taking such a course. While it is true that this book devotes the first two or three pages of each chapter to a description of the statistical procedure that follows, these descriptions are designed to refresh the reader's memory, *not* to instruct the novice. While it is certainly possible for the novice to follow the steps in each chapter and get SPSS to produce pages of output, a fundamental grounding in statistics is important for an understanding of which procedures to use and what all the output means. In addition, while the first 16 chapters should be understandable by individuals with limited statistical background, the final 12 chapters deal with much more complex and involved types of analyses. These chapters require substantial grounding in the statistical techniques involved.

Computer knowledge: Your knowledge of the computer may be quite limited. The following, however, are necessary. You must:

- Have access to a personal computer that has
  - Windows 98, Windows 2000, ME, NT 4.0, or XP installed
  - SPSS for Windows Release 13 installed
- Know how to turn the computer on
- Have a working knowledge of the keys on the keyboard and how to use a mouse

This book will take you the rest of the way. If you are using SPSS on a network of computers (rather than your own PC) the steps necessary to *access* SPSS for Windows may vary slightly from the single step shown in the pages that follow.

SCOPE OF COVERAGE

SPSS for Windows is a complex and powerful statistical program by any standards. The software occupies about 200MB of your hard drive, and requires at least 128MB of RAM to operate adequately. If you didn't understand the last sentence, don't worry! Despite its size and complexity, SPSS has created a program that is not only powerful but is very user friendly (you're the user, the program tries to be friendly). By creating the windows version, SPSS has done for data analysis what Henry Ford did for the automobile: made it available to the masses. SPSS is able to perform essentially any type of statistical analysis ever used in the social sciences, in the business world, and in other scientific disciplines.

This book was written for Version 13 of SPSS for Windows. More specifically, the screen shots and output are based on Version 13.0. With some exceptions, what you see here will be similar to SPSS Version 7.0 and higher. Because only a few parts of SPSS are changed with each version, most of this book will apply to previous versions. It's 100% up-to-date with version 13.0, but it will only lead you astray about 3% of the time if you're using version 12.0, and it's still 85% accurate for version 7.0 (if you can find a computer that old).

Our book covers the statistical procedures present in the three *modules* created by SPSS that are most frequently used by researchers. A module (within the SPSS context) is simply a set of different statistical operations. We include the **Base System Module**, the module covering **Advanced Models**, and the module that addresses **Regression Models**—all described in greater detail later in this chapter. To support their program, SPSS has created a set of comprehensive manuals that cover all procedures these three modules are designed to perform. To a person fluent in statistics and data analysis, the manuals are well written and intelligently organized. To anyone less fluent, however, the organization is often undetectable, and the comprehensiveness (about 3,000 pages of fine-print text) is overwhelming. Our book is about 380 pages long. Clearly we cannot cover in 380 pages as much material as the manuals do in 3,000, but herein lies our major advantage.

The purpose of this book is to make the fundamentals of most types of data analysis clear. To create this clarity requires the omission of much (often unnecessary) detail. Despite brevity, we have been keenly selective in what we have included and believe that the material presented here is sufficient to provide simple instructions that cover 95% of analyses ever conducted by social science researchers. Although we cannot substantiate that exact number, our time in the manuals suggests that at least 2,000 of the 3,000 pages involve detail that few researchers ever consider. How often do you really need 7 different methods of extracting and 6 methods of rotating factors in factor analysis, or 18 different methods for post-hoc comparisons after a one-way ANOVA? (By the way, that last sentence should be understood by statistical geeks only.)

We are in no way critical of the manuals; they do well what they are designed to do and we regard them as important adjuncts to the present book. When our space limitations prevent explanation of certain details, we often refer our readers to the SPSS manuals. Within the context of presenting a statistical procedure, we often show a window that includes several options but describe only one or two of them. This is done without apology except for the occasional "description of these options extends beyond the scope of this book" and cheerfully refer you to the appropriate SPSS manual. The ultimate goal of this format is to create clarity without sacrificing necessary detail.

OVERVIEW

This chapter introduces the major concepts discussed in this book and gives a brief overview of the book's organization and the basic tools that are needed in order to use it.

If you want to run a particular statistical procedure, have used SPSS for Windows before, and already know which analysis you wish to conduct, you should read the Typographical and Formatting Conventions section in this chapter (pages 6-8) and then go to the appropriate chapter in the last portion of the book (Chapters 6 through 28). Those chapters will tell you exactly what steps you need to perform to produce the output you desire.

If, however, you are new to SPSS for Windows, then this chapter will give you important background information that will be useful whenever you use this book.

THIS BOOK'S ORGANIZATION, CHAPTER BY CHAPTER

This book was created to describe the crucial concepts of analyzing data. There are three basic tasks associated with data analysis:

A. You must type data into the computer, and organize and format the data so both SPSS and you can identify it easily,
B. You must tell SPSS what type of analysis you wish to conduct, and
C. You must be able to interpret what the SPSS output means.

After this introductory chapter, Chapter 2 deals with basic operations such as types of SPSS windows, the use of the toolbar and menus, saving, viewing and editing the output, printing output, and so forth. While this chapter has been created with the beginner in mind, there is much SPSS-specific information that should be useful to anyone. Chapter 3 addresses the first step mentioned above—creating, editing, and formatting a data file. The SPSS data editor is an instrument that makes the building, organizing, and formatting of data files wonderfully clear and straightforward.

Chapters 4 and 5 deal with two important issues—modification and transformation of data (Chapter 4), and creation of graphs or charts (Chapter 5). Chapter 4 deals specifically with different types of data manipulation, such as creating new variables, reordering, restructuring, merging files, or selecting subsets of data for analysis. Chapter 5 introduces the basic procedures used when making a number of different graphs; some graphs, however, are described more fully in the later chapters.

Chapters 6 through 28 then address Steps B and C—analyzing your data and interpreting the output. It is important to note that each of the analysis chapters is self-contained. If the beginner, for example, were instructed to conduct *t* tests on certain data, Chapter 11 would give complete instructions for accomplishing that procedure. In the Step by Step section, Step 1 is always "start the SPSS program" and refers the reader to Chapter 2 if there are questions about how to do this. The second step is always "create a data file or edit (if necessary) an already existing file," and the reader is then referred to Chapter 3 for instructions if needed. Then the steps that follow explain exactly how to conduct a *t* test.

As mentioned previously, this book covers three basic modules produced by SPSS: **Base System**, **Advanced Models**, and **Regression Models**. Since some computers at colleges or universities may not have all of these modules (the base-system module is always present), we organize the book according to the structure SPSS has imposed: In this book we cover ALL procedures included in the Base System module and then selected procedures from the more complex Advanced Models and Regression Models. Chapters 6-22 deal with processes included in the Base System module. Chapters 23-27 deal with procedures in the Advanced and Regression Models, and Chapter 28, the analysis of residuals, draws from all three modules.

Base System Module: Chapters 6 through 10 describe the most fundamental data analysis methods available, including frequencies, bar charts, histograms, and percentiles (Chapter 6); descriptive statistics such as means, medians, modes, skewness, and ranges (Chapter 7); crosstabulations and chi-square tests of independence (Chapter 8); subpopulation means (Chapter 9); and correlations between variables (Chapter 10).

The next group of chapters (Chapters 11 through 17) explains ways of testing for differences between subgroups within your data or showing the strength of relationships between a dependent variable and one or more independent variables through the use of *t* tests (Chapter 11), ANOVAs (Chapters 12, 13, and 14); linear, curvilinear, and multiple regression analysis (Chapters 15 and 16); and the most common forms of nonparametric tests are discussed in Chapter 17.

Reliability analysis (Chapter 18) is a standard measure used in research that involves multiple response measures, multidimensional scaling is designed to identify and model the structure and dimensions of a set of stimuli from dissimilarity data (Chapter 19), then factor analysis (Chapter 20), cluster analysis (Chapter 21), and discriminant analysis (Chapter 22) all occupy stable and important niches in research conducted by scientists.

Advanced and Regression Models: The next series of chapters deals with analyses that involve multiple *dependent* variables (SPSS calls these procedures General Linear Models; they are also commonly called MANOVAs or MANCOVAs). Included under the heading General Linear Model are simple and general factorial models and multivariate models (Chapter 23), and models with repeated measures or within-subjects factors (Chapter 24).

The next three chapters deal with procedures that are only infrequently performed, but they are described here because when these procedures are needed they are indispensable. Chapter 25 describes logistic regression analysis and Chapters 26 and 27 describe hierarchical and nonhierarchical log-linear models, respectively. As mentioned previously, Chapter 28 on residuals closes out the book.

AN INTRODUCTION TO THE EXAMPLE

A single example is used in 17 of the first 19 chapters of this book. For more complex procedures it has been necessary to select different examples to reflect the particular procedures that are presented. Examples are useful because often, things that appear to be confusing in the SPSS documentation become quite clear when you see an example of how they are done. Although only the most frequently used example is described here, there are a total of twelve data sets that are used to demonstrate procedures throughout the book, in addition to datasets used in the exercises. Data files are available for download at **www.ablongman.com/george6e**. These files can be of substantial benefit to you as you practice some of the processes presented here without the added burden of having to input the data. We suggest that you make generous use of these files by trying different procedures and then comparing your results with those included in the output sections of different chapters.

The example has been designed so that it may be used to demonstrate most of the statistical procedures presented here. It consists of a single data tile used by a teacher who teaches three sections of a class with approximately 35 students in each section. For each student, the following information is recorded:

- ID number
- Name
- Gender
- Ethnicity
- Year in school

- Upper- or lower-division classperson
- Previous GPA
- Section
- Whether or not he or she attended review sessions or did the extra credit
- The scores on five 10-point quizzes and one 75-point final exam

In Chapter 4 we describe how to create four new variables. In all presentations that follow (and on the data file available on the website), these four variables are also included:

- The total number of points earned
- The final percent
- The final grade attained
- Whether the student passed or failed the course

The example data file (the entire data set is displayed at the end of Chapter 3) will also be used as the example in the introductory chapters (Chapters 2 through 5). If you enter the data yourself and follow the procedures described in these chapters, you will have a working example data file identical to that used through the first half of this book. Yes, the same material is recorded on the downloadable data files, but it may be useful for you to practice data entry, formatting, and certain data manipulations with this data set. If you have your own set of data to work with, all the better.

One final note: All of the data in the **grades** file are totally fictional, so any findings exist only because we created them when we made the file.

TYPOGRAPHICAL and FORMATTING CONVENTIONS

Chapter organization: Chapters 2 through 5 describe SPSS for Windows formatting and procedures, and the material covered dictates each chapter's organization. Chapters 6 through 28 (the analysis chapters) are, with only occasional exceptions, organized identically. This format includes:

1. The **Introduction** in which the procedure that follows is described briefly but concisely. These introductions vary in length from one to seven pages depending on the complexity of the analysis being described.
2. The **Step by Step** section in which the actual steps necessary to accomplish particular analyses are presented. Most of the typographical and formatting conventions described below refer to the Step by Step sections.
3. The **Output** section, in which the results from analyses described earlier are displayed—often abbreviated. Text clarifies the meaning of the output, and all of the critical output terms are defined.

The screens: Due to the very visual nature of SPSS, every chapter contains pictures of *screens* or *windows* that appear on the computer monitor as you work. The first picture from Chapter 6 (following page) provides an example. These pictures are labeled "Screens" despite the fact that sometimes what is pictured is a screen (everything that appears on the monitor at a given time) and other times is a *portion* of a screen (a window, a dialog box, or something smaller). If the reader sees reference to Screen 13.3, she knows that this is simply the third picture in Chapter 13. The screens are typically positioned within breaks in the text (the screen icon and a title are included) and are used for sake of reference as procedures involving that screen are described. Sometimes the screens are separate from the text and labels identify certain characteristics of the screen (see the inside front cover for an example). Because screens take up a lot of space, frequently used screens are included on the inside front and back covers of this book. At other times, within a particular chapter, a screen from a different chapter may be cited to save space.

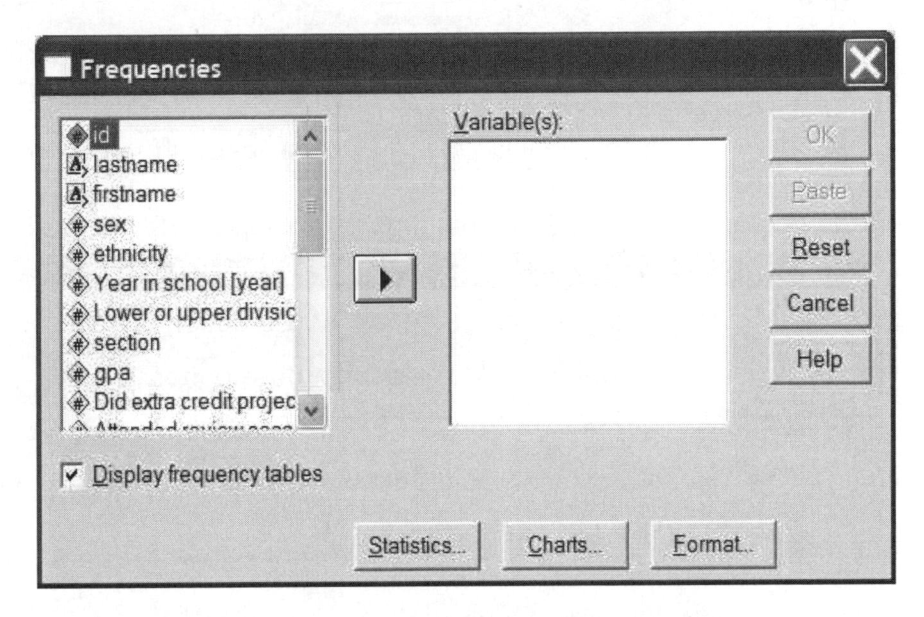

Sometimes a portion of a screen or window is displayed (such as the menu bar included here)

and is embedded within the text without a label.

The Step by Step boxes: Text that surrounds the screens *describes* a procedure, but it is the step by step boxes that identify exactly what must be done to *do* a procedure. The following box illustrates:

In Screen	Do This		Step 3 (sample)
Front1	**File** → **Open** → **Data**	[or]	
Front2	type `grades.sav` → **Open**	[or **grades.sav**]	Data

Sequence Step 3 means: "Beginning with Screen 1 (displayed on the inside front cover), click on the word **File,** move the cursor to **Open**, then click the word **Data**. At this point a new window will open (Screen 2 on the inside front cover); type the words `grades.sav` then click the **Open** button, at which point a screen with your data file opens." Notice that within brackets shortcuts are sometimes suggested: Rather than the **File** → **Open**,→ **Data** sequence, it is quicker to click the icon. Instead of typing `grades.sav` then clicking **Open**, it is quicker to double click on the **grades.sav** file name. Items within Step-by-Step boxes include:

Screens: A small screen icon will be placed to the left of each group of instructions that are based on that screen. There are three different types of screen icons:

Type of Screen Icon	Example Icon	Description of Example
Inside Cover Screens	Front1	Screen #1 on the inside front cover
General Screens	Menu	Any screen with the menu bar across the top
	Graph	Any screen that displays a graph or chart
Chapter Screens	4.3	The third screen in Chapter 4
	21.4	The fourth screen in Chapter 21

Other Images with special meaning inside of Step by Step boxes include:

Image	What it Means
(single mouse click icon)	A single click of the left mouse button
(double mouse click icon)	A double-click of the left mouse button
type	A type icon appears before words that need to be typed
press	A press icon appears when a button such as the **TAB** key needs to be pressed
→	Proceed to the next step.

Sometimes fonts can convey information, as well:

Font	What it Means
`Monospaced font (Courier)`	Any text within the boxes that is rendered in the `Courier` font represents text (numbers, letters, words) to be *typed* into the computer (rather than being clicked or selected).
Italicized text	*Italicized text* is used for information or clarifications within the Step by Step boxes.
Bold font	The **bold font** is used for words that appear on the computer screen.

The groundwork is now laid. We wish you a pleasant trip through the exciting and challenging world of data analysis!

SPSS Windows Processes

WE MENTIONED in the introductory chapter that it was necessary for the user to understand how to turn the computer on and get as far as the Windows desktop. This chapter will give you the remaining skills required to use SPSS for Windows: How to use the mouse, how to navigate using the taskbar, what the various buttons (on the toolbar and elsewhere) do, and how to navigate the primary windows used in SPSS.

If you are fluent with computers, you may not need to read this chapter as carefully as if you are new to computers. But every one should read at least portions of this chapter carefully; it contains a great deal of information unique to SPSS for Windows.

THE MOUSE

Most computers today have a mouse attached to them. Not since the days of the Pied Piper of Hamelin (A.D. 1376) have mice proliferated so rapidly. Three different types of mouse operations are used frequently in this book:

- **Point and click** (or single click) means to position the point of the arrow on the word, symbol, or icon that you desire, then press and release the left mouse button.
- **Double click** refers to holding the mouse so the arrow is positioned correctly and rapidly pressing the left mouse button twice.
- **Dragging** refers to selecting several items with one mouse operation. If, for instance, there is a list of words and you wish to select several of them, you may drag the mouse over all of the words. To do this, position the arrow on the first word and press the left mouse button, continue to hold the button down while you move the arrow down to the last desired word, and then release the mouse button. This will highlight all the words or text you wish for further operations.

Mice usually have at least two buttons (left and right). Many have a third (center) button, and some have dials as well. Within the context of this book, the left mouse button is used so extensively that a click or a double click always refers to a left-mouse-button operation. On the occasion when the right button is clicked, we will specify that in words. As mentioned in the first chapter a small picture of a mouse ( ) always represents the mouse click within Step by Step boxes.

Feel free to experiment with the right mouse button. In SPSS, when you position the cursor on an object or word, a right click may do nothing but often one of two different things occur:

- It provides a brief summary or definition of whatever you right clicked. This is particularly useful when you do not know the definition of a term or when you need to quickly access variable and value-label information. If you can't remember how you coded gender or ethnicity or can't recall the meaning of the variable named "cntbusb", a right click can provide quick relief.
- It provides a short pop-up menu of commands you can select, such as cut, copy, paste, delete, or others.

There are a few occasions in which moving the mouse cursor over something on the screen will provide either more information (in the case of toolbar icons), or will produce a submenu (in the case of the taskbar) that will cause additional information to pop up.

THE TASKBAR AND START MENU

Once you have arrived at the Windows desktop, your screen should look something like that shown below. It will certainly not look *exactly* like this, but it will be similar. There will typically be a number of icons along the left side of the screen, and a bar across the bottom (or top) of the screen (with the word "Start" on the left, and the time on the right).

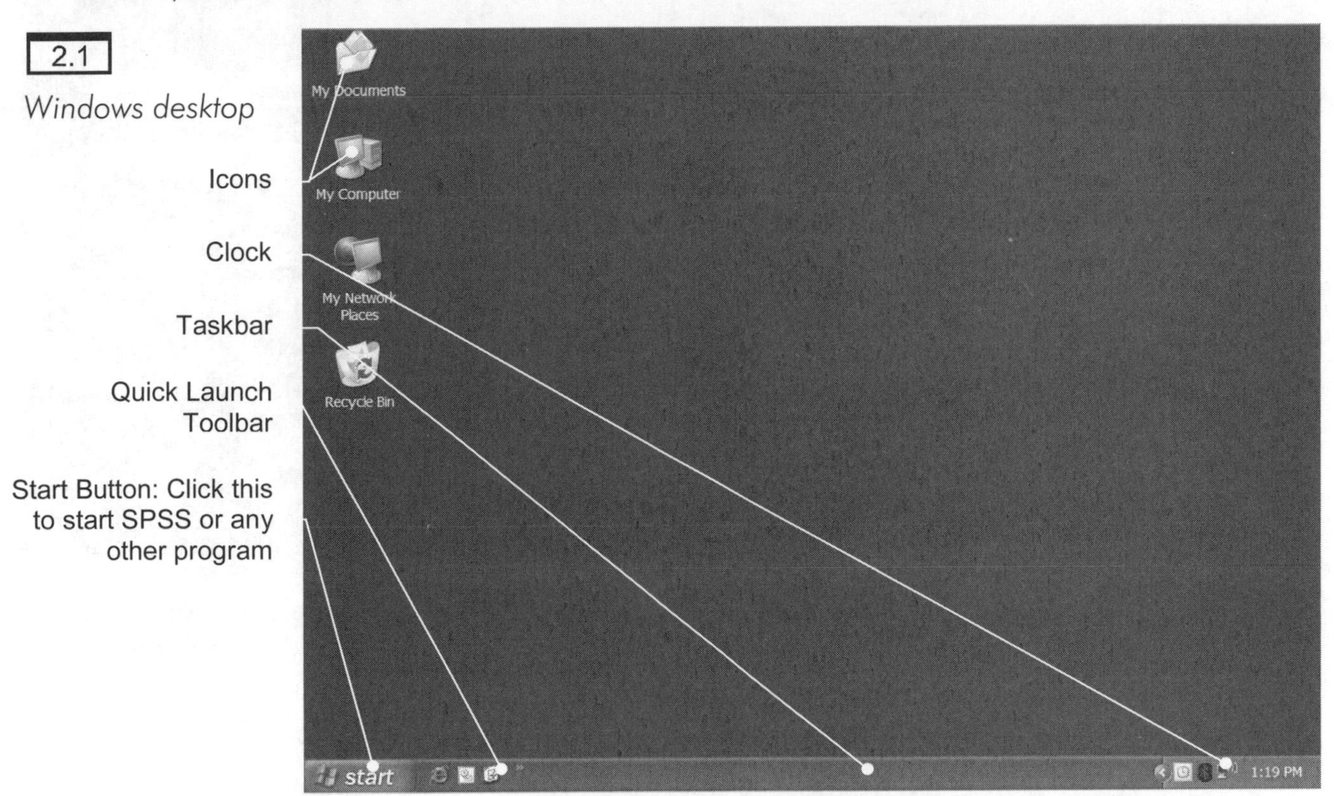

2.1

Windows desktop

There are two main types of icons on the Windows desktop: **Program** icons represent a particular program, while **folder** icons actually contain other icons (usually several programs that are related in some way).

The most important thing you need to know about the Windows desktop (at least as long as you are reading this book) is how to start the SPSS program. To do this on most computers, you need to click the start button, move the cursor over the All Programs menu folder, and then over the SPSS for Windows program icon. Once SPSS 13.0 for Windows emerges, click on the icon, and SPSS will begin. On most computers, the Windows desktop will look similar to that shown in Screen 2.2 (following page) immediately before you click on SPSS 13.0 for Windows.

One word of warning: On some computers, the SPSS program icon may be in a different location within the Start menu. You may have to move the cursor around the Start menu (look especially for any folders labeled "SPSS"). Don't worry, though; if you look around enough you will find the icon. Occasionally, the icon is on the Windows desktop (along the left side), and you don't have to use the Start button at all.

In addition to starting the SPSS program, the other important required skill that you must possess when using the Taskbar is changing between programs. This is especially important because SPSS is actually a collection of several programs. When you first start the SPSS system, the Data Editor program is started, but as soon as you perform some statistics on your data, another program (in another window) is started: the Output program. Depending on your SPSS settings and what you are doing, you may have other SPSS windows open, but the Data Editor and Output windows are the primary ones you will use.

2.2

View of the Windows desktop Start menu immediately before clicking to start the SPSS program.

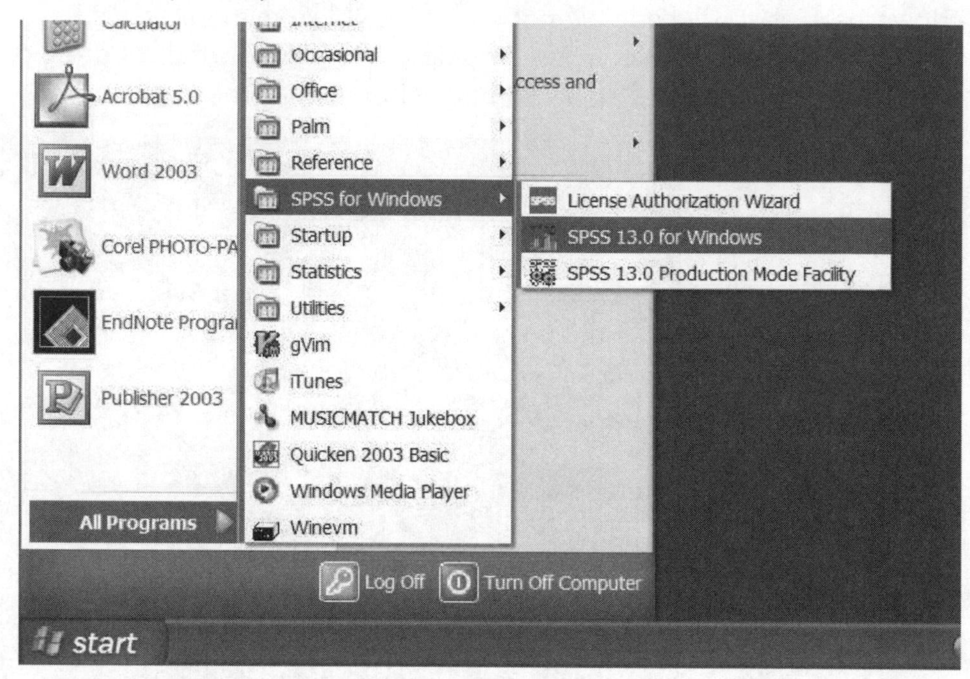

Sometimes, SPSS will change between windows automatically; other times, you may need to change between the windows yourself. This is the other option that the taskbar provides.

When several programs are running at the same time, each program will have a button on the taskbar. Here's what the taskbar looks like when the SPSS data editor and SPSS output programs are both running:

If you want to change from one program window to the other, simply move the cursor down to the taskbar and click the appropriate button.

COMMON BUTTONS

A number of buttons occur frequently within screens and around the borders of windows and screens. The standard push buttons have the same function in any context. The most frequently used ones are identified and explained below:

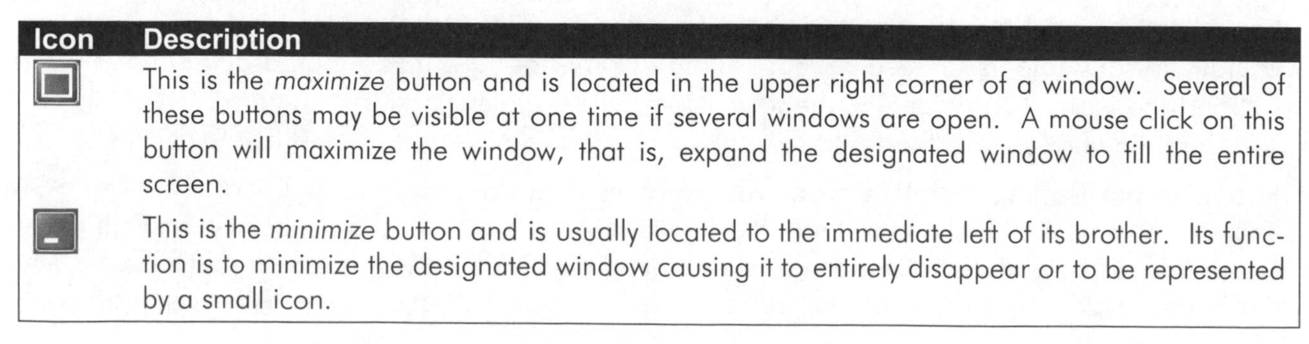

Icon	Description
[maximize icon]	This is the *maximize* button and is located in the upper right corner of a window. Several of these buttons may be visible at one time if several windows are open. A mouse click on this button will maximize the window, that is, expand the designated window to fill the entire screen.
[minimize icon]	This is the *minimize* button and is usually located to the immediate left of its brother. Its function is to minimize the designated window causing it to entirely disappear or to be represented by a small icon.

Icon	Description
	This is called the *restore* button. A click on this button will reduce the size of the window with which it is associated. Sometimes this reduction will result in additional windows becoming visible. This allows you to see and then click on whichever window you wish to use.
	Within the SPSS context, this button will usually paste (move text from one location to another) a designated variable from a list in one location to an *active box* in another location. Such a move indicates that this variable is to be used in an analysis.
	This button is the reverse. It will move a variable from the active box back to the original list.
	These are up, down, right, and left scroll arrows that are positioned on each end of a scroll bar. To illustrate, see Screen 2.3 in which both vertical and horizontal scroll bars are identified. Most data or output files are longer than the open window and in many instances also wider. The scroll arrows and scroll bar allow you to view either above, below, to the right, or to the left of what is visible on the screen. A quicker method of scrolling is to click and drag on the small box within the scroll bar.
	A click on a down arrow will reveal a menu of options (called a drop-down menu) directly below. Screen 2.4 (page 16) shows such buttons under the title **Files of type**.

THE DATA WINDOW and OTHER COMMONLY USED WINDOWS

What follows are pictures and descriptions of the most frequently used windows or screens in SPSS processes. These include:

1. The initial screen on entry into SPSS that includes detail concerning the meaning of each icon and a brief description of the function of each command,
2. The **Open File** dialog which identifies several different ways to access a previously created file, and
3. A main dialog box, which, although different for each procedure, has certain similarities that will be highlighted.

Following these three presentations the initial window that appears following completion of data analysis (the Output Window) and instructions dealing with how to edit and manipulate output prior to printing results are presented. This chapter concludes with a description of a number of printing options. We now begin with the screen that appears when you first enter SPSS.

THE INITIAL SCREEN, ICON DETAIL, and MEANING OF COMMANDS

To access the initial SPSS screen from the Windows display, perform the following sequence of steps:

Screen 2.3 (shown on the following page) is a slightly modified version of Screen 1 on the inside front cover. When you start the SPSS program, Screen 2.3 is the first screen to appear:

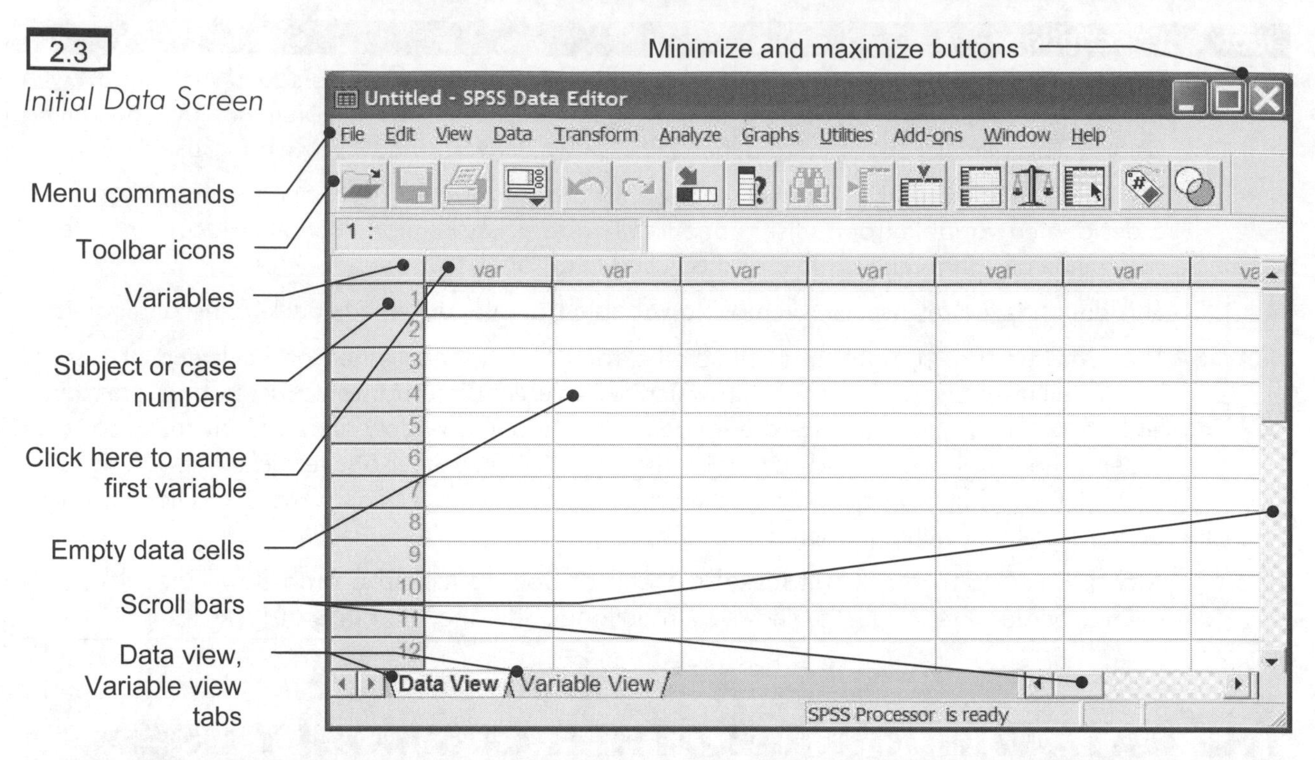

2.3

Initial Data Screen

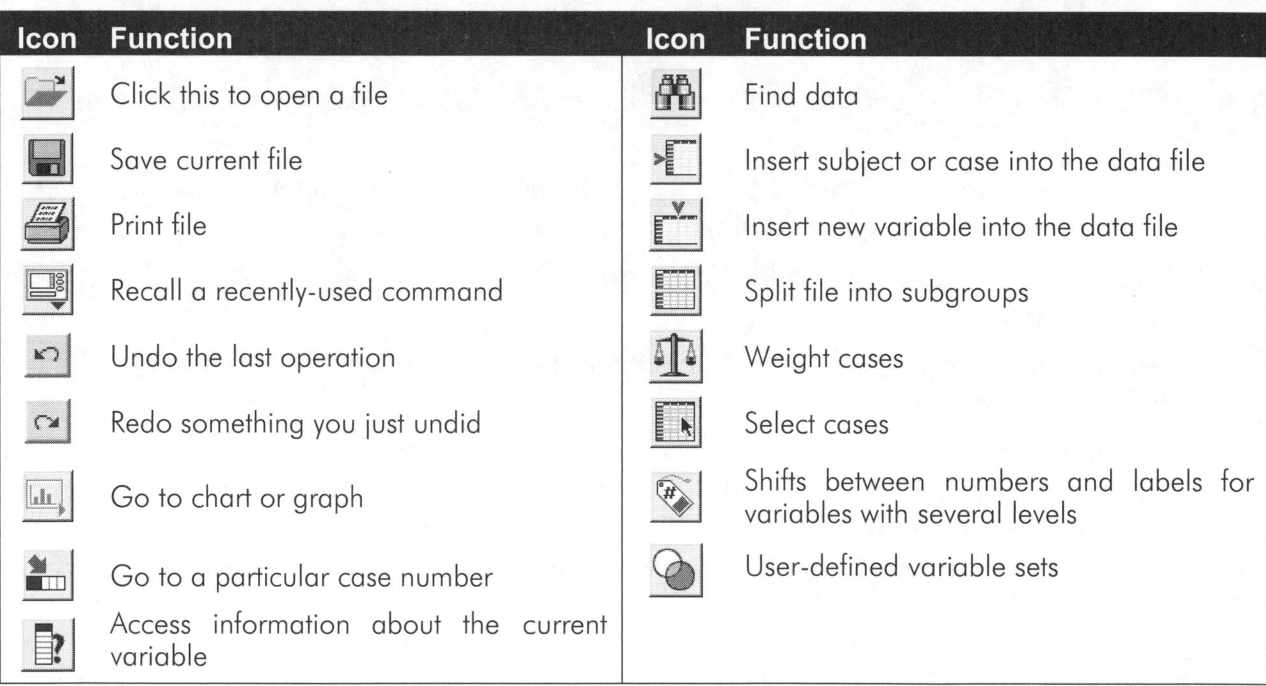

Icon	Function	Icon	Function
	Click this to open a file		Find data
	Save current file		Insert subject or case into the data file
	Print file		Insert new variable into the data file
	Recall a recently-used command		Split file into subgroups
	Undo the last operation		Weight cases
	Redo something you just undid		Select cases
	Go to chart or graph		Shifts between numbers and labels for variables with several levels
	Go to a particular case number		User-defined variable sets
	Access information about the current variable		

Screen 2.3 pictures a full-screen image of the data editor, with a detailed breakdown of the toolbar buttons below. The menu bar (the commands) and the toolbar are located at the top of the screen and are described below. When you start SPSS, there are no data in the data editor. To fill the data editor window, you may type data into the empty cells (see Chapter 3) or access an already existing data file (described later in this chapter).

Toolbar: The toolbar icons are located below the menu bar at the top of the screen. The icons were created specifically for ease of point-and-click mouse operations. It must be noted that even in SPSS applications the format of the icon bar may vary slightly. The toolbar shown above applies to the data editor window; a different toolbar is shown (page 20) that applies to the output window. Also note that some of the icons are bright and clear and others are "grayed". Grayed icons are those that are not currently available. Note for instance that the Print File icon is grayed because there are no data to print. When data are entered into the data editor, then these icons become clear because they are now available. The best way to learn how the icons work is to click on them and see what happens.

The menu bar: The menu bar (just above the toolbar) displays the commands that perform most of the operations that SPSS provides. You will become well acquainted with these commands as you spend time in this book. Whenever you click on a particular command, a series of options appears below and you will select the one that fits your particular need. The commands are now listed and briefly described:

- **File**: Deals with different functions associated with files including opening, reading, and saving, as well as exiting SPSS.
- **Edit**: A number of editing functions including copying, pasting, finding, and replacing.
- **View**: Several options that affect the way the screen appears; the option most frequently used is **Value Labels**.
- **Data**: Operations related to defining, configuring, and entering data; also deals with sorting cases, merging or aggregating files, and selecting or weighting cases.
- **Transform**: Transformation of previously entered data including recoding, computing new variables, reordering, and dealing with missing values.
- **Analyze**: All forms of data analysis begin with a click of the Analyze command.
- **Graphs**: Creation of graphs or charts can begin either with a click on the Graphs command or (often) as an option while other statistics are being performed.
- **Utilities**: Utilities deal largely with fairly sophisticated ways of making complex data operations easier. Most of these commands are for advanced users, and will not be described in this book.
- **Add-ons**: If you want to do some advanced statistics that aren't already in SPSS, these menu options will direct you to other programs and services that SPSS can sell you.
- **Windows**: Deals with the position, status, and format of open windows. This menu may be used instead of the taskbar to change between SPSS windows.
- **Help**: A truly useful aid with search capabilities, tutorials, and a statistics coach that can help you decide what type of SPSS procedure to use to analyze your data.

THE OPEN DATA FILE DIALOG WINDOW

The **Open File** dialog window provides opportunity to access previously created data files. On the following page we show the Open File window with key elements identified. To access this window, it is necessary to perform the following sequence of steps.

From Screen 2.3, perform the following procedure to access the **Open File** *window.*

To assist in description of the functions of this window, we will make use of the data file presented in the first chapter (and described in greater detail in the Chapter 3). The name of the data file is **grades.sav**. All data files created in the SPSS for Windows data editor are followed by a period (.) and the 3-letter extension, **sav**; depending on your Windows settings, the **.sav** may or may not appear in the Open File dialog window.

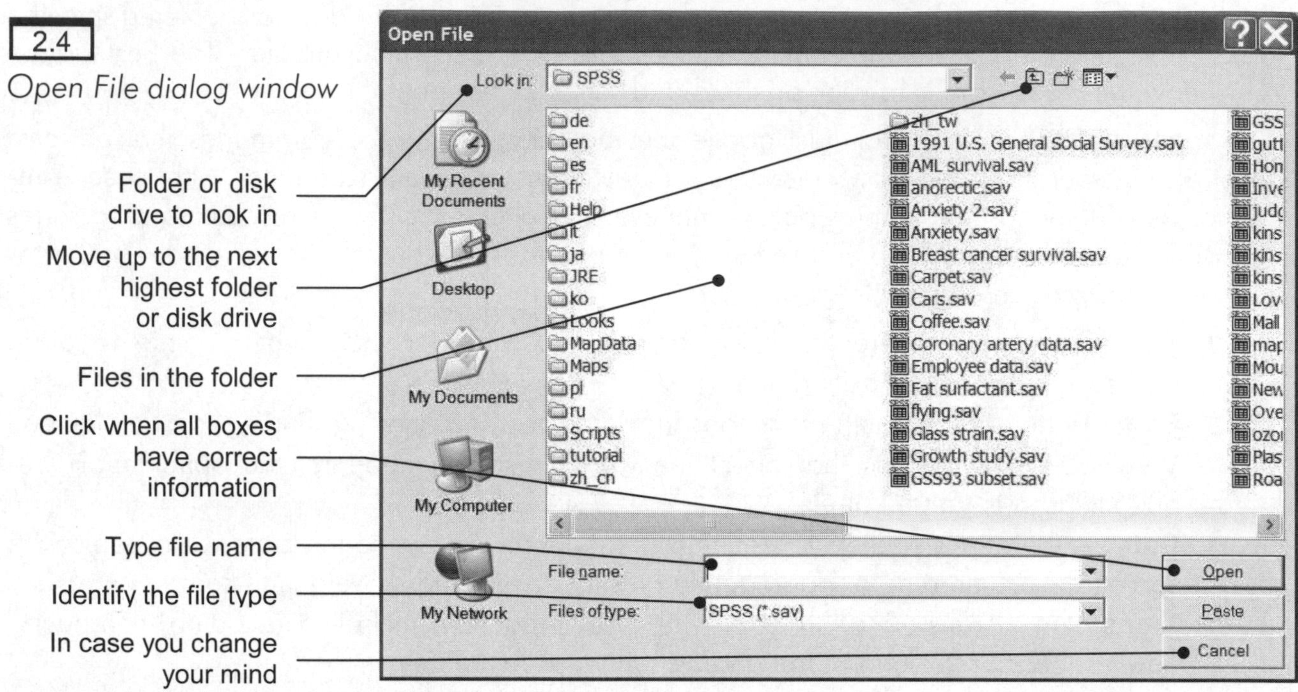

2.4
Open File dialog window

If you wish to open a data file, there are several ways this may be accomplished.

Often the name of the file will be listed in the files box. If the **grades.sav** *file is there,*

In Screen | Do This | Step 3
2.4 | **grades.sav** | Menu

A second option is to simply type the file name in the **File name** *box, making sure the folder and disk drive (above) are correct, then click on the* **OK** *button*

In Screen | Do This | Step 3a
2.4 | type `grades.sav` → **Open** | Menu

If you are reading from a floppy disk and the file is an SPSS for Windows file, then in the place where the caption says "type file name"

In Screen | Do This | Step 3b
2.4 | type `a:\grades.sav` → **Open** | Menu

*If you are reading a data file that was not created on the SPSS for Windows data editor, you need to identify the file type before SPSS can read it accurately. If, for instance, you are reading an Excel file (***grades.xls** *in this case) from a data disk, perform the following steps to read this file accurately.*

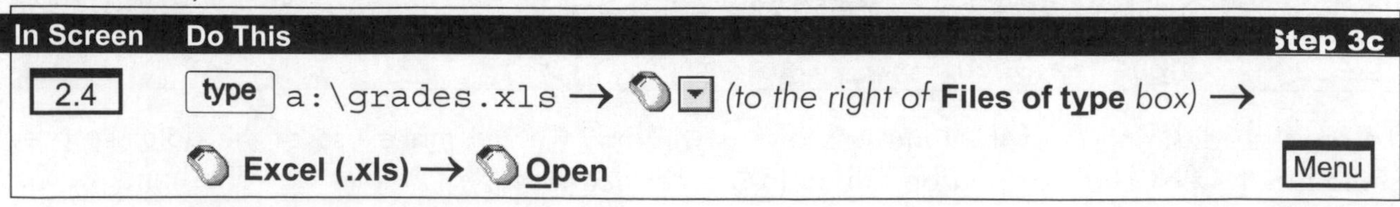

In Screen | Do This | Step 3c
2.4 | type `a:\grades.xls` → ▼ *(to the right of* **Files of type** *box)* → **Excel (.xls)** → **Open** | Menu

The result of any of these four operations is to produce a screen with the menu of commands across the top and the data from the **grades.sav** file in the data editor window. We note the menu screen as the outcome of each of these operations because the menu of commands across the top of the screen is what allows analyses to take place.

AN EXAMPLE OF A STATISTICAL-PROCEDURE DIALOG WINDOW

There are as many statistical procedure dialog windows as there are statistical procedures. Despite the fact that each is different from its fellow, there are fundamental similarities in each one. We illustrate by use of the dialog window for Frequencies (Chapter 6). In all such boxes, to the left will be a list of available variables from the data file. To the right will be another box or even two or three boxes. Although the title of this (or these) box(es) will differ, they are designed to show which variables will be processed in the analysis. Between the boxes will be ▸ and/or ◂ buttons to move variables from one box to another.

2.5

Sample statistical procedure dialog window

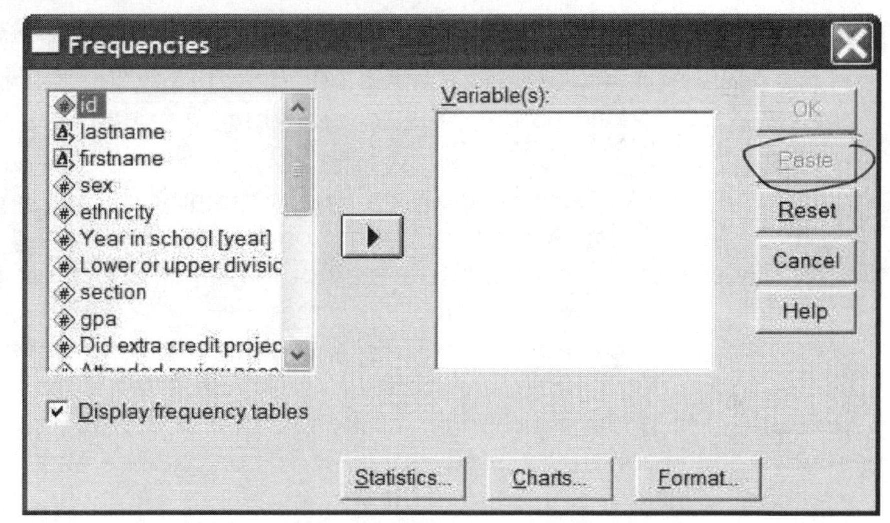

All statistical procedure dialog boxes have the same five buttons, usually to the right but sometimes below:

- **OK**: When all variables and specifications are selected, click the **OK** button to compute the statistics.
- **Paste**: A click of the paste button will open a syntax window with the specifications identified in the dialog boxes pasted into the syntax box in a command-file format. Those acquainted with computer programming will enjoy this option. The reason this option is available is because the windows format (although simpler) has certain limitations that make it less flexible than the command file format. If you wish to tailor your program to fit your specific needs then it is often useful to work with a command file. For those unfamiliar with programming in a command file format, read the hefty *SPSS Syntax Reference Guide* before attempting this—or better yet, pretend this button doesn't exist.
- **Reset**: One characteristic of these main dialog boxes is called *persistence*. That means if one analysis is completed and the researcher wishes to conduct another, upon returning to the dialog box, all the same variables and specifications remain from the

previous analysis. If that is what you wish, fine. If not, click the **Reset** button and the variables will all return to the variable list and all specifications will return to the defaults (defaults are the automatic settings). There is one other way of returning *all* statistical procedure dialog windows to the default settings: open a different data file (or close and re-open SPSS and open the same data file again).

- **Cancel**: For immediate exit from that procedure.
- **Help**: Similar to the **Help** on the main command menu except that this help is context specific. If you are working with frequencies, then a click on the help button will yield information about frequencies.

Finally the three buttons along the lower edge of the window (**Statistics**, **Charts**, and **Format**) represent different procedural and formatting options. Most main-dialog boxes have one or more options similar to these. Their contents, of course, are dependent on the statistical procedure you are performing.

KEYBOARD PROCESSING, CHECK BOXES, and RADIO BUTTONS

Some people cannot use mice and most software packages (including SPSS) may be operated entirely from the keyboard. We'll not go into great detail, but will give the basics of keyboard maneuvers.

When using the keyboard exclusively, the point and click with a mouse is replaced by highlighting a particular icon, word, command, or phrase, then hitting the **Enter** key. Highlighting is generally shown by color enhancement of the selected item or a border surrounding it, and is accomplished by moving around the screen using the **TAB** key, the space bar and the cursor keys. Once the correct element is highlighted, press the **Enter** key to select it. Another keyboard option deals with commands in which one letter of that option is underlined (e.g., **Edit**, **Options**). The underlining appears when you press the **ALT** key, then you press the underlined letter on the keyboard to select that option. For instance, to select **Edit,** press the **ALT** key then the letter **E**. The menu under **Edit** will immediately appear. Even for mouse users, the keystrokes are often faster than mouse clicks once you have learned them. Selecting (or highlighting) text is often much easier and quicker with keystrokes than using the click and drag of the mouse. To highlight a block of cells, one positions the cursor in a corner cell of the desired block of cells, presses down the shift key, then presses the cursor keys to highlight to the right, to the left, above or beneath the original insertion point. There are many more keyboard operations but just this brief paragraph should be sufficient to allow you to run SPSS without the use of a mouse.

Finally, there are two different types of selections of options that may take place within dialog boxes. Note the window (following page) from Chapter 10. Notice the **Tests of Significance** box near the bottom; it contains **radio buttons**. They are called radio buttons because if one is selected the others are automatically deselected (as in a car stereo—you select one channel, and you deselect the other channels). Not to insult your intelligence, but a circle with a black dot in the center (◉) represents a selected option.

Check boxes are shown under the **Correlation Coefficients** heading and at the very bottom of the dialog box. With check boxes you simply select those you desire. You may select all three, two, one, or none of them. A selection is indicated when a ✔ appears in the box to the left of the option. For radio buttons, click on a different option to *de*select the present option. In check boxes, click on the box a second time to deselect it.

2.6

Sample dialog window with radio buttons and check boxes.

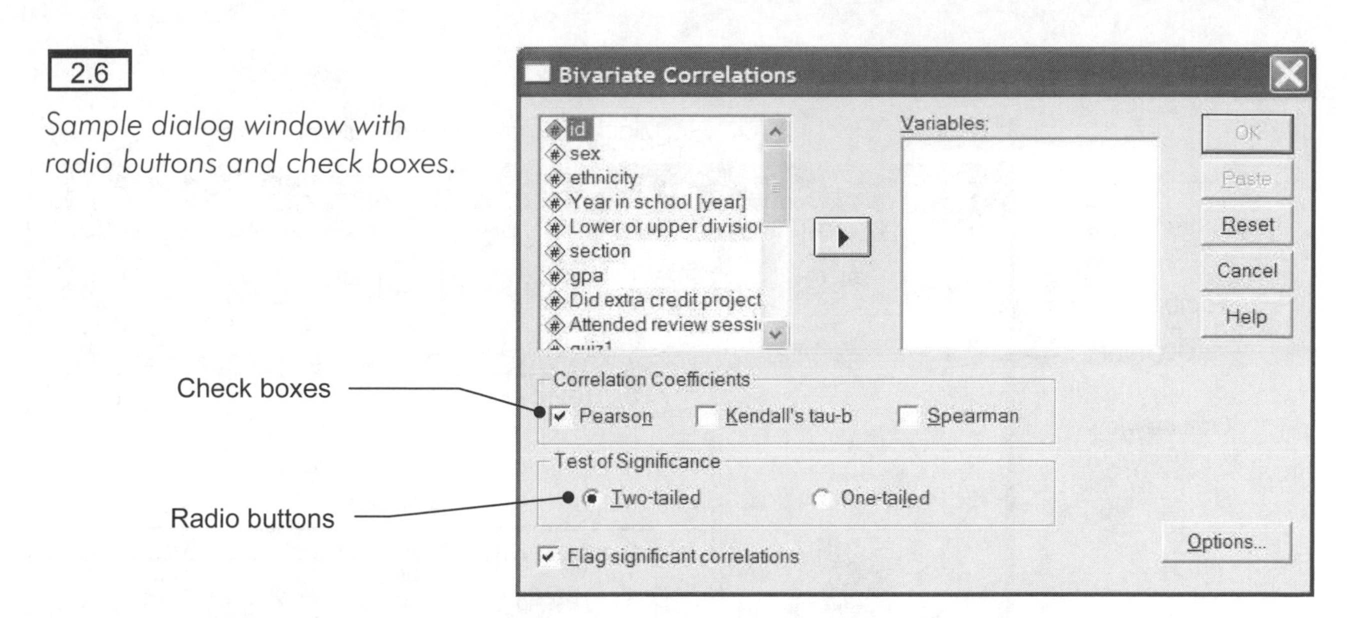

THE OUTPUT WINDOW

The "Output" is the term used to identify the results of previously conducted analyses. It is the objective of all data analysis. SPSS has a long and storied history of efforts to create a format of output that is clear yet comprehensive. The current version uses a tables-with-borders format. When utilizing options described below, this is relatively clear, but output can still be very awkward and occupy many pages. It is hoped that the information that follows will maximize your ability to identify, select, edit, and print out the most relevant output. If you don't read it now, do come back to it before you first press the "Print" button.

The initial output screen is shown on the inside back cover. Each chapter that involves output will give instructions concerning how to deal with the results displayed on this screen. Our focus is to explain how to *edit* output so that, when printed, it will be reproduced in a format most useful to you. Of course, you do not have to edit or reorganize output before you print, but there are often advantages to doing so:

- Extensive outputs will often use/waste many pages of paper.
- Most outputs will include some information that is unnecessary.
- At times a large table will be clearer if it is reorganized.
- You may wish to type in comments or titles for ease or clarity of interpretation.

This chapter will explain the SPSS Output window (and toolbar), how to delete output that you no longer need, how to add comments to your file, how to re-arrange the order of the output, and how to save the output.

Screen 2.7 (following page) summarizes the key elements of the output window, as well as providing a detailed description of each toolbar item on the output window. In this sample, several procedures have been performed on the **grades.sav** file; don't worry about the content of the analysis, just pay attention to what SPSS will allow you to do with the output.

You will notice that several of the toolbar icons are identical to those in the SPSS Data Editor window; these buttons do the same thing that they do in the Data Editor, but with the output instead of the data. For example, clicking on the print icon prints the output instead of the data.

2.7 *The SPSS Output Navigator Window*

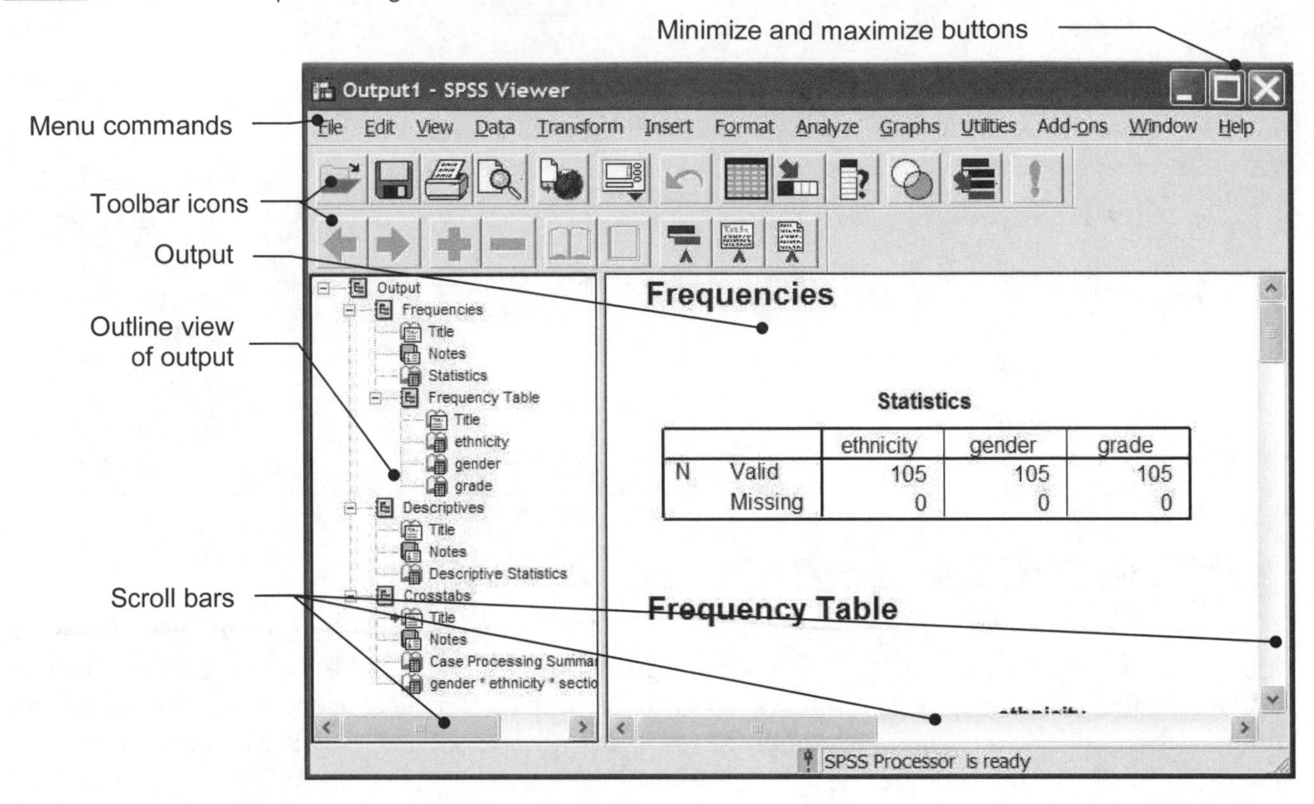

Icon	Function	Icon	Function
	Click to open a file		Go to the bottom of most recent output
	Click to save file		Send future output to this window
	Print output		Make the currently selected subheading into a heading
	Print preview (see what the printout will look like)		Make the currently selected heading into a subheading
	Export output to text or web (HTML) file		Display everything under the currently selected heading
	Recall a recently used command		Hide everything under the currently selected heading
	Undo the last operation		Display the currently selected object (if it is hidden)
	Go to SPSS Data editor (frequently used!)		Hide the currently selected object (visible in outline view only)
	Go to a particular case number		Insert heading (above titles, tables, and charts)
	Get information about variables		Insert title after the currently selected object
	User-defined sets		Insert text after the currently selected object

One of the most important things to learn about the SPSS Output window is the use of the outline view on the left of the screen. On the right side of the window is the output from the SPSS procedures that were run, and on the left is the outline (like a table of contents without page numbers) of that output. The SPSS output is actually composed of a series of *output objects;* these objects may be titles (e.g., "Frequencies"), tables of numbers, or charts, among other things. Each of these objects is listed in the outline view:

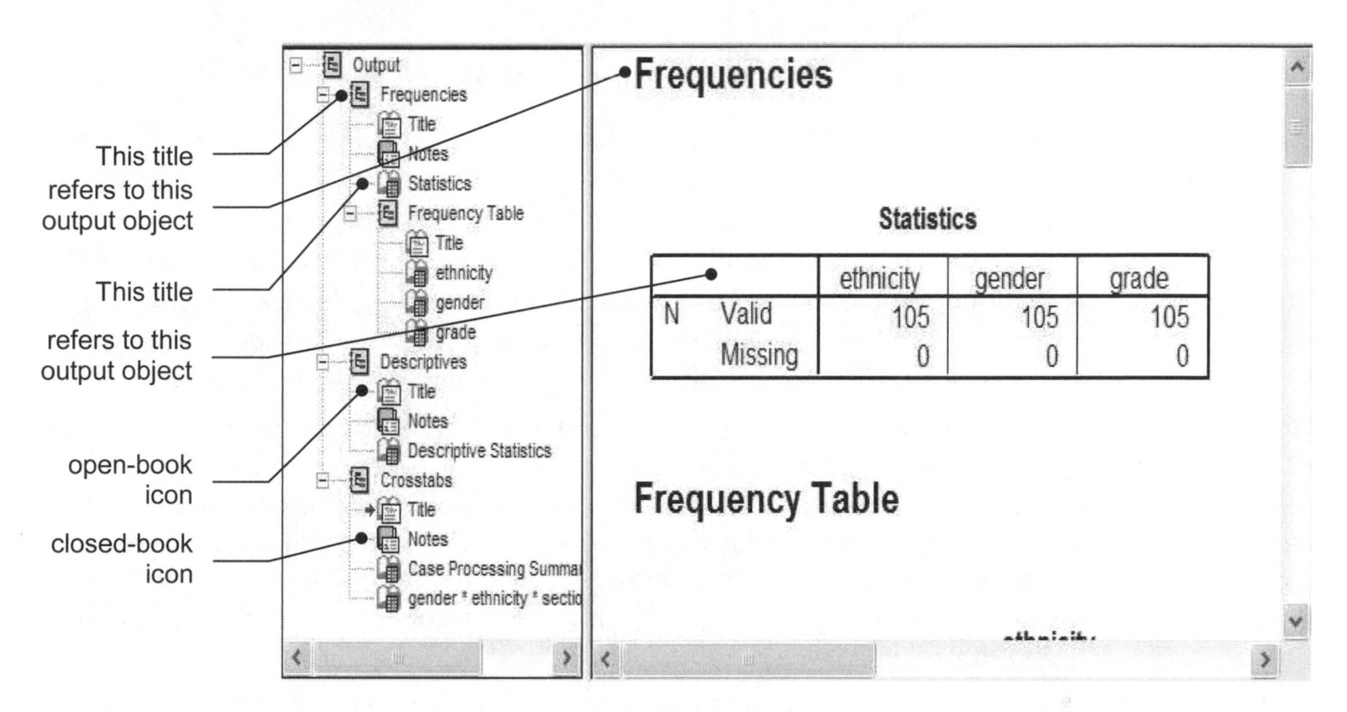

Note: You will notice that there is no "notes" section in the output window to correspond with the "notes" title in the outline view. That's because the notes are (by default) hidden. If you want to see the notes, just double click on the closed book icon to the left of the **notes** title. The closed book icon will then become an open-book icon and the notes will materialize in the window to the right.

The outline view makes navigating the output easier. If you want to move to the Crosstabs output, for example, you merely need to click on the word "Crosstabs" in the outline view, and the crosstabs will appear in the output window. If you want to delete the Descriptives section (perhaps because you selected an incorrect variable), simply click on the word "Descriptives" and select menu item **Edit** then click **Delete**. If you want to move some output from one section to another (to re-arrange the order), you can select an output object (or a group of output objects), and select **Edit** then click **Cut**. After that, select another output object below which you want to place the output object(s) you have cut. Then select **Edit** and click **Paste After**.

If you have been working with the same data file for a while, you may produce a lot of output. So much output may be produced, in fact, that it becomes difficult to navigate throughout the output even with the outline view. To help with this problem, you can "collapse" a group of output objects underneath a heading. To do this, click on the minus sign to the left of the heading. For example, if you want to collapse the Frequencies output (to get it out of the way and come back to it later) simply click on the minus sign to the left of the word **Frequencies**. If you want to expand the frequencies heading later, all you have to do is click on the plus sign to the left of the title **Frequencies**. This operation is illustrated on the top of the following page.

Clicking on the minus sign...

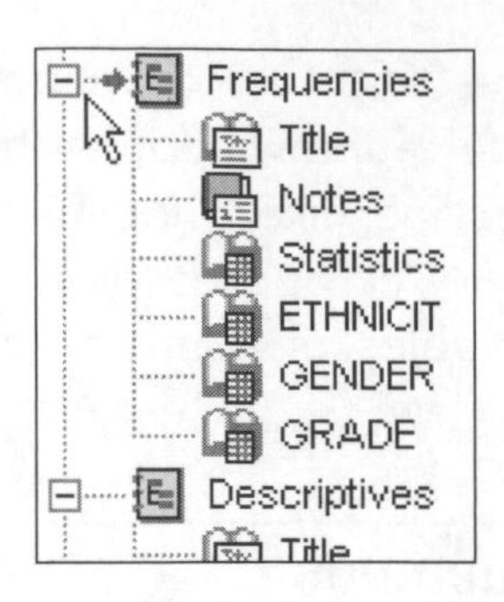

...will collapse the frequency objects to this:

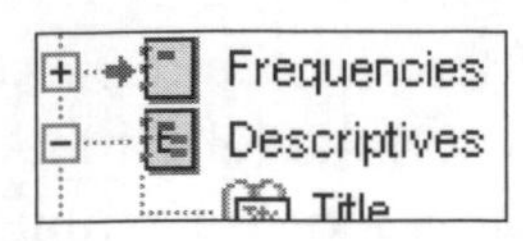

One particularly useful command when you are working with output is the insert text command ( ). When you click this button, an SPSS Text object is inserted. In this box, you can type comments to remind yourself what is interesting about the SPSS output (for example, "The next chart would look good in the results section," or "Oh, no, my hypothesis was not supported!"). Once you have typed your comments, click on another SPSS object to de-select the SPSS Text object. Note: If you type more than one line of comments, you will have to click on the square black box on the bottom of the SPSS Text object, and drag it down to make the box larger.

Now that you have learned the basics of getting around within SPSS output, there is one more major concept to learn. When SPSS produces output, whether it be charts or tables, you can make changes in the format of that output. Although this book will not describe all of the procedures that you can perform to modify the appearance of charts and tables, some basic operations will be described that will allow you to make your charts and tables more attractive or easier to read. The basics of editing charts and graphs are described in Chapter 5; the basic processes of modifying tables are described here.

First, we will start with some background and theory of how tables work. It is important to realize that, although SPSS automatically arranges tables for you—it decides what to put in the columns and rows—the arrangement of a table is somewhat arbitrary. For example, in the crosstabulation of **gender** x **ethnicity** x **section** (described in Chapter 8), by default SPSS places the ethnicity in the columns, and section and gender in rows:

gender * ethnicity * section Crosstabulation

Count

			ethnicity					
section			Native	Asian	Black	White	Hispanic	Total
1	gender	Female		6	2	9	3	20
		Male		1	5	7	0	13
	Total			7	7	16	3	33
2	gender	Female	3	5	6	11	1	26
		Male	1	4	1	7	0	13
	Total		4	9	7	18	1	39
3	gender	Female	1	2	6	6	3	18
		Male	0	2	4	5	4	15
	Total		1	4	10	11	7	33

Although there is nothing wrong with this arrangement, you might want to rearrange the table, depending on what exactly you were trying to demonstrate. It is important to remember that you are *not* changing the content of the statistics present within the table; you are only rearranging the *order* or the format in which the data and statistics are displayed.

You will notice in the table above, there are both *rows* and *columns*. The columns contain the various categories of **ethnicity**. The rows contain two different levels of headings: In the first level is the course **section**, and in the second level is the **gender** of the students. You will notice that each level of the second heading (each gender) is displayed for each level of the first heading (each section). This table has one *layer*: **Count** (the number of subjects in each category) is visible in the upper left hand corner of the table. In most tables, there is only one layer, and we will focus our discussion here on arranging rows and columns.

In order to edit and change the format of an output table, you must first double click on the output object in the output view (the right portion of Screen 2.7). When you do that, the toolbars will disappear, and in the menu bar will appear a new item: **Pivot**. In addition to this, some of the options available on the various menus change. We will describe some of the options available on the **Pivot** and **Format** menus, including: **Rotate inner column labels**, **Rotate outer row labels**, **Transpose rows and columns**, and **Pivoting trays**.

You will notice in the sample table shown on the previous page that the columns are much wider than they need to be to display the numbers. This is so the columns will be wide enough to display the labels. Clicking on **Format** followed by **Rotate inner column labels** will fix this problem by making the column labels much taller and narrower, as shown to the right. If you decide you prefer the previous format, simply click on **Format** followed by **Rotate inner column labels** once again, and the old format will be restored. It is also possible to change the orientation of the row titles. This would make the section numbers tall and narrow. But this also makes the heading "Section" tall and narrow, and does render the table itself narrower.

ethnicity					Total
Native	Asian	Black	White	Hispanic	
	6	2	9	3	20
	1	5	7	0	13
	7	7	16	3	33
3	5	6	11	1	26
1	4	1	7	0	13
[illegible]	[illegible]	[illegible]	[illegible]	[illegible]	[illegible]

When SPSS produced the sample **gender** x **ethnicit** x **section** table, it placed ethnicity in the columns, and course section and gender in the rows. A researcher may want these variables to be displayed differently, perhaps with course section in the rows, and both gender and ethnicity in the columns. In order to make this change, you may use either **Transpose rows and columns** or **Pivoting trays**. Clicking **Pivot** followed by **Transpose rows and columns** is the simplest way to rearrange rows and columns. If we were to choose this command on our sample output table, then the rows would become columns and the columns rows, as shown below:

gender * ethnicity * section Crosstabulation

Count

		section								
		1			2			3		
		gender			gender			gender		
		Female	Male	Total	Female	Male	Total	Female	Male	Total
ethnicity	Native				3	1	4	1	0	1
	Asian	6	1	7	5	4	9	2	2	4
	Black	2	5	7	6	1	7	6	4	10
	White	9	7	16	11	7	18	6	5	11
	Hispanic	3	0	3	1	0	1	3	4	7
Total		20	13	33	26	13	39	18	15	33

A somewhat more complex and sophisticated way of rearranging columns and rows is to use **Pivoting trays**. Selecting this option (a subcommand on the **Pivot** menu) will produce a new window that allows you to rearrange rows and columns (Screen 2.8). This window will also allow you to rearrange layers, but because the arrangement of layers usually doesn't need to be changed, we won't worry about layers in this discussion.

2.8

Pivoting Trays Window

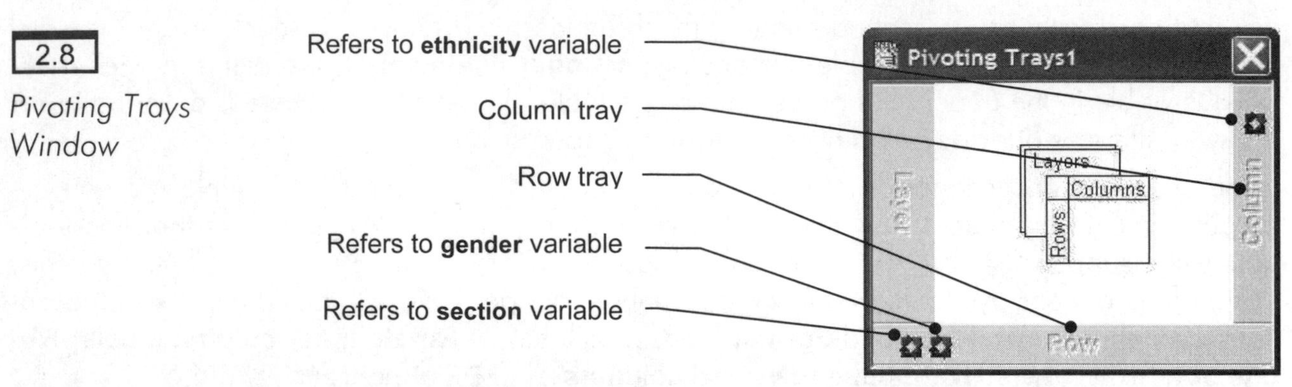

The pivoting tray window consists of different trays: a tray for row variables, and a tray for column variables. In our example (the original output, not the modified output), ethnicity is in the column tray, with section and gender both in the row tray.

Each variable icon (■) may be dragged within each tray, or between the two trays. For example, if you want to rearrange the order of the section and gender variables, you could click and drag the gender variable to the left of the section variable. This would rearrange the rows so that the gender was the main heading level, with three levels of each section for each of the two levels of gender.

You can also rearrange variables between rows and columns, by dragging the appropriate variable icon from one tray to another. For example, if you wanted to make section a major column heading with levels of ethnicity displayed for each section, you would drag the section variable icon to just above the ethnicity variable.

As you can see, there are many different combinations of rows and columns possible. The best way to understand pivoting tables is simply to experiment with them. Because of the persistence function, it's not at all difficult to run the analysis again if you get too confused.

Once you have done all of this work making your output look the way you want it to look (not to mention doing the statistical analyses to begin with), you may want to save your work. This, by the way, is quite different from saving the data file (always followed by a ".sav" suffix). Output is generally saved with a ".spo" suffix. The sequence step required to do that follows:

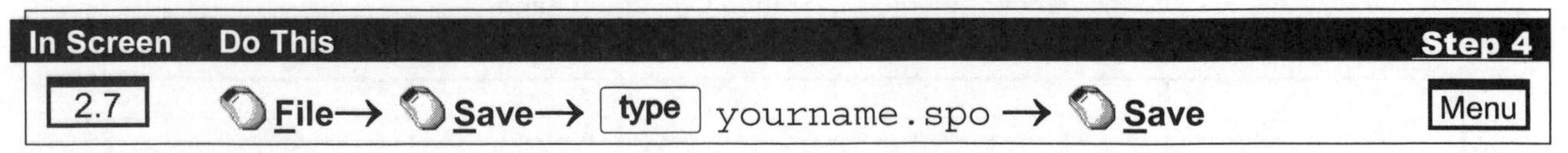

In Screen	Do This	Step 4
2.7	**File**→ **Save**→ type `yourname.spo` → **Save**	Menu

PRINTING OUTPUT

Toward the end of the Step by Step section of each chapter skeletal information is provided concerning printing of output and/or graphs. Every time this is done the reader is referred to this section in this chapter for greater detail. If you desire more complete information than provided here, please refer to the *SPSS* manuals.

The first time you print, it is a good idea to select the **File**, then click the **Page setup** option. This very intuitive dialog box lets you specify whether the orientation of printed material will be portrait (longer dimension is vertical) or landscape (longer dimension is horizontal), the margins, and the size and source of your paper.

In order to print your output, the material that you wish to print must be *designated*—that is, you must select the portions of your output that you want printed. If you know that you want all of your output printed, then you don't have to designate the material prior to printing. We don't, however, recommend making this a habit: Save the trees. (You don't have to hug them if you don't want, but at least think about what you are going to print before you start.)

To print a section of the output you must highlight the portion of the file you wish to reproduce *before* beginning the print sequence. This may be accomplished by

1. Clicking on the folder or output object you want to print on the left side of the SPSS output window;
2. Clicking on a folder or output object on the left side of the SPSS output window, then holding down the **Shift key** while selecting another folder or output object (this will select both objects you clicked and all the objects in between); or
3. Clicking on a folder or object on the left side of the SPSS output window, then holding down the **Control** key while clicking on one or more additional folders or objects (this will select only the folders or objects that you click on).

The process always starts with a click on the **File** command followed by a click on the **Print** option. This causes a small dialog box to open (Screen 2.9) allowing you to choose which printer you want to use (if there is more than one option), how many copies to print, and whether to print just the designated (selected) part of the output, or the entire output.

Once all of the options in the **Print** dialog have been specified as desired, click the **OK** button and watch your output pour from the printer.

2.9

Print dialog box

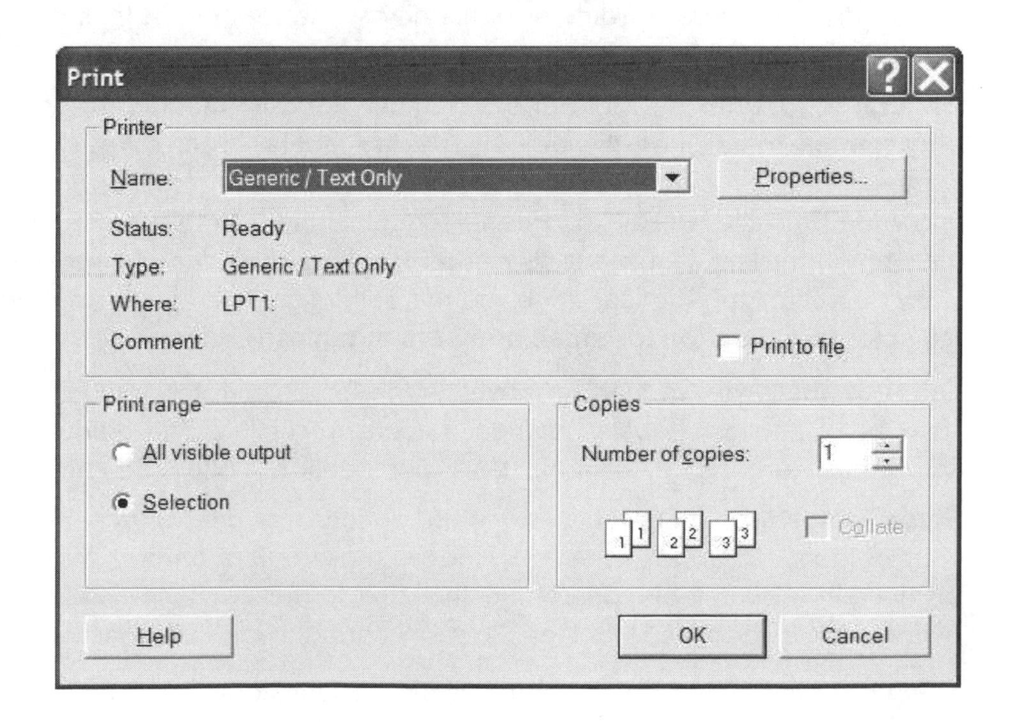

THE "Options..." OPTION: CHANGING THE FORMATS

Under the **Edit → Options...** sequence lies a wide variety of alternatives to SPSS defaults. An awareness of the available options allows you to format many features associated with the appearance of the SPSS screen, dialog windows, and output. We reproduce the dialog window here and then identify the selections that we have found to be the most useful.

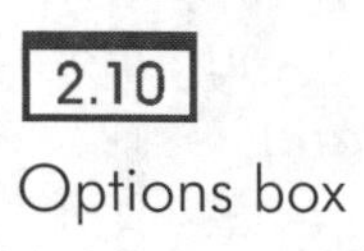

Options box

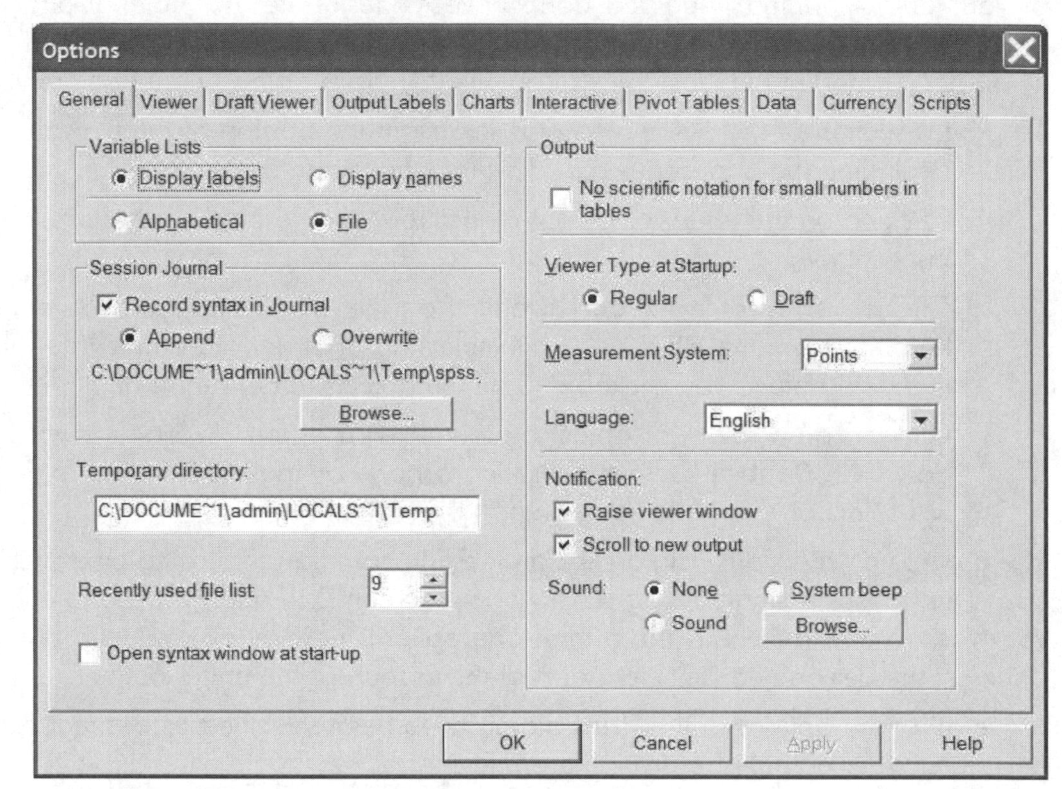

Notice that there are 10 different tabs across the top; each deals with a different set of functions. Here are the options we use most frequently:

General tab (currently visible above): Under **Variable Lists**, you may want to change to **Display names** rather than the default **Display labels**. Labels are typically much too long and cumbersome to fit neatly in the box of variables visible to the left in the procedure box. If you know the data file well, select **File** to sequence the list of variables (in every procedure box) in the original order. If you are dealing with a large data file or one that you are not so familiar with, then the **Alphabetical** might be the better solution. To the right of the screen make sure **No scientific notation for small numbers in tables** is selected.

Output Labels tab: In the four boxes to the left be sure to change to a sequence (from top to bottom) of **Names**, **Labels**, **Names**, **Labels** rather than the default of all **Labels**. Then click **Apply** to secure the new setting. The default option is often too cumbersome to be useful.

Pivot Tables tab: There are 54 different options that deal with the format of the tables used in your outputs. You may selectively choose those output formats that best suit your needs. For instance in a large table, one of the "narrow" formats will allow you to get more information on a single page, and the "Hot Dog" format is good if you want to give yourself a headache.

There are others, many others, but this gives you a start.

Creating and Editing a Data File

THIS CHAPTER describes the first step necessary to analyze data: Typing the data into the computer. SPSS uses a spreadsheet for entering and editing data. Variable names are listed across the top columns, case or subject numbers are listed along the left rows, and cells in the middle are available for entering data. Vertical and horizontal scroll bars allow you to maneuver rapidly through even a large file, and there are several automatic *go-to* functions that allow you to identify or find a particular case number or variable quickly.

This is the first chapter in which we use a step-by-step sequence to guide you through the data entry procedure. A data file was introduced in the first chapter that will be used to illustrate the data-entry process. The name of the file is **grades.sav**, and it involves a course with 105 students taught in three different sections. For each student a variety of information has been recorded, such as quiz scores and the final grade for the course. The content of this entire file is listed on the last pages of this chapter and provides an opportunity to enter the file into your own computer. This file is available for download (along with the other data files used in examples and exercises) at **www.ablongman.com/george6e**. Because of this, you don't have to enter these data yourself if you don't want to, but you may wish to do so (or at least enter the first few students) to gain practice in the data entry procedure.

RESEARCH CONCERNS and STRUCTURE of the DATA FILE

Before creating the data file, you should think carefully to ensure that the structure of your file allows all the analyses you desire. Actually, this careful consideration needs to occur much earlier than the data entry phase. We have both had individuals come to us for help with data analysis. They are often working on some large project (such as a dissertation) and have collected data from hundreds of subjects that assess hundreds of variables. Clearly the thought of what to actually *do* with these data has never occurred to them. If such individuals are wealthy and not our students, we're in for a nice little financial benefit as we attempt to tidy up the mess they have created. On other occasions, their lack of planning makes it impossible to make important comparisons or to find significant relationships. Examples of common errors include: failure to include key variables (such as gender or age) when such variables are central to the study; requesting yes-no answers to complex personal questions (irritating the subjects and eliminating variability); including many variables without a clear *dependent* variable to identify the objective of the study; having a clear dependent variable but no independent variables that are designed to influence it; and a variety of other outrages.

If the original research has been carefully constructed, then creating a good data file is much easier. For instance, many studies use **ordinal data** (that is, it has an intrinsic order such as seven levels of income from low to high, or levels of education from junior high to Ph.D.). Ordinal data should be coded with numbers rather than letters to allow SPSS to conduct analyses with such data. If you have missing values (as almost all studies do), it is far better to deal with these at the data entry stage rather than resorting to SPSS defaults for automatic handling of such. A simple example illustrates one way to handle missing data: If you have an ethnicity variable, several subjects may not answer such a question or may be of an ethnic background that doesn't fit the categories you have created. If you have provided 4 categories such as White, Black, Asian, and Hispanic (coded as 1, 2, 3, and 4), you might create an additional category (5 = Other or Decline to State) when you enter the data. Ethnicity then has no missing values and you can make any type of ethnic comparisons you wish. The issue of missing values is dealt with in greater detail in Chapter 4.

The order in which you enter variables also deserves some thought before you enter the data. One of us had an individual come with a data file 150 variables wide and the demographics (such as gender, age, and marital status) were positioned at the end of the file. Since demographics were important to the interpretation of the data, they had to be moved back to the beginning of the file where they were more accessible. In determining the order for data entry, IDs and demographics are usually placed at the beginning of the file, then entry of other variables should follow some sort of logical order.

It is also important to think about the way your data were collected, and how that influences the way your file should be structured. An important rule to remember is that each **row** in your data file represents one **case**. Usually, one case is one person or subject, but sometimes a case may be more than one subject (for example, if a husband and wife rate each other, that may be one case). So, if you collect data from one person for several different variables (for a correlational analysis, paired-samples *t* test, or within-subjects ANOVA) then all of the data from that person should be entered in one row. Specifically, each **column** represents a separate **variable** and the variable labels at the top of the columns should make it clear what you are measuring.

The **grades.sav** sample data file illustrates: There are 105 cases in 105 rows representing 105 students. Case #1 in row #1 is the first student (Maria Ross). Case #2 in row #2 is the second student (Mihaela Shima) and so forth. The variables in the **grades.sav** are listed across the top in a useful order. For instance the five quizzes are listed in order (**quiz1**, **quiz2**, **quiz3**, **quiz4**, **quiz5**) for ease of access when we are performing analyses.

If, however, you collect data from different subjects in different conditions (for an independent-samples *t* test or between-subjects ANOVA, for example) each subject, as above, should be listed on a different row, but a variable in a column should make it clear which condition is involved for each subject. For instance, In the sample data file (as in pretty much all data files) the **gender** variable categorizes subjects into the "female" category or the "male" category. A similar principle applies to the section variable (**section**) which separates subjects into three different categories, "section 1", "section 2", and "section 3".

STEP BY STEP

To enter SPSS, a click on **Start** *in the taskbar (bottom of screen) activates the start menu:*

Your screen should now look something like Screen 3.1 (following page). Critical elements important for data entry are labeled.

With a simple click on the Variable View tab (at the bottom of the screen) the second screen (Screen 3.2, also following page) emerges that allows you to create your data file. The step-by-step box (requiring only one click) follows.

In Screen	Do This	Step 2
3.1	**Variable View** tab	3.2

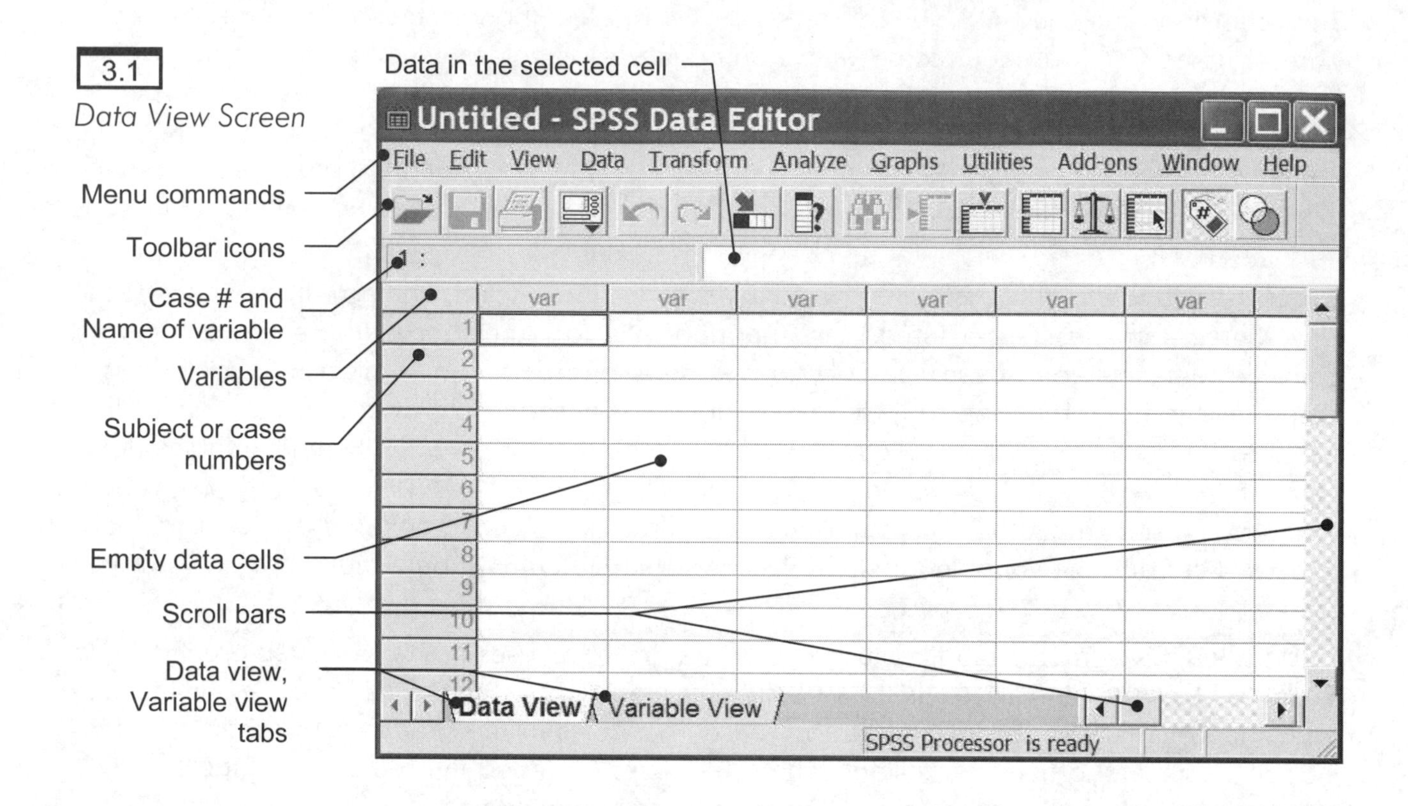

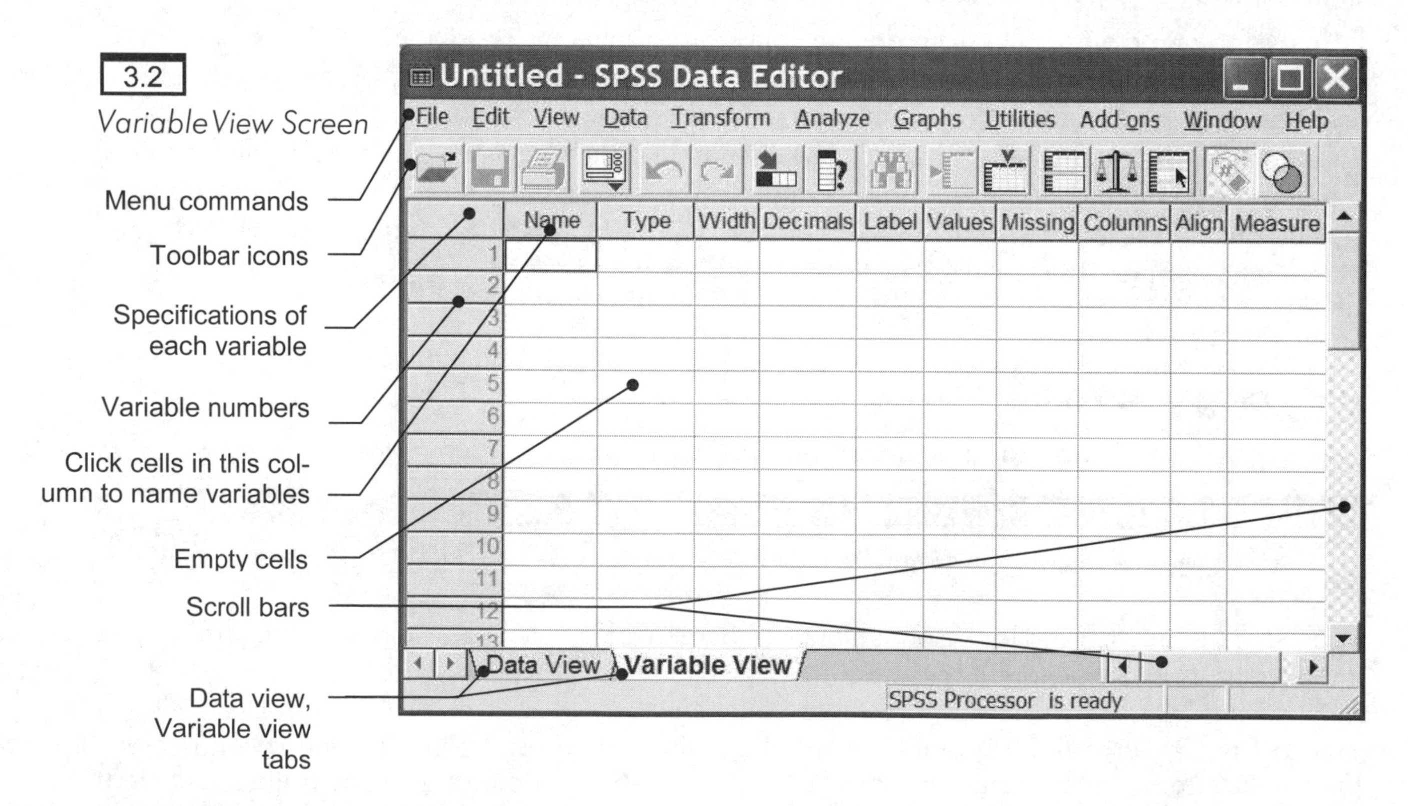

The two screens serve distinct and complementary purposes: **Screen 3.1** is designed to enter data after the data file has been created. **Screen 3.2** is designed to name, label, and determine specifications for each variable. Notice the 10 words in the row near the top of Screen 3.2 labeled "Specifications of each variable". In case you left your spectacles at school (and thus can't read the fine print) they are: **Name**, **Type**, **Width**, **Decimals**, **Label**, **Values**, **Missing**, **Columns**, **Align**, and **Measure**. This is the order in which we will present the various processes associated with setting up a data file.

We now take you step-by-step through the procedures necessary to enter the data included in the **grades.sav** file. The entire file is listed beginning on page 39 for sake of reference. While there are a number of different *orders* in which you might execute the steps to create a data file, for the sake of simplicity we will have you first enter all variable names, then format the variables, and then enter all of the data. If you want to conduct data analysis before all of the data are entered, however, there is nothing wrong with entering the names and formats for several variables, entering the data for those variables, performing data analysis, and then continuing with the next set of variables. A file of class records provides an excellent example of this kind of data entry: You would enter data by variables (quizzes, homework assignments, etc.) as they are completed.

Name

Beginning with the variable view screen (Screen 3.2), simply type the names of your variables one at a time in the first column. After a name is typed, you may use the cursor keys or the tab to move to the next cell to type the next variable name. There are 17 variable names in the original **grades.sav** data file. You may begin the process by typing all 17 names in the first 17 rows of the Variable View screen. Here's step-by-step on the first five variables.

We continue on from sequence step 2 (above) beginning with Screen 3.2.

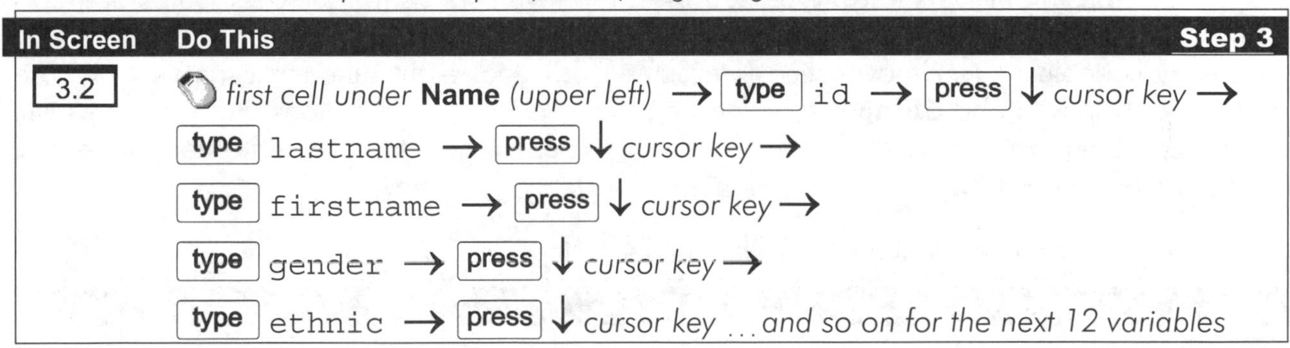

In Screen	Do This	Step 3
3.2	*first cell under* **Name** *(upper left)* → **type** `id` → **press** ↓ *cursor key* →	
	type `lastname` → **press** ↓ *cursor key* →	
	type `firstname` → **press** ↓ *cursor key* →	
	type `gender` → **press** ↓ *cursor key* →	
	type `ethnic` → **press** ↓ *cursor key* ... *and so on for the next 12 variables*	

You will notice that as you press the down cursor key that to the right of the variable name just typed, eight of the other nine columns fill in with default information. The default settings will typically need to be altered, but that can be done easily after you have typed the names of each of the (in this case) 17 variables.

Each variable name you use must adhere to the following rules. There was a time when these rules made sense. Now, you just have to follow them.

- Each variable may be any length but shorter than 10 characters is usually desirable.
- It must begin with a letter, but after that any letters, numbers, a period, or the symbols @, #, _, or $ may be used. However, the name may not end with a period.
- All variable names must be unique; duplicates are not allowed.
- Variable names are not sensitive to upper or lower case. **`ID`**, **`Id`**, and **`id`** are all identical to SPSS.
- Because they have a unique meaning in SPSS, certain variable names may not be used, including: **all, ne, eq, to, le, lt, by, or, gt, and, not, ge,** and **with**.

You now have all 17 variable names in the left column. We will now specify how to configure each variable to facilitate the most effective use of your data.

Type

When you click on a cell in any row under **Type** you notice a small grayed box (...) to the right of the word **Numeric**. A click on the box opens up Screen 3.3.

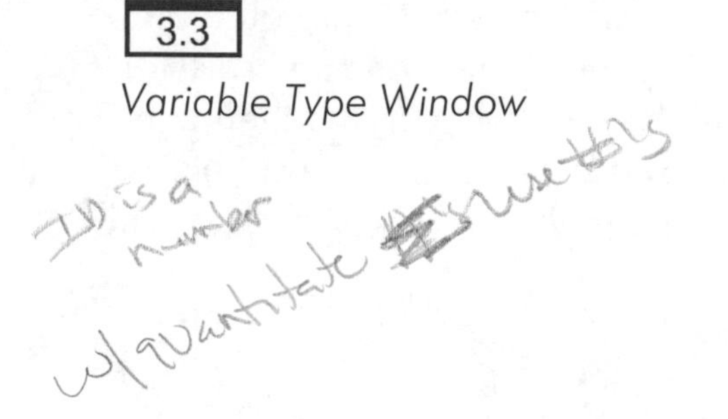

3.3

Variable Type Window

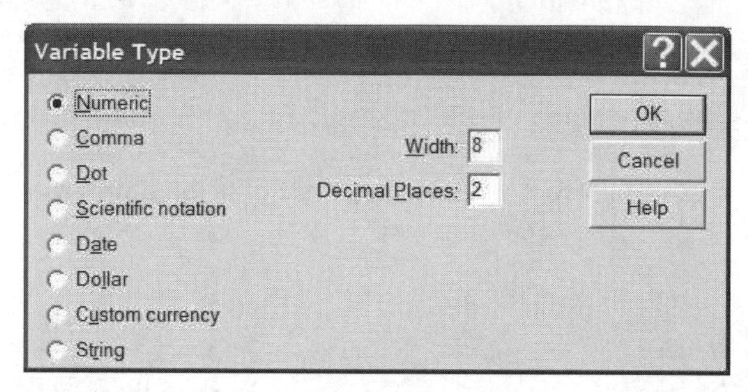

Notice that **Numeric** is selected. Most of your variables WILL be numeric and you will allow the default setting to stand. There are two non-numeric variables in the original **grades.sav** file, **lastname** and **firstname**. For those two variables you will click on the grayed box to select **String.**

A variable that contains letters (rather than only numbers) is called a *string* variable. String variables may contain numbers (e.g., **type 2**, **JonesIII**) or even consist of numbers, but SPSS treats strings as nonnumeric, and only a very limited number of operations may be conducted with strings as variables. You may notice (see Screen 3.3) that eight different variable types are available. **Numeric** and **String** are by far the most frequently used options and the others will not be considered here. What follows is a step-by-step of how to change the **lastname** and **firstname** to string variables.

We continue on from sequence step 3 (above) beginning with Screen 3.2.

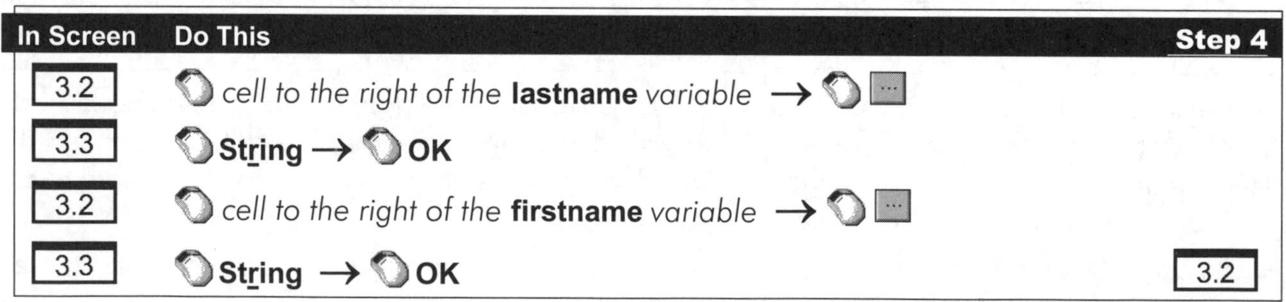

In Screen	Do This	Step 4
3.2	*cell to the right of the* **lastname** *variable* → ...	
3.3	**String** → **OK**	
3.2	*cell to the right of the* **firstname** *variable* → ...	
3.3	**String** → **OK**	3.2

Width

In the **Width** column you determine the largest number or longest string that will occur for each variable. For **id** the width will be 6 because all ID numbers in the data set have six digits. For **lastname** we might select 10. While there may be a student with a last name longer than 10 letters, that is typically long enough for identification. For **gpa** the width will be 4: one digit left of the decimal, two digits to the right of the decimal, and one more space for the decimal point.

When you click on the cell under **Width** to the right of any variable name the cell is highlighted and a small appears in the right margin of the cell. Use this to increase or decrease width from the default of 8 characters, or simply highlight the original number and type in the number you desire.

We continue on from sequence step 4 (above) beginning with Screen 3.2.

In Screen	Do This	Step 5
3.2	**Width** *cell to the right of* **id** → *lower arrow of* *until 6 appears*	
	Width *cell to the right of* **lastname** → *upper arrow of* *until 10 appears*	
	Width *cell to the right of* **firstname** → *upper arrow of* *until 10 appears*	
	Width *cell to the right of* **gender** → *lower arrow of* *until 1 appears*	

Decimals

In the **Decimals** column you identify the number of decimal places desired to the right of the decimal point. The default is two places. For string variables the number of decimals is 0 by default. In the **grades.sav** file only the **gpa** variable has decimals—two digits—which is also the default. Since the decimal defaults are already correct for **lastname**, **firstname**, and **gpa**, we show in the following step-by-step box how to adjust four of the variables to 0 digits to the right of the decimal point. Note: if you attempt to reduce a **Width** (immediately above) to 2 or less while the decimal default of 2 remains, a warning will flash informing you that it isn't possible. You need to change the decimal value to 0 before creating a width of 2 or less.

We continue on from sequence step 5 (above) beginning with Screen 3.2.

In Screen	Do This	Step 6
3.2	**Decimals** *cell to the right of* **id** → *lower arrow of* *until 0 appears*	
	Decimals *cell to the right of* **gender** → *lower arrow of* *until 0 appears*	
	Decimals *cell to the right of* **ethnic** → *lower arrow of* *until 0 appears*	
	Decimals *cell to the right of* **year** → *lower arrow of* *until 0 appears*	

Label

The **Label** column allows you to label any variable whose meaning is not clear from the variable name. Many times the meaning is clear from the variable name itself (e.g., **id, gender, quiz1, quiz2**) and no label is required. Other times the meaning is NOT clear and a label is very useful. When a label is entered, in the Output section following any analyses, the label may be listed next to the variable name to assist in clarity and interpretation.

The cells in the **Label** column are simply text boxes and you type the label you desire. The maximum length is 256 characters, but something that long is very cumbersome. Twenty or 30 characters is usually plenty to get your point across.

As you type a label, the width of the **Label** column will increase to allow room for the label. If it becomes too wide, simply position the cursor at the line to the right of **Label** in the top row. The cursor will become a ✛ and you can adjust the column back to a shorter (or longer) width. Use this procedure freely to custom format the Variable View screen.

We continue on from sequence step 6 (above) and create labels for two variables.

In Screen	Do This	Step 7
3.2	**Label** *cell to the right of* **year** → type `year in school`	
	Label *cell to the right of* **lowup** → type `lower or upper division`	

Values

Value labels allow you to identify *levels* of a variable (e.g., **gender:** 1=female, 2=male; **marital:** 1=married, 2=single, 3=divorced, 4=widowed). Entering value labels for variables that have several distinct groups is absolutely critical for clarity of interpretation of output. SPSS can do the arithmetic whether or not you include value labels, but you'll never remember whether a three means single or divorced (or maybe it was widowed? I'll bet you had to look!). Another advantage of value labels is that SPSS can display these labels in your data file and in Output following analyses. The pleasure of having labels included in the data file is something that quickly becomes addictive. SPSS allows up to 60 characters for each value label, but clearly something shorter is more practical.

Just like the **Type** option, a click on any cell in the **Values** column will produce the small grayed box (...). A click on this box will produce a dialog box (Screen 3.4, following page) that will allow you to create value labels.

In the step-by-step box below we show how to use this dialog box to create labels for **gender** *and* **year**. *Again we begin at Screen 3.2.*

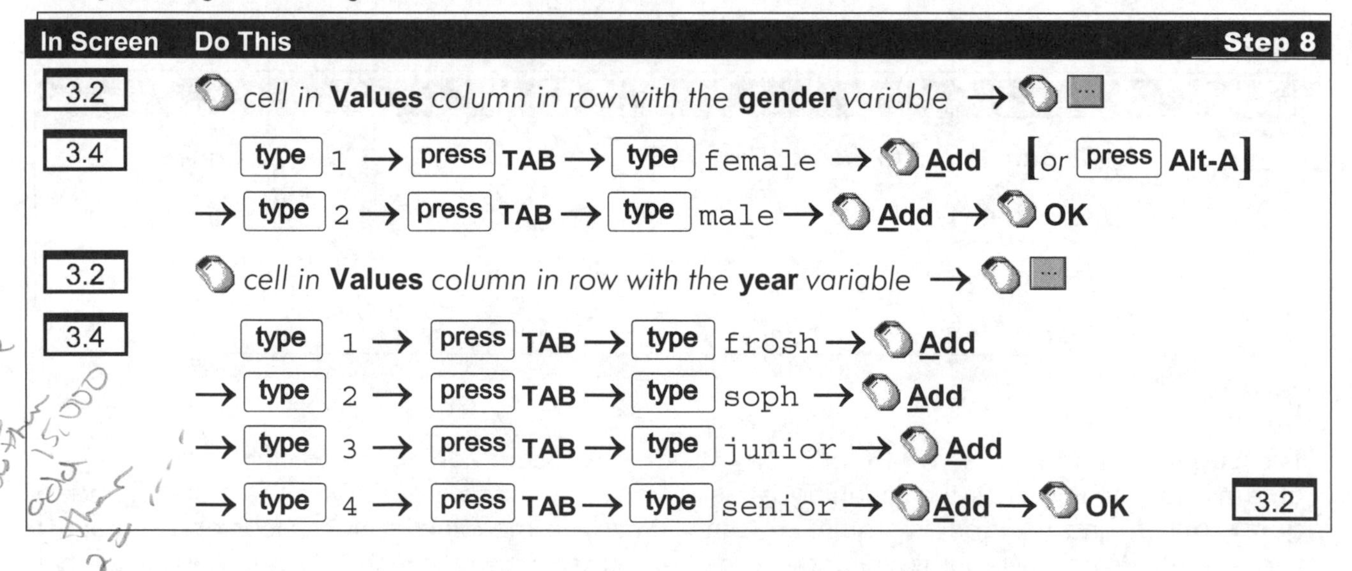

In Screen	Do This	Step 8
3.2	*cell in* **Values** *column in row with the* **gender** *variable* → ...	
3.4	type `1` → press TAB → type `female` → **Add** [or press **Alt-A**] → type `2` → press TAB → type `male` → **Add** → **OK**	
3.2	*cell in* **Values** *column in row with the* **year** *variable* → ...	
3.4	type `1` → press TAB → type `frosh` → **Add** → type `2` → press TAB → type `soph` → **Add** → type `3` → press TAB → type `junior` → **Add** → type `4` → press TAB → type `senior` → **Add** → **OK**	3.2

Missing

The **Missing** column is rarely used. Its purpose is to designate different types of missing values in your data. For instance, subjects who refused to answer the ethnicity question might be coded 8 and those who were of different ethnicity than those listed might be coded 9. If you have entered these values in the **Missing** value column, then you may designate that 8s and 9s not enter into any of the analyses that follow. A small dialog box (not reproduced here) opens when you click the ... that allows you to specify which values are user-missing.

3.4

The Value Labels Window

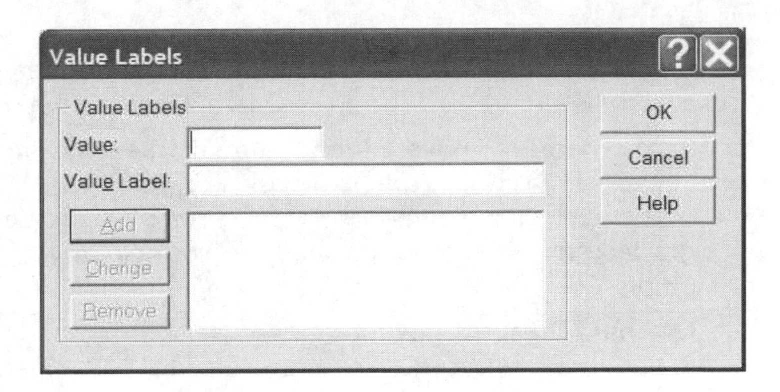

Columns

The **Columns** column, by contrast, is used with most variables. This allows you to identify how much room to allow for your data and labels. The pros and cons on wide versus narrow columns are clear: If you have wide columns you can see the entire variable name and thus seem less crowded. If you have narrow columns, you have the advantage of getting many variables visible on the screen at one time but you may have to truncate variable names to do so.

For instance if you have columns that are 3 characters wide you may be able to fit 33 variables into the visible portion of the data screen (depending on your monitor); if the columns are all 8 characters wide (the default), you can fit only 12. Anyone who has spent a good deal of time working with data files knows how convenient it is to be able to view a large portion of the data screen without scrolling. But many variable names would be truncated and thus unintelligible with only a 3-character width. The same rationale applies to the value labels. Note that for the **ethnic** variable it might be more precise to create labels of "American Native or Inuit, Asian or Asian American, African or African American/Canadian, Caucasian, Hispanic", but the brief labels of "Native, Asian, Black, White, and Hispanic" gets the point across without clutter.

Once again the ⬍ becomes visible in the right margin when you click a cell in the **Columns** column. You may either click number higher or lower, or type in the desired number after highlighting the default number, 8.

We enter the column width for the first five variables in the step-by-step box below. Notice how the width reflects the name of the variable. The starting point, again is Screen 3.2

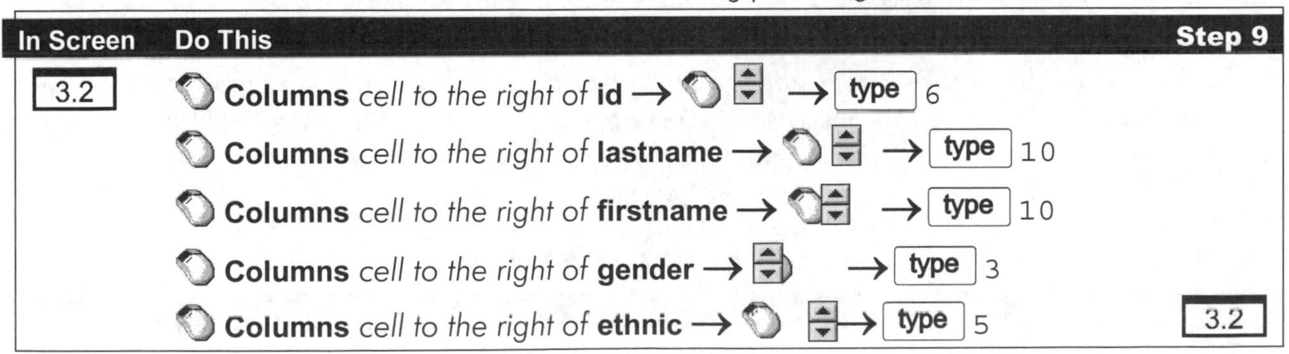

In Screen	Do This	Step 9
3.2	**Columns** *cell to the right of* **id** → ⬍ → **type** 6	
	Columns *cell to the right of* **lastname** → ⬍ → **type** 10	
	Columns *cell to the right of* **firstname** → ⬍ → **type** 10	
	Columns *cell to the right of* **gender** → ⬍ → **type** 3	
	Columns *cell to the right of* **ethnic** → ⬍ → **type** 5	3.2

Align

The **Align** column provides a drop-down menu that allows you to align the data in each cell either right, left, or center. By default, numeric variables align to the right, string variables align to the left. You may select otherwise if you wish.

Measure

The **Measure** column also provides a drop-down menu that allows you to select three options based on the nature of your data: **Scale**, **Ordinal**, and **Nominal**.

- **Scale** measures have intrinsic numeric meaning that allow typical mathematical manipulations. For instance, age is a scale variable: 16 is twice as much as 8, 4 is half as much as 8, the sum of a 4 and 8 is 12, and so forth. **Scale** is default for numeric variables. In the **grades.sav** file, scale variables would include year, the quizzes, the final, total score, and others.
- **Ordinal** measures have intrinsic order but mathematical manipulations are typically meaningless. On an aggression scale of 1 to 10, someone higher on the scale is more aggressive than someone lower on the scale, but someone who rates 4 is not twice as aggressive as someone who rates 2. Interestingly, dichotomous variables (that have exactly two options, such as male/female, pass/fail, true/false) are considered ordinal for reasons explained in Chapter 10 on correlations.
- **Nominal** measures are used for identification but have no intrinsic (lesser to greater) order such as ethnicity, marital status, and any string variables. Nominal data may be used for categorization but cannot be used in most analyses.

Sometimes it can be difficult to choose between scale and ordinal. If so, don't worry too much. In all analysis, SPSS handles both ordinal and scale variables identically.

Each variable is now named and formatted; the next step is to enter data.

ENTERING DATA

After naming and formatting the variables, entering data is a simple process. We first give you the sequence of steps for saving the data file—an event that should occur frequently during the creation of the file and entry of data. After that, we describe two different ways of entering the data; depending on the particular data file you are creating, either may be appropriate.

When you save the file for the first time you need to enter the file name.

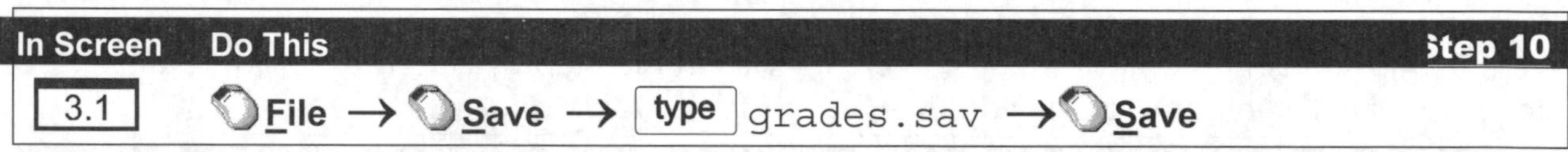

Note: These instructions assume that you want to save the file in the default folder (usually C:\Program Files\SPSS or A:\, but depending on your particular computer setup it could be a different folder). If you don't want to save your work in the default folder, you will need to click the up-one-level button ( ) one or more times, and double click on the disk drives, network paths, or folders in which you want to save your file.

After the first time you save the file, saving your changes requires just one or two clicks.

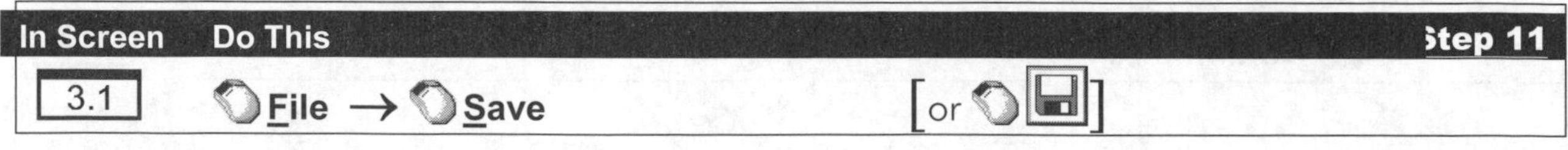

There are two different ways of entering data that we will describe here: by variable and by case or subject. We will describe both methods of entering data; depending on the format of the data you are entering, either method may be easier.

Enter data by variable: Click on the first empty cell under the first variable, type the number (or word), press the **Down-arrow** key or **Enter** key, then type the next number/word, press the **Down-arrow** or **Enter** key, and so forth. When you finish one variable, scroll up to the top of the file and enter data for the next variable in the same manner.

Enter data by case or subject: Click on the first empty cell for the first subject under the first variable, then type the first number/word, press the **Right-arrow** key or **TAB** key, type the next number, press the **Right-arrow** or **TAB** key, and so forth. When you finish one subject, scroll back to the first column and enter data for the next subject.

EDITING DATA

Just as data entry is a simple procedure, so also is editing previously entered data. The following options are available:

Changing a cell value: Simply click on the cell of interest, type the new value, then hit **Enter**, **TAB**, or any of the **Arrow keys**.

Inserting a new case: If you wish to insert data for a new subject or case, click on the case number above which you would like the new case to be. Then click on the insert-case toolbar icon ( ) and a new line will open and push all other cases down by exactly one line. You may then enter data for the new subject or case.

Inserting a new variable: To insert a new variable, click on the variable to the right of where you would like the new variable to be located, click on the insert-variable icon, ( ) and a new column will open and push all other variables exactly one column to the right. You may then name and format the new variable and enter data in that column.

To copy or cut cells: To copy rows or columns of data, first highlight the data you want to copy by a click on the variable name (to highlight all entries for one variable) or click on the case number (to highlight all the entries for one subject or case) then use the shift and arrow keys to highlight larger blocks of data. Once the cells you want to copy are highlighted, click on the **Edit** menu command followed by the **Copy** command if you wish to *leave* the cells where they are but put them onto the clipboard for future pasting. Click the **Cut** option if you wish to *delete* the cells from the data editor but place them onto the clipboard.

To paste cells: At first glance, pasting cells seems simply a matter of clicking on the upper-left cell of the group of cells you want to paste, then **Edit**, then **Paste**. This process is actually somewhat tricky, however, for two reasons: 1) If you paste data into already existing cells, you will erase any data that is already in those cells, and 2) SPSS lets you paste data copied from one variable into another variable (this can cause confusion if, for example, you copy data from the **gender** variable into the **marital** status variable). Here are some tips to avoid problems when pasting cells:

- Save your data file before doing major cutting and pasting, so that you can go back to an earlier version if you make a mistake.
- Create space for the new data by inserting new cases and variables (see instructions on the previous page) to make room for the cells you are going to paste.
- In most cases you will cut and paste entire rows (cases) or columns (variables) at the same time. It requires careful attention to paste *segments* of rows or columns.

- ❒ Be careful to align variables appropriately when pasting, so that you don't try pasting string variables into numeric variables or other problematic variable mismatches.

To search for data: One of the handiest editing procedures is the **Find** function. A click on the **Edit** command followed by a click on the **Find** option (or click on the toolbar icon) opens up a screen (Screen 3.5) that allows you to search for a particular word or data value. This function is most frequently used for two different purposes:

- ❒ If you have a large file that includes names, you can quickly find a particular name that is embedded within the file.
- ❒ If you discover errors in your data file (e.g., a GPA of 6.72), the search function can quickly find those errors for correction.

3.5

The Find Data Window

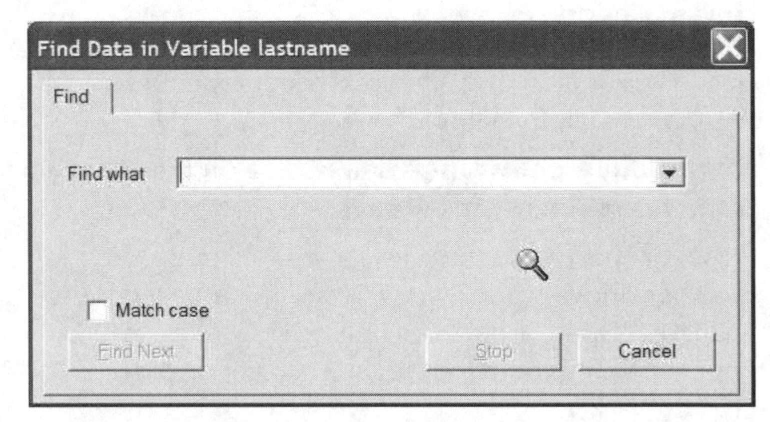

If for instance we wish to find a student with the last name Osborne, execute the following sequence of steps:

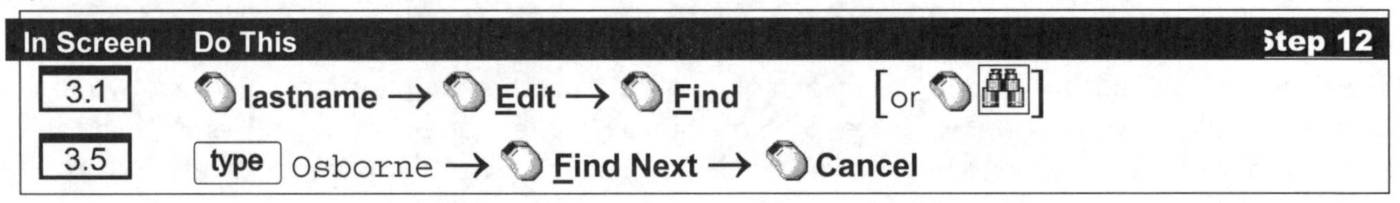

In Screen	Do This	Step 12
3.1	lastname → Edit → Find [or]	
3.5	type Osborne → Find Next → Cancel	

If you are working with a data file with a large number of subjects, and notice (based on frequency print-outs) that you have three subjects whose scores are outside the normal range of the variable (e.g., for **anxiety** *coded 1 to 7 you had three subjects coded 0), and you wish to locate these errors, perform the following sequence of steps.*

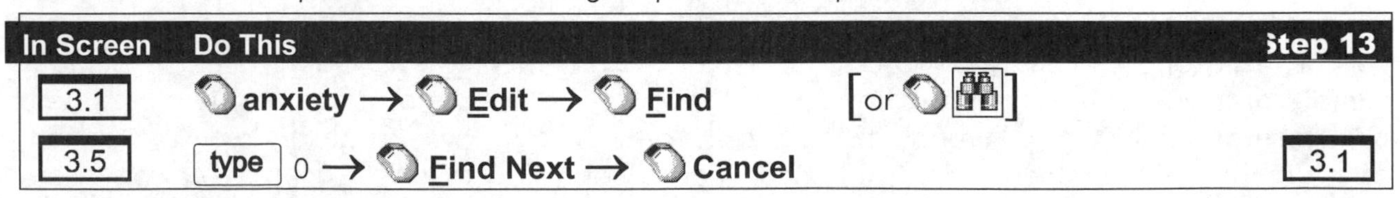

In Screen	Do This	Step 13
3.1	anxiety → Edit → Find [or]	
3.5	type 0 → Find Next → Cancel	3.1

To make subject identifiers always visible: When a file has many variables, it is often useful to make some of the columns (those on the left, if you set the file up as we recommend) always visible, even when you are scrolling over to view or enter variables along the right.

This is easy to do, once you find where to put the cursor. Near the bottom right of the screen, look for a small area just to the right of the horizontal scroll bar.

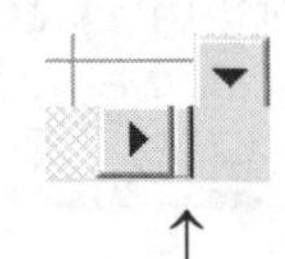

↑

This is what you're looking for.

Drag that line to the left, until it is immediately to the right of any variables you want to always be ready to see. For example, in the **grades.sav** file, you might move it just to the right of the **lastname** variable. Then, your file would look like this:

	id	lastname	gender	ethnicity	year
1	106484	VILLARRUZ	Male	Asian	Sc
2	108642	VALAZQUEZ	Male	White	Jun
3	127285	GALVEZ	Female	White	Sen
4	132931	OSBORNE	Female	Black	Sc
5	140219	GUADIZ	Female	Asian	Sen

This part is always visible — This part scrolls

This makes it *much* easier to work with a "wide" file with many variables.

GRADES.SAV: THE SAMPLE DATA FILE

The data file is the raw data for calculating the grades in a particular class. The example consists of a single file, used by a teacher who teaches three sections of a class with approximately 35 students in each section. From left to right, the variables that are used in the data file are:

Variable	Description
ID	Six-digit student ID number
LASTNAME	The last name of the student
FIRSTNAME	First name of the student
GENDER	Gender of the student: 1=female, 2=male
ETHNIC	Ethnicity of the student: 1=Native (Native American or Inuit), 2=Asian (or Asian American), 3=Black, 4=White (non-Hispanic), 5=Hispanic
YEAR	Year in school; 1=Frosh (1st year), 2=Soph (2nd year), 3=Junior (3rd year), 4=Senior (4th year)
LOWUP	Lower or upper division student: 1=Lower (1st or 2nd year), 2=Upper (3rd or 4th year)
SECTION	Section of the class (1 through 3)
GPA	Cumulative GPA at the beginning of the course
EXTRCREDIT	Whether or not the student did the extra credit project: 1=No, 2=Yes
REVIEW	Whether or not the student attended the review sessions: 1=No, 2=Yes
QUIZ1 to QUIZ5	Scores out of 10 points on five quizzes throughout the term
FINAL	Final exam worth 75 points.

ID	Lastname	Firstname	Gender	Ethnic	Year	Lowup	Section	GPA	Extrcredit	Review	Quiz1	Quiz2	Quiz3	Quiz4	Quiz5	Final
973427	ROSS	MARIA	1	4	4	2	1	3.19	1	2	9	7	10	9	7	65
390203	SHIMA	MIHAELA	1	2	3	2	2	2.28	1	2	6	7	9	6	8	61
703740	SUNYA	DALE	2	5	3	2	3	3.58	1	2	10	9	10	10	7	62
354601	CARPIO	MARY	1	2	2	1	1	2.03	1	2	10	10	10	10	9	71
979028	NEUHARTH	JIM	2	4	3	2	3	1.80	1	2	3	6	3	4	5	49
768995	DUMITRESCU	STACY	2	4	4	2	2	2.88	1	1	7	10	8	9	10	60
574170	HURRIA	WAYNE	2	1	2	1	2	3.84	1	1	4	5	6	6	6	48
380157	LUTZ	WILLIAM	2	4	3	2	2	2.25	2	2	10	9	10	10	8	61
167664	SWARM	MARK	2	4	3	2	3	2.35	1	2	8	10	10	10	9	71

ID	Lastname	Firstname	Gender	Ethnic	Year	Lowup	Section	GPA	Extrcredit	Review	Quiz1	Quiz2	Quiz3	Quiz4	Quiz5	Final
245473	DAYES	ROBERT	2	4	3	2	1	2.74	1	1	8	9	6	7	10	48
436413	PANG	SUZANNE	1	2	3	2	1	2.66	1	2	8	6	7	8	7	60
515586	FIALLOS	LAUREL	1	4	2	1	2	3.90	1	1	7	8	8	6	6	63
106484	VILLARUIZ	ALFRED	2	2	2	1	2	1.18	1	2	6	5	7	6	3	53
725987	BATILLER	FRED	2	2	2	1	2	1.77	1	2	6	7	7	7	5	60
870810	REYNO	NICHOLAS	2	4	3	2	3	3.66	2	1	10	8	10	10	10	68
127285	GALVEZ	JACKIE	1	4	4	2	2	2.46	2	2	10	7	8	9	7	57
623857	CORTEZ	VIKKI	1	3	4	2	3	2.56	1	2	5	7	6	5	6	58
905109	JENKINS	ERIC	2	3	2	1	3	2.84	1	1	6	8	6	6	10	64
392464	DOMINGO	MONIKA	1	4	3	2	3	3.02	2	1	10	10	10	9	9	55
447659	GLANVILLE	DANA	1	5	4	2	3	2.77	1	1	6	8	9	5	8	63
958384	RONCO	SHERRY	1	4	2	1	1	2.30	1	2	10	9	10	10	7	60
414775	RATANA	JASON	2	2	3	2	1	2.38	1	2	8	9	10	10	9	50
237983	LEE	JONATHAN	2	2	4	2	2	1.66	2	2	5	7	4	7	6	63
818528	CARRINGTON	JYLL	1	4	3	2	1	1.95	1	2	9	10	10	8	8	53
938881	YEO	DENISE	1	1	3	2	3	3.53	1	2	7	10	9	8	9	72
944702	LEDESMA	MARTINE	1	4	3	2	2	3.90	1	2	6	7	7	5	9	67
154441	LIAN	JENNY	1	5	2	1	1	3.57	1	2	10	9	10	10	10	71
108642	VALAZQUEZ	SCOTT	2	4	3	2	2	2.19	2	1	10	10	7	6	9	54
985700	CHA	LILY	1	4	2	1	1	2.43	2	2	10	9	10	10	7	63
911355	LESKO	LETITIA	1	3	2	1	3	3.49	1	2	10	9	10	10	8	71
249586	STOLL	GLENDON	2	4	3	2	2	2.51	1	1	5	9	5	6	10	63
164605	LANGFORD	DAWN	1	3	3	2	2	3.49	2	1	10	10	9	10	10	75
419891	DE CANIO	PAULA	1	4	3	2	2	3.53	1	2	6	7	7	9	9	54
615115	VASENIUS	RUSS	2	3	3	2	3	1.77	1	2	6	7	6	8	6	59
192627	MISCHKE	ELAINE	1	4	1	1	2	2.90	1	1	3	8	4	6	8	55
595177	WILLIAMS	OLIMPIA	1	3	3	2	3	1.24	1	1	7	6	7	10	5	53
434571	SURI	MATTHEW	2	2	3	2	2	2.80	1	1	7	6	9	8	8	60
506467	SCARBROUGH	CYNTHIA	1	4	3	2	2	1.33	1	2	8	5	6	4	7	58
546022	HAMIDI	KIMBERLY	1	5	3	2	1	2.96	1	1	7	7	6	9	8	61
498900	HUANG	JOE	2	5	3	2	3	2.47	1	1	0	5	0	2	5	40
781676	WATKINS	YVONNE	1	3	4	2	1	4.00	1	2	9	9	10	10	9	70
664653	KHAN	JOHN	2	4	3	2	3	1.24	1	2	3	8	5	2	7	59
908754	MARQUEZ	CHYRELLE	1	4	1	1	2	1.85	1	2	4	8	5	7	9	57
142630	RANGFIO	TANIECE	1	4	3	2	3	3.90	1	2	10	10	10	9	9	74
175325	KHOURY	DENNIS	2	4	3	2	1	2.45	1	1	8	8	10	10	6	69
378446	SAUNDERS	TAMARA	1	1	2	1	2	2.80	1	2	4	6	5	4	5	57
289652	BRADLEY	SHANNON	1	4	3	2	1	2.46	1	2	6	9	8	9	9	68
466407	PICKERING	HEIDI	1	3	3	2	3	2.38	1	1	4	7	6	4	7	56
898766	RAO	DAWN	1	2	3	2	1	3.90	1	2	8	10	10	8	9	73
302400	JONES	ROBERT	2	3	4	2	3	1.14	1	2	2	5	4	5	6	43
157147	BAKKEN	KREG	2	4	3	2	1	3.95	2	2	10	10	10	10	9	74
519444	RATHBUN	DAWN	1	4	4	2	2	3.90	1	1	10	9	10	10	8	74
420327	BADGER	SUSAN	1	4	3	2	3	2.61	1	2	10	10	10	10	10	53
260983	CUSTER	JAMES	2	4	4	2	1	2.54	1	1	10	9	10	10	7	60
554809	JONES	LISA	1	3	3	2	3	3.35	1	1	7	8	8	9	6	69
777683	ANDERSON	ERIC	2	5	4	2	3	2.40	1	1	3	6	3	2	6	50
553919	KWON	SHELLY	1	2	3	2	1	3.90	2	2	10	10	10	10	8	75
479547	LANGFORD	BLAIR	2	3	3	2	1	3.42	2	2	10	10	10	9	10	75
755724	LANGFORD	TREVOR	2	4	3	2	2	2.96	1	2	8	9	9	9	8	62

ID	Lastname	Firstname	Gender	Ethnic	Year	Lowup	Section	GPA	Extrcredit	Review	Quiz1	Quiz2	Quiz3	Quiz4	Quiz5	Final
337908	UYEYAMA	VICTORINE	1	1	3	2	2	2.34	2	1	10	8	10	10	7	63
798931	ZUILL	RENEE	1	4	3	2	1	2.22	2	2	10	9	10	10	8	62
843472	PRADO	DON	2	5	3	2	3	3.54	1	2	9	9	10	8	9	68
762308	GOUW	BONNIE	1	4	2	1	3	3.90	1	2	8	7	9	10	8	57
700978	WEBSTER	DEANNA	1	3	2	1	3	3.90	1	2	8	9	9	10	10	67
721311	SONG	LOIS	2	2	3	2	3	1.61	1	1	6	9	9	7	10	64
417003	EVANGELIST	NIKKI	1	2	3	2	2	1.91	1	2	9	8	10	10	6	66
153964	TOMOSAWA	DANIEL	2	2	3	2	3	2.84	2	1	10	9	10	10	10	63
765360	ROBINSON	ERIC	2	3	3	2	2	2.43	1	2	8	8	7	8	10	65
463276	HANSEN	TIM	2	4	3	2	1	3.84	2	2	10	10	10	9	10	74
737728	BELTRAN	JIM	2	3	3	2	1	2.57	1	1	6	8	9	5	7	62
594463	CRUZADO	MARITESS	1	4	4	2	2	3.05	1	2	9	8	10	8	8	65
938666	SUAREZ-TAN	KHANH	1	2	3	2	3	2.02	2	2	10	8	10	10	7	52
616095	SPRINGER	ANNELIES	1	4	3	2	1	3.64	1	2	10	10	10	10	10	72
219593	POTTER	MICKEY	1	5	3	2	3	2.54	1	2	5	8	6	4	10	61
473303	PARK	SANDRA	1	3	4	2	2	3.17	1	2	8	8	8	10	9	70
287617	CUMMINGS	DAVENA	1	5	3	2	3	2.21	1	2	9	10	9	9	9	52
681855	GRISWOLD	TAMMY	1	4	3	2	2	1.50	1	2	5	7	8	5	8	57
899529	HAWKINS	CATHERINE	1	3	4	2	2	2.31	1	1	10	8	9	10	7	49
576141	MISHALANY	LUCY	1	4	3	2	1	3.57	1	2	0	3	2	2	2	42
273611	WU	VIDYUTH	1	2	2	1	2	3.70	1	2	3	6	2	6	6	55
900485	COCHRAN	STACY	2	4	3	2	2	2.77	2	2	10	9	10	10	9	61
211239	AUSTIN	DERRICK	2	4	3	2	3	2.33	1	2	5	5	7	6	4	52
780028	ROBINSON	CLAYTON	2	4	3	2	1	3.90	1	2	10	10	10	9	10	73
896972	HUANG	MIRNA	1	2	3	2	1	2.56	1	1	7	6	10	8	7	57
280440	CHANG	RENE	1	2	3	2	2	3.90	1	2	10	8	10	10	8	68
920656	LIAO	MICHELLE	1	2	2	1	2	3.28	2	2	10	9	10	10	9	72
490016	STEPHEN	LIZA	1	5	3	2	2	2.72	1	2	8	9	9	8	10	60
140219	GUADIZ	VALERIE	1	2	4	2	1	1.84	1	1	7	8	9	8	10	66
972678	KAHRS	JANNA	1	4	4	2	2	2.37	1	2	10	10	10	10	10	53
988808	MCCONAHA	CORA	1	4	3	2	3	3.06	1	2	7	8	9	8	7	68
132931	OSBORNE	ANN	1	3	2	1	2	3.98	1	1	7	8	7	7	6	68
915457	SHEARER	LUCIO	2	3	3	2	1	2.22	1	2	10	10	10	9	8	52
762813	DEAL	IVAN	2	3	2	1	1	2.27	2	2	10	9	10	10	10	62
897606	GENOBAGA	JACQUELIN	1	2	3	2	3	2.92	1	2	8	9	8	8	7	68
164842	VALENZUELA	NANCY	1	1	4	2	2	2.32	1	1	7	8	6	7	10	59
822485	VALENZUELA	KATHRYN	1	4	1	1	1	3.90	1	2	8	9	10	10	8	66
978889	ZIMCHEK	ARMANDO	2	4	4	2	1	3.90	1	2	4	8	6	6	9	64
779481	AHGHEL	BRENDA	1	5	3	2	1	3.01	1	2	3	5	3	2	4	49
921297	KINZER	RICHARD	2	4	3	2	2	2.73	1	2	7	9	9	7	8	67
983522	SLOAT	AARON	2	3	3	2	3	2.11	1	1	4	5	6	6	6	50
807963	LEWIS	CARL	2	3	2	1	1	2.56	2	1	8	5	6	4	7	62
756097	KURSEE	JACKIE	1	3	3	2	2	3.13	1	2	9	6	8	7	10	66
576008	BULMERKA	HUSIBA	1	4	4	2	3	3.45	2	1	10	8	7	9	7	68
467806	DEVERS	GAIL	1	3	3	2	1	2.34	1	1	7	6	8	7	9	59
307894	TORRENCE	GWEN	1	3	2	1	2	2.09	2	2	6	5	4	7	6	62

EXERCISES

Answers to selected exercises are available for download at www.abacon.com/george6e.

1. Set up the variables described above for the **grades.sav** file, using appropriate variable names, variable labels, and variable values. Enter the data for the first five students into the data file.
2. Perhaps the instructor of the classes in the **grades.sav** dataset teaches these classes at two different schools. Create a new variable in this dataset named **school**, with values of 1 and 2. Create variable labels, where 1 is the name of a school you like, and 2 is the name of a school you don't like. Save your dataset with the name **gradesme.sav**.
3. Which of the following variable names will SPSS accept, and which will SPSS reject? For those that SPSS will reject, how could you change the variable name to make it "legal"?
   - **age**
   - **firstname**
   - **@edu**
   - **sex.**
   - **grade**
   - **not**
   - **anxeceu**
   - **date**
   - **iq**
4. Using the **grades.sav** file, make the **gpa** variable values (which currently have two digits after the decimal place) have no digits after the decimal point. You should be able to do this without retyping any numbers. *Note that this won't actually round the numbers, but it will change the way they are displayed and how many digits are displayed after the decimal point for statistical analyses you perform on the numbers.*
5. Using **grades.sav**, search for a student who got 121 on the final exam. What is his or her name?
6. Why is each of the following variables defined with the measure listed? Is it possible for any of these variables to be defined as a different type of measure?

ethnicity	Nominal
extrcred	Ordinal
quiz4	Scale
grade	Nominal

7. Ten people were given a test of balance while standing on level ground, and ten other people were given a test of balance while standing on a 30° slope. Their scores follow. Set up the appropriate variables, and enter the data into SPSS.
   - Scores of people standing on level ground: 56, 50, 41, 65, 47, 50, 64, 48, 47, 57
   - Scores of people standing on a slope: 30, 50, 51, 26, 37, 32, 37, 29, 52, 54
8. Ten people were given two tests of balance, first while standing on level ground and then while standing on a 30° slope. Their scores follow. Set up the appropriate variables, and enter the data into SPSS.

Participant:	1	2	3	4	5	6	7	8	9	10
Score standing on level ground:	56	50	41	65	47	50	64	48	47	57
Score standing on a slope:	38	50	46	46	42	41	49	38	49	55

Managing Data

THE MATERIAL covered in this chapter, when mastered, will provide a firm grounding in data management that will serve you well throughout the remaining chapters of this book. Rarely will you complete an analysis session without making use of some of the features in this chapter. Occasionally you may have a single data file and want to run a single operation. If so, reference to this chapter may be unnecessary; just use the data as you originally entered and formatted them. Chapter 3 tells how to create and format a data file, and Chapters 6-27 provide lucid step-by-step instructions about how to analyze data. However, it is common to want to analyze variables that are not *in* the original data file but are derived *from* the original variables (such as total points, percentage of possible points, or others). It is also sometimes desirable to display the data differently (e.g., alphabetically by student last names, by total points from high to low, or class records listed by sections), to code variables differently (e.g., 90-100 coded A, 80-89 coded B, and so forth), to perform analyses on only a certain portion of the data set (e.g., scores for females, GPAs for Whites, final percentages for Sophomores, final grades for Section 2, etc.), or if the sections were in three different files, to merge the files to gain a clearer picture of characteristics of the entire class.

All these operations may at times be necessary if you wish to accomplish more than a single analysis on a single set of data. Some operations presented here are complex, and it is possible that you may encounter difficulties that are not covered in this book. The different functions covered in this chapter are considered one at a time, and we present them as clearly and succinctly as space allows. You can also find additional help by selecting **Help → Topics** within SPSS; this will allow you to browse the documentation, or use the index to look up a particular key term that you are having problems with.

Despite the potential difficulty involved, a thorough understanding of these procedures will provide the foundation for fluency in the use of other SPSS operations. Once you have learned to manage data effectively, many statistical processes can be accomplished with almost ridiculous ease. We will present here seven different types of data management tools:

1. The judicious use of the **Case Summaries** option to assist in proofing and editing your data.
2. Using the **Replace Missing Values** subcommand for several options that deal with replacing missing data.
3. Computing and creating new variables from an already existing data set by using the **Transform** command and the **Compute** subcommand.
4. Using the **Recode** subcommand to change the coding associated with certain variables.
5. Employing the **Select Cases** procedure to allow statistical operations on only a portion of your data.
6. Utilizing the **Sort Cases** subcommand to reorder your data.
7. Adding new cases or new variables to an already existing file by use of the **Merge Files** option.

Beginning with this chapter we introduce the convention that allows all chapters that follow to be self-contained. The three steps necessary to get as far as reading a data file and accessing a screen that will allow you to conduct analyses will begin the Step by Step section in each chapter. This will be followed by the steps necessary to run a particular analysis. A description of how to print the results and exit the program will conclude each section. The critical steps designed to perform a particular procedure for each chapter will typically involve a sequence step 4 that accesses and sets up the procedure, and one or more versions of sequence step 5 that describes the specifics of how to complete the procedure(s). If there is more than one step-5 version, the steps will be designated 5, 5a, 5b, 5c, and so forth.

This chapter will be formatted somewhat differently from the analysis chapters (6-27) because we are presenting seven different types of computations or manipulations of data. We will begin the Step by Step section with the three introductory steps mentioned above and then present each of the seven operations with its own title, introduction, and discussion. After the seventh of these presentations, we will conclude with the final steps. Unlike the analysis chapters, there are no Output sections in this chapter.

STEP BY STEP

Manipulation of Data

To enter SPSS, a click on **Start** *in the taskbar (bottom of screen) activates the start menu:*

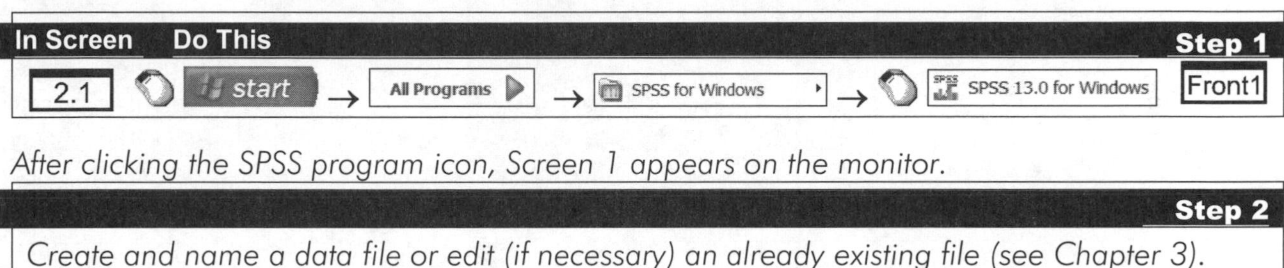

Screens 1 and 2 (displayed on the inside front cover) allow you to access the data file used in conducting the analysis of interest. The following sequence accesses the **grades.sav** *file for further analyses:*

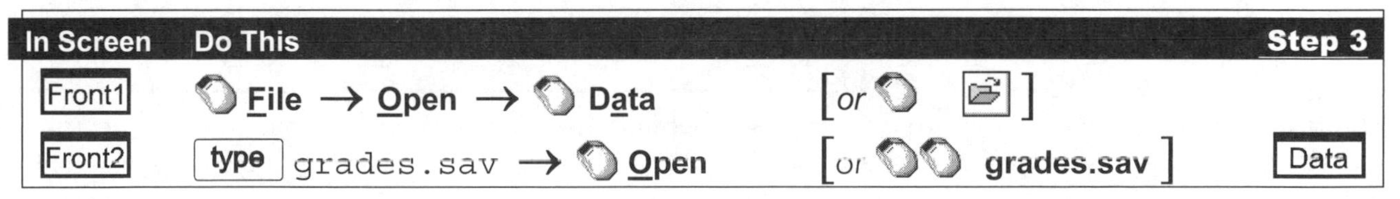

Whether first entering SPSS or returning from earlier operations the standard menu of commands across the top is required (shown below). As long as it is visible you may perform any analyses. It is not necessary for the data window to be visible.

This menu of commands disappears or modifies when using pivot tables or editing graphs. To uncover the standard menu of commands simply click on the ▬ or the ▣ icon.

The CASE SUMMARIES Procedure

Listing All or a Portion of Data

After completion of Step 3, a screen with the desired menu bar appears. When you click a command (from the menu bar) a series of options will appear (usually) below the selected command. With each new set of options, click the desired item. To access the **Case Summaries** procedure, begin at any screen that shows the standard menu of commands.

One of the best ways to ensure that you have entered data correctly or that SPSS has followed your instructions and created, formatted, or arranged your data is to make generous use of the **Case Summaries** operation. The **Case Summaries** command allows you to list an entire data file or a subset of that file, either grouped or in the order of the original data. After executing the steps listed above, a new window opens (Screen 4.1) that allows you to specify which variables and/or cases you wish to have listed.

4.1

The Summarize Cases Window

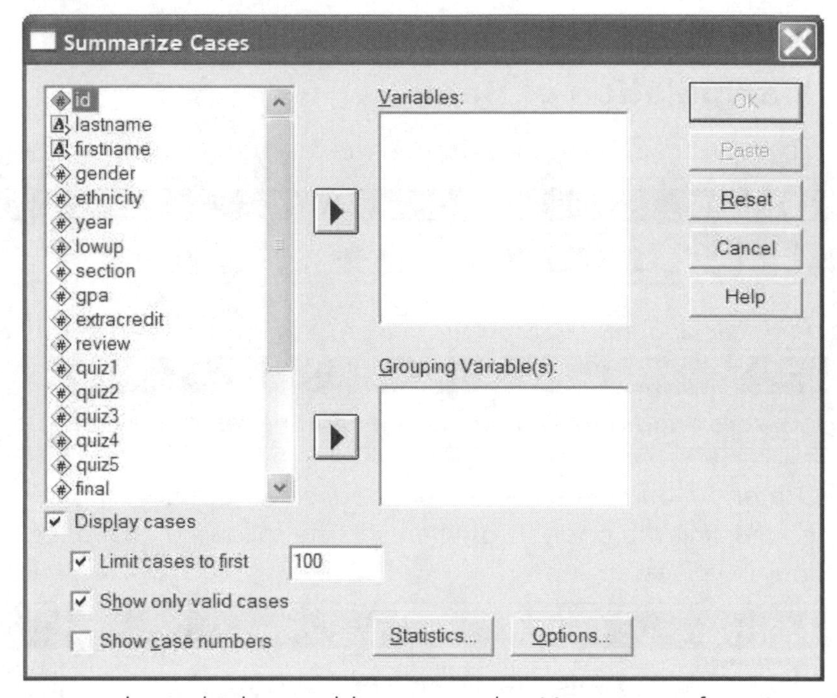

The variable list on the left allows you to select which variables you wish. You may, of course, select all of them. In selecting the variables you have the option of choosing the order in which they appear. The order of selection in the **Variables** box will be the same order presented in the output. Several options allow you to select and format both the content and structure of the output.

- **Variables**: The list of variables selected for listing all or a subset of values.
- **Grouping Variables**: Will create an order for the listing of variables. If there are no grouping variables, cases will be listed in the order of the data file. If, for instance, gender were included under grouping variables, then the output would list all women (in the order of the data file) then all men, also in order. Women would be listed first because in our data set, women are coded 1 and men are coded 2. If you selected section, the output would list all students in section 1, then students in section 2, then students in section 3. You may include more than one grouping variable. If you included section and gender, the output would first list women in section 1, then men in section 1, then women in section 2, and so forth.
- **Display cases**: This is selected by default (and is the major function of the Case Summaries procedure). If deselected, then output will list only the number of data points in each category.

- **Limit Cases to first**: The first 100 cases is default value. If you wish more or less, delete the **100** and type in the number you wish.
- **Show only valid cases**: This option is selected by default. It ensures that when variables are listed if there are three subjects who did not include their age, those three subjects would not be listed in the output. This option will often be deselected because we are often quite interested in where missing values occur.
- **Show case numbers**: This is not default but is almost routinely selected because you may wish to identify a particular subject or case. This would be difficult if the case number were not included.
- **Statistics**: A wide array of descriptive statistics is available. Most of these are described in Chapter 7. The **number of cases** is included by default. Any other statistics you wish may be clicked into the **Cell Statistics** box.
- **Options**: Allows you to exclude any subjects who have missing values in their data or to include or edit titles or subtitles. Since you may edit titles in the output anyway, this option is generally ignored.

The steps below request a listing of all variables (ordered alphabetically) for the first 20 cases with cases numbered. We begin this procedure where sequence step 4 ended, Screen 4.1.

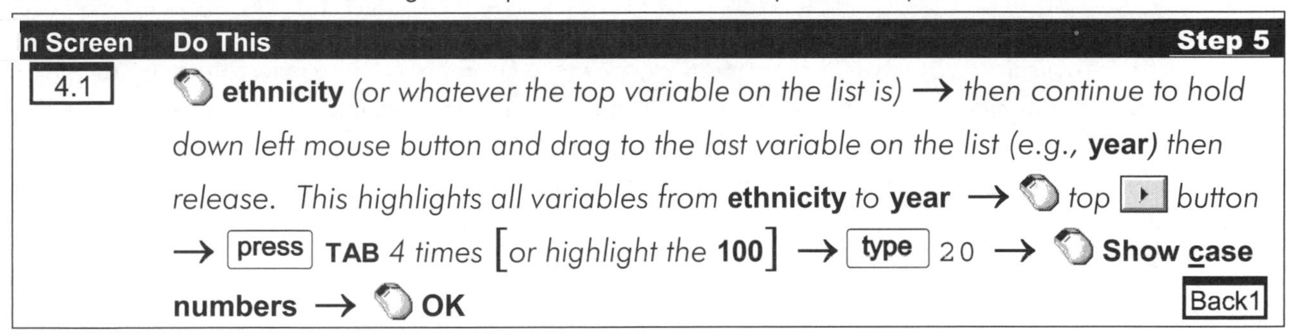

In Screen	Do This	Step 5
4.1	**ethnicity** *(or whatever the top variable on the list is)* → *then continue to hold down left mouse button and drag to the last variable on the list (e.g.,* **year***) then release. This highlights all variables from* **ethnicity** *to* **year** → *top* ▸ *button* → press **TAB** *4 times* [*or highlight the* **100**] → type 20 → **Show case numbers** → **OK**	Back1

Note: **ethnicity** will be the first variable only if your variable list is ordered alphabetically. The **Edit** → **Options** sequence allows you to select an alphabetic ordering or a variable list ordering. This sequence will produce a listing of the first 20 subjects with all variables included in the order displayed in the original variable list.

To select the variables **id, total,** *and* **grade,** *grouped by* **section** *and* **gender,** *with cases numbered for all subjects, perform the following sequence of steps:*

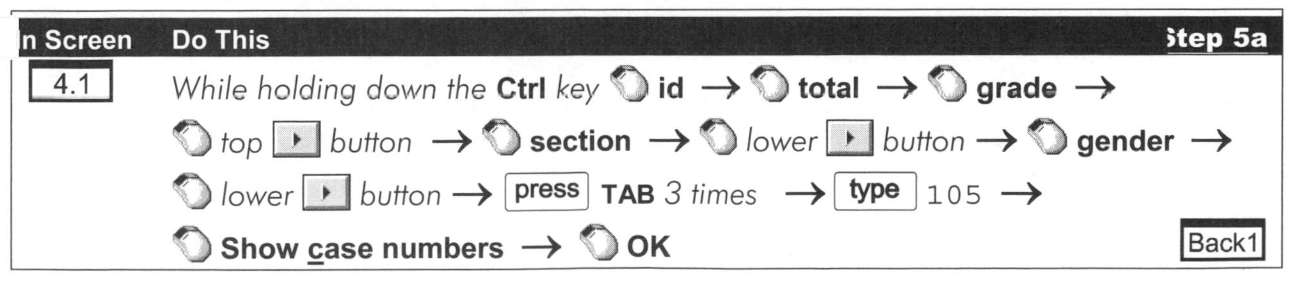

In Screen	Do This	Step 5a
4.1	*While holding down the* **Ctrl** *key* **id** → **total** → **grade** → *top* ▸ *button* → **section** → *lower* ▸ *button* → **gender** → *lower* ▸ *button* → press **TAB** *3 times* → type 105 → **Show case numbers** → **OK**	Back1

This operation will produce a listing of student ids, total points, and grades; first for women in section 1, then men in section 1, then women in section 2, and so forth; each case will be numbered.

The REPLACING MISSING VALUES Procedure

During the course of data entry you will discover soon enough that several subjects refused to answer the ethnicity question, three subjects did not take the first quiz, and other gaps in your data file appear. These are called *missing values*. Missing values are not only an irritant but can influence your analyses in a number of undesirable ways. Missing values often make your data file more difficult to work with. There are a number of SPSS procedures that delete any case (or subject) that has any missing values from consideration prior to analysis. If, for instance, you are computing correlations, and 13 of your 35 subjects have one or more missing values in the data file, SPSS computes correlations for only the 22 subjects that have *no* missing values. This results in a distressing loss of legitimate information that should be available.

In many procedures SPSS allows you to delete subjects **listwise** (if one or more missing value(s) exist for a subject, *all* data for that subject are removed before any analyses), or to delete subjects **pairwise** (if for any calculation a necessary data point is missing for a subject, the *calculation* will be conducted without the influence of that subject). In other words, for listwise deletion, a subject with missing values will be involved in all analyses except those for which critical values are missing. The authors suggest that you, to the greatest extent possible, deal with missing values during the data-coding and data-entry phase rather than relying on SPSS-missing-value defaults. For categorical data, this is easy, but for some continuous data, the SPSS procedures may be appropriate.

1. If you have missing values in *categorical data* (e.g., subjects didn't answer the ethnicity or level-of-income questions), you can create an additional level for that variable and replace the missing values with a different number. For instance, if you have five levels of ethnicity coded for five different ethnic groups, you could create a sixth level (coded 6) labeled "unknown". This would not interfere with any analyses you wished to conduct with the ethnicity variable, and there would no longer be missing values to contend with.

2. For *continuous* data a frequent procedure is to replace missing values with the mean score of all other subjects for that variable. SPSS has a procedure that allows this type of replacement to occur. Within this context, SPSS also has several other options (e.g., replace with the median, replace with the mean of surrounding points) that are presented in the Step by Step section. Although replacing *many* missing values by these techniques can sometimes bias the results, a small number of replacements has little influence on the outcome of your analyses. An often-used rule of thumb suggests that it is acceptable to replace up to 15% of data by the mean of the distribution (or equivalent procedures) with little damage to the resulting outcomes. If a particular subject (or case) or a certain variable has more than 15% missing data, it is recommended that you drop that subject or variable from the analysis entirely.

3. A more sophisticated way to replace missing values in continuous data is to create a regression equation with the variable of interest as the dependent (or criterion) variable and replace missing values with the *predicted* values. This requires a fair amount of work (SPSS has no automatic procedures to accomplish this) and a substantial knowledge of multiple regression analysis (Chapter 16); but it is considered one of the better methods for dealing with missing data. If you make use of predicted values, you will not be able to use the procedure described below. It is necessary to compute the new values separately and insert them into the original data file.

Concerning research ethics: If you replace missing data in research that you plan to publish, it is required that you state in the article how you handled any missing values that may influence your outcome. The reviewers and the readers then know what procedures you have used and can interpret accordingly.

If you spend time in the SPSS manuals, you will often see the phrases *system-missing values* and *user-missing values*. Just to clarify, system-missing values are simply omissions in your data set and are represented by a period in the cell where the missing value occurs. A user-missing value is a value that is specified *by the researcher* as a missing value for a particular level (or levels) of a variable. For instance, in an educational setting subjects might be coded 1, 2, or 3 based on which of three different forms of a test they took. Some subjects could be coded 8 for did-not-complete-test, or 9 for did-not-attend. The researcher could specify codings 8 and 9 as user-missing values.

Begin with any screen that shows the menu of commands across the top. To access the missing values procedure perform the following sequence of steps:

In Screen	Do This	Step 4b
Menu	Transform → Replace Missing Values	4.2

At this point a dialog box will open (Screen 4.2, below) that provides several options for dealing with missing values. The list of available variables appears to the left, and you will paste into the **New Variable(s)** box variables designated for missing-values replacement. You may designate more than one variable, and, by use of the **Change** button, can even select different methods for replacing missing values for each variable. We illustrate this procedure using the **grades.sav** file even though this file contains no missing values. Under **Method** in the dialog box shown below, SPSS provides five different techniques for replacing missing values:

4.2

The Replace Missing Values Window

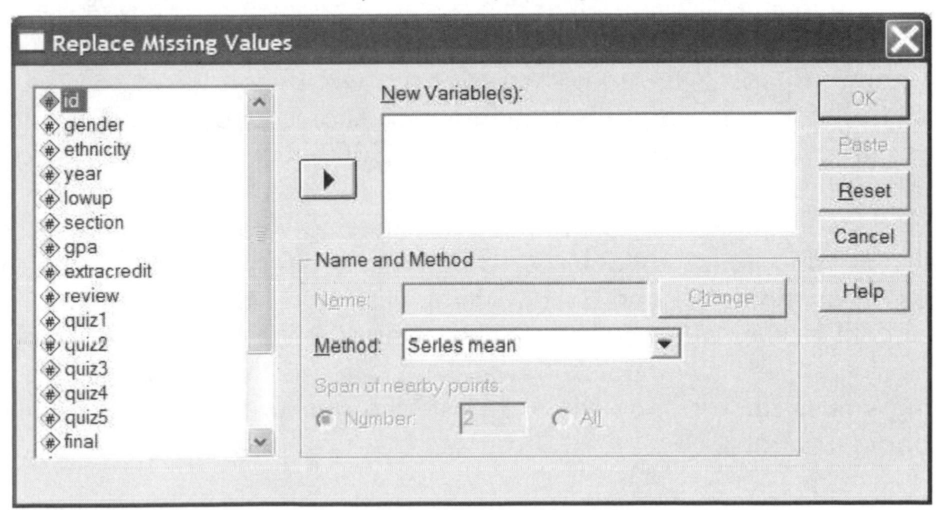

- **Series mean**: Missing values are replaced by the mean (or average) value of all other values for that variable. This is the default procedure.
- **Mean of nearby points**: Missing values are replaced by the mean of surrounding values (that is, values whose SPSS case numbers are close to the case with a missing value). You may designate how many values to use under **Span of nearby points**.
- **Median of nearby points**: Missing values are replaced by the median (see glossary) of surrounding points. You may designate how many surrounding values to use.

- **Linear interpolation**: Missing values are replaced by the value midway between the surrounding two values.
- **Linear trend at point**: If values of the variable tend to increase or decrease from the first to the last case, then missing values will be replaced by a value consistent with that trend.

If you select one of the two options that is based on a span of points, you will designate how many values by clicking **Number** and typing in the number of surrounding values you wish. Of the five techniques, **Series mean** is by far the most frequently used method. The other four are usually applied in time series analyses when there is some intrinsic trend of values for the cases over time. For instance if your data were constructed so that the cases represented 10 different times rather than 10 different subjects, then you might anticipate changes in your variables over time. This does not frequently occur with subjects unless they are ordered based on some intrinsic quality that relates to the independent variables in the study, such as IQ from low to high (on a study of academic achievement) or body weight from low to high (for a study dealing with weight loss).

In the example that follows we will replace missing values for previous **GPA** and **quiz1** with the mean value.

Beginning with Screen 4.2, perform the following sequence of steps.

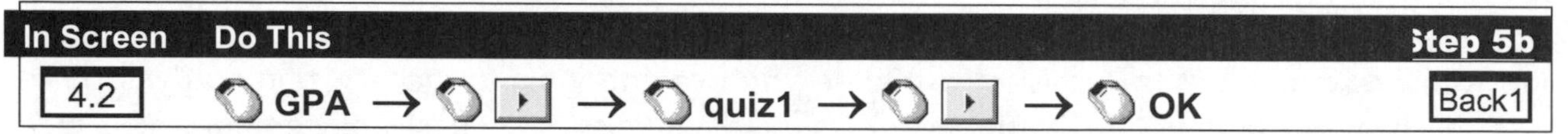

In Screen	Do This	Step 5b
4.2	GPA → ▸ → quiz1 → ▸ → OK	Back1

Some comments concerning this very simple sequence of steps is required: (1) After clicking the **OK**, the output will list the number of values replaced, the value that replaced the missing values, and the first and last case number of cases considered. (2) The sequence of steps is simple because the default option **(Series mean)** is the desired option, and usually will be. To select a different option, after you paste a variable into the **New Variable(s)** box, click the ▾ to the right of the **Method** label, click whichever option you desire from the menu, then click the **Change** button. You may then click the **OK** button to run the program. (3) Despite the fact that SPSS can perform the operations illustrated above, there is little logical reason for doing so. If you wanted to replace a missing GPA you would probably do so based on other indicators of that person's academic ability (such as performance on tests and quizzes) rather than a class average. (4) When you replace missing values by standard SPSS procedures for a particular variable (or variables), SPSS creates a new variable with the old variable name followed by an underline (_) and a "1". If you wish to replace a missing quiz score, an average of the student's other quizzes makes more sense than a class average for that quiz. You may then keep both sets of variables (the original with missing values and the new variable without missing values) or you may cut and paste the data without missing values under the name of the original variable.

The COMPUTE Procedure: Creating Variables

In the current data set, it would be quite normal for the teacher to instruct the computer to create new variables that calculate total points earned as the sum of the quizzes and final exam, and to determine the percent of total points for each student. The sequence of steps that follows will compute two new variables called **total** and **percent**. As noted earlier, the **grades.sav** file, available for download at www.ablongman.com/george6e, already contains the four new variables computed in this chapter, **total**, **percent**, **grade**, and **passfail**.

Beginning with a screen that shows the menu of commands, perform the following step to access the **Compute Variable** *window.*

At this point, a new window opens (Screen 4.3, below) that allows you to compute new variables. The list of variables in the active data file are listed in a box to the left of the screen. Above these is the **Target Variable** box. In this box you will type the name of the new variable you wish to compute. To the right is the **Numeric Expression** box. In this box you will type or paste (usually a combination of both) the expression that will define the new variable. Three options are then provided to assist you in creating the expression defining the new variable.

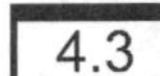

The Compute Variable Window

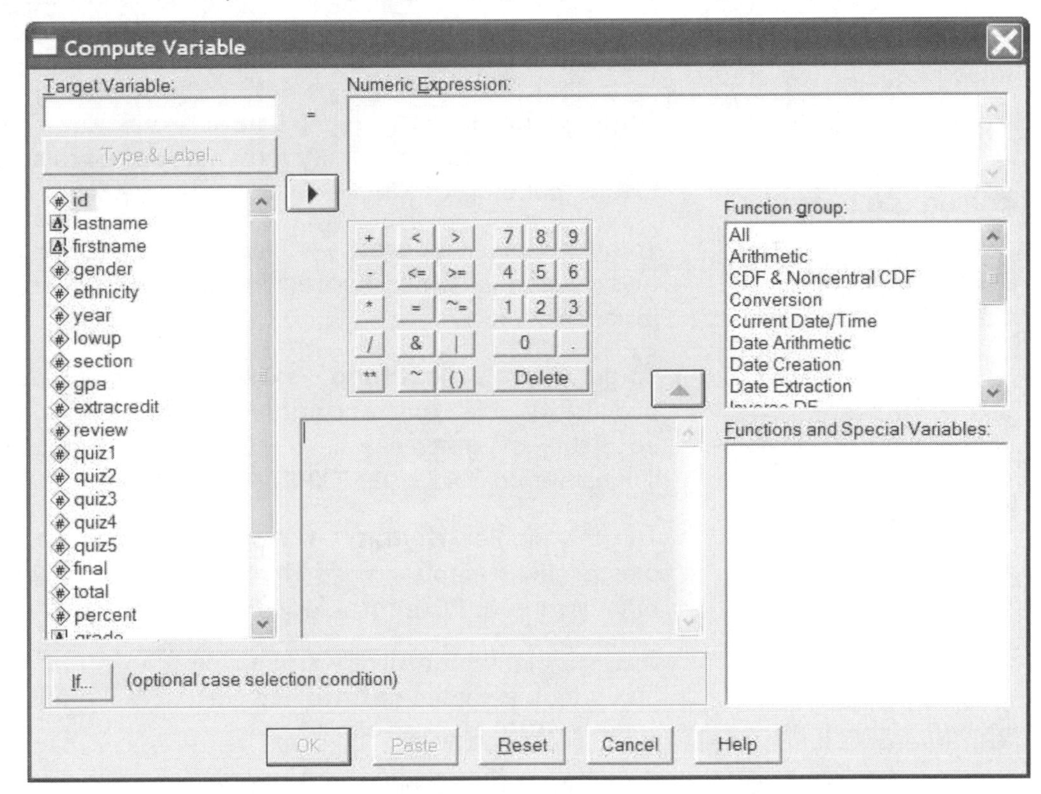

- *The calculator pad*: On this pad are all single-digit numbers and a decimal point. Each of these may be entered into the **Numeric Expression** by a mouse click on the screen button or typing the same on the keyboard. In addition there are a number of *operation* buttons. If a similar button occurs on the keyboard (e.g., <, >, +, etc.) you may type that symbol rather than click the screen button. What follow are the operation keys and their meanings:

Arithmetic operations		Rational operations		Logical operations	
+	add	<	less than	&	and: both relations must be true
–	subtract	>	greater than		
*	Multiply	<=	less than or equal	|	or: either relation may be true
/	Divide	>=	greater than or equal		
**	raise to power	=	equal	~	negation: true becomes false, false becomes true
( )	order of operations	~=	not equal		

- *The* **Functions** *box*: A terrifying array of over 180 different functions, many of which even Einstein (were he alive) couldn't identify. We present only 9 of these functions, the 9 we feel are most likely to be used. If you wish to see all 180 defined, consult the *SPSS Base 13.0 User's Guide*. The term **numexpr** represents a *numerical expression* and we give an example of a sample **Target Variable**, **Numeric Expression**, and how it would compute in your file. Please refer back to Screen 4.3 for visual reference.

Expression	Illustration
ABS(numexpr) (absolute value)	Target variable: **zpositiv** → Numeric expression: **ABS(zscore).** Creates a variable named **zpositiv** that calculates the absolute value of a variable named **zscore** for each subject.
RND(numexpr) (round to the nearest integer)	Target variable: **simple** → Numeric expression: **RND(gpa).** Computes a variable named **simple** by rounding off each subject's **gpa** to the nearest integer.
TRUNC(numexpr) (truncates decimal portion of a number)	Target variable: **easy** → Numeric expression: **TRUNC(gpa).** Computes a variable named **easy** that truncates each subject's **gpa** (*truncate* means "cut off"). This is like rounding, but it *always* rounds down.
SQRT(numexpr) (square root)	Target variable: **scorert** → Numeric expression: **SQRT(score).** Creates a variable named **scorert** by taking the square root of each **score** for each subject.
EXP(numexpr) (exponential (*e*) raised to a power)	Target variable: **confuse** → Numeric expression: **EXP(gpa).** Computes a variable named **confuse** that calculates the value of e raised to each subject's **gpa**'s power. $e \approx 2.721$ (it is irrational). For a particular subject with a 3.49 **gpa**, **EXP**(3.49) = $(2.721\ldots)^{3.49} \approx 32.900506\ldots$
LG10(numexpr) (base 10 logarithm)	Target variable: **rhythm** → Numeric expression: **LOG10(total).** Creates a new variable named **rhythm** that calculates the base 10 logarithm for the variable **total** for each subject.
LN(numexpr) (natural logarithm)	Target variable: **natural** → Numeric expression: **LN(total).** Creates a new variable named **natural** that calculates the natural logarithm for the variable **total** for each subject.
RV.NORMAL(mean, stddev)	Computes random numbers based on a normal distribution with a user-specified mean (**mean**) and standard deviation (**stddev**). Target variable: **randnorm** → Numeric expression: **RV.NORMAL(5,3).** This function will generate random numbers based on a normal distribution of values with a mean of 5 and a standard deviation of 3. One of these numbers will be assigned randomly to each subject (or case) in your data file under the variable name **randnorm**.
RV.UNIFORM(min, max)	Computes random numbers based on equal probability of selection for a range of numbers between a user-specified minimum (**min**) and maximum (**max**). Target variable: **random** → Numeric expression: **UNIFORM(1,100).** Random numbers from 1 to 100 will be assigned to each subject (or case) in your data file.

- *The **I̲f** pushbutton*: A click on this button opens a new screen so similar to Screen 4.3 that we will not reproduce it here. The only differences are an **Include a̲ll cases** option paired with an **Include i̲f case satisfies condition**, and a **Continue** button at the bottom of the window.

All of these resources will be demonstrated in several examples that follow. The first example computes two new variables, **total** and **percent**. **Total** will be the sum of the five quizzes and the final exam. **Percent** will be each subjects **total** score, divided by 125 (possible points), multiplied by 100 (to make it a percent rather than a decimal) and rounded to the nearest integer. You will need to perform two separate sequences to create these two variables.

The starting point for this operation is Screen 4.3. To compute the new variable **total** *perform the following sequence of steps:*

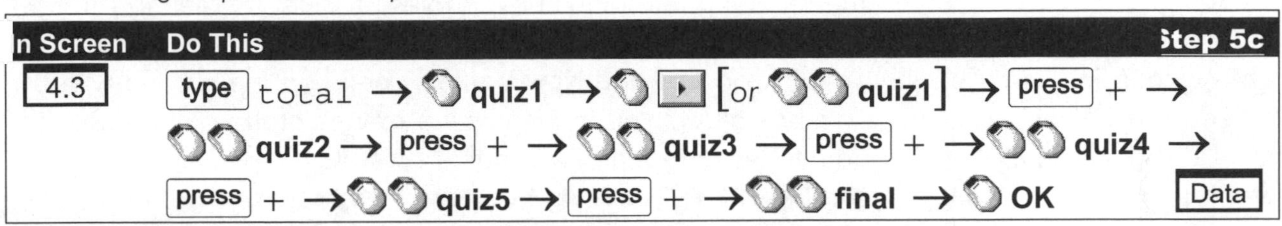

In Screen	Do This	Step 5c
4.3	type `total` → quiz1 → ▸ [or quiz1] → press + → quiz2 → press + → quiz3 → press + → quiz4 → press + → quiz5 → press + → final → OK	Data

With the click of the **OK**, the new variable is computed and entered in the data file in the last column position. If you wish to move this new variable to a more convenient location, you may cut and paste it to another area in the data file. The **total** variable is already in the downloaded file, so if you wish to practice this procedure use a different variable name, such as **total2**.

The starting point for the next operation is also Screen 4.3. To compute the new variable **percent** *and save in the data file the two variables we have created, perform the following sequence of steps:*

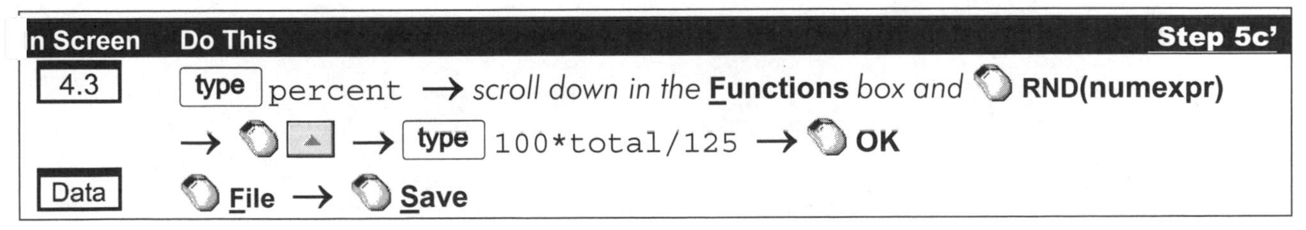

In Screen	Do This	Step 5c'
4.3	type `percent` → *scroll down in the* **F̲unctions** *box and* **RND(numexpr)** → ▲ → type `100*total/125` → **OK**	
Data	**F̲ile** → **S̲ave**	

The final step saves the newly created data. If you wish to create a variable only for the present session, then don't save the changes and when you next open the data file it will NOT have the new variable.

Upon completion of a compute operation, SPSS returns the screen to the data file and you may use the scroll bar to check the accuracy of the new variable(s). You may, of course, use several of these program functions in a single command. The example in step 5c demonstrates a computation that includes rounding, multiplying by 100, and dividing by 125. In creating more complex computations, be sure to adhere strictly to the basic algebraic rules of orders of operations. If you wish to do an operation on a complex expression, be sure to include it within parentheses.

The RECODE INTO DIFFERENT VARIABLES Procedure: Creating New Variables

This procedure also creates new variables, not by means of calculations with already existing variables but rather by dividing a pre-existing variable into different categories and coding each category differently. This procedure is ideally suited for a data file that includes class records because at the end of the term the teacher wants to divide scores into different grades. In the present example this coding is based on the class **percent** variable in which percents between 90 and 100 = A, between 80 and 89 = B, between 70 and 79 = C, between 60 and 69 = D, and between 0 and 59 = F. The **percent** variable can also be used to classify subjects into pass/fail categories with codings of greater or equal to 60 = P and less than 60 = F.

Two different windows control the recoding-into-different-variables process. Screen 4.4 (below) allows you to identify the **Input Variable** (**percent** in the present example) and the **Output Variable** (**grade** in this case), then a click on **Old and New Values** opens up the second window, Screen 4.5 (following page). This box allows you to identify the ranges of the old variable (90 to 100, 80 to 89, etc.) that code into levels of the new variable (A, B, C, D, and F).

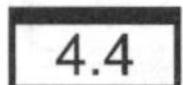

4.4

The Recode Into Different Variables Window

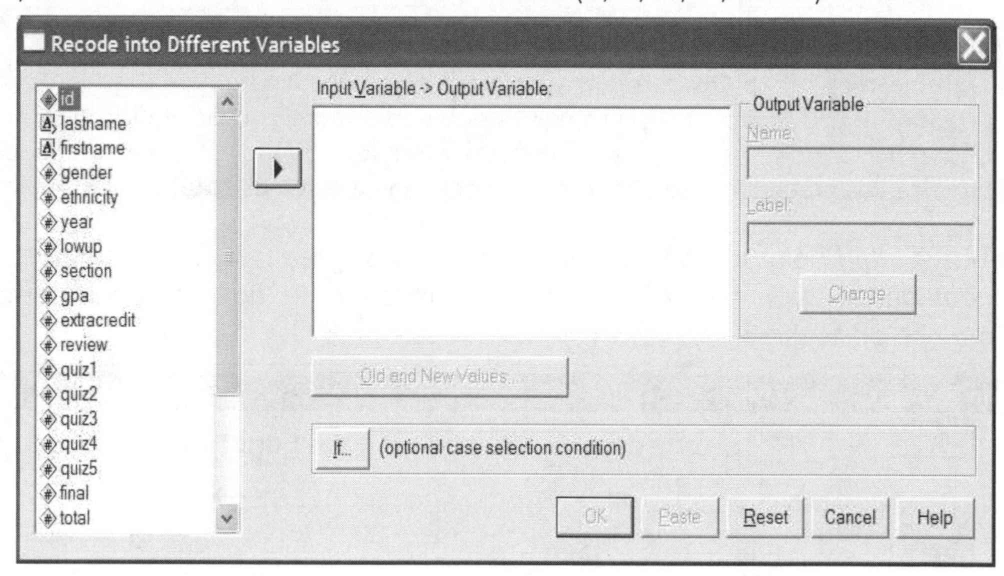

Notice that in Screen 4.4 all available variables are listed to the left and the five function buttons are at the bottom of the dialog box. The initial stage (in the present example) is to click on **percent** and paste it into the **Input Variable → Output Variable** box. Then click in the box beneath the **Output Variable** title and type the name of the new variable, **grade**. A click on the **Change** button will move the **grade** variable back to the previous box. The entry in the active box will now read **percent -> grade**. A click on the **Old and New Values** pushbutton opens the next window.

The dialog box represented by Screen 4.5 allows you to identify the values or range of values in the input variable (**percent**) that is used to code for levels of the Output variable (**grade**). Since we want a *range* of values to represent a certain grade, click on the top **Range** option and indicate the range of values (90-100) associated with the first level (A) of the new variable (**grade**). When the first range is entered, click in the **Value** box and type the letter A. A click on the **Add** button will paste that range and grade into the **Old -> New** box. We are now ready to perform a similar sequence for each of the other four grades. When all five levels of the new variable are entered (with the associated ranges and letter grades) then click **Continue** and Screen 4.4 will reappear. A click on the **OK** button will create the new variable.

4.5

The Old and New Values Window

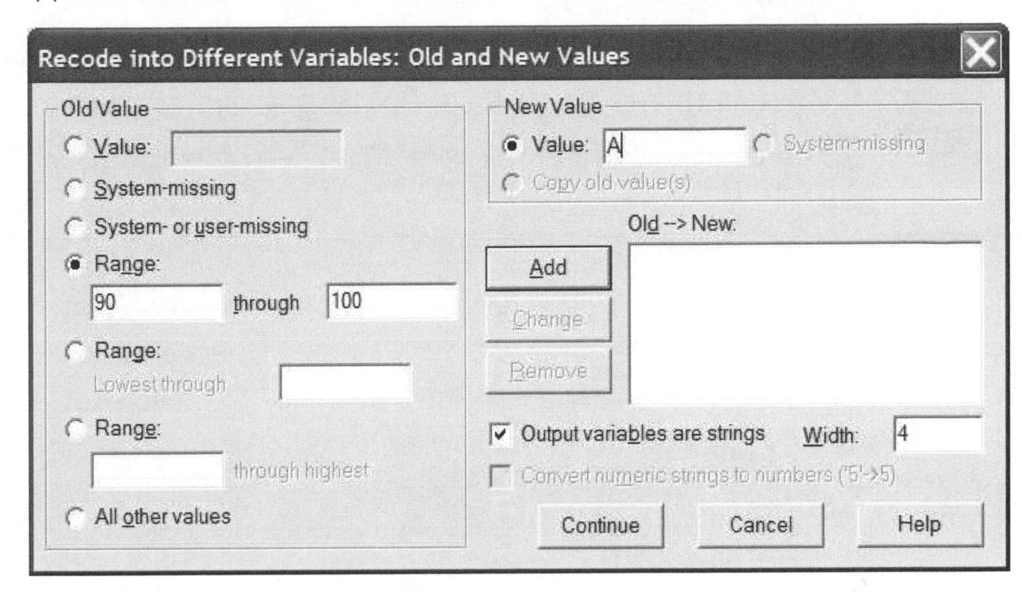

A similar process will be conducted for creating the **passfail** variable. When these two new variables are created they will be listed in the last two columns of the data file. Once the new variable is created, you may type in value or variable labels, change the format, or move the variables to a new location in the data file by techniques described in Chapter 3.

The creation of the two new variables is described below in the step-by-step boxes. Remember that these variables are already in the data available by download. If you want to practice with that file, create variables with slightly different names (e.g., **grade2**) and then you can compare your results with those already present.

To create the new variable **grade** *the starting point is any menu screen. Perform the following sequence of steps to create this new variable.*

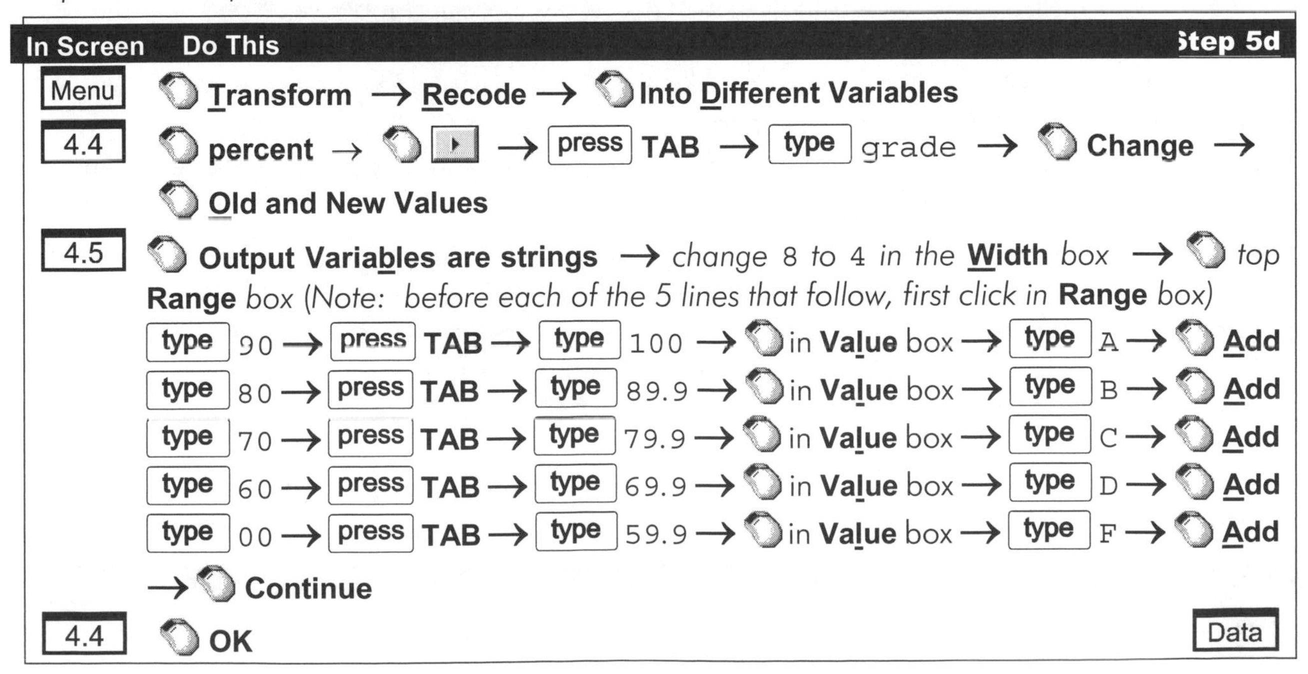

In Screen	Do This	Step 5d
Menu	**Transform** → **Recode** → **Into Different Variables**	
4.4	**percent** → ▸ → press TAB → type grade → **Change** → **Old and New Values**	
4.5	**Output Variables are strings** → *change* 8 *to* 4 *in the* **Width** *box* → *top* **Range** *box (Note: before each of the 5 lines that follow, first click in* **Range** *box)*	
	type 90 → press TAB → type 100 → in **Value** box → type A → **Add**	
	type 80 → press TAB → type 89.9 → in **Value** box → type B → **Add**	
	type 70 → press TAB → type 79.9 → in **Value** box → type C → **Add**	
	type 60 → press TAB → type 69.9 → in **Value** box → type D → **Add**	
	type 00 → press TAB → type 59.9 → in **Value** box → type F → **Add**	
	→ **Continue**	
4.4	**OK**	Data

To create the new variable **passfail** *the starting point is any menu screen. Perform the following sequence of steps to create this new variable.*

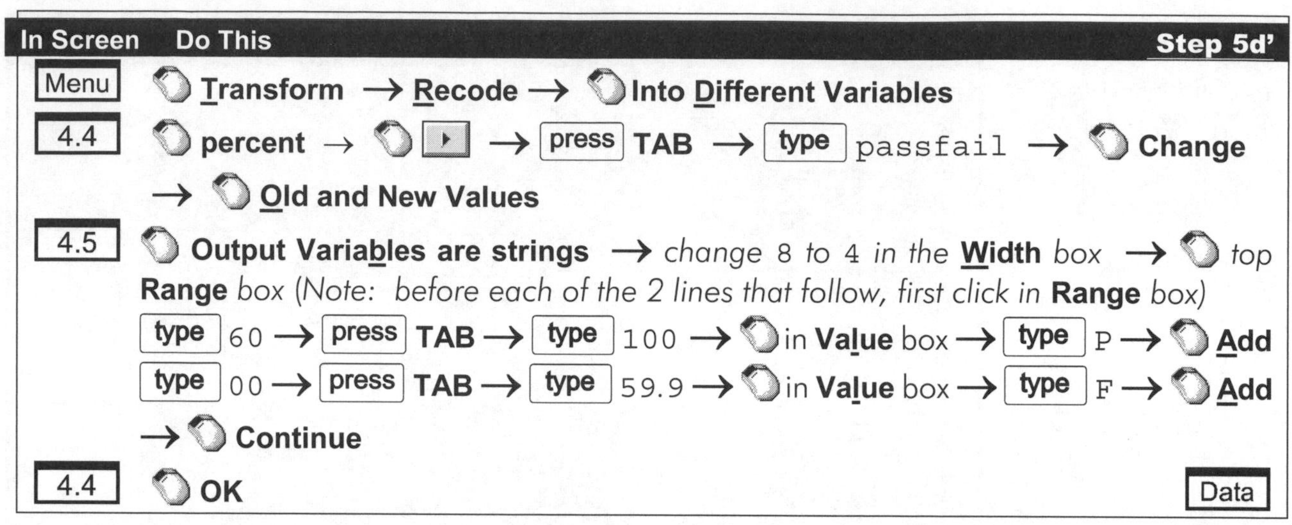

In Screen	Do This	Step 5d'
Menu	Transform → Recode → Into Different Variables	
4.4	percent → ▸ → press TAB → type passfail → Change → Old and New Values	
4.5	Output Variables are strings → *change* 8 *to* 4 *in the* **Width** *box* → *top* **Range** *box (Note: before each of the 2 lines that follow, first click in* **Range** *box)* type 60 → press TAB → type 100 → in **Value** box → type P → Add type 00 → press TAB → type 59.9 → in **Value** box → type F → Add → Continue	
4.4	OK	Data

The RECODE Option:

Changing the Coding of Variables

There are times when you may wish to change the coding of your variables. There are usually two reasons why it is desirable to do so. The first is when you wish to reverse coding to be consistent with other data files, or when one coding pattern makes more sense than another. For instance if you had a file in which the gender variable was coded male = 1 and female = 2, you may have other files where it is coded female = 1, male = 2. It would be desirable to recode the files so that they are consistent with each other. Or perhaps you have an **income** variable that has >$100,000 coded 1 and <$10,000 coded 7. If you felt that it would make more sense to have the lower incomes coded with the lower numbers, the **Recode** command could accomplish that easily for you.

The second reason for recoding is to group variables that are ungrouped in the original data. For instance, you might have a variable named **marital** that is coded 1 = never married, 2 = divorced, 3 = widowed, 4 = separated, 5 = married. You could use the **Recode** procedure to change the marital variable so that 1 = single (a combination of 1, 2, and 3) and 2 = married (a combination of 4 and 5).

The starting point for recoding variables is a screen that shows the menu of commands across the top. Select the following options to access the Recode screen.

In Screen	Do This	Step 4e
Menu	Transform → Recode → Into Same Variable	4.6

4.6

The Recode into Same Variable Window

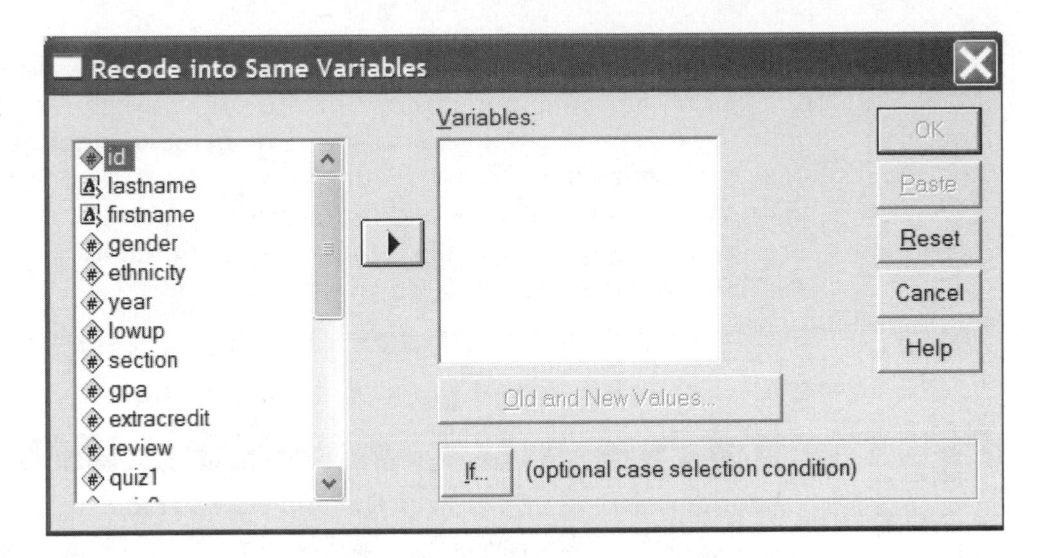

Two different dialog boxes control (1) the selection of variable(s) to be recoded (Screen 4.6 above), and (2) clarification of old and new coding values (Screen 4.7, below). It is also possible to recode a string variable (such as **grade**) into numeric coding. A researcher might wish to recode the letters A, B, C, D, and F into 4, 3, 2, 1, and 0, so as to run correlations of **grade** with other variables. The **Recode** option can accommodate this comfortably. You can recode several variables in one setting but would need to switch back and forth between the two screens to do so.

In Screen 4.6 (above) the first step is to select the variable(s) of interest and paste them into the active box to the right (titled **Variables**). A click on **Old and New Values** opens up a new window (Screen 4.7).

4.7

The Old and New Values Window

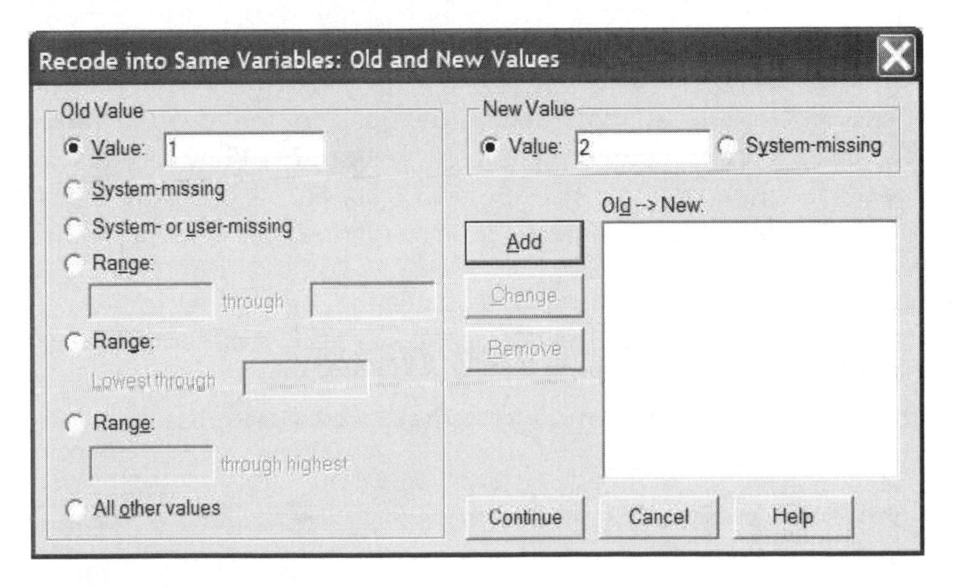

Options for the **Old Value** coding include:

- A single value
- System-missing values (blanks in your data set)
- User-missing values (values you have designated as missing for various reasons)
- A range of values (internal range, lowest to specified value, specified value to highest)
- All other values (useful for complex combinations)

For each variable you specify the old value(s), the new value(s), then paste them into the **Old →New** box one level at a time. To demonstrate both a single switch recode and a recoding that involves a range of values, we will reverse the coding for the **gender** variable; then, even though it's not in the **grades.sav** file, recode the fictitious **marital** variable into levels 1, 2, and 3 representing unmarried, and levels 4 and 5 representing married. If you wish to practice the latter procedure using the **grades.sav** file, simply use the **ethnicity** variable for recoding. Like the fictional **marital** variable, it also has five levels.

The starting point is Screen 4.6. Perform the following sequence of steps to reverse the coding on the **sex** *variable, and to recode the mythical* **marital** *variable.*

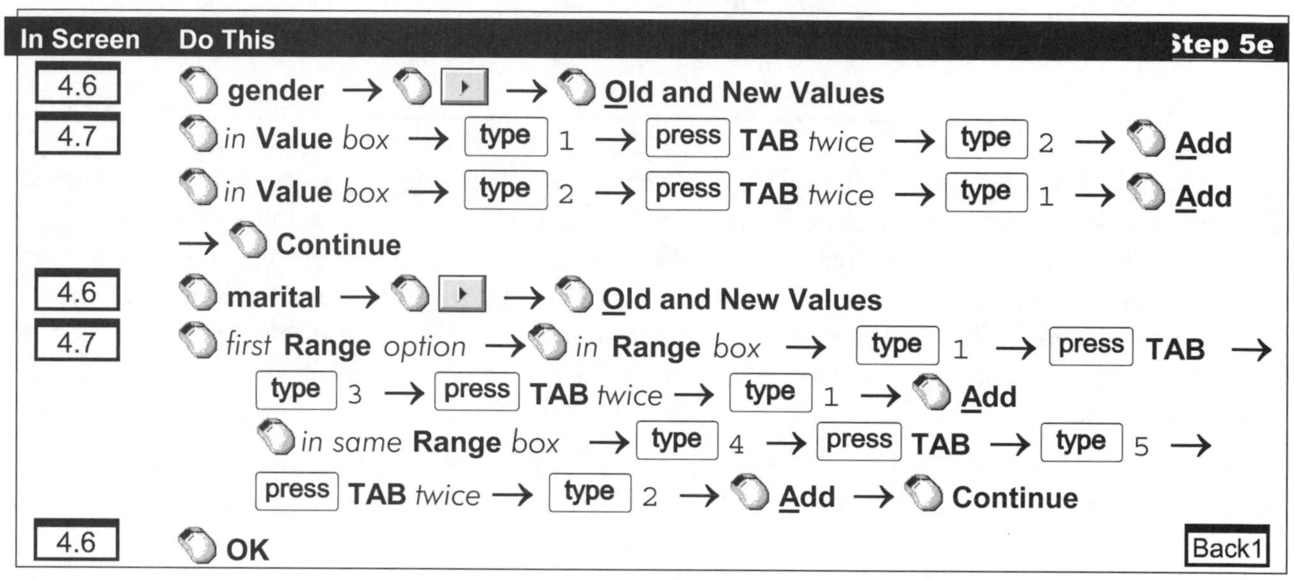

In Screen	Do This	Step 5e
4.6	**gender** → ▸ → **Old and New Values**	
4.7	*in* **Value** *box* → type 1 → press **TAB** *twice* → type 2 → **Add**	
	in **Value** *box* → type 2 → press **TAB** *twice* → type 1 → **Add** → **Continue**	
4.6	**marital** → ▸ → **Old and New Values**	
4.7	*first* **Range** *option* → *in* **Range** *box* → type 1 → press **TAB** → type 3 → press **TAB** *twice* → type 1 → **Add** *in same* **Range** *box* → type 4 → press **TAB** → type 5 → press **TAB** *twice* → type 2 → **Add** → **Continue**	
4.6	**OK**	Back1

To create new value labels for the recoded variables is an urgent concern. Otherwise Robert will be coded female, Bill Cosby will be coded Asian, and E. Taylor will be coded never married. If you wish to create new value labels for the recoded variables, you will need to perform the appropriate operation (click on the **Variable View** tab and the appropriate cell in the **Values** column) described in Chapter 3. Further, if you want these changes to be permanent in your data file, you will need to go through the save data procedure (click **File** → click **Save**).

The SELECT CASES Option:

Selecting a Portion of the Data for Analysis

To access the necessary initial screen, select the following options:

In Screen	Do This	Step 4f
Menu	**Data** → **Select Cases**	4.8

The purpose of the **Select Cases** option is to allow the user to conduct analyses on only a subset of data. This will be a frequently used function. Often we want to know what the mean **total** score is for females, or the average **total** score for Section 3, or the mean **total** score for Sophomores. **Select Cases** enables you to accomplish this type of selection. It chooses from the data a subset of those data for analysis. Once analyses have been conducted with a subset of data, you may revert to the entire file by clicking **All cases** in the **Select Cases** dialog box. If you wish to create a file that consists only of the selected cases, you must first delete unselected cases and then save the file (click **File** → click **Save**). Then only the selected cases (or subjects) will remain. For ease of identifying cases that are selected and those that are not, in the data file a diagonal line is placed through the row number of nonselected cases. The example to the right illustrates females selected. The lines through case numbers represent males in the sample. Then, when certain cases are selected, SPSS creates a new variable named **filter_$** that codes selected cases as 1 and nonselected cases as 0. You may keep that variable for future reference (if you wish to make comparisons between the two groups), or you may delete it.

1	106484	VILLARRUZ	ALFRED
2	108642	VALAZQUEZ	SCOTT
3	127285	GALVEZ	JACKIE
4	132931	OSBORNE	ANN
5	140219	GUADIZ	VALERIE
6	142630	RANGIFO	TANIECE
7	153964	TOMOSAWA	DANIEL
8	154441	LIAN	JENNY
9	157147	BAKKEN	KREG
10	164605	LANGFORD	DAWN
11	164842	VALENZUELA	NANCY
12	167664	SWARM	MARK

In the **Select Cases** box (Screen 4.8) due to space constraints, we discuss only two options. The **All cases** option (mentioned above) simply turns off the select option, and further analyses will be conducted on all subjects. To access the working screen for creating the conditional statements (Screen 4.9) click on the **If condition is satisfied** circle, then click on the **If** button.

4.8

The Select Cases Window

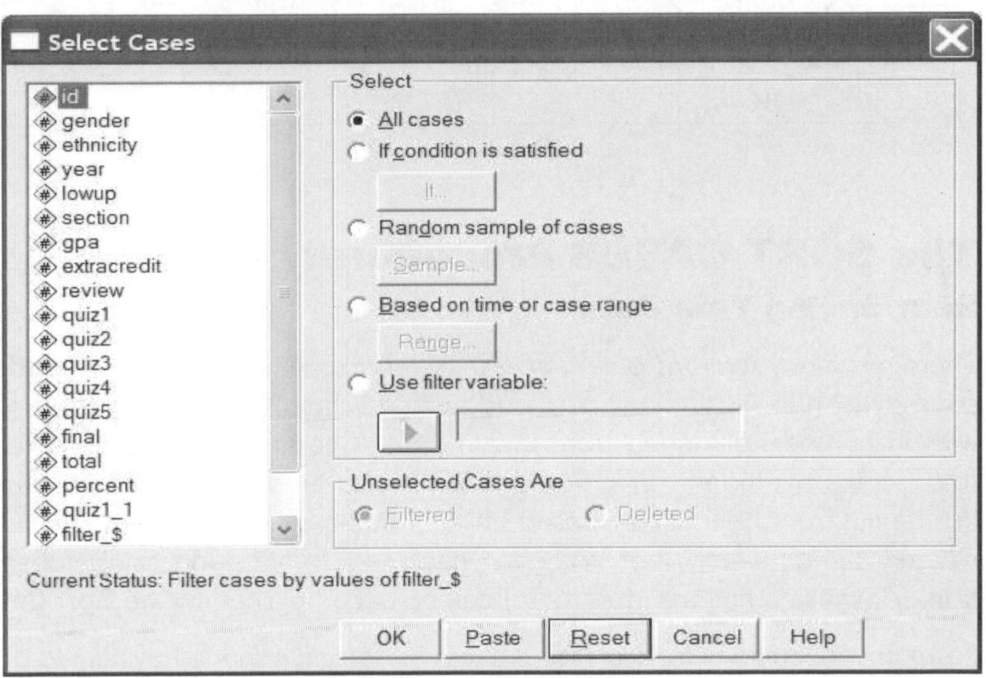

Screen 4.3 now opens. The actual screen that opens is not identical to Screen 4.3, but it is so similar that we'll not take up space by reproducing it here. In this screen is an active box where you may either type or click-and-paste relevant information. Below is the key pad, and to the right, the 180 functions lurking behind an innocent looking window. Quite a variety of conditional statements may be created. Several examples follow:

- *women*: **gender = 1**
- *third section*: **section = 3**
- *juniors and seniors*: **year >= 3**
- *sophomores and juniors*: **year >= 2 & year <= 3** or **year > 1 & year < 4**
- *first-year students and seniors*: **year = 1 | year = 4**

To select a subset of variables the process is to click the desired variable then paste it into the active box. You can then use the key pad (or simply type in the conditional statements) to create the selection you desire. In the sequences that follow we will demonstrate how to select women for an analysis. Then in another sequence we will show how to select sophomores and juniors.

*The starting point is Screen 4.8. Below we select females (**gender = 1**) for future analyses.*

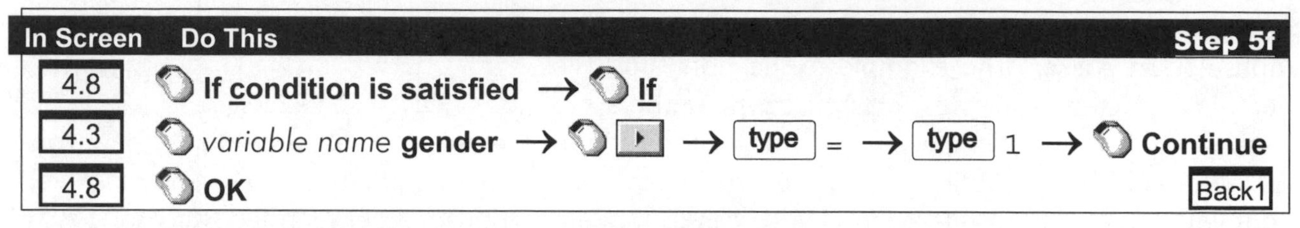

In Screen	Do This	Step 5f
4.8	**If condition is satisfied** → **If**	
4.3	*variable name* **gender** → ▸ → type = → type 1 → **Continue**	
4.8	**OK**	Back1

*To select sophomores and juniors (**year = 2 and 3**) also begin at Screen 4.8.*

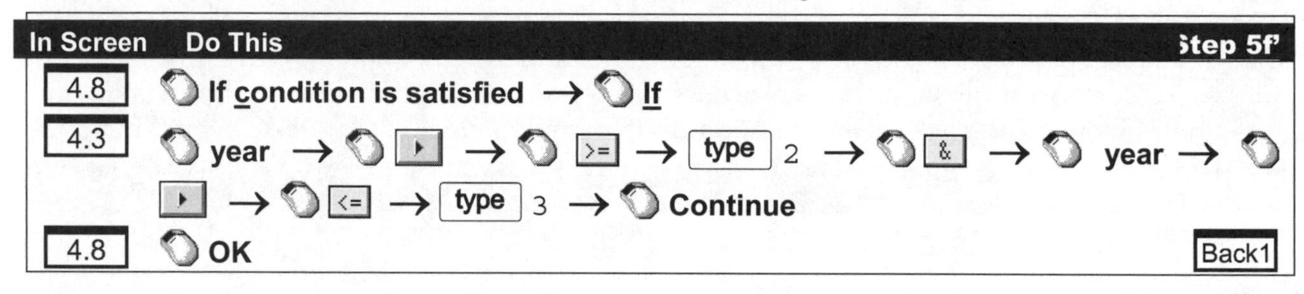

In Screen	Do This	Step 5f'
4.8	**If condition is satisfied** → **If**	
4.3	**year** → ▸ → >= → type 2 → & → **year** → ▸ → <= → type 3 → **Continue**	
4.8	**OK**	Back1

The SORT CASES Procedure:

Rearranging Your Data

There are many reasons one may wish to rearrange data. With the **grades.sav** file, the professor may wish to list final grades by **id** number. If so, it would be useful to be able to list scores with **id** numbers arranged from low to high. She may also be interested in the distribution of total scores of students in order to make grade breakdowns. Then it might be appropriate to list the file from high to low based on total points or total percentage. A class list is usually arranged alphabetically, but it is often necessary to list students alphabetically within each section. All these functions (and more) can be accomplished by the **Sort Cases** procedure.

From any menu screen begin the process by selecting the following two options.

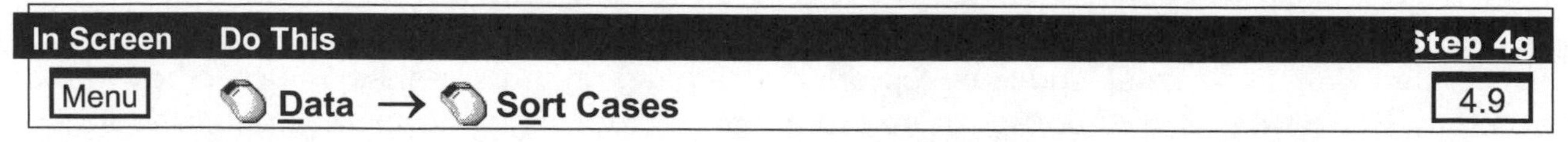

In Screen	Do This	Step 4g
Menu	**Data** → **Sort Cases**	4.9

To sort cases a simple dialog box (Screen 4.9) handles the entire process. You simply select a variable of interest and specify if you wish the sort to be **Ascending** (small to large for numbers, alphabetic for string variables) or **Descending** (the opposite). The only slight twist to the whole procedure is that you can sort by more than one variable at a time. This is useful in sorting names. If you have several students with the same last name, then when sorting by **lastname**, **firstname**, if subjects have identical last names, it will sort alphabetically by first name. When a sort is completed, then the **Cases Summaries** procedure may also prove useful.

4.9

The Sort Cases Window

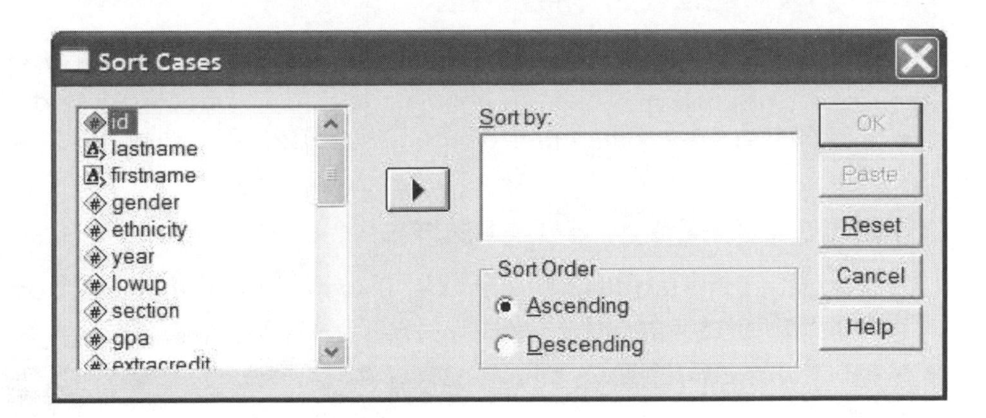

To sort subjects by **total** *points from high to low, execute the following steps:*

To sort subjects by **lastname** *and* **firstname** *alphabetically, execute the following steps.*

Note that **Ascending** is the default order so it doesn't need to be clicked here.

MERGING FILES:

Adding Blocks of Variables or Cases

Merging files is a procedure that can baffle the most proficient of computer experts at times. Files can be constructed in a variety of different formats and be created in different data-analysis or word-processing programs. What we show here, however, are the simplest steps of merging files created in the SPSS for Windows data editor, in the same format, in the same order, and with identical variable names (when we are adding more cases) or identical number and order of subjects (when we are adding more variables). While the last few sentences may sound apologetic, they actually contain sound advice for merging files: Prepare the foundation before you attempt the procedure. If possible perform the following steps before you begin:

- format files in the same data editor (i.e., SPSS for Windows),
- create identical formats for each variable,
- make sure that matching variables have identical names,
- make sure that cases are in the same order (if you are adding new variables), and
- make sure that variables are in the same order (if you are adding new cases).

To be sure, there will be times that you are not able to do all of these things, but the more preparation beforehand, the less difficulty you will have in the merging process. For example, the present procedure allows you to merge an SPSS for Windows file with an SPSS/PC+ file. Having used SPSS/PC+ many times I have a number of files in that format. If I wish to merge a PC+ file with a Windows file, even though I know it can be done, I will convert the PC+ file to SPSS for Windows format before I attempt to merge them. Be especially careful to insure that all variables have the same names and formats and are in the same order in the data file (the same order of cases is *required* when adding new variables). Adding new cases and adding new variables are two different procedures in SPSS. We will begin with adding cases.

ADDING CASES or SUBJECTS

After accessing the **grades.sav** *file (Step 3 early in this chapter), from the menu of commands select the following options.*

At this point a new window will open (Screen 4.10, below) titled **Add Cases: Read File**. This screen provides opportunity to read the file you wish to merge (called the *external data file*) with the already active file (called the *working data file*). If the external file is visible in the **File Names** box, you may simply click on the name to access that file. For the sake of illustration, we have created a file named **graderow.sav** that contains 10 new subjects with identical variables and formats as the **grades.sav** file. **grades.sav** in this case is the working file and **graderow.sav** is the external file. Often you will be merging with a file on a floppy disk. If your disk is in drive a, then the simplest way to access it is to type in the file name indicating the source (e.g., **a:\filename.sav**). Once the file name is selected, click the **Open** button to proceed with the process.

4.10

The Add Cases: Read File Window

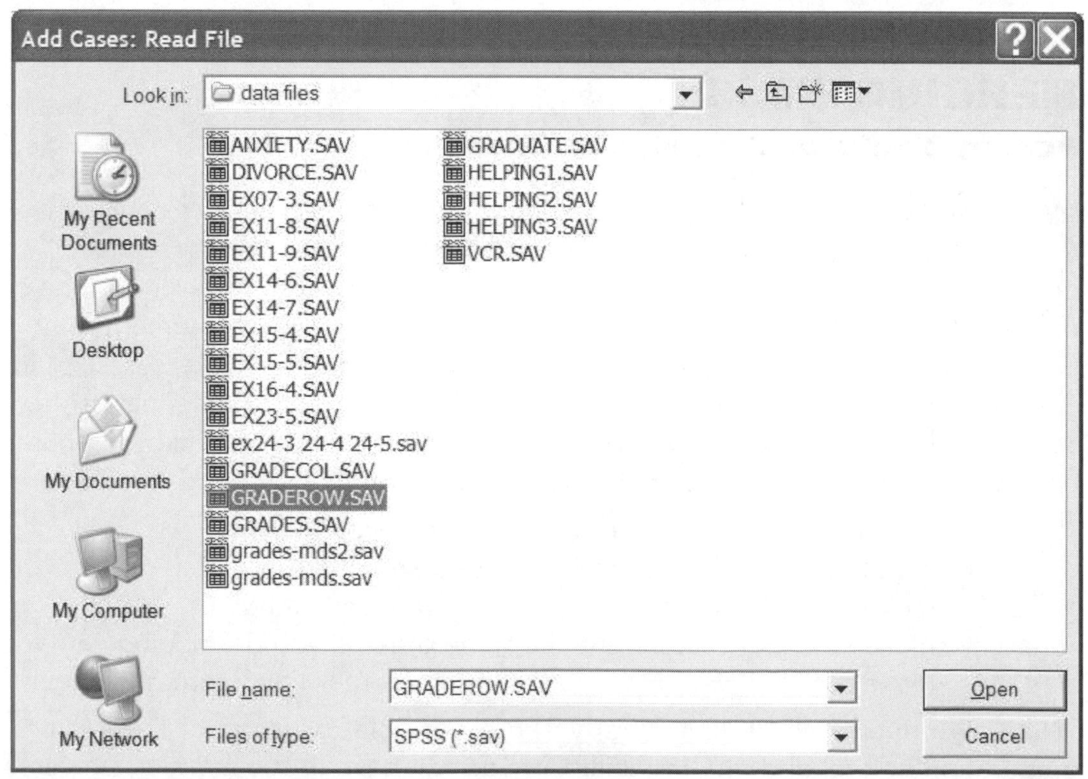

The next screen to appear is the **Add Cases from [ ]** box (Screen 4.11, following page) with the name and source of your external file included in the title. Note in the title of Screen 4.11 the source of the external file (**D** drive), the directory (**spss data**), and the name of the file name (**graderow.sav**) are all included. There are no **Unpaired Variables** in this example. If there were, unpaired variables would appear in the box to the left. Those from the working data file would be marked with an asterisk (*), those from the external data file would be marked with a plus (+). If you think the variable structures of the two files are identical and variable names *do* appear in the window to the left, then you need to examine those variables to make sure that names and formats are identical. Two simple operations are available before you click the **OK** to approve the merger.

4.11

The Add Cases from Drive, Program, and File Name Window

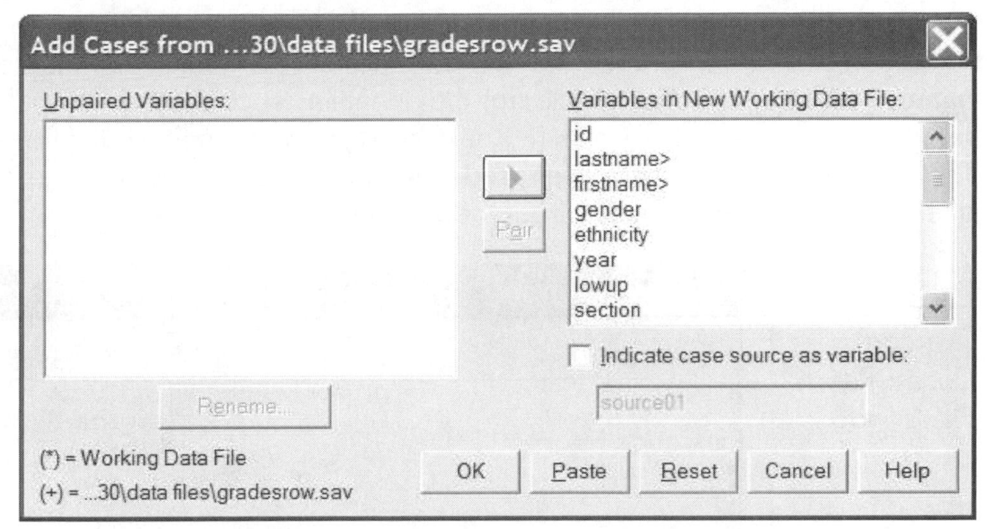

- Any variables that are common to both files (paired variables, shown in the **Variables in the New Working Data File** window) that you *do not* wish to have in the merged file you may click (to highlight) and delete by pressing the delete button.
- All unpaired variables are *excluded* from the merged file by default. If you wish any of these variables to be *included* click the desired variable(s) and press the ▸ to paste them into the **Variables in New Working Data File** box. Selected variable(s) will appear in the new file with blank cells where data are missing. These blank cells will be treated as missing values.

To merge the **graderow.sav** *file with the* **grades.sav** *file, perform the following sequence of steps. The starting point is Screen 4.10.*

In Screen	Do This	Step 5h
4.10	graderow.sav → Open	
4.11	OK	Back1

What will result is a new data file that will need to be named and saved. The merger does not alter in any way the original two files.

ADDING VARIABLES

After accessing the **grades.sav** *file (Step 3 early in this chapter), from the menu of commands select the following options.*

In Screen	Do This	Step 4i
Menu	Data → Merge Files → Add Variables	4.12

The procedure for adding variables is in many ways similar to that for adding cases or subjects. The initial entry procedure is the same. After clicking **Data**, **Merge Files**, *and* **Add Variables**, the same screen opens (Screen 4.10, previous page) as the screen used for adding cases. The purpose of this screen is to select an external file to merge with the working file. Upon clicking **Open**, however, a new screen opens (Screen 4.12) that looks quite different from the screen for adding cases. The title is similar in that it identifies the drive, directory and name of the external file to be merged. Note the new external file is **gradecol.sav**. This is a file that has the same subjects as the **grades.sav** file, includes several identical variables (**id**, **firstname**, **lastname**, **gpa**), and a new variable (**iq**) that identifies each subject's IQ. These variables you will see listed to the left with a plus (+) next to each one to indicate they are from the external file. The variables in the working file, **grades.sav**, are listed in a box to the right with an asterisk (*) after each to indicate they are from the working file.

4.12

The Add Cases: Read File Window

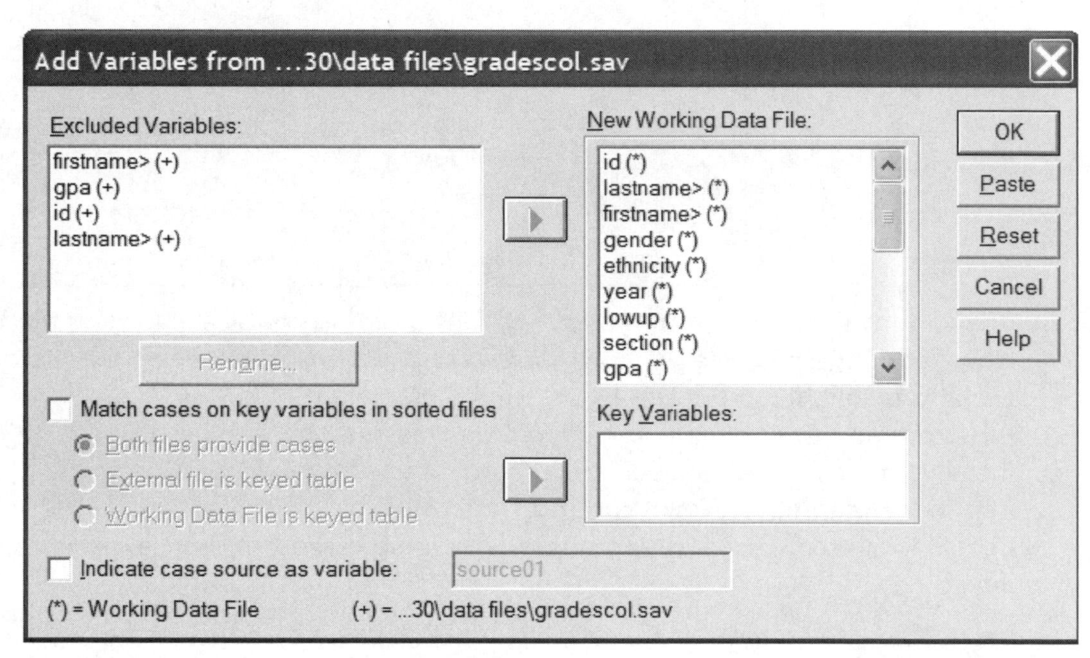

The procedure is to click on a matching variable in the external data file. It is required that this matching variable be in the same order in both files. Desirable matching variables might be **id** (ordered from low to high value), or **lastname** (ordered alphabetically). This requires first that you make sure that the matching variable is ordered identically in each file. Then, click the **Match cases on key variable in sorted files** option, select the matching variable in the box to the left, click the lower ▶ button, and click **OK**. Below is the step-by-step sequence for merging a working file (**grades.sav**) with an external file (**gradecol.sav**). In this example, the matching variable will be **id**, ordered from low to high value.

The starting point for this sequence is Screen 4.10. Perform the following steps to merge the two files mentioned above.

In Screen	Do This	Step 5i
4.10	**gradecol.sav** → **Open**	
4.12	**Match cases on key variables in sorted files** → **id** *in variable list to the left* → *lower* ▶ → **OK**	Back1

If the process doesn't work, check that the matching variables in the two files are in identical order.

PRINTING RESULTS

Results of the analysis (or analyses) that have just been conducted requires a window that displays the standard commands (**File Edit Data Transform Analyze** . . .) across the top. A typical print procedure is shown below beginning with the standard output screen (Screen 1, inside back cover).

To print results, from the Output screen perform the following sequence of steps:

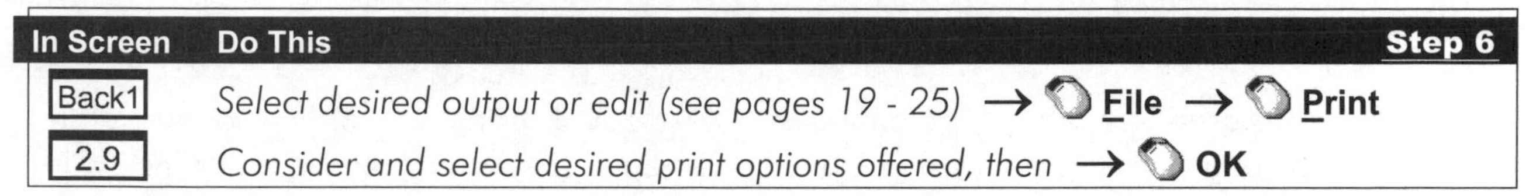

In Screen	Do This	Step 6
Back1	*Select desired output or edit (see pages 19 - 25)* → **File** → **Print**	
2.9	*Consider and select desired print options offered, then* → **OK**	

To exit you may begin from any screen that shows the **File** *command at the top.*

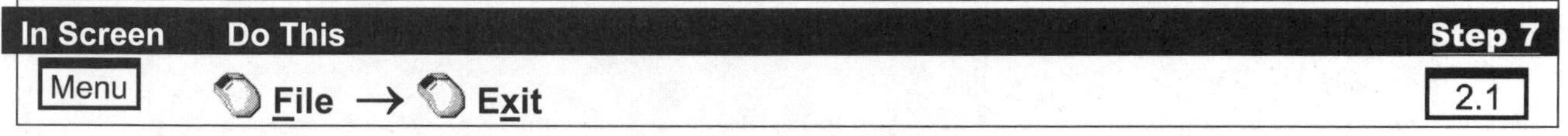

In Screen	Do This	Step 7
Menu	**File** → **Exit**	2.1

Note: After clicking **Exit**, there will frequently be small windows that appear asking if you wish to save or change anything. Simply click each appropriate response.

EXERCISES

Answers to selected exercises are available for download at **www.ablongman.com/george6e**. Some of the exercises that follow change the original data file. If you wish to leave the data in their original form, don't save your changes.

Case Summaries

1. Using the **grades.sav** file, list variables (in the original order) from **id** to **quiz5**, first 30 students consecutive, number cases, fit on one page by editing.
2. Using the **helping3.sav** file, list variables **hclose**, **hseveret**, **angert**, **hcontrot**, **sympathi**, **worry**, **obligat**, **hcopet**, first 30 cases, number cases, fit on one page by editing.
3. List the first 20 students in the **grades.sav** file, with the lower division students listed first, followed by the upper division students.

Missing Values

4. Using the **grades.sav** file delete the **quiz1** scores for the cases selected in exercise 3, above. Replace the (now) missing scores with the average score for all other students in the class.

Computing Variables

5. Now that you have changed the **quiz1** scores (in exercise 4), recalculate **total** (the sum of all five quizzes and the final) and **percent** (100 times the total divided by the points possible, 125).

6. Using the **divorce.sav** file compute a variable named **spirit** (spirituality) that is the mean of **sp8** through **sp57** (there should be 18 of them). Print out **id, sex**, and the new variable **spirit**, first 30 cases, edit to fit on one page.
7. Using the **grades.sav** file, compute a variable named **quizsum** that is the sum of **quiz1** through **quiz5**. Print out variables **id**, **lastname**, **firstnam**, and the new variable **quizsum**, first 30, all on one page.

Recode Variables

8. Using the **grades.sav** file, compute a variable named **grade2** according to the instructions on page 55. Print out variables **id, lastname, firstnam, grade** and the new variable **grade2**, first 30, edit to fit all on one page. If done correctly, **grade** and **grade2** should be identical.
9. Recode the **passfail** variable so that D's and F's are failing, and A's, B's, and C's are passing.
10. Using the **helping3.sav** file, redo the coding of the ethnic variable so that **Black** = 1, **Hispanic** = 2, **Asian** = 3, **Caucasian** = 4, and **Other/DTS** = 5. Now change the value labels to be consistent with reality (that is the coding numbers are different but the labels are consistent with the original ethnicity). Print out the variables **id** and **ethnic**, first 30 cases.

Selecting Cases

11. Using the **divorce.sav** file select females (**sex** = 1); print out **id** and **sex**, first 40 subjects, numbered, fit on one page.
12. Select all of the students in the **grades.sav** file whose previous **GPA**'s are less than 2, and whose **percent**ages for the class is greater than 85.
13. Using the **helping3.sav** file, select females (**gender** = 1) who give more than the average amount of help (**thelplnz** > 0). Print out **id**, **gender**, **thelplnz**, first 40 subjects, numbered, fit on one page.

Sorting Cases

14. Alphabetize the **grades.sav** file by **lastname**, **firstnam**, first 40 cases.
15. Using the **grades.sav** file, sort by **id** (ascending order). Print out **id**, **total**, **percent**, and **grade**, first 40 subjects, fit on one page.

GRAPHS: Creating and Editing Graphs and Charts

SPSS for Windows possesses impressive and dramatic graphics capabilities. It produces high quality graphs and charts, and editing and enhancement options are extensive. Furthermore, there are two complete sets of graphing procedures available in SPSS: **Regular graphs** and **Interactive graphs**.

Regular graphs are simpler, usually have fewer editing options, and provide adequate graphical representation in most cases. If you have a reasonably simple procedure and/or if you wish graphical representation for your own perusal, then regular graphs may be the best option. Further, graphs that are accessed within specific statistical procedures (you will encounter these a number of times throughout the chapters that follow) are typically regular graphs and thus it is important to have a reasonable working knowledge of how to read and edit them. A section discussing the main ways to edit regular graphs is included at the end of this chapter. There are other instances when specific procedures are not available through interactive graphs. For example, if you want to produce a pareto chart (combining the features of a bar chart and a line graph with the bars generally representing the number of cases or subjects within particular categories and the lines showing cumulative frequencies), you will need to use the regular chart procedures instead of the interactive graphs procedures. You access regular graphs through clicking the **Graphs** command and then any subcommand other than **Interactive**.

Interactive graphs provide a wider array of useful features (for instance, including regression lines in scatter plots), have a more sophisticated interface, provide greater flexibility and, in some cases, create a more professional looking final product. You access interactive graphs through clicking the **Graphs** command and then the **Interactive** subcommand. Then select whichever interactive graph you wish.

Both regular and interactive graphs have different menu options, dialog boxes, and toolbars. Both sets of procedures can produce the graphs discussed in this chapter. We are limited, however, in terms of how much we can present. By way of perspective, this chapter, at 15 pages, is one of the longer in the book, but it is dwarfed by the more than 300 pages devoted to graphic options in the SPSS manuals. In short, we are unable to even hint at comprehensiveness, however, we hope that what we have chosen to include in this chapter will address the graphing needs of most of our readers.

In pages that follow, we first consider the general procedure for accessing graphs and include the dialog boxes and the available options. After that, seven particularly useful types of graphs are presented and discussed. Next, editing procedures are presented, and we finish with a page that addresses editing regular (non-interactive) graphs. Types of graphs covered in this chapter include:

- **Bar graphs**: Bar graphs are used most often to display the distribution of subjects or cases in particular categories, such as the number of A, B, C, D, and F grades in a particular class.
- **Line graphs**: Line graphs may be used to display trends in data and multiple line charts are often employed to demonstrate ANOVA interactions.
- **Pie charts**: Pie charts, like bar charts, are another popular way of displaying the number of subjects or cases within different subsets of categorical data.
- **Box plots**: Box plots are based on percentiles and provide an excellent vehicle to display the distribution of your data.

- **Error bar charts:** Produces a chart that includes error bars indicating the standard error of measurement, or indicating the confidence interval. This is useful for visually clarifying which differences are significant and which are not significant.
- **Histograms**: Histograms look similar to bar graph but are used more often to indicate the number of subjects or cases in ranges of values for a continuous variable, such as the number of students who scored between 90 and 100, between 80 and 89, between 70 and 79, and so forth on a final exam.
- **Scatter plots** (simple and overlay): Scatter plots are a popular way of displaying the nature of correlations between variables.

The Sample Graph

In the graph that follows, we attempt to crowd as many SPSS graphics options into one graph as possible. This provides a clear reference as you read through this chapter. When you encounter a word such as *"category axis"*, and don't know what it means, a quick glance back at this chart provides an example of a category axis. We cannot in one graph illustrate all terms, but we display most of them in the chart below. Be aware that some terminology is idiosyncratic to the SPSS graph editor.

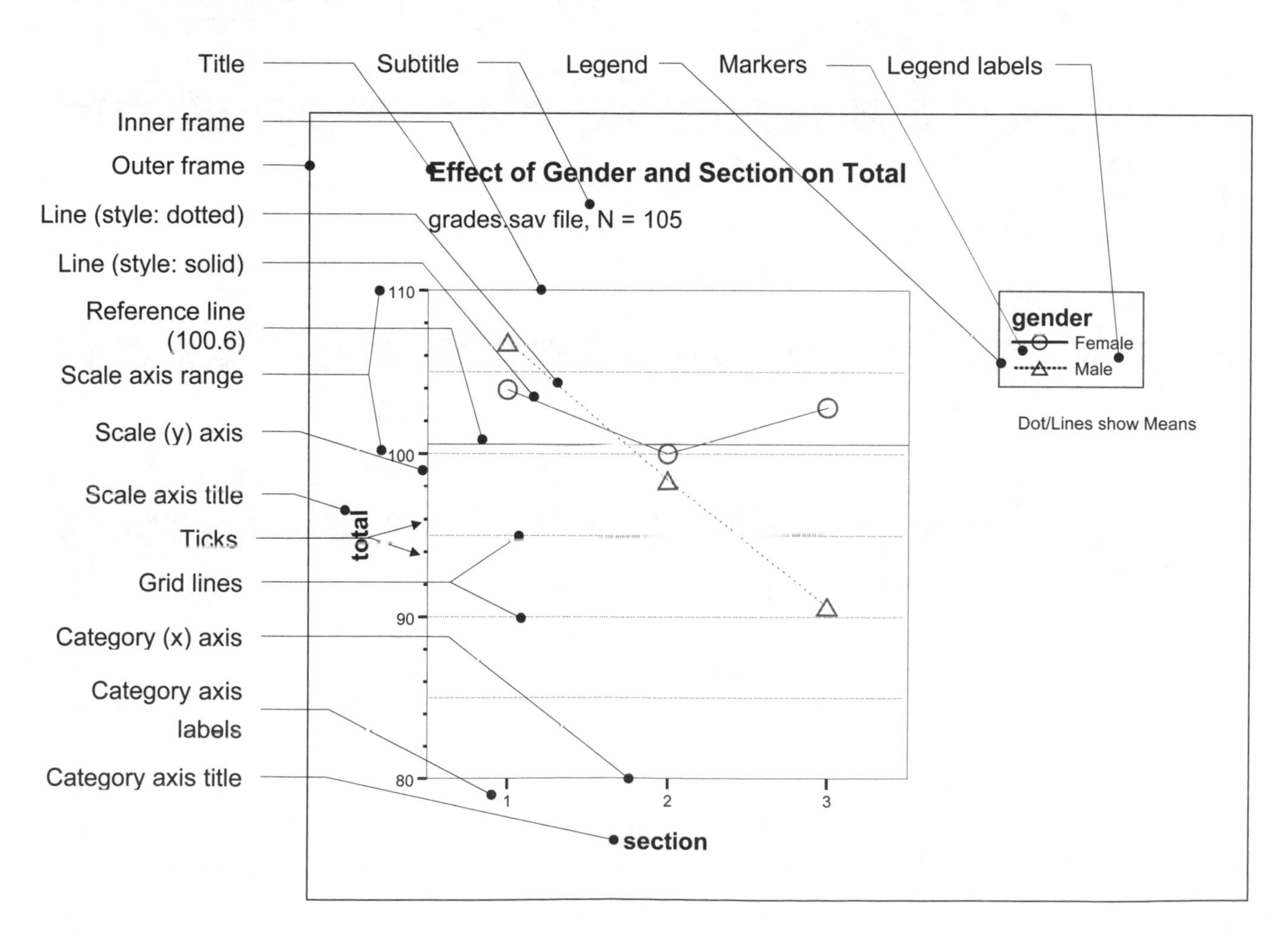

Producing Graphs and Charts

Although there are a wide variety of graphs, there are certain features they have in common. They almost all have an X-axis and a Y-axis (with a certain variable graphed on the X-axis, and a different variable graphed on the Y-axis), and something in the data region of the graph (dots, lines, bars, etc.) to indicate the relation between the X-axis variable and the Y-axis variable.

The following steps indicate the general procedure to produce a graph. In this operation, a line graph is produced (similar to the sample graph, previous page). The steps here focus on the general operations involved in creating any graph; later in the chapter, a description of some of the key additional options for creating particular kinds of graphs are discussed. A note for advanced users: For many procedures, you can select certain values in the SPSS output, right-click, and select **Create Graph** and then select kind of graph that you want to create. If you are not an advanced user already familiar with the SPSS graphing procedures, this can go very wrong or be misinterpreted. The usual steps to access the interactive graphs follow:

To enter SPSS, a click on **Start** *in the taskbar (bottom of screen) activates the start menu:*

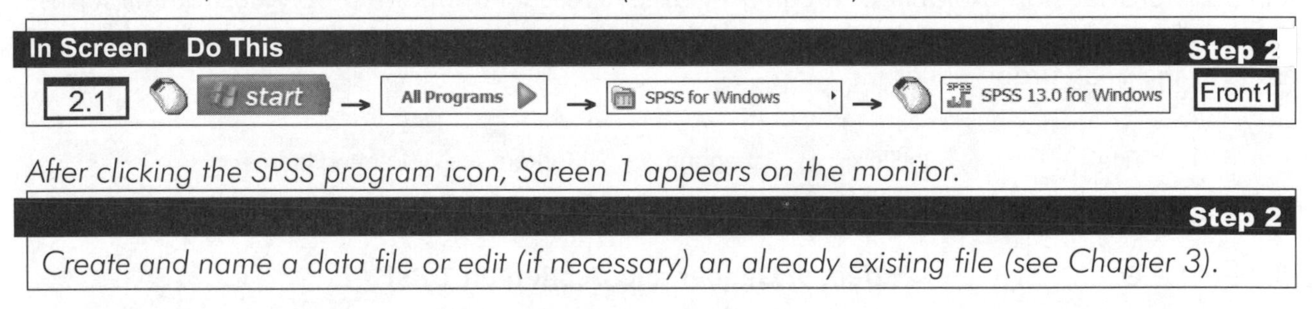

Screens 1 and 2 (displayed on the inside front cover) allow you to access the data file used in conducting the analysis of interest. The following sequence accesses the **grades.sav** *file for further analyses:*

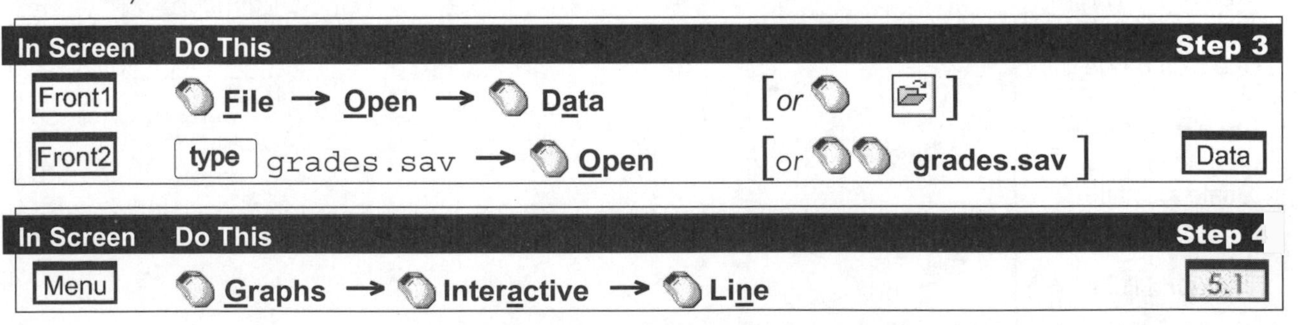

You may select any kind of graph available in Step 4; for the sake of example, we have chosen a line graph. You could also choose to create a **Bar** chart, **Pie** chart, **Boxplot**, **Error bar** graph, **Histogram**, or **Scatterplot**.

Once Screen 5.1 appears (following page), you have a dizzying array of options. First, note that there is a series of tabs across the top of the window: **Assign Variables**, **Dots and Lines**, **Error Bars**, **Titles**, and **Options**. When you start, the **Assign Variables** tab is selected. If you click on a different tab, then you will get an entirely different set of options. (With five tabs and a total of 87 options for the line graph procedure, you can see why we have had to select only the most common options to describe here!) We will describe the **Assign Variables** and **Titles** tabs here, and mention the key elements of the second tab from the left later as each graph is discussed in greater depth. (The title of the second tab, here labeled **Dots and Lines**, is different for each graph.)

In the box at the left of Screen 5.1 is the list of variables that you can select to graph. In addition to the usual list of variables, you will notice two new variables here: **$case** and **$pct**. A third new variable, **$count**, currently sits in the box provided for the vertical (X) axis. Sometimes this variable is also listed along with the others. These variables are computed by SPSS at the time the graph is created:

- **$case** refers to the case number (the ones listed on the left side of the Data View window (see page 30). Creating graphs using case numbers makes sense only if you have a very small data file, or if your cases are in a particular order (chronological, for example).
- **$count** refers to the number of cases in a particular category. For example, if you select gender for your X-axis variable and $count for your Y-axis variable, you will get a graph indicating how many females and how many males are in the data.
- **$pct** refers to the percentage of cases in a particular category. For example, if you select gender for your X-axis variable and $pct for your Y-axis variable, you will get a graph indicating the percentages of females and males in the data.

5.1

The Create Lines Window (Assign Variables Tab)

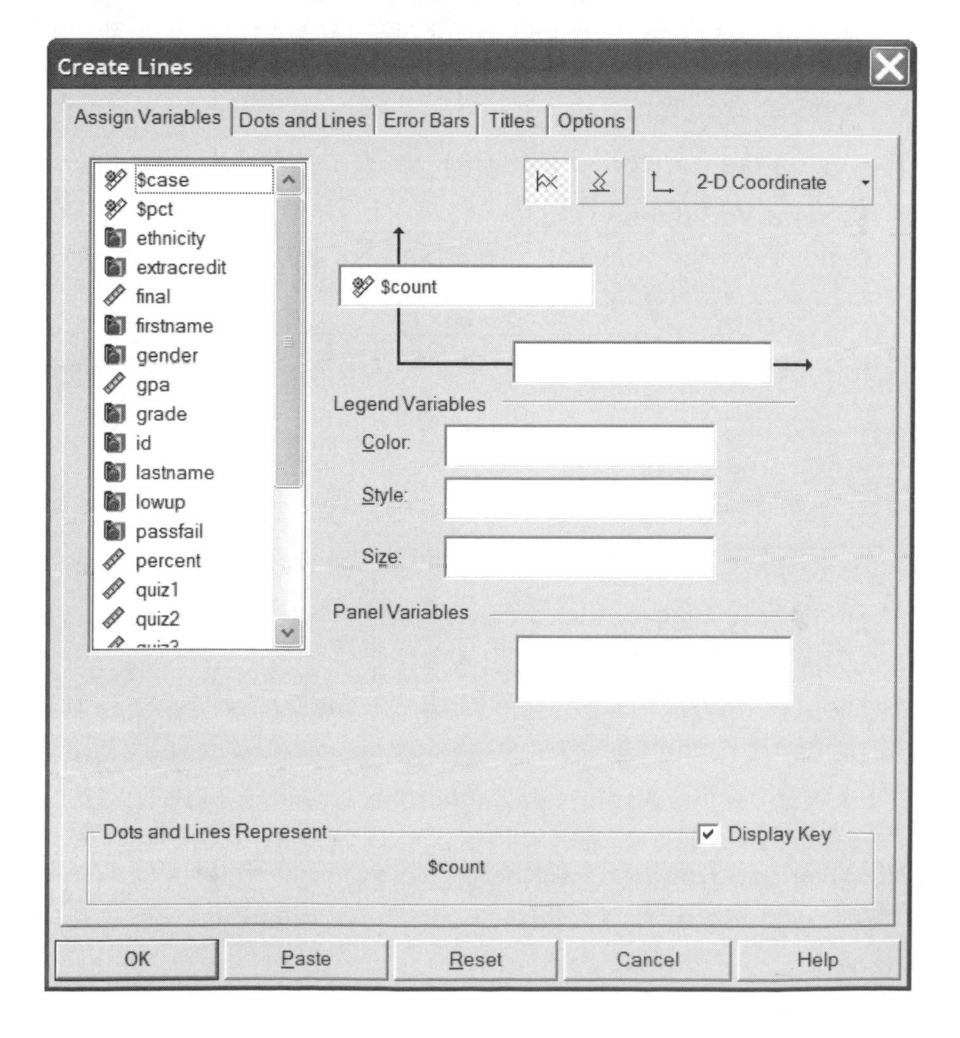

There are small icons to the left of the variable names. These icons indicate whether the variables are a categorical () or a continuous scale (). SPSS bases this on the measure type set up when you defined the variable (see page 36): Scale variables have the scale icon (), and nominal or ordinal variables have the categorical icon (). If you try to produce a graph with a scale variable where SPSS expects a categorical variable (or a categorical variable where a scale variable is expected), SPSS will ask you if you want to change the variable from one kind to the other. When this happens, this usually means that you made a mistake (when selecting the variable, or when setting it up in your data file), or you want an unusual graph.

Unlike most of the SPSS dialog boxes (where you push the button to move variables from the variable list at the left over to the active box(es) on the right), for interactive graphs you click and drag the variable from the list on the left to the appropriate box to the right.

Here is a brief description of the most important elements in Screen 5.1:

- **2-D Coordinate**: This is selected by default to produce a 2-dimentional graph (with two variables, one on the X-axis and one on the Y-axis). You can click the small down-arrow to the right of this box and select 3-D Coordinate if you want to also create a Z-axis. Because computer screens are still typically two-dimensional, it is usually hard to interpret a three-dimensional graph; stick with two dimensions if possible.
- **Box on the vertical axis (↑)**: Drag a variable to this box and it will be plotted on the Y-axis. In this case, the default variable (**$count**) starts in that box; you can drag it to the left variable-list box if you want to use a different variable. If that sounds like too much effort, then just drag a new variable into the box and the **$count** will scurry back on its own.
- **Box on the horizontal axis (→)**: Drag a variable to this box to plot it on the X-axis.
- **Legend Variables**: These options tell SPSS to draw data differently in the data region depending on the value of a variable. For example, in the sample graph (page 69), the lines are plotted differently for females (a solid line with circular markers) and males (a dotted line with triangular markers). These variables are called legend variables because you have to look at the legend to figure out what they mean: You can't tell from looking at the X- or Y-axis. Using legend variables, you can have SPSS control:
    - The **Color** of the lines or dots plotted in the data region. Note that if you want to print your output on a black-and-white printer, this is probably not a useful option.
    - The **Style** of the lines (solid and a variety of dotted patterns).
    - The **Size** of the lines or dots. For example, you could have cases with higher test scores be plotted with larger dots.
- **Display Key**: If this box is checked, a key (called the legend) indicating what dots and lines represent is plotted.

In addition to the **Assign Variables** tab, you will usually want to visit the **Titles** tab. Three large boxes on this screen allow you to type in a **Chart Title**, **Chart Subtitle**, and **Caption**. Whatever you type here will be displayed in the output (as shown in the sample graph, page 69).

This sequence step produces a line graph (similar to the Sample Chart) with total points on the Y-axis, class section on the X-axis, and separate lines (with different styles) for females and males. Begin after sequence step 4 (p. 70).

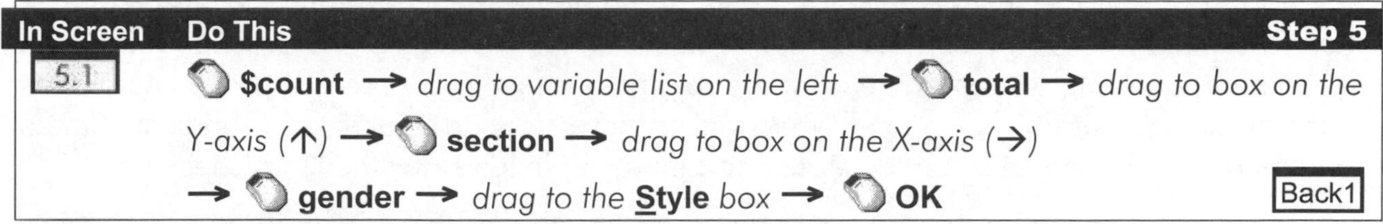

In Screen	Do This	Step 5
5.1	**$count** → *drag to variable list on the left* → **total** → *drag to box on the Y-axis (↑)* → **section** → *drag to box on the X-axis (→)* → **gender** → *drag to the* **Style** *box* → **OK**	Back1

The interactive graph will now appear on the output screen. It is likely that you will wish to edit the graph before printing. A number of editing options are described in the pages that follow. Printing instructions are the same for all graphs and will be presented at the end of the chapter.

SPECIFIC GRAPHS SUMMARIZED

The previous section describes in general terms how to create graphs. Those instructions apply to any graph you want to produce. This section focuses on certain graphs, addressing particularly useful or effective options that are specific to those graphs.

Bar Charts

Bar graphs are used most often to display the distribution of subjects or cases in particular categories, such as the number of A, B, C, D, and F grades in a particular class. To create a bar chart, select **Graphs** → **Interactive** → **Bar**. A dialog box very similar to Screen 5.1 appears. Some of the options available for bar charts (not available for other charts) include:

Clicking on the icon will produce bars going from the bottom to the top of the data region (oriented vertically). Clicking on the icon will produce bars going from the left of the data region to the right (oriented horizontally).

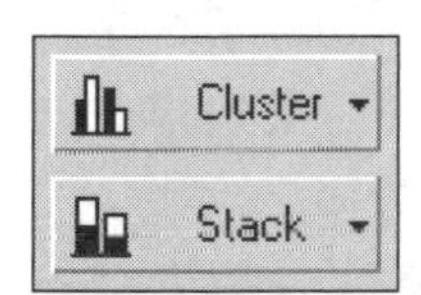

Legend variables in bar charts may be clustered, in which each category of a variable has its own bar (with a unique color or style), or stacked, in which each bar is composed of several sections (one on top of the other). In a stacked bar, there are several sections of the bar (each with its own color or style) indicating how much of each bar is made up of each category of the legend variable. To change a legend variable from cluster to stacked (or stacked to cluster), click on the ▾ to the right of the button and select the new format for the bars.

The sequence step below produces a bar graph with the class grades on the X-axis and the number of students with each grade on the Y-axis. The class section is set as a stacked legend variable.

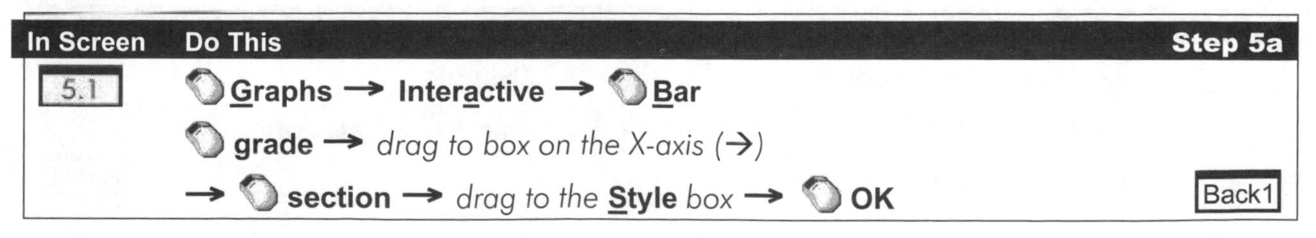

In Screen	Do This	Step 5a
5.1	**Graphs** → **Interactive** → **Bar** **grade** → *drag to box on the X-axis (→)* → **section** → *drag to the* **Style** *box* → **OK**	Back1

Line Graphs

Line graphs may be used to display trends in data and multiple line charts are often employed to demonstrate ANOVA interactions. To create a line graph, select **Graphs** → **Interactive** → **Line**. The dialog box in Screen 5.1 appears; because this procedure was described earlier in the chapter (page 73), little more needs to be said here. One option unique to the line graphs procedure is quite useful however: If you want to have dots on your lines to indicate where group mean is graphed, click on the **Dots and Lines** tab, followed by the **Dots** check box.

The sequence step below produces a line graph (similar to the Sample Chart) with total points on the Y-axis, class section on the X-axis, and separate lines (with different styles) for females and males. Dots are included at each data point.

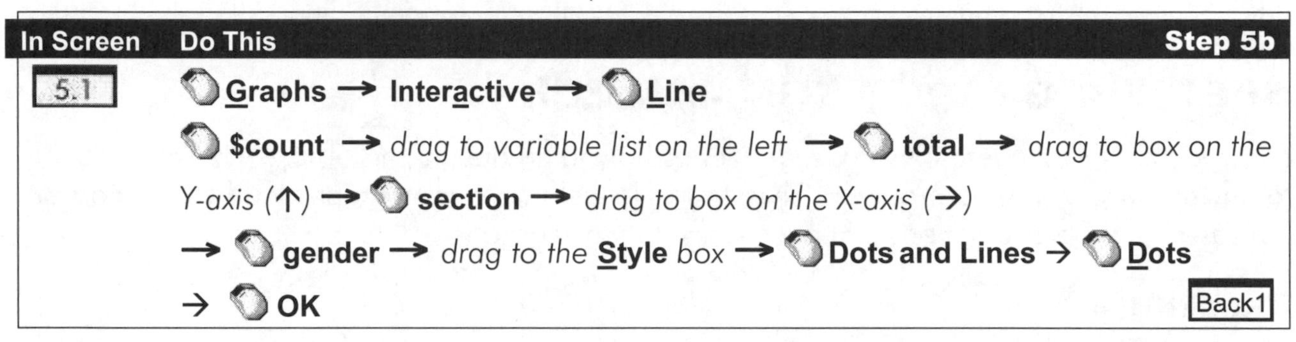

In Screen	Do This	Step 5b
5.1	**Graphs** → **Interactive** → **Line** **$count** → *drag to variable list on the left* → **total** → *drag to box on the Y-axis (↑)* → **section** → *drag to box on the X-axis (→)* → **gender** → *drag to the* **Style** *box* → **Dots and Lines** → **Dots** → **OK**	Back1

To see an interpretation of a line graph (in the context of ANOVA), examine the output section of Chapters 13 and 14.

Pie Charts

Pie charts, like bar charts, are another popular way of displaying the number of subjects or cases within different subsets of categorical data. To create a pie chart, select **Graphs** → **Interactive** → **Pie** → **Simple**. A dialog box similar to Screen 5.1 appears. Because pie charts do not have X- and Y-axes, it is very easy to create a pie chart: First, drag the categorical variable that you want to use to "slice the pie" to the **Slice By** box, and then select either the **Color** or **Style** radio button. It is usually easier to read a pie chart if you select **Color** (or leave it selected, as it is the default), as there will be different colors for each slice of the pie. If you are going to print on a black-and-white printer, though, you should probably select **Style**.

If you want to label your pie slices, click on the **Pies** tab. Here you can choose to add **Category** labels (the value labels for each pie slice), **Count** values (the number of cases in each pie slice), or **Percent**ages for each pie slice (summing to 100% of the pie).

The sequence step below produces a pie chart with a different slice of the pie chart representing a grade in the class. Slices are labeled with category labels and percentages.

In Screen	Do This	Step 5c
5.1	**Graphs** → **Interactive** → **Pie** → **Simple** **grade** → *drag to* **Slice By** *box* → **Pies** → **Category** → **Percent** → **OK**	Back1

The graph now appears on the Output screen, ready to edit or print.

Boxplots

Boxplots are based on percentiles and provide an excellent vehicle to display the distribution of your data. To create a boxplot, select **Graphs → Interactive → Boxplot**. A dialog box very similar to Screen 5.1 appears. For boxplots, you can also **Label Cases By** a variable (for example, by student ID number). This will not label all cases in the data, but it will label outliers farther than 1.5 times the interquartile range away from the median so that you can give these cases special treatment (for example, by examining these cases and be sure they are not mistakes).

The sequence step below produces a graph with total points on the Y-axis, class section on the X-axis, and separate boxplots for females and males. Boxplots are graphed at each data point.

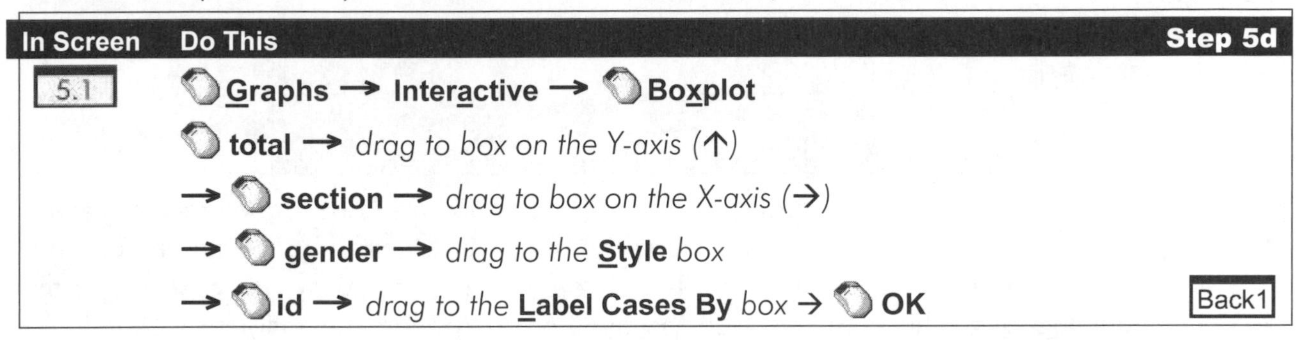

In Screen	Do This	Step 5d
5.1	**Graphs → Interactive → Boxplot**	
	total → *drag to box on the Y-axis (↑)*	
	→ section → *drag to box on the X-axis (→)*	
	→ gender → *drag to the* **Style** *box*	
	→ id → *drag to the* **Label Cases By** *box* **→ OK**	Back1

If you want a boxplot of a single variable on the Y-axis without any variables on the X-axis, you may leave the X-axis blank. If you want to produce different colored boxplots instead of boxplots with different styles, drag gender to **Color** instead of to **Style**.

The following boxplot was produced by sequence step 5d; the leftmost boxplot has been labeled.

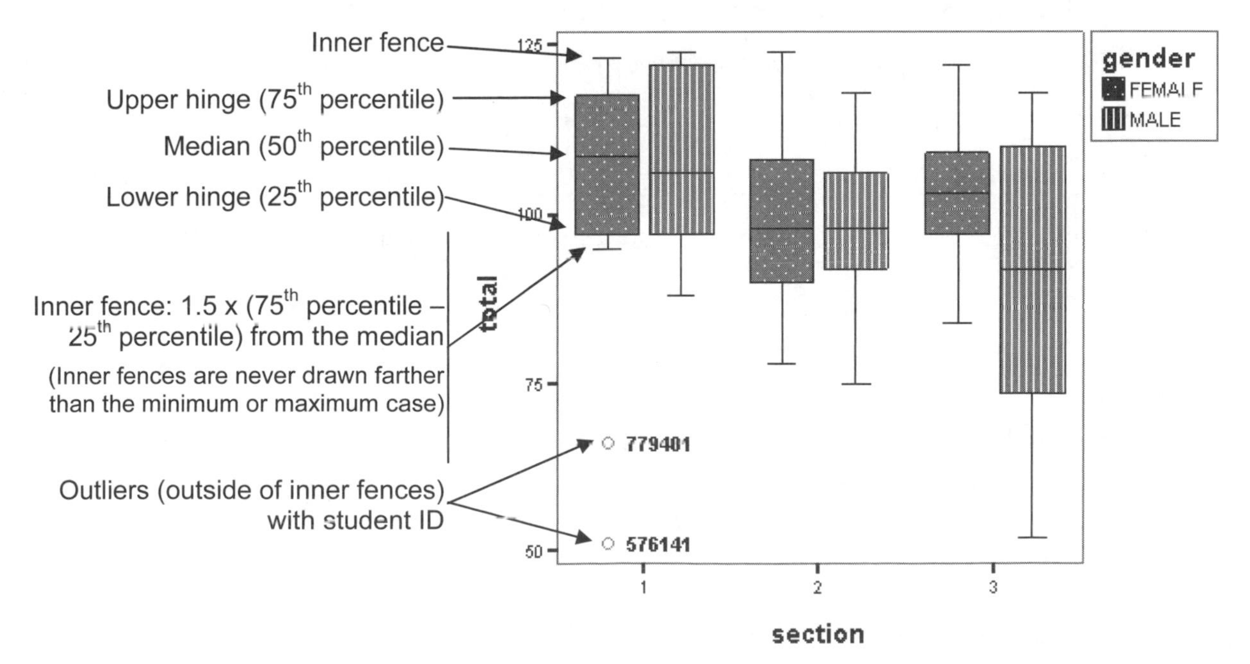

Error Bar Charts

Error bars indicate the confidence interval, the standard deviation, or the standard error of the mean. This is useful for visually clarifying which differences are significant and which are not significant.

To create an error bar chart, select **Graphs** → **Interactive** → **Error Bar**. A dialog box very similar to Screen 5.1 appears. Options specific to error bar charts let you specify what type of error bars you want to produce. You can request error bars that are long enough to represent a certain confidence interval of the mean (typically, a 95% confidence interval; this is the default), a certain number of standard deviations around the mean (for example, the error bars could be one or two standard deviations long), or a certain number of standard errors of the mean long.

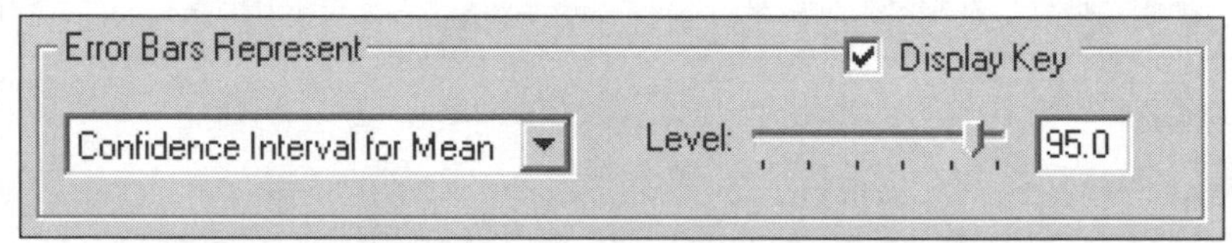

To select a different confidence interval, select the "95.0" and type a new confidence interval. To have SPSS produce error bars based on a certain number of standard deviations or standard errors of the mean, click on the ▾, choose the type of error bar you want, select the "2.0" that will appear to the right and type the number of standard deviations or confidence intervals you desire.

The sequence step below produces a graph with total points on the Y-axis, class section on the X-axis, and separate 95% confidence interval error bars for females and males.

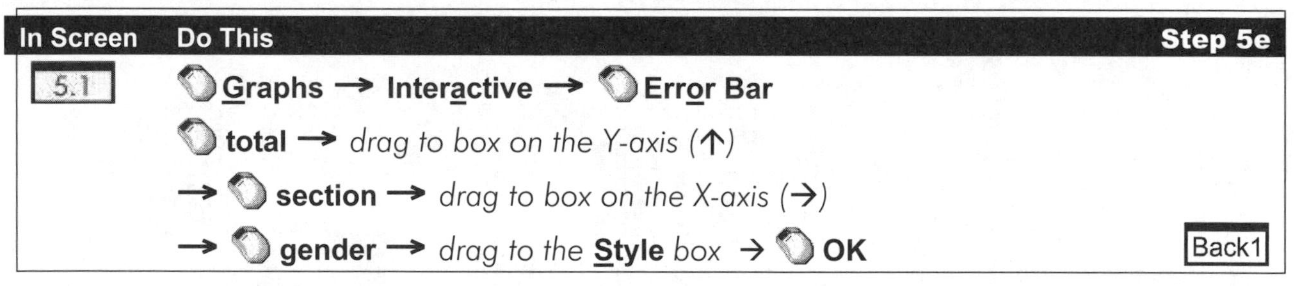

In Screen	Do This	Step 5e
5.1	**Graphs** → **Interactive** → **Error Bar**	
	total → *drag to box on the Y-axis (↑)*	
	→ **section** → *drag to box on the X-axis (→)*	
	→ **gender** → *drag to the* **Style** *box* → **OK**	Back1

If you want to produce different colored error bars instead of error bars with different styles, drag gender to **Color** instead of to **Style**.

Histograms

Histograms look similar to bar graphs but are used more often to indicate the number of subjects or cases in ranges of values for a continuous variable, such as the number of students who scored between 90 and 100, between 80 and 89, between 70 and 79, and so forth for a final class percentage. To create an error bar chart, select **Graphs** → **Interactive** → **Histogram**. A dialog box very similar to Screen 5.1 appears. There are several additional options that you can access by clicking on the Histogram tab, shown on Screen 5.2 below.

- **Normal curve**: If this box is checked, a line representing a normal curve with the mean and standard deviation of your data will be drawn in front of the histogram bars. This makes it possible to easily examine how normally your data is distributed.

- **Set interval size automatically**: If this box is checked, then SPSS will automatically determine how many categories to create for the histogram bars. It is checked by default; if you uncheck the box, you can manually specify the number or size of the histogram bars.
- **Number of intervals**: You can type here the number of bars you want in your histogram.
- **Width of intervals**: You can type here the width of each bar you want in your histogram. Remember that the number you type here is in the units of the variable you are graphing—so if you are graphing GPA's, you will want a much smaller interval width than if you are graphing class percentages, for example.

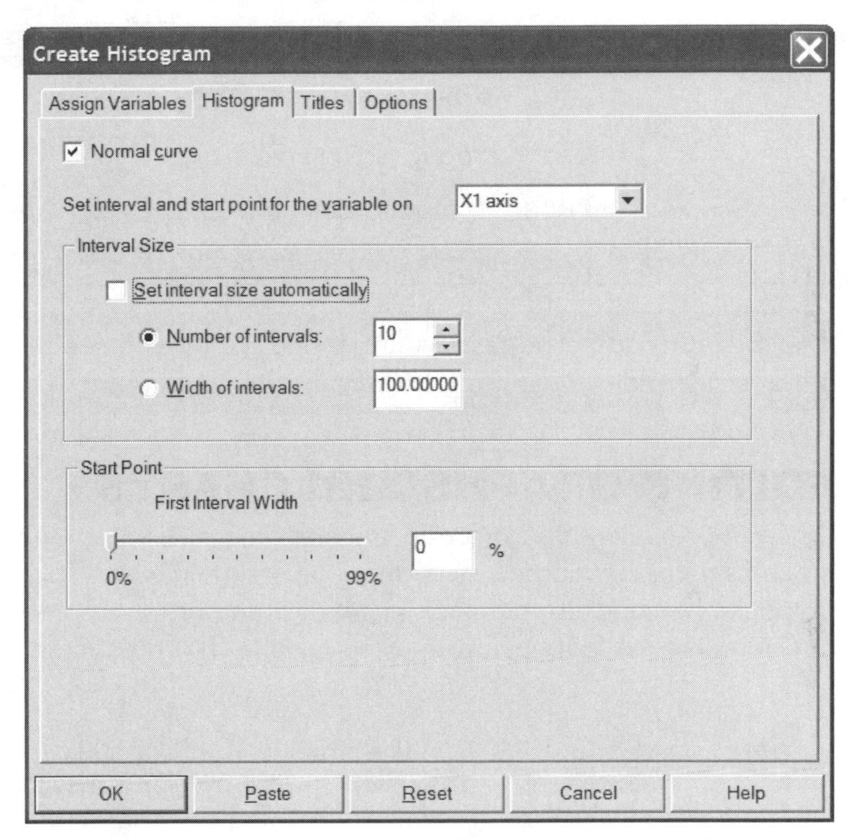

5.2

The Create Histogram Window (Histogram Tab)

The sequence step below produces a histogram of previous GPA with bars .25 GPA's wide. A normal curve will be superimposed on the graph.

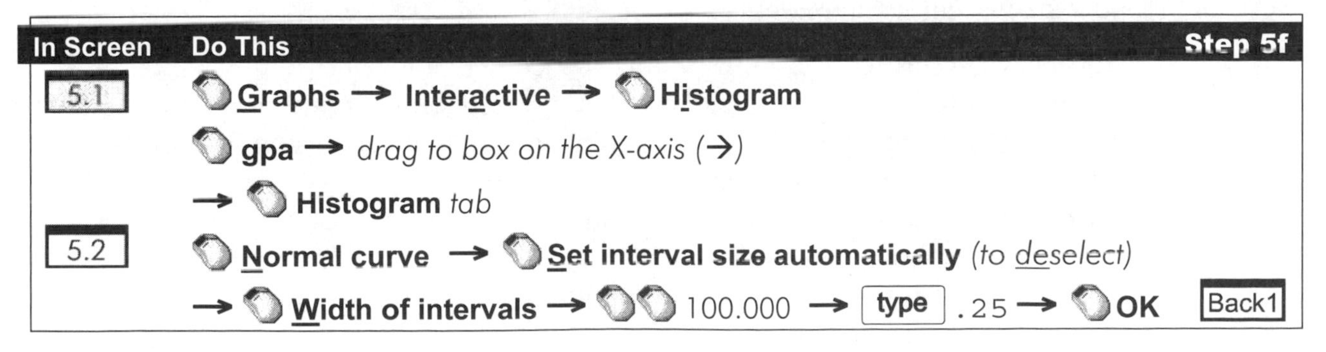

In Screen	Do This	Step 5f
5.1	**Graphs** → **Interactive** → **Histogram**	
	gpa → *drag to box on the X-axis (→)*	
	→ **Histogram** *tab*	
5.2	**Normal curve** → **Set interval size automatically** *(to deselect)*	
	→ **Width of intervals** → 100.000 → type .25 → **OK**	Back1

For an example and explanation of a histogram, see page 92.

Scatterplots

Scatterplots are a popular way of displaying the nature of correlations between variables. One dot is drawn for each case in the data file; both the X- and Y-axis variables are scale variables. To create a scatterplot, select **Graphs** → **Interactive** → **Scatterplot**. A dialog box very much like Screen 5.1 appears where you can select what variables you want plotted on your X- and Y-axes.

The sequence step below produces a scatterplot with total points on the Y-axis, and previous GPA on the X-axis. Different shapes will be used for dots representing females and males.

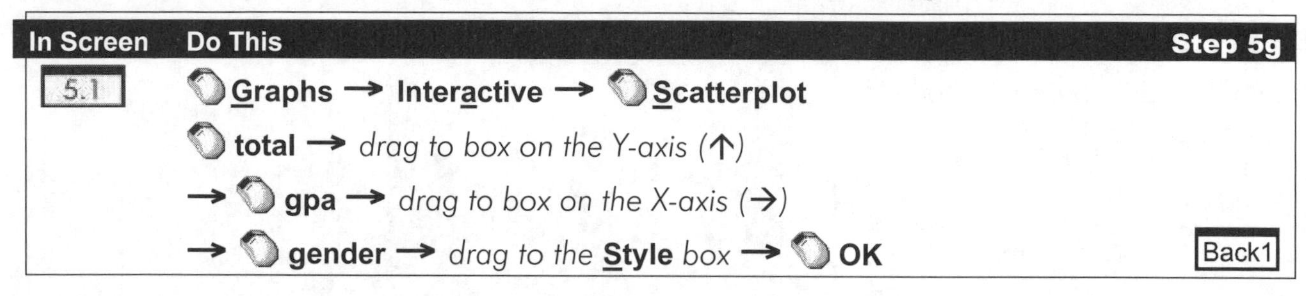

In Screen	Do This	Step 5g
5.1	**Graphs** → **Interactive** → **Scatterplot**	
	total → *drag to box on the Y-axis (↑)*	
	→ **gpa** → *drag to box on the X-axis (→)*	
	→ **gender** → *drag to the* **Style** *box* → **OK**	Back1

If you want to add a regression line to the scatterplot, see regression fit (below). For an example of scatterplots in use, see the introduction to Chapter 10 (on correlations) and Chapter 15.

EDITING GRAPHS AND CHARTS

Once you have created a graph or chart, you may want to modify it to change its appearance, make its meaning clearer, or format it in a certain way. To do this, go to the graph in the SPSS output viewer and double-click on the graph. Some of the menu options will change, and toolbars will appear around the graph. The most common editing options are described here:

Clicking on this icon brings up a dialog that looks almost exactly like Screen 5.1 (with different tabs). Here, you may add variables to the chart legend by dragging new variables from the variable list on the left to the appropriate legend variable box. You can also change your X- or Y-axis variables by dragging the current variable(s) to the variable list on the left, and then dragging the new variables to the X- and Y-axis boxes on the right.

If you click on this icon, a list of things you can add to your graph pops up. The options you are most likely to use are summarized here; note that if you select an option and nothing happens, that probably means that it isn't allowed for the particular kind of chart you are editing.

- **Error Bar**: Replaces each mean in the chart with an error bar. For a description of error bars, see page 76.
- **Box**: Replaces each mean in the chart with a boxplot. For a description of boxplots, see page 75.
- **Mean Fit**: Inserts a labeled reference line at the mean of all of the data displayed in the data region.
- **Regression Fit**: Inserts a labeled regression line going through all of the points. This is typically used in scatterplots. The label for the regression line will

include the R^2 and the regression equation. If you want to move the label, you can click and drag it somewhere else on the graph.

- **Title**, **Subtitle**, and **Caption**: If you didn't include a title, subtitle, or caption when you created the graph, you can insert one later by selecting this option. You will probably want to change these to something more informative by double-clicking on the text and editing it—the default text when you insert a new title, for example, is the word "title."

Clicking on the icon will move the X-axis variable to the Y-axis, and the Y-axis variable to the X-axis. Clicking on the icon will move the variables back where they started.

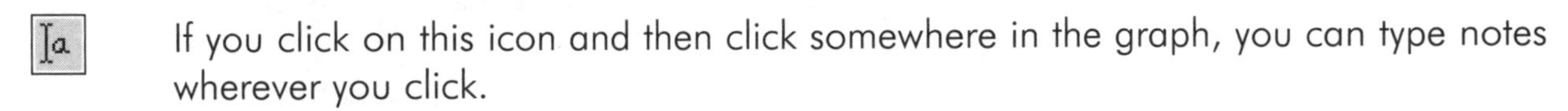

If you click on this icon and then click somewhere in the graph, you can type notes wherever you click.

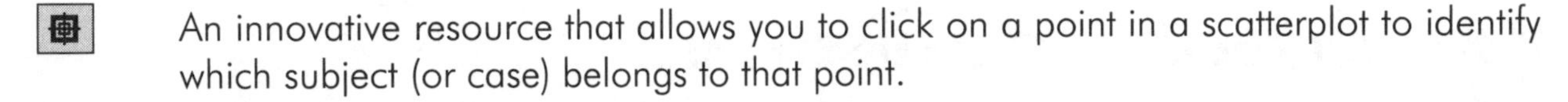

An innovative resource that allows you to click on a point in a scatterplot to identify which subject (or case) belongs to that point.

In addition to these icons, one menu option is particularly useful when editing graphs: **Format → Axis → Scale Axis.** This menu item brings up Screen 5.3. This allows you to change the characteristics of the X- and/or Y-axes. The name of the variable on each axis will be listed next to **Scale Axis**, so there is no doubt which axis you are formatting.

5.3

The Scale Axis Window (Scale Tab)

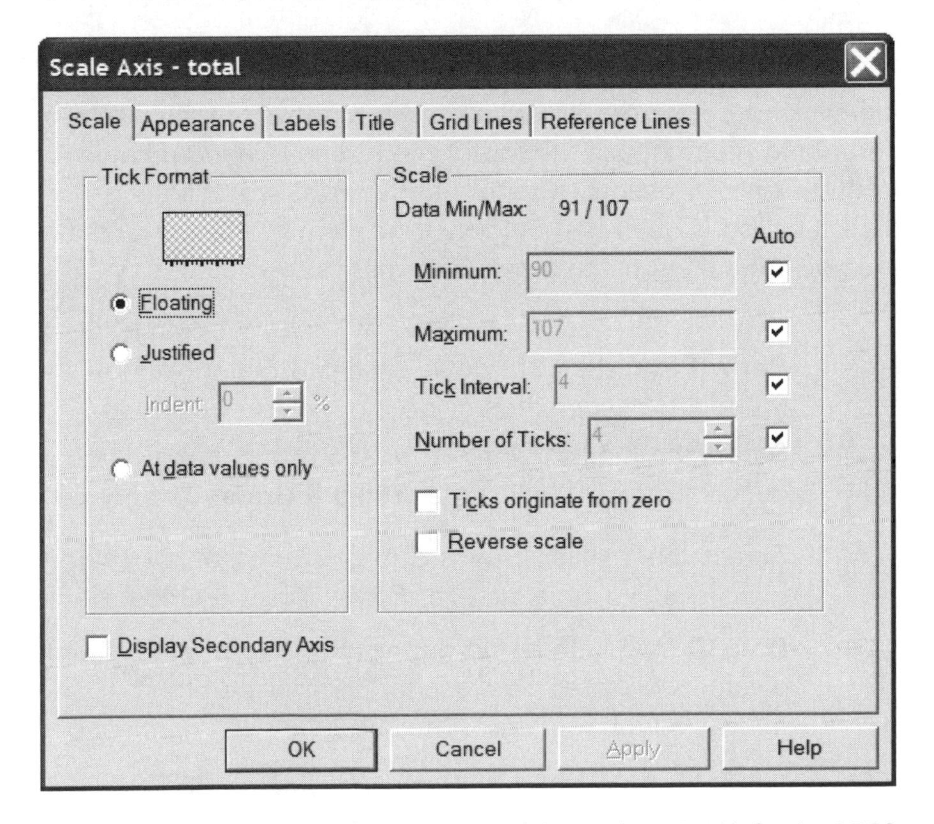

The most useful options in this screen are in the **Scale** portion to the right. By default, SPSS automatically selects the range of possible values for each scale made of a continuous variable. If you want to select your own range, you can do so by clicking on the **Auto** check-mark and typing a new value for the **Minimum** or **Maximum**. For example, SPSS chose a scale range of 90 to 107 by default for total on the sample graph. We defined the **Minimum** at 80 and the **Maximum** at 110 (just because they're nice, round numbers).

You can also modify the **Tick Interval** (the distance between ticks) and the **Number of Ticks** on the scale. It is possible to tell SPSS to do something that is impossible—in which case, SPSS will change something to make it work. For example, if you say you want a minimum of 1 and a maximum of 25 and a tick interval of 5, you can't have 10 ticks (they just won't fit). If you say this is what you want, SPSS will ignore something you said and decide to calculate it automatically.

By default, the X-axis is at the minimum value. If you want the X-axis to be at 0, then select **Ticks originate from 0**. If you want to display minor (smaller, unlabeled) ticks in addition to the major (bigger, labeled) ticks on the scale, click on the **Appearance** tab, followed by clicking on the **Minor ticks** check box.

Editing (Regular) Output Graphs

Many statistical procedures can produce graphs automatically, without having to use the procedures involved in this chapter to specify exactly what the graphs should look like. Unfortunately, if you want to edit these graphs you have to use a completely different set of procedures than used for editing graphs that you request with the interactive graphs procedures. This final section of the chapter describes the most important things you need to know to edit an output graph.

You begin to edit an output graph the same way you would edit an interactive graph—by double clicking on the graph in the output window. At this point, a window opens that includes a command menu across the top of the screen offering different options, and also a bar of icons to allow quicker access to certain frequently-used options. The most frequently used icons and their function is described below. Once you are aware of the function of each of these icons, then actual editing of the graph becomes fairly intuitive.

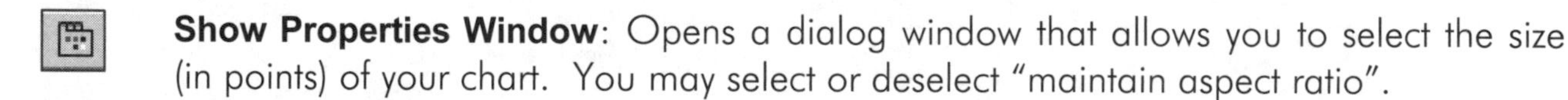

Show Properties Window: Opens a dialog window that allows you to select the size (in points) of your chart. You may select or deselect "maintain aspect ratio".

Select the X axis: Opens a dialog box that allows you to format the X (horizontal) axis. Functions available include: **Ticks & Grids, Categories,** and **Axis Labels.**

Select the Y axis: Opens a dialog box that allows you to format the Y (vertical) axis. Functions available include: **Scale, Ticks & Grids, Number Format,** and **Axis Labels**.

Data ID Mode: This little puppy allows you, with a simple click, to identify particular individuals or cases on a scatter chart or barplot, but also allows you to identify the frequency (number of cases) for bar charts, histograms, or pie graphs.

Insert an Annotation: This option opens a dialog box that allows you to manually type information into your graph and then position and format the information you have typed.

Insert a Text Box: Similar to the Annotation box, but is primarily designed to insert titles and sub titles while the annotation box allows identification of objects within the chart.

Transpose Chart Coordinate system: A click on this button switches the horizontal and vertical axes. A second click returns the chart to its original orientation.

Add Interpolation Line: Within the line-graph procedure this opens a dialog box that allows you to select how you wish to connect your points. Options include: **Straight, Step, Jump,** and **Spline.**

Add fit line: When a scatter plot is displayed this option opens a dialog box that allows you to include in the chart a best-fit line. Choices include: Mean of Y, Linear (the widely used regression line), Quadratic, Cubic, and Loess.

Show Data Labels: A click on this option will (for bar charts, line charts, histograms, and pie charts) automatically identify frequency within each category. At the same time a dialog box will open that allows you to format the size, font, style, and color of those labels.

Show/Hide legend: A click causes to show or hide the legend. The legend might be, for instance, the box that identifies females by a solid line and males by a dotted line.

Show Line Markers: Opens a dialog box that allows you to select the style (e.g., circle, square, triangle, asterisk, etc.), size, and color of markers on line or scatter plots.

Explode slice: In a pie chart, this button allows one or several of the slices to explode. Fortunately the explosion does not usually damage your computer.

PRINTING RESULTS

Results of the analysis (or analyses) that have just been conducted requires a window that displays the standard commands (**File Edit Data Transform Analyze** . . .) across the top. A typical print procedure is shown below beginning with the standard output screen (Screen 1, inside back cover).

To print results, from the Output screen perform the following sequence of steps:

In Screen	Do This	Step 6
Back1	*Select desired output or edit (see pages 19 - 25)* → **File** → **Print**	
2.9	*Consider and select desired print options offered, then* → **OK**	

To exit you may begin from any screen that shows the **File** *command at the top.*

In Screen	Do This	Step 7
Menu	**File** → **Exit**	2.1

Note: After clicking **Exit**, there will frequently be small windows that appear asking if you wish to save or change anything. Simply click each appropriate response.

EXERCISES

Answers to selected exercises are available for download at **www.ablongman.com/george6e**.

All of the following exercises use the **grades.sav** sample data file.

1. Using a bar chart, examine the number of students in each section of the class along with whether or not student attended the review session. Does there appear to be a relation between these variables?
2. Using a line graph, examine the relationship between attending the **review** session and **section** on the **final** exam score. What does this relationship look like?
3. Create a boxplot of **quiz 1** scores. What does this tell you about the distribution of the quiz scores? Create a boxplot of **quiz 2** scores. How does the distribution of this quiz differ from the distribution of quiz 1? Which case number is the outlier?
4. Create an error bar graph highlighting the 95% confidence interval of the mean for each of the three **section**s' **final** exam scores. What does this mean?
5. Based on the examination of a histogram, does it appear that students' previous GPA's are normally distributed?
6. Create the scatterplot described in Step 5g. What does the relationship appear to be between **gender** and academic performance (**total**)? Add a regression line to this scatterplot. What does this regression line tell you?

FREQUENCIES

THIS CHAPTER deals with frequencies, graphical representation of frequencies (bar charts and pie charts), histograms, and percentiles. Each of these procedures is described below. **Frequencies** is one of the SPSS commands in which it is possible to access certain graphs directly (specifically, bar charts, pie charts, and histograms) rather than accessing them through the **Graph** command. Greater detail about editing these graphs is treated in some detail in Chapter 5. Bar charts or pie charts are typically used to show the number of cases ("frequencies") in different categories. As such they clearly belong in a chapter on frequencies. Inclusion of histograms and percentiles seems a bit odd because they are most often used with a continuous distribution of values and are rarely used with categorical data. They are included here because the **Frequencies** command in SPSS is configured in such a way that, in addition to frequency information, you can also access histograms for continuous variables, certain descriptive information, and percentiles. The **Descriptives** command and descriptive statistics are described in Chapter 7, however, that procedure does not allow access to histograms or percentiles.

FREQUENCIES

Frequencies is one of the simplest yet one of the most useful of all SPSS procedures. The **Frequencies** command simply sums the number of instances within a particular category: There were 56 males and 37 females. There were 16 Whites, 7 Blacks, 14 Hispanics, 19 Asians, and 5 others. There were 13 A's, 29 B's, 37 C's, 7 D's, and 3 F's. Using the **Frequencies** command, SPSS will list the following information: Value labels, the value code (the number associated with each level of a variable, e.g., female = 1, male = 2), the frequency, the percent of total for each value, the valid percent (percent after missing values are excluded), and the cumulative percent. These are each illustrated and described in the Output section.

BAR CHARTS

The **Bar chart(s)** option is used to create a visual display of frequency information. A bar chart should be used only for categorical (not continuous) data. The gender, ethnicity, and grade variables listed in the previous paragraph represent categorical data. Each of these variables divides the data set into distinct categories such as male, female; A, B, C, D, F; and others. These variables can be appropriately displayed in a bar chart. Continuous data contain a series of numbers or values such as scores on the final exam, total points, finishing times in a road race, weight in pounds of individuals in your class, and so forth. Continuous variables are typically represented graphically with histograms, our next topic.

HISTOGRAMS

For continuous data, the **Histogram(s)** option will create the appropriate visual display. A histogram is used to indicate frequencies of a *range* of values. A histogram is used when the number of instances of a variable is too large to want to list all of them. A good example is the breakdown of the final point totals in a class of students. Since it would be too cumbersome to list *all* scores on a graph, it is more practical to list the number of subjects within a *range* of values, such as how many students scored between 60 and 69 points, between 70 and 79 points, and so forth.

PERCENTILES

The **Percentile** option will compute any desired percentiles for continuous data. Percentiles are used to indicate what percent of a distribution lies below (and above) a particular value. For instance if a score of 111 was at the 75th percentile, this would mean that 75% of values are lower than 111 and 25% of values are higher than 111. Percentiles are used extensively in educational and psychological measurement.

The file we use to illustrate frequencies, bar charts, histograms, and percentiles (pie charts are so intuitive we do not present them here) is the example described in the first chapter. The file is called **grades.sav** and has an $N = 105$. This analysis computes frequencies, bar charts, histograms, and percentiles utilizing the **gender**, **ethnicity**, **grade**, and **total** variables.

STEP BY STEP

Frequencies

*To enter SPSS, a click on **Start** in the taskbar (bottom of screen) activates the start menu:*

After clicking the SPSS program icon, Screen 1 (inside front cover) appears on the monitor.

	Step 2
Create and name a data file or edit (if necessary) an already existing file (see Chapter 3).	

*Screens 1 and 2 (inside front cover) allow you to access the data file used in conducting the analysis of interest. The following sequence accesses the **grades.sav** file for further analyses:*

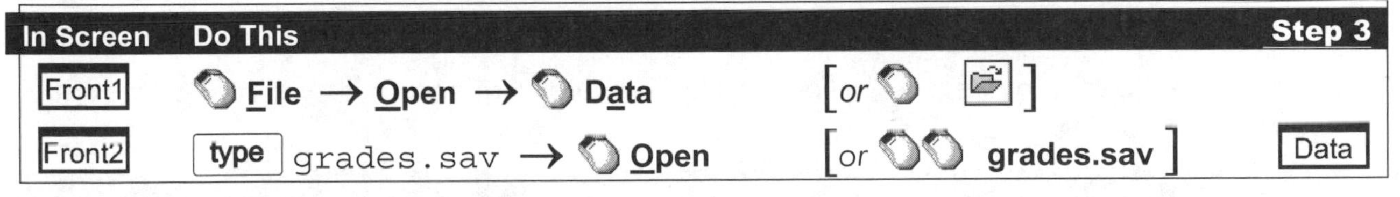

Whether first entering SPSS or returning from earlier operations the standard menu of commands across the top is required (shown below). As long as it is visible you may perform any analyses. It is not necessary for the data window to be visible.

This menu of commands disappears or modifies when using pivot tables or editing graphs. To uncover the standard menu of commands simply click on the or the icon.

After completion of Step 3 a screen with the desired menu bar appears. When you click a command (from the menu bar), a series of options will appear (usually) below the selected command. With each new set of options, click the desired item. The sequence to access frequencies begins at any screen with the menu of commands visible:

FREQUENCIES

A new screen now appears (below) that allows you to select variables for which you wish to compute frequencies. The procedure involves clicking the desired variable name in the box to the left and then pasting it into the **Variables(s)** (or "active") box to the right by clicking the right arrow (▸) in the middle of the screen. If the desired variable is not visible, use the scroll bar arrows (▴ ▾) to bring it to view. To deselect a variable (that is, to move it from the **Variable(s)** box back to the original list), click on the variable in the active box and the ▸ in the center will become a ◂. Click on the left arrow to move the variable back. To clear all variables from the **Variables(s)** box, click the **Reset** button.

6.1

The Frequencies Window

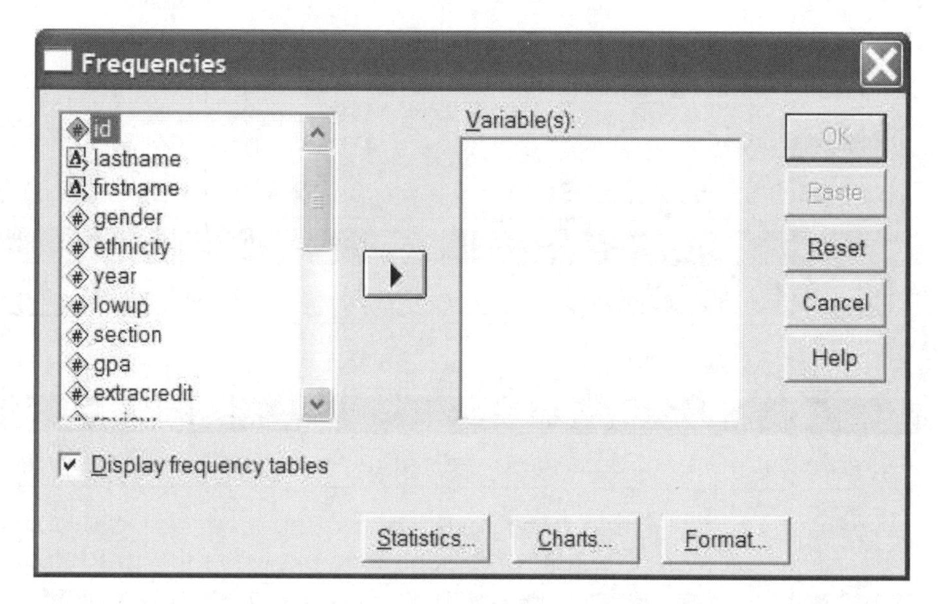

The following sequence of steps will allow you to compute frequencies for the variables **ethnicity**, **gender**, *and* **grade**.

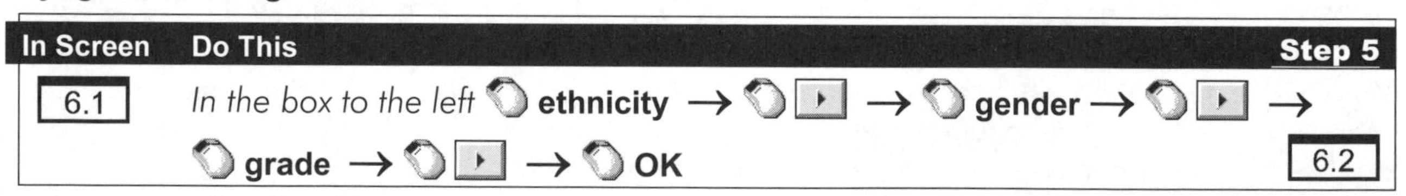

In Screen	Do This	Step 5
6.1	*In the box to the left* **ethnicity** → ▸ → **gender** → ▸ → **grade** → ▸ → **OK**	6.2

You have now selected the three variables associated with gender, ethnicity, and grades. By clicking the **OK** button, SPSS proceeds to compute frequencies. After a few moments the output will be displayed on the screen. The Output screen will appear every time an analysis is conducted (labeled Screen 6.2), and appears on the following page.

The results are now located in a window with the title **Output1 – SPSS Viewer** at the top. To view the results, make use of the up and down arrows on the scroll bar (▴ ▾). Partial results from the procedure described above are found in the Output section. More complete information about output screens, editing output, and pivot charts are included in Chapter 2 (pages 19-25). If you wish to conduct further analyses with the same data set, the starting point is again Screen 6.1. Perform whichever of Steps 1-4 (usually Step 4 is all that is necessary) are needed to arrive at this screen.

6.2

SPSS Output Navigator

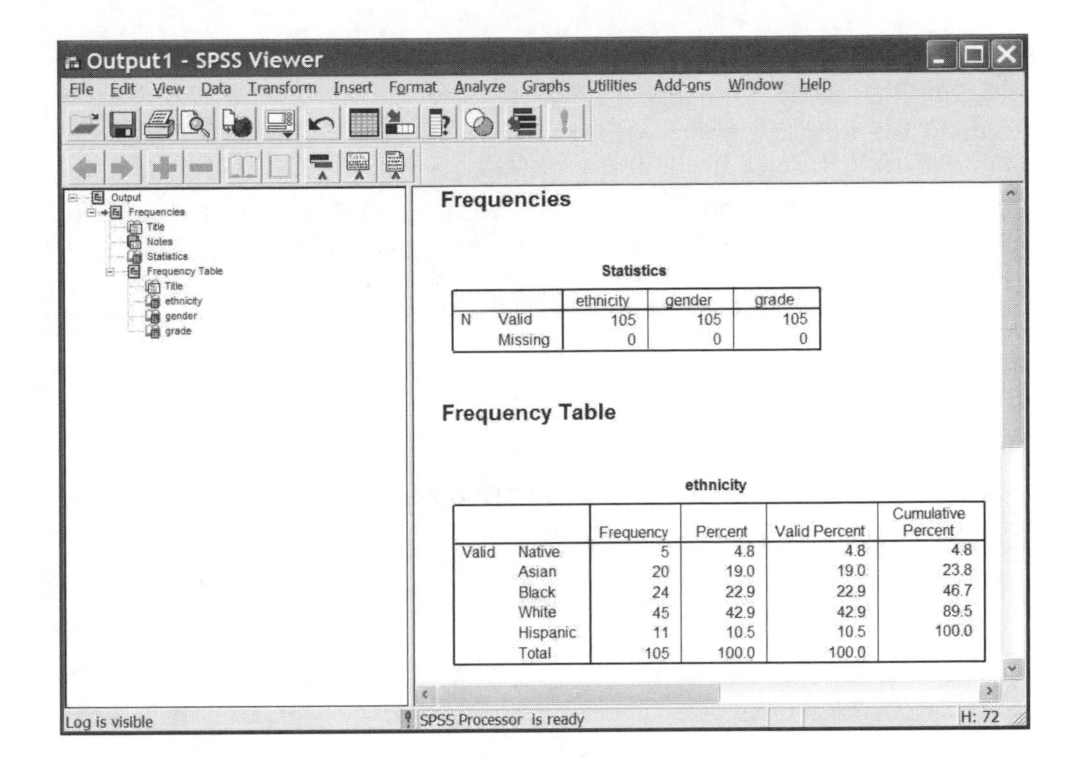

Statistics

		ethnicity	gender	grade
N	Valid	105	105	105
	Missing	0	0	0

ethnicity

		Frequency	Percent	Valid Percent	Cumulative Percent
Valid	Native	5	4.8	4.8	4.8
	Asian	20	19.0	19.0	23.8
	Black	24	22.9	22.9	46.7
	White	45	42.9	42.9	89.5
	Hispanic	11	10.5	10.5	100.0
	Total	105	100.0	100.0	

Bar Charts

To create bar charts of categorical data, the process is identical to sequence step 5 (above), except that instead of clicking the final OK, you will click the **Charts** option (see Screen 6.1). At this point a new screen (Screen 6.3, below) appears: **Bar charts**, **Pie charts**, and **Histograms** are the types of charts offered. For categorical data you will usually choose **Bar charts**. The choice of **Frequencies** (the number of instances within each category) or **Percentages** (the percent of total for each category) depends on your preference. After you click **Continue**, the Charts box disappears leaving Screen 6.1. A click on **OK** completes the procedure.

6.3

The Frequencies: Charts Window

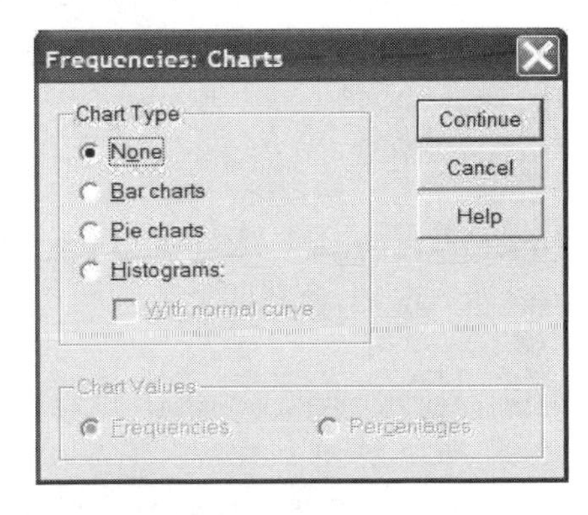

Screen 6.1 is also the starting point for this procedure. Notice that we demonstrated a double click of the variable name to paste it into the active box (rather than a click on the ▸ button).

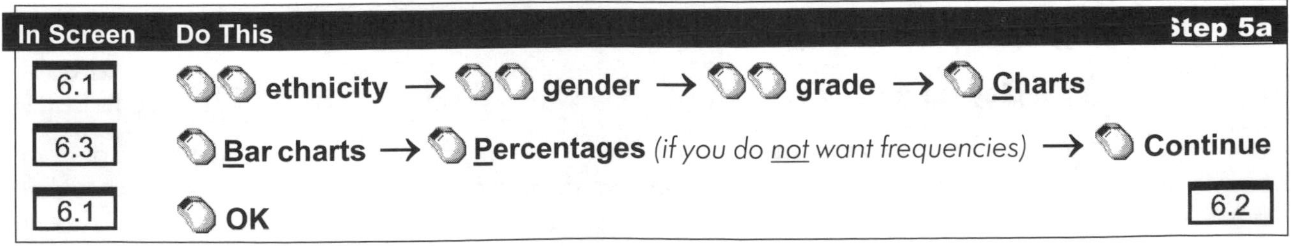

In Screen	Do This	Step 5a
6.1	ethnicity → gender → grade → **Charts**	
6.3	**Bar charts** → **Percentages** *(if you do not want frequencies)* → **Continue**	
6.1	**OK**	6.2

After a few moments of processing time (several hours if you are working on a typical university network) the output screen will emerge. A total of three bar charts have been created, one describing the ethnic breakdown, another describing the gender breakdown, and a third dealing with grades. To see these three graphs simply requires scrolling down the output page until you arrive at the desired graph. If you wish to edit the graphs for enhanced clarity, double click on the graph and then turn to Chapter 5 to assist you with a number of editing options. The chart that follows (Screen 6.4) shows the bar chart for ethnicity.

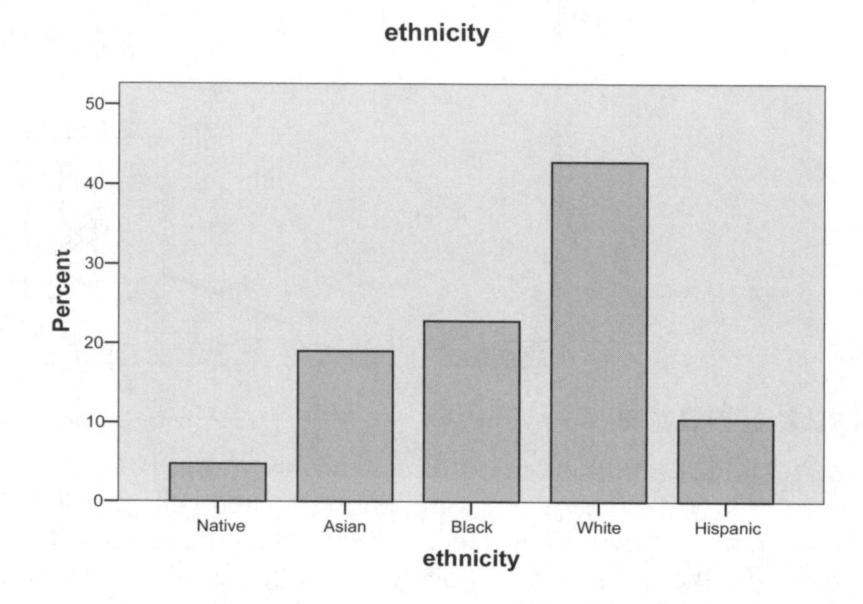

6.4

A Sample Bar Chart

HISTOGRAMS

Histograms may be accessed in the same way as bar charts. The distinction between bar charts and histograms is that histograms are typically used for display of continuous (not categorical) data. For the variables used above, (gender, ethnicity, and grades) histograms would not be appropriate. We will here make use of a histogram to display the distribution for the total points earned by students in the class. Perform the following sequence of steps to create a histogram for total. Refer, if necessary, to Screens 6.1, 6.2, and 6.3 on previous pages for visual reference. The histogram for this procedure is displayed in the Output section.

This procedure begins at Screen 6.1. Perform whichever of steps 1-4 (pp. 85-86) are necessary to arrive at this screen. You may also need to click the **Reset** *button before beginning.*

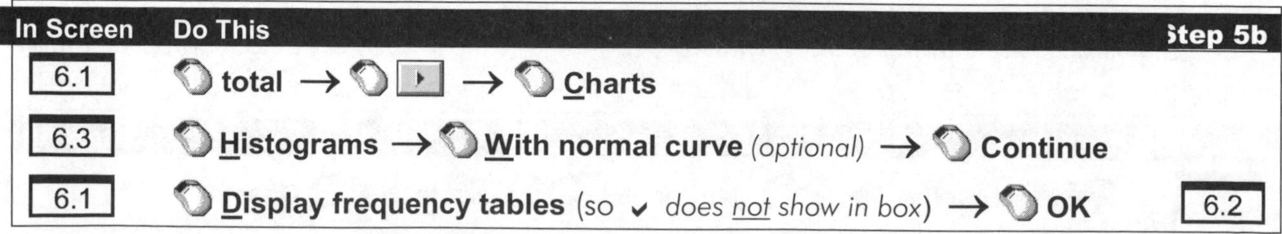

In Screen	Do This	Step 5b
6.1	**total** → ▸ → **Charts**	
6.3	**Histograms** → **With normal curve** *(optional)* → **Continue**	
6.1	**Display frequency tables** (so ✔ *does not show in box*) → **OK**	6.2

Note the step where you click **Display frequency tables** to deselect that option. For categorical data, you will always keep this option since it constitutes the entire non-graphical output. For continuous data (the **total** variable in this case), a display of frequencies would be a list about 70 items long indicating that 1 subject scored 45, 2 subjects scores 47 and so on up to the number of subjects who scored 125. This is rarely desired. If you click this option prior to requesting a histogram, a warning will flash indicating that there will be no output. The **With normal curve** option allows a normal curve to be superimposed over the histogram.

PERCENTILES and DESCRIPTIVES

Descriptive statistics are explained in detail in Chapter 7. Using the **Frequencies** command, under the **Statistics** option (see Screen 6.1), descriptive statistics and percentile values are available. When you click on the **Statistics** option, a new screen appears (Screen 6.5, below) that allows access to this additional information. Three different step sequences (below and on the following page) will explain (a) how to create a histogram and access descriptive data, (b) how to calculate a series of percentiles with equal spacing between each value, and (c) how to access specific numeric percentiles. All three sequences will utilize the **total** points variable.

6.5

The Frequencies: Statistics Window

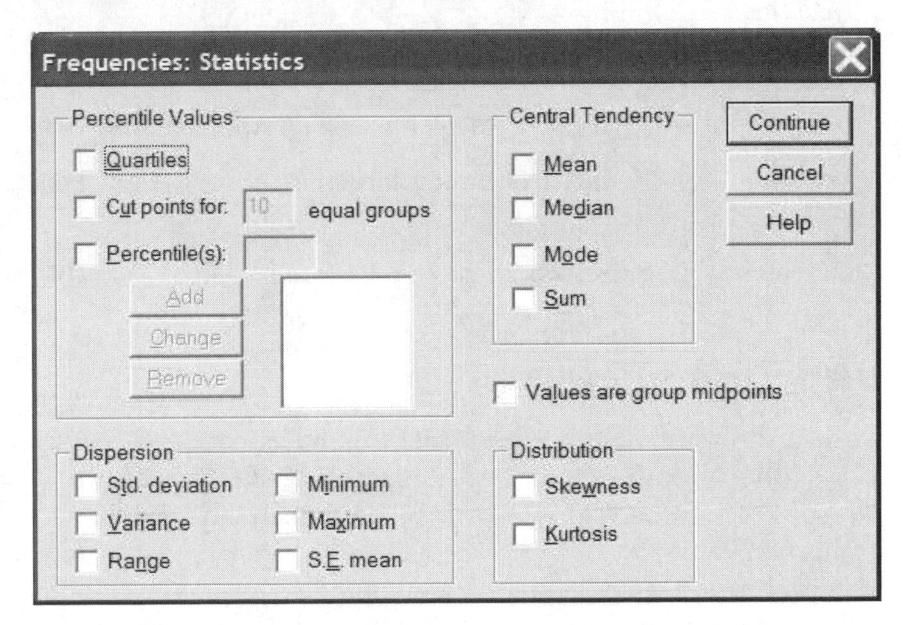

For any of the three procedures below, the starting point is Screen 6.1. Perform whichever of steps 1-4 (pp. 85-86) are necessary to arrive at this screen. Step 5c gives steps to create a histogram for total points and also requests the mean of the distribution, the standard deviation, the skewness, and the kurtosis. Click the **Reset** *button before beginning if necessary.*

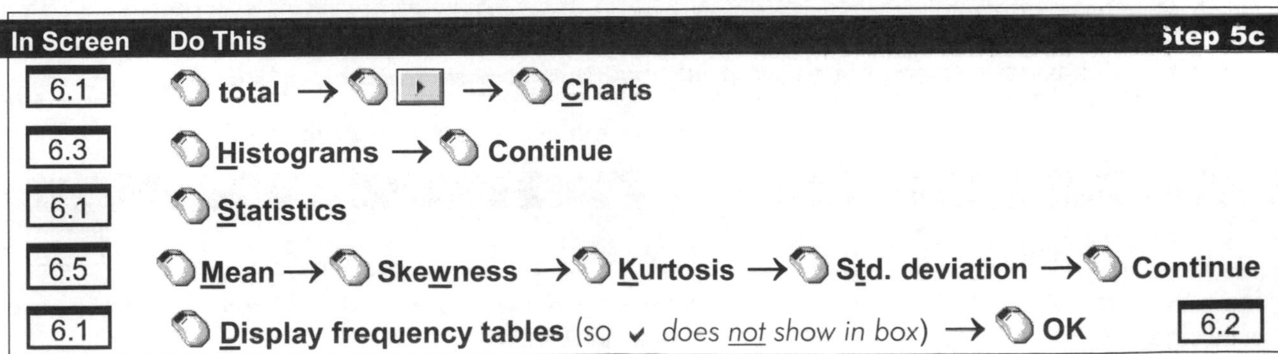

In Screen	Do This	Step 5c
6.1	**total** → ▸ → **Charts**	
6.3	**Histograms** → **Continue**	
6.1	**Statistics**	
6.5	**Mean** → **Skewness** → **Kurtosis** → **Std. deviation** → **Continue**	
6.1	**Display frequency tables** (so ✔ *does not show in box*) → **OK**	6.2

To calculate percentiles of the **total** *variable for every 5th percentile value (e.g., 5th, 10th, 15th, etc.) it is necessary to divide the percentile scale into 20 equal parts. Click* **Reset** *if necessary.*

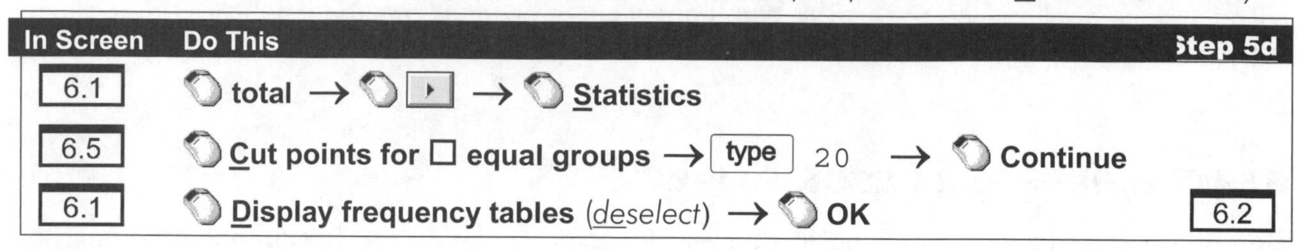

Note that when you type the 20 (or any number) it automatically writes over the default value of 10, already showing.

Finally, to designate particular percentile values (in this case, 2, 16, 50, 84, 98) perform the following sequence of steps. Click **Reset** *if necessary.*

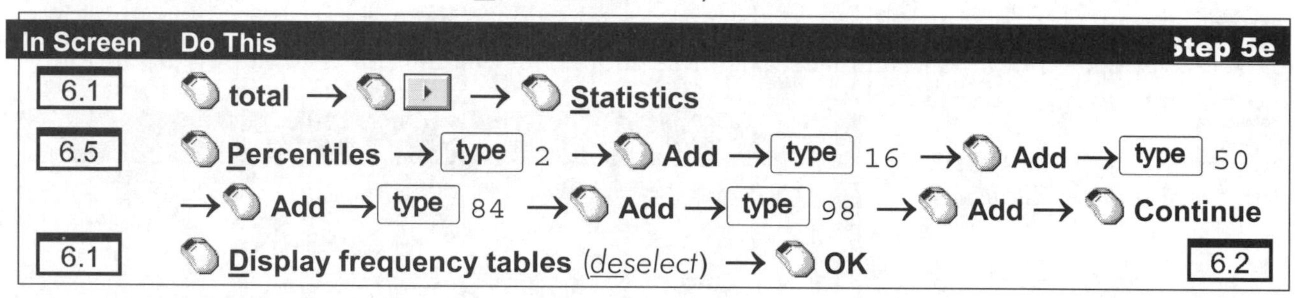

Note: Quartile values (the 25th, 50th, and 75th percentiles) may be obtained quickly by clicking the **Quartiles** box (see screen 6.5), clicking **Continue**, then clicking **OK** (see Screen 6.1).

PRINTING RESULTS

Results of the analysis (or analyses) that have just been conducted requires a window that displays the standard commands (**File Edit Data Transform Analyze** . . .) across the top. A typical print procedure is shown below beginning with the standard output screen (Screen 1, inside back cover). Screen 1 on the inside back cover, incidentally, is the same as Screen 6.2 in this chapter. They differ only in that the cover screen is larger and provides more detail. We have chosen to include the output screen in Chapters 6 and 7, but the remainder of the chapters in this book simply refer to the back-cover screen.

To print results, from the Output screen perform the following sequence of steps:

In Screen	Do This	Step 6
Back1	*Select desired output or edit (see pages 19 - 25)* → **File** → **Print**	
2.9	*Consider and select desired print options offered, then* → **OK**	

To exit you may begin from any screen that shows the **File** *command at the top.*

In Screen	Do This	Step 7
Menu	**File** → **Exit**	2.1

Note: After clicking **Exit**, there will frequently be small windows that appear asking if you wish to save or change anything. Simply click each appropriate response.

OUTPUT

Frequencies, Histograms, Descriptives, and Percentiles

In the output, due to space constraints, we often present results of analyses in a more space-conserving format than is typically done by SPSS. We use identical terminology as that used in the SPSS output and hope that minor formatting differences do not detract from understanding.

FREQUENCIES

What follows is partial results (and slightly different format) from sequence step 5, page 86.

Variable	Value Label	Frequency	Percent	Valid Percent	Cum Percent
ETHNICITY	Native	5	4.8	4.8	4.8
	Asian	20	19.0	19.0	23.8
	Black	24	22.9	22.9	46.7
	White	45	42.9	42.9	89.5
	Hispanic	11	10.5	10.5	100.0
	TOTAL	105	100.0	100.0	
GENDER	Female	64	61.0	61.0	61.0
	Male	41	39.0	39.0	100.0
	TOTAL	105	100.0	100.0	

The number of subjects in each category is self-explanatory. Definitions of other terms follow:

Term	Definition/Description
Value label	Names for levels of a variable.
Value	The number associated with each level of the variable (just in front of each label).
Frequency	Number of data points for a variable or level.
Percent	The percent for each component part, including missing values. If there were missing values, they would be listed in the last row as **missing** along with the frequency and percent of missing values. The total would still sum to 100.0%.
Valid percent	Percent of each value excluding missing values.
Cum percent	Cumulative percentage of the **Valid percent.**

HISTOGRAMS

What follows is output from sequence step 5b on page 89.

Note that on the horizontal axis (graph on the following page) the border values of each of the bars are indicated. This makes for clear interpretation since it is easy to identify that, for instance, 11 students scored between 90 and 95 points, 20 students scored between 95 and 100 points, and 8 students scored between 100 and 105 points. The graph has been edited to create the 5-point increments for bars. For creation of an identical graph several of the editing options would need to be applied. Please see Chapter 5 to assist you with this. A normal curve is superimposed on the graph due to selecting the **With normal curve** option.

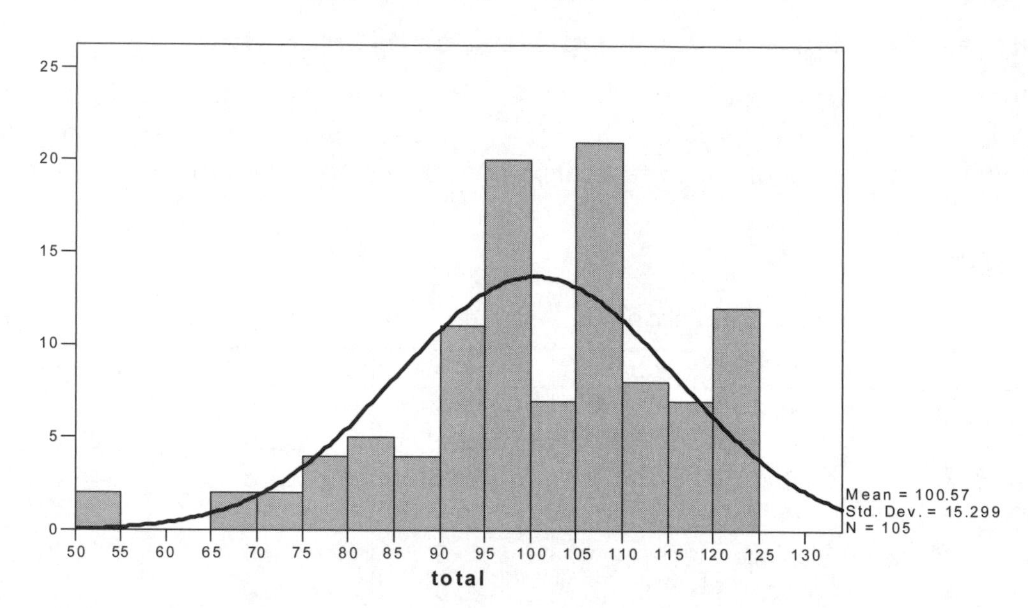

DESCRIPTIVES and PERCENTILES

What follows is complete output (slightly different format) from sequence step 5d on page 90.

Descriptives

Variable	Mean	Std Deviation	Kurtosis	S E Kurtosis	Skewness	S E Skewness
TOTAL	100.571	15.299	.943	.467	-.837	.236

Percentiles

Percentile	Value	Percentile	Value	Percentile	Value	Percentile	Value
5.00	70.00	30.00	95.80	55.00	105.00	80.00	113.00
10.00	79.60	35.00	97.00	60.00	106.60	85.00	118.00
15.00	86.70	40.00	98.00	65.00	108.00	90.00	120.00
20.00	90.00	45.00	98.70	70.00	109.00	95.00	122.70
25.00	92.00	50.00	103.00	75.00	111.00		

Descriptive information is covered in Chapter 7 so we will not discuss those terms here. Note that when the skewness and kurtosis are requested, the standard errors of those two measures are also included.

For Percentiles: For the **total** points variable, 5% of values fall below 70 points and 95% of values are higher than 70 points. 10% of values fall below 79.6 points and 90% are higher, and so forth.

EXERCISES

Answers to selected exercises are available for download at **www.ablongman.com/george6e**.

Notice that data files other than the **grades.sav** file are being used here. Please refer to the **Data Files** section starting on page 365 to acquire all necessary information about these files and the meaning of the variables. As a reminder, all data files are downloadable from the web address shown above.

1. Using the **divorce.sav** file display frequencies for **gender**, **ethnicity**, **status**. Print output to show frequencies for all three; edit output so it fits on one page. Include three bar graphs of these data and provide labels to clarify what each one means.

2. Using the **graduate.sav** file display frequencies for **motiv**, **stable**, **hostile**. Print output to show frequencies for all three; edit output so it fits on one page. Note: this type of procedure is typically done to check for accuracy of data. Motivation (**motiv**), emotional stability (**stable**), and hostility (**hostile**) are scored on 1 to 7 scales. You are checking to see if you have, by mistake, entered any 0s or 8s or 77s.

3. Using the **helping3.sav** file compute percentiles for **thelplnz** (time helping, measured in z scores), **tqualitz** (quality of help measured in z scores). Use percentile values 2, 16, 50, 84, 98. Print output and circle values associated with percentiles for **thelplnz**; box percentile values for **tqualitz**.

4. Using the **helping3.sav** file compute percentiles for **age**. Compute every 10th percentile (10, 20, 30, etc.). Edit (if necessary) to fit on one page.

5. Using the **graduate.sav** file display frequencies for **gpa**, **areagpa**, **grequant**. Compute quartiles for these three variables. Edit (if necessary) to fit on one page.

6. Using the **grades.sav** file create a histogram for **final**. Create a title for the graph that makes clear what is being measured.

DESCRIPTIVE Statistics

Descriptives is another frequently used SPSS procedure. Descriptive statistics are designed to give you information about the distributions of your variables. Within this broad category are measures of central tendency (**Mean**, **Median**, **Mode**), measures of variability around the mean (**Std deviation** and **Variance**), measures of deviation from normality (**Skewness** and **Kurtosis**), information concerning the spread of the distribution (**Maximum**, **Minimum**, and **Range**), and information about the stability or sampling error of certain measures including standard error (S.E.) of the mean (**S.E. mean**), S.E. of the kurtosis, and S.E. of the skewness (included by default when skewness and kurtosis are requested). Using the **Descriptives** command, it is possible to access all of these statistics or any subset of them. In this introductory section of the chapter, we begin with a brief description of statistical significance (included in all forms of data analysis) and the normal distribution (because most statistical procedures require normally distributed data). Then each of the statistics identified above is briefly described and illustrated.

STATISTICAL SIGNIFICANCE

All procedures in the chapters that follow involve testing the significance of the results of each analysis. Although statistical significance is not employed in the present chapter it was thought desirable to cover the concept of statistical significance (and normal distributions in the section that follows) early in the book.

Significance is typically designated with words such as "significance", "statistical significance", or "probability". The latter word is the source of the letter that represents significance, the letter "p". The *p* value identifies the likelihood that a particular outcome may have occurred by chance. For instance, group A may score an average of 37 on a scale of depression while group B scores 41 on the same scale. If a *t* test determines that group A differs from group B at a $p = .01$ level of significance, it may be concluded that there is a 1 in 100 probability that the resulting difference happened by chance, and a 99 in 100 probability that the discrepancy in scores is a reliable finding.

Regardless of the type of analysis the *p* value identifies the likelihood that a particular outcome occurred by chance. A Chi-square analysis identifies whether observed values differ significantly from expected values; a *t* test or ANOVA identifies whether the mean of one group differs significantly from the mean of another group or groups; correlations and regressions identify whether two or more variables are significantly related to each other. In all instances a significance value will be calculated identifying the likelihood that a particular outcome is or is not reliable. Within the context of research in the social sciences, nothing is ever "proved". It is demonstrated or supported at a certain level of likelihood or significance. The smaller the *p* value, the greater the likelihood that the findings are valid.

Social scientists have generally accepted that if the *p* value is less than .05 then the result is considered **statistically significant**. Thus, when there is less than a 1 in 20 probability that a certain outcome occurred by chance, then that result is considered statistically significant. Another frequently observed convention is that when a significance level falls between .05 and .10, the result is considered **marginally significant**. When the significance level falls far below .05 (e.g., .001, .0001, etc.) the smaller the value the greater confidence the researcher has that his or her findings are valid.

When one writes up the findings of a particular study, certain statistical information and *p* values are always included. Whether or not a significant result has occurred is the key focus of most studies that involve statistics.

THE NORMAL DISTRIBUTION

Many naturally occurring phenomena produce distributions of data that approximate a normal distribution. Some examples include the height of adult humans in the world, the weight of collie dogs, the scoring averages of players in the NBA, and the IQs of residents of the United States. In all of these distributions, there are many mid-range values (e.g., 60-70 inches, 22-28 pounds, 9-14 points, 90-110 IQ points) and few extreme values (e.g., 30 inches, 80 pounds, 60 points, 12 IQ points). There are other distributions that approximate normality but deviate in predictable ways. For instance, times of runners in a 10-kilometer race will have few values less than 30 minutes (none less than 26:20), but many values greater than 40 minutes. The majority of values will lie above the mean (average) value. This is called a *negatively skewed distribution*. Then there is the distribution of ages of persons living in the United States. While there are individuals who are 1 year old and others who are 100 years old, there are far more 1-year-olds, and in general the population has more values below the mean than above the mean. This is called a *positively skewed distribution*. It is possible for distributions to deviate from normality in other ways, some of which are described in this chapter.

A normal distribution is symmetric about the mean or average value. In a normal distribution, 68% of values will lie between plus-or-minus (±) 1 standard deviation (described below) of the mean, 95.5% of values will lie between ± 2 standard deviations of the mean, and 99.7% of values will lie between ± 3 standard deviations of the mean. A normal distribution is illustrated in the figure below.

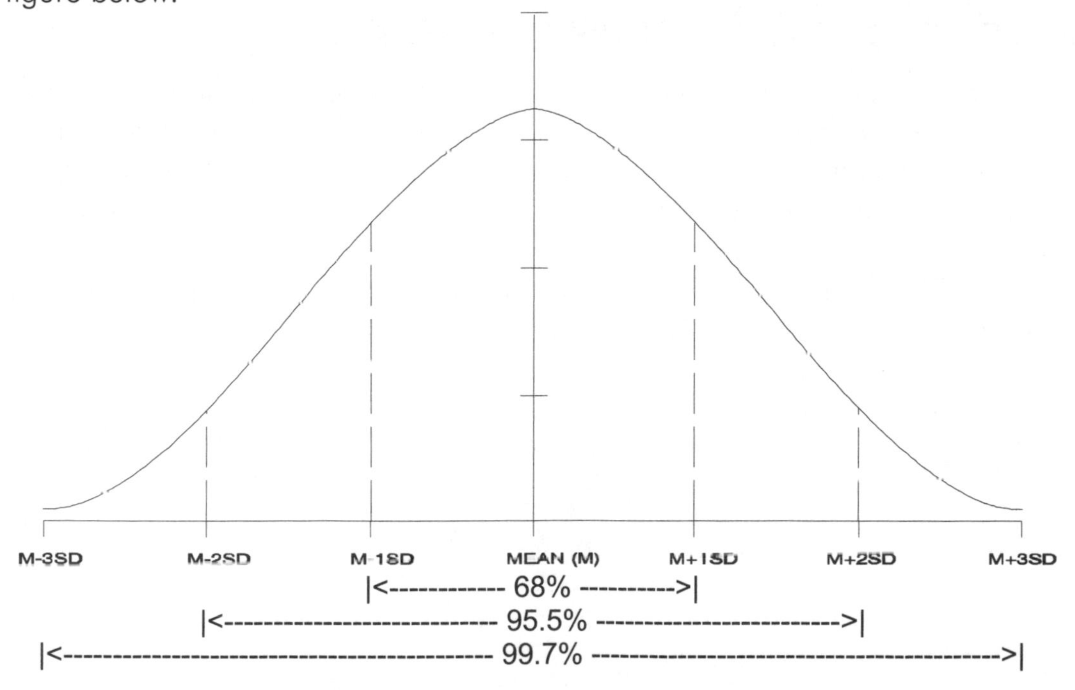

A final example will complete this section. The average (or mean) height of an American adult male is 69 inches (5 ft 9 in.) with a standard deviation of 4 inches. Thus, 68% of American men are between 5 ft 5 in. and 6 ft 1 in. (69 ± 4), 95.5% of American men are between 5 ft 1

in. and 6 ft 5 in. (69 ± 8), and 99.7% of American men are between 4 ft 9 in. and 6 ft 9 in. (69 ± 12) in height (don't let the NBA fool you).

MEASURES OF CENTRAL TENDENCY

The **Mean** is the average value of the distribution, or, the sum of all values divided by the number of values. The mean of the distribution [3 5 7 5 6 8 9] is

$$(3 + 5 + 7 + 5 + 6 + 8 + 9)/7 = \underline{6.14}.$$

The **Median** is the middle value of the distribution. The median of the distribution [3 5 7 5 6 8 9], is 6, the middle value (when reordered from small to large, 3 5 5 **6** 7 8 9).

The **Mode** is the most frequently occurring value. The mode of the distribution [3 **5** 7 **5** 6 8 9] is 5, because the 5 occurs most frequently (twice, all other values occur only once).

MEASURES OF VARIABILITY AROUND THE MEAN

The **Variance** is the sum of squared deviations from the mean divided by $N - 1$. The variance for the distribution [3 5 7 5 6 8 9] is

$$((3-6.14)^2 + (5-6.14)^2 + (7-6.14)^2 + (5-6.14)^2 + (6-6.14)^2 + (8-6.14)^2 + (9-6.14)^2)/6 = \underline{4.1429}$$

Variance is used mainly for computational purposes. Standard deviation is the more commonly used measure of variability.

The **Standard deviation** is the positive square root of the variance. For the distribution [3 5 7 5 6 8 9], the standard deviation is the square root of 4.1429, or 2.0354.

MEASURES OF DEVIATION FROM NORMALITY

Kurtosis is a measure of the "peakedness" or the "flatness" of a distribution. A kurtosis value near zero (0) indicates a shape close to normal. A positive value for the kurtosis indicates a distribution more peaked than normal. A negative kurtosis indicates a shape flatter than normal. An extreme negative kurtosis (e.g., < -5.0) indicates a distribution where more of the values are in the tails of the distribution than around the mean. A kurtosis value between ±1.0 is considered excellent for most psychometric purposes, but a value between ±2.0 is in many cases also acceptable, depending on the particular application.

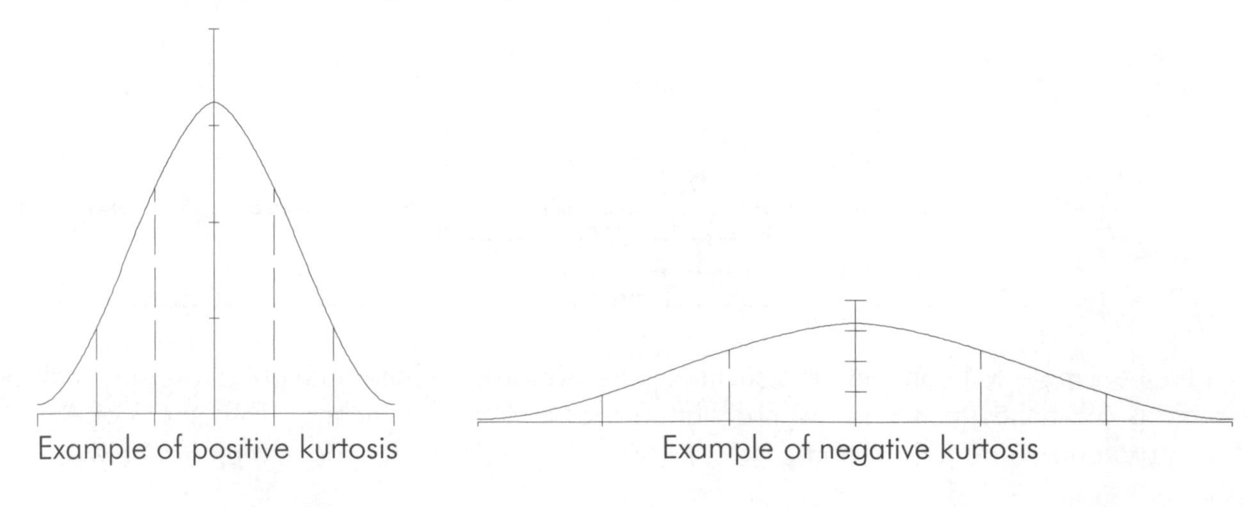

Skewness measures to what extent a distribution of values deviates from symmetry around the mean. A value of zero (0) represents a symmetric or evenly balanced distribution. A positive skewness indicates a greater number of *smaller* values (sounds backward, but this is correct). A negative skewness indicates a greater number of *larger* values. As with kurtosis, a skewness value between ±1.0 is considered excellent for most psychometric purposes, but a value between ±2.0 is in many cases also acceptable, depending on your particular application.

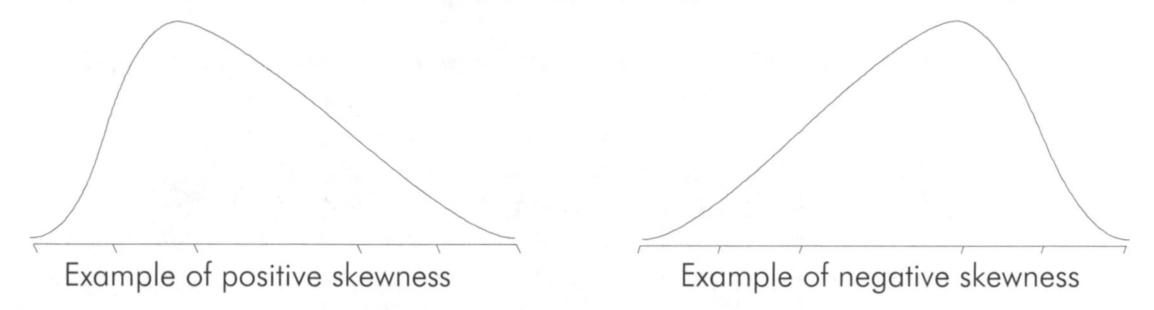
Example of positive skewness   Example of negative skewness

MEASURES FOR SIZE OF THE DISTRIBUTION

For the distribution [3 5 7 5 6 8 9], the **Maximum** value is 9, the **Minimum** value is 3, and the **Range** is 9 - 3 = 6. The **Sum** of the scores is 3 + 5 + 7 + 5 + 6 + 8 + 9 = 43.

MEASURES OF STABILITY: STANDARD ERROR

SPSS computes the **Standard errors** for the mean, the kurtosis, and the skewness. As indicated above, standard error is designed to be a measure of stability or of sampling error. The logic behind standard error is this: If you take a random sample from a population, you can compute the mean, a single number. If you take another sample of the same size from the same population you can again compute the mean—a number likely to be slightly different from the first number. If you collect many such samples, the standard error of the mean is the standard deviation of this sampling distribution of means. A similar logic is behind the computation of standard error for kurtosis or skewness. A small value (what is "small" depends on the nature of your distribution) indicates *greater* stability or *smaller* sampling error.

The file we use to illustrate the **Descriptives** command is our example described in the first chapter. The data file is called **grades.sav** and has an $N = 105$. This analysis computes descriptive statistics for variables **gpa**, **total**, **final**, and **percent**.

STEP BY STEP

Descriptives

To enter SPSS, a click on **Start** *in the taskbar (bottom of screen) activates the start menu:*

After clicking the SPSS program icon, Screen 1 (inside front cover) appears on the monitor.

Step 2
Create and name a data file or edit (if necessary) an already existing file (see Chapter 3).

Screens 1 and 2 (inside front cover) allow you to access the data file used in conducting the analysis of interest. The following sequence accesses the **grades.sav** *file for further analyses:*

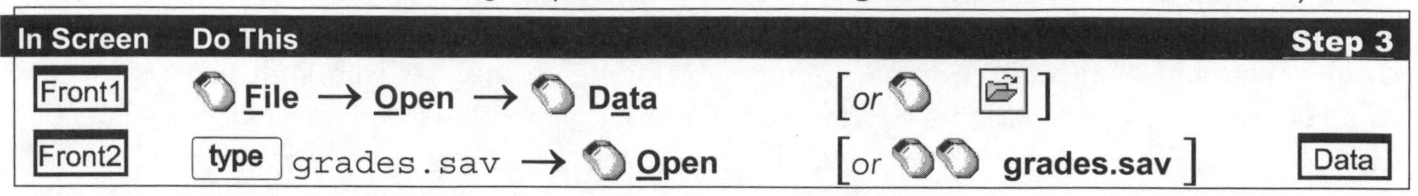

In Screen	Do This		Step 3
Front1	File → Open → Data	[or]	
Front2	type grades.sav → Open	[or grades.sav]	Data

Whether first entering SPSS or returning from earlier operations the standard menu of commands across the top is required (shown below). As long as it is visible you may perform any analyses. It is not necessary for the data window to be visible.

This menu of commands disappears or modifies when using pivot tables or editing graphs. To uncover the standard menu of commands simply click on the or the icon.

After completion of Step 3 a screen with the desired menu bar appears. When you click a command (from the menu bar), a series of options will appear (usually) below the selected command. With each new set of options, click the desired item. The sequence to access Descriptive Statistics begins at any screen with the menu of commands visible:

In Screen	Do This	Step 4
Menu	Analyze → Descriptive Statistics → Descriptives	7.1

A new screen now appears (below) that allows you to select variables for which you wish to compute descriptives. The procedure involves clicking the desired variable name in the box to the left and then pasting it into the **Variable(s)** (or "active") box to the right by clicking the right arrow ( ) in the middle of the screen. If the desired variable is not visible, use the scroll bar arrows ( ) to bring it to view. To deselect a variable (that is, to move it from the **Variable(s)** box back to the original list), click on the variable in the active box and the in the center will become a . Click on the left arrow to move the variable back. To clear all variables from the active box, click the **Reset** button.

7.1

The Descriptives Window

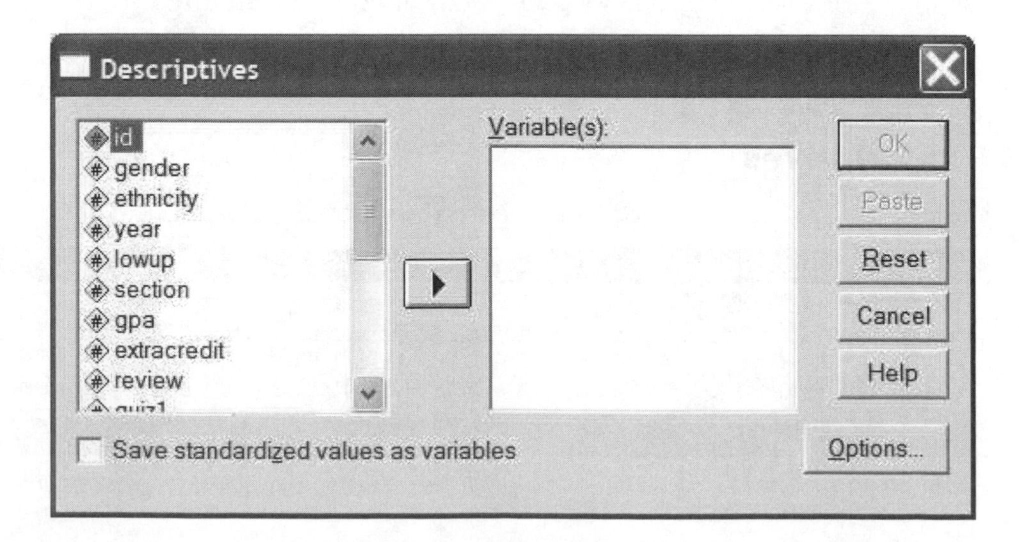

The only check box on the initial screen, **Save standardized values as variables**, will convert all designated variables (those in the **Variable(s)** box) to z scores. The original variables will remain, but new variables with a "z" attached to the front will be included in the list of variables. For instance, if you click the **Save standardized values as variables** option, and the variable **final** was in the **Variable(s)** box, it would be listed in two ways: **final** in the original scale, and **zfinal** for the same variable converted to z scores. You may then do analyses with either the original variable or the variable converted to z scores. Recall that z scores are values that have been mathematically transposed to create a distribution with a mean of zero and a standard deviation of one. See the glossary for a more complete definition. Also note that for non-mouse users, the SPSS people have cleverly underlined the "z" in the word "standardized" as a gentle reminder that standardized scores and z scores are the same thing.

To create a table of the default descriptives (mean, standard deviation, maximum, minimum) for the variables **gpa** *and* **total**, *perform the following sequence of steps:*

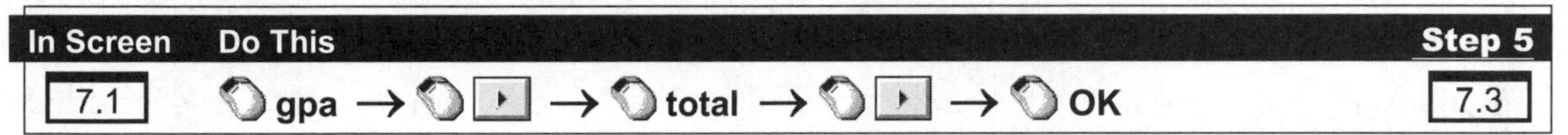

If you wish to calculate more than the four default statistics, after selecting the desired variables, before clicking the **OK**, it is necessary to click the **Options** button (at the bottom of screen 7.1). Here every descriptive statistic presented earlier in this chapter is included with a couple of exceptions: Median and mode are accessed through the Frequencies command only. See Chapter 6 to determine how to access these values. Also, the standard errors ("S.E.") of the kurtosis and skewness are not included. This is because when you click either kurtosis or skewness, the standard errors of those values are automatically included. To select the desired descriptive statistics, the procedure is simply to click (so as to leave an ✔ in the box to the left of the desired value) the descriptive statistics you wish. This is followed by a click of **Continue** and **OK**. The **Display order** options include (a) **Variable list** (the default—in the same order as displayed in the data editor), (b) **Alphabetic** (names of variables ordered alphabetically), (c) **Ascending means** (ordered from smallest mean value to largest mean value in the output), and (d) **Descending means** (from largest to smallest).

7.2

The Descriptives: Options Window

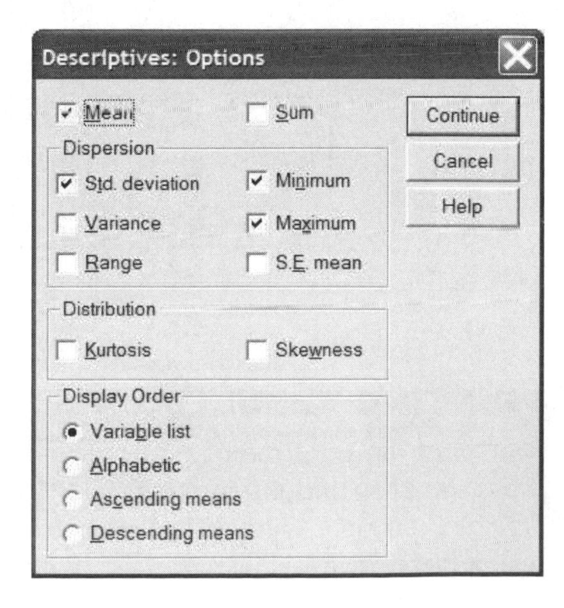

To select the variables **final**, **percent**, **gpa**, *and* **total**, *and then select all desired descriptive statistics, perform the following sequence of steps; Press the* **Reset** *button if there are undesired variables in the active box.*

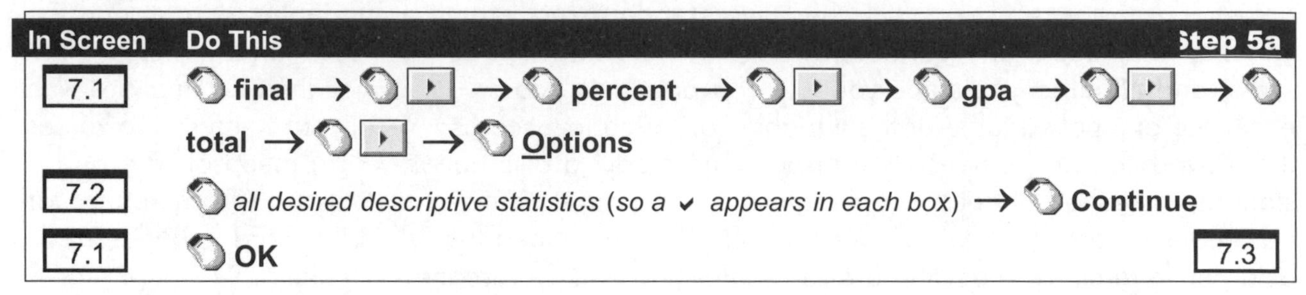

In Screen	Do This	Step 5a
7.1	final → ▸ → percent → ▸ → gpa → ▸ → total → ▸ → Options	
7.2	*all desired descriptive statistics (so a ✓ appears in each box)* → Continue	
7.1	OK	7.3

Upon completion of either step 5 or step 5a, Screen 7.3 will appear (below). The results of the just-completed analysis are included in the top window labeled **Output1 – SPSS Viewer**. Click on the ☐ to the right of this title if you wish the output to fill the entire screen, then make use of the arrows on the scroll bar (▲▼▶◀) to view the results. Even when viewing output, the standard menu of commands is still listed across the top of the window. Further analyses may be conducted without returning to the data screen. Partial output from this analysis is included in the Output section.

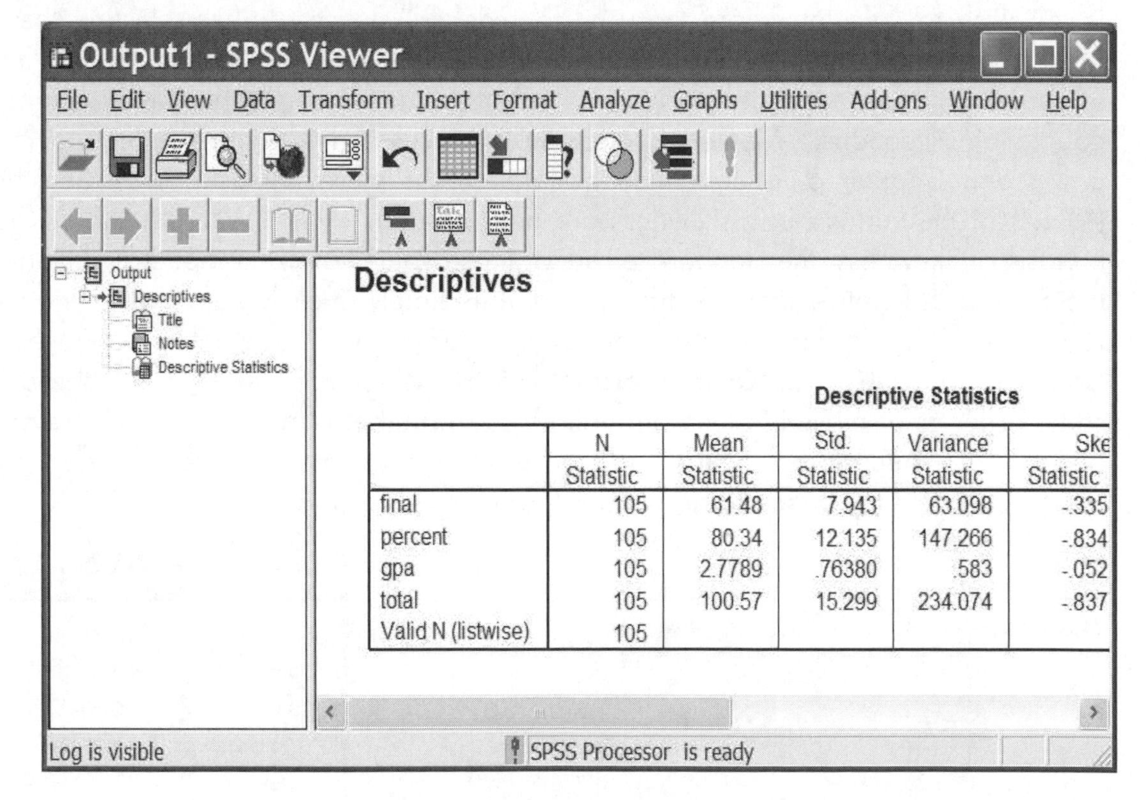

	N	Mean	Std.	Variance	Ske
	Statistic	Statistic	Statistic	Statistic	Statistic
final	105	61.48	7.943	63.098	-.335
percent	105	80.34	12.135	147.266	-.834
gpa	105	2.7789	.76380	.583	-.052
total	105	100.57	15.299	234.074	-.837
Valid N (listwise)	105				

7.3 *The Output1 – SPSS Output Viewer Window*

PRINTING RESULTS

Results of the analysis (or analyses) that have just been conducted requires a window that displays the standard commands (**File Edit Data Transform Analyze** . . .) across the top. A typical print procedure is shown below beginning with the standard output screen (Screen 1, inside back cover).

To print results, from the Output screen perform the following sequence of steps:

In Screen	Do This	Step 6
Back1	*Select desired output or edit (see pages 19 - 25)* → File → Print	
2.9	*Consider and select desired print options offered, then* → OK	

To exit you may begin from any screen that shows the **File** *command at the top.*

In Screen	Do This	Step 7
Menu	File → Exit	2.1

Note: After clicking **Exit**, there will frequently be small windows that appear asking if you wish to save or change anything. Simply click each appropriate response.

OUTPUT

Descriptive Statistics

What follows is output from sequence step 5a, page 102. Notice that the statistics requested included the N, the Mean, the Standard Deviation, the Variance, the Skewness, and the Kurtosis. The Standard Errors of the Skewness and Kurtosis are included by default.

SPSS for Windows: Descriptive Statistics

	N	Mean	Std. Deviation	Variance	Skewness		Kurtosis	
	Statistic	Statistic	Statistic	Statistic	Statistic	Std. Error	Statistic	Std Error
FINAL	105	61.48	7.94	63.098	-.335	.236	-.332	.467
PERCENT	105	80.381	12.177	148.272	-.844	.236	.987	.467
GPA	105	2.779	.763	.583	-.052	.236	-.811	.467
TOTAL	105	100.57	15.30	234.074	-.837	.236	.943	.467
Valid N (listwise)	105							

First observe that in this display the entire output fits neatly onto a single page or is entirely visible on the screen. This is rarely the case. When more extensive output is produced, make use of the up, down, left, and right scroll bar arrows to move to the desired place. You may also use the index in the left window to move to particular output more quickly. Notice that all four variables fall within the "excellent" range as acceptable variables for further analyses; the skewness and kurtosis values all lie between ± 1.0. All terms are identified and described in the introductory portion of the chapter. The only undefined word is **listwise**. This means that any subject that has a missing value for any variable has been deleted from the analysis. Since in the **grades.sav** file there are no missing values, all 105 subjects are included.

EXERCISES

Answers to selected exercises are available for download at **www.ablongman.com/george6e**.

Notice that data files other than the **grades.sav** file are being used here. Please refer to the **Data Files** section starting on page 365 to acquire all necessary information about these files and the meaning of the variables. As a reminder, all data files are downloadable from the web address shown on the previous page.

1. Using the **grades.sav** file select all variables except **lastname**, **firstname**, **grade**, **passfail**. Compute descriptive statistics including **mean**, **standard deviation**, **kurtosis**, **skewness**. Edit so that you eliminate "S.E. Kurt" and "S.E. Skew" and your chart is easier to interpret, and the output fits on one page.
   - Draw a line through any variable for which descriptives are meaningless (either they are categorical or they are known to not be normally distributed)
   - Place an "*" next to variables that are in the ideal range for both skewness and kurtosis
   - Place an **x** next to variables that are acceptable but not excellent
   - Place a ψ next to any variables that are not acceptable for further analysis

2. Using the **divorce.sav** file select **all** variables except the indicators (for spirituality, **sp8 – sp57**, for cognitive coping, **cc1 – cc11**, for behavioral coping, **bc1 – bc12**, for avoidant coping, **ac1 – ac7**, and for physical closeness, **pc1 – pc10**). Compute descriptive statistics including **mean**, **standard deviation**, **kurtosis**, **skewness**. Edit so that you eliminate "S.E. Kurt" and "S.E. Skew" and your chart is easier to interpret, and the output fits on two pages.
   - Draw a line through any variable for which descriptives are meaningless (either they are categorical or they are known to not be normally distributed)
   - Place an "*" next to variables that are in the ideal range for both skewness and kurtosis
   - Place an **x** next to variables that are acceptable but not excellent
   - Place a ψ next to any variables that are not acceptable for further analysis

3. Create a practice data file that contains the following variables and values:
   - VAR1: 3 5 7 6 2 1 4 5 9 5
   - VAR2: 9 8 7 6 2 3 3 4 3 2
   - VAR3: 10 4 3 5 6 5 4 5 2 9

   Compute: the mean, the standard deviation, and variance and print out on a single page.

CROSSTABULATION and χ^2 Analyses

THE PURPOSE of crosstabulation is to show in tabular format the relationship between two or more categorical variables. Categorical variables include those in which distinct categories exist such as gender (female, male), ethnicity (Asian, White, Hispanic), place of residence (urban, suburban, rural), responses (yes, no), grade (A, B, C, D, F), and many more. Crosstabulation can be used with continuous data only if such data are divided into separate categories, such as age (0-19 years, 20-39 years, 40-59 years, 60-79 years, 80-99 years), total points (0-99, 100-149, 150-199, 200-250), and so on. While it is acceptable to perform crosstabulation with continuous data that has been categorized, it is rare to perform chi-square analyses with continuous data because a great deal of useful information about the distribution is lost by the process of categorization. For instance, in the total points distribution (above), two persons who scored 99 and 100 points, respectively, would be in the first and second categories and would be considered identical to two persons who scored 0 and 149 points, respectively. Nonetheless, crosstabulation with continuous data is often used for purposes of data description and display. The SPSS command **Crosstabs** and the subcommands **Cells** and **Statistics** are used to access all necessary information about comparisons of frequency data.

CROSSTABULATION

While the **Frequencies** command can tell us there are 5 Natives, 20 Asians, 24 Blacks, 45 Whites, and 11 Hispanics (and that there are 64 females and 41 males) in our **grades.sav** file, it cannot tell us how many female Asians or male Whites there are. This is the function of the **Crosstabs** command. It would be appropriate to "cross" two variables (**ethnicity** by **gender**) to answer the questions posed above. This would produce a table of 10 different cells with associated frequencies inserted in each cell by crossing two (2) levels of **gender** with five (5) levels of **ethnicity**. It is possible to cross three or more variables, although only with a very large data set would a researcher be likely to perform a crosstabulation with three variables because there would be many low-count and empty cells if the number of subjects was not sufficient. For the present sample, an **ethnicity** by **gender** by **grade** crosstabulation would probably not be recommended. This procedure would create a 5 (**ethnicity**) × 2 (**gender**) × 5 (**grade**) display of frequencies—a total of 50 cells to be filled with only 105 subjects. A large number of low-count or empty cells would be guaranteed. If such a crosstabulation were created with a larger *N*, SPSS would produce five different 5 × 2 tables to display these data.

CHI-SQUARE (χ^2) TESTS OF INDEPENDENCE

In addition to frequencies (or the *observed values*) within each cell, SPSS can also compute the expected value for each cell. *Expected value* is based on the assumption that the two variables are independent of each other. A simple example demonstrates the derivation of expected value. Suppose there is a group of 100 persons in a room and that 30 are male and 70 are female. If there are 10 Asians in the group, it would be anticipated (expected)—if the two variables are independent of each other—that among the 10 Asians there would be 3 males and 7 females (the same proportion as is observed in the entire group). However, with the same group of 100, if 10 of them were football players we would *not* expect 3 male football players and 7 female football players. In American society, most football players are male, and the two categories (gender and football players) are *not* independent of each other. If there were no additional information given, it would be expected that all 10 of the players would be male. The purpose of a chi-square test of independence is to determine whether the observed values for the cells deviate significantly from the corresponding expected values for those cells.

The chi-square statistic is computed by summing the squared deviations [observed value (f_o) minus expected value (f_e)] divided by the expected value for each cell:

$$\chi^2 = \Sigma[(f_o - f_e)^2/f_e]$$

As you can see, if there is a large discrepancy between the observed values and the expected values, the χ^2 statistic would be large, suggesting a significant difference between observed and expected values. Along with this statistic, a probability value is computed. With $p < .05$, it is commonly accepted that the observed values differ significantly from the expected values and that the two variables are *NOT* independent of each other. More complete descriptions and definitions will be included in the Output section of this chapter.

An additional concern addresses the fact that a chi-square statistic is often thought of as a test of association (the opposite of independence) between variables. This invalid assumption can create difficulty because a chi-square value is largely dependent on the number of dimensions and sample size, and thus comparisons of one chi-square value with another are often misleading. To control for this difficulty, Pearson suggested the phi (ϕ) statistic, which divides the chi-square value by *N* and then takes the positive square root of the result. The purpose was to standardize a measure of association to values between 0 and 1 (with 0 indicating completely independent variables and a value close to 1 indicating a strong association between variables). However, if one of the dimensions of the crosstabulation is larger than 2, ϕ may attain a value larger than 1.0. To control for this, Cramér's V was introduced (the positive square root of $\chi^2/[N(k-1)]$, where *k* is the smaller of the number of rows and columns). This measure *does* vary between 0 and 1.0 and is a commonly used measure of the strength of association between variables in a chi-square analysis.

The file we use to illustrate **Crosstabs** is our example described in the first chapter. The data file is called **grades.sav** and has an $N = 105$. This analysis creates crosstabulations and calculates chi-square statistics for **gender** by **ethnicity**.

STEP BY STEP

Cross Tabulation and Chi-Square Tests of Independence

To enter SPSS, a click on **Start** *in the taskbar (bottom of screen) activates the start menu:*

After clicking the SPSS program icon, Screen 1 (inside front cover) appears on the monitor.

Step 2

Create and name a data file or edit (if necessary) an already existing file (see Chapter 3).

Screens 1 and 2 (inside front cover) allow you to access the data file used in conducting the analysis of interest. The following sequence accesses the **grades.sav** *file for further analyses:*

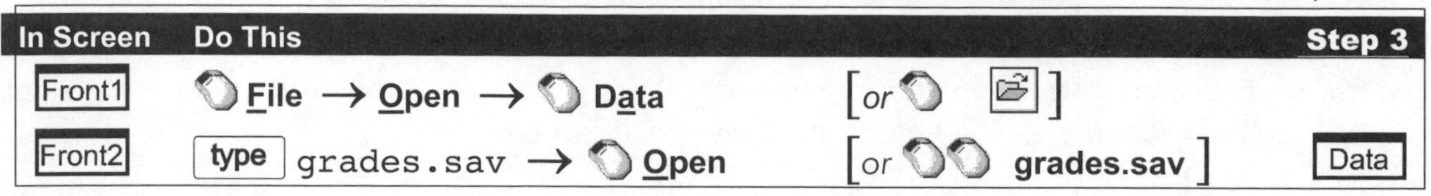

In Screen	Do This		Step 3
Front1	File → Open → Data	[or]	
Front2	type grades.sav → Open	[or grades.sav]	Data

Whether first entering SPSS or returning from earlier operations the standard menu of commands across the top is required (shown below). As long as it is visible you may perform any analyses. It is not necessary for the data window to be visible.

This menu of commands disappears or modifies when using pivot tables or editing graphs. To uncover the standard menu of commands simply click on the or the icon.

After completion of Step 3 a screen with the desired menu bar appears. When you click a command (from the menu bar), a series of options will appear (usually) below the selected command. With each new set of options, click the desired item. The sequence to access chi-square statistics begins at any screen with the menu of commands visible:

In Screen	Do This	Step 4
Menu	Analyze → Descriptive Statistics → Crosstabs	8.1

A new window now appears (Screen 8.1, next page) that provides the framework for conducting a crosstabs analysis. The procedure is to click on the desired variable in the list to the left (**gender** in this example), then click the uppermost of the right arrows ( ) to indicate that we wish gender to be the row variable. Then click a second variable (**ethnicity** in this example) and click the middle right arrow (to indicate that we wish ethnicity to be the column variable). This is all that is necessary to create a cross tabulation of two variables. This will create a 2 (**gender**) by 5 (**ethnicity**) table that contains 10 cells.

The lowest box in the window allows for crosstabulation of three or more variables. If, for instance, we wanted to find the gender by ethnicity breakdown for the three sections, you would click the **section** variable in the list of variables then click the lowest of the three right arrows. This would result in three tables: A gender by ethnicity breakdown for the first section, a gender by ethnicity breakdown for the second section, and a gender by ethnicity breakdown for the third section. The **Previous** and **Next** to the left and right of **Layer 1 of 1** are used if you wanted a gender by ethnicity analysis for more than one variable. For instance, if you wanted this breakdown for both **section** and **year** (year in school), you would click **section**, click the lowest right arrow, click **Next**, click **year**, then click the lowest right arrow again. This would produce three 2 × 5 tables for **section** and four 2 × 5 tables for **year**. The sequence of steps below shows how to create a **gender** by **ethnicity** by **section** crosstabulation.

8.1

The Crosstabs Window

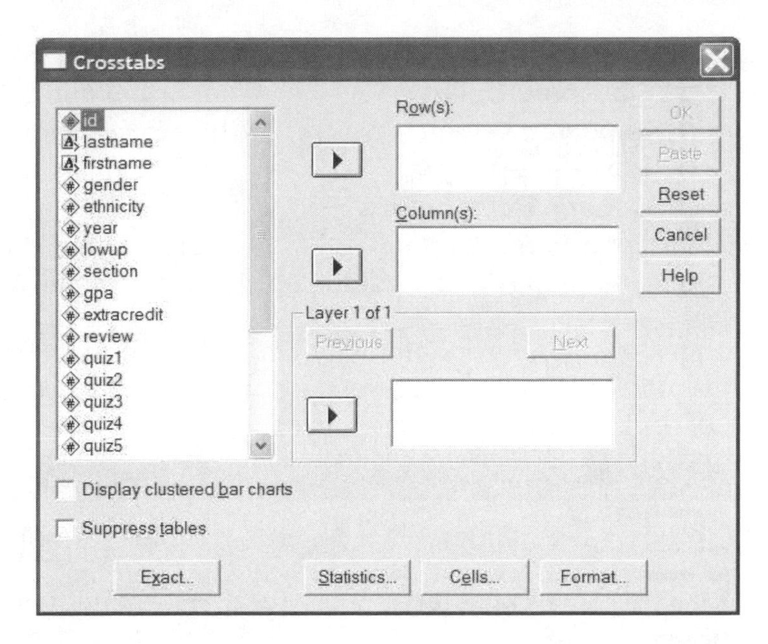

The following sequence begins from Screen 8.1. To arrive at this screen, perform whichever of steps 1-4 (pages 107-108) are necessary.

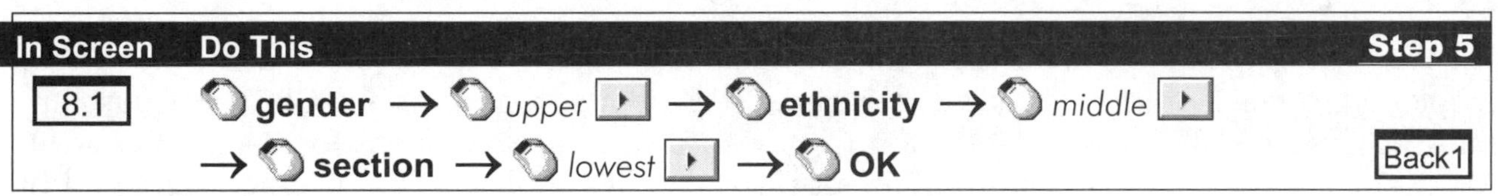

In Screen	Do This	Step 5
8.1	**gender** → *upper* ▶ → **ethnicity** → *middle* ▶ → **section** → *lowest* ▶ → **OK**	Back1

For a simple **gender** by **ethnicity** crosstabulation, omit the two clicks that involve the **section** variable.

It is rare for a researcher to want to compute only cell frequencies. In addition to frequencies, it is possible to include within each cell a number of additional options. Those most frequently used are listed below with a brief definition of each. When you press the **Cells** button (Screen 8.1), a new screen appears (Screen 8.2, below) that allows you to select a number of options. The **Observed** count is selected by default. The **Expected** count (more frequently referred to as the *expected value*) is in most cases also desired. Inclusion of other values depends on the preference of the researcher.

8.2

The Crosstabs: Cell Display Window

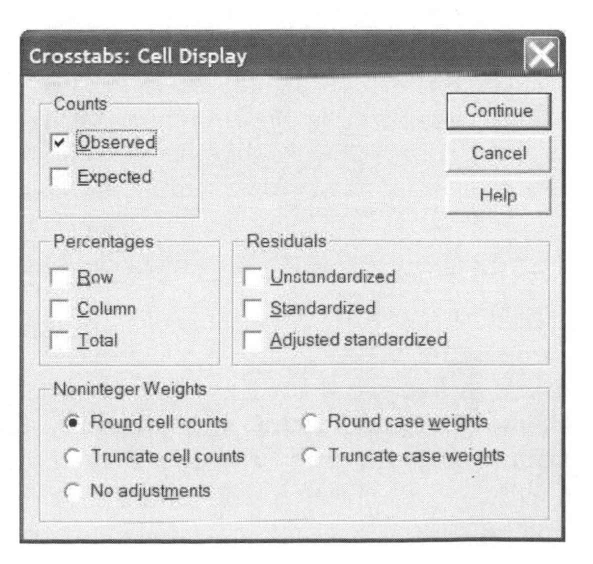

- ❐ **Observed Count**: The actual number of subjects or cases within each cell.
- ❐ **Expected Count**: The expected value for each cell (see page 106).
- ❐ **Row Percentages**: The percent of values in each cell for that row.
- ❐ **Column Percentages**: The percent of values in each cell for that column.
- ❐ **Total Percentages**: The percent of values in each cell for the whole table.
- ❐ **Unstandardized Residuals**: Observed value minus expected value.

To create a **gender** *by* **ethnicity** *crosstabulation that includes observed count, expected count, total percentages, and unstandardized residuals within each cell, perform the following sequence of steps. We begin at Screen 8.1; press* **Reset** *if variables remain from a prior analysis.*

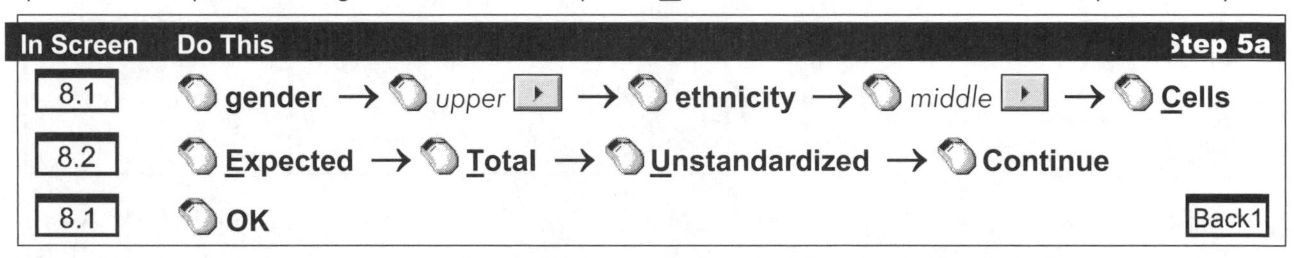

In Screen	Do This	Step 5a
8.1	**gender** → *upper* ▸ → **ethnicity** → *middle* ▸ → **Cells**	
8.2	**Expected** → **Total** → **Unstandardized** → **Continue**	
8.1	**OK**	Back1

Note: We don't click on **Observed**, because this option is already selected by default.

Thus far we have only created tabulations of numbers within cells. Usually, along with cross tabulation, a chi-square analysis is conducted. This requires a click on the **Statistics** pushbutton (see Screen 8.1). When this button is clicked, a new window opens (Screen 8.3, below). Many different tests of independence or association are listed here. Only **Chi-square**, and **Phi and Cramér's V** (see page 107) and **Correlations** (explained in Chapter 10) will be considered here. As in the **Cells** window, the procedure is to click in the small box to the left of the desired statistic before returning to the previous screen to conduct the analysis. See the *SPSS for Windows Base System Users Guide* for a description of the other statistics included in this chart.

8.3

The Crosstabs: Statistics Window

Crosstabs: Statistics
Chi-square
Correlations
Continue
Cancel
Help
Nominal
Contingency coefficient
Phi and Cramér's V
Lambda
Uncertainty coefficient
Ordinal
Gamma
Somers' d
Kendall's tau-b
Kendall's tau-c
Nominal by Interval
Eta
Kappa
Risk
McNemar
Cochran's and Mantel-Haenszel statistics
Test common odds ratio equals: 1

In the sequence below, we create a **gender** *by* **ethnicity** *crosstabulation, request* **Observed count**, **Expected count**, *and* **Unstandardized Residuals** *within each cell, and include the* **Chi-square**, *and* **Phi and Cramér's V** *as forms of analysis. We do not include correlations because they have no meaning when there is no intrinsic order to the associated variables. The starting point is Screen 8.1. Complete results from this analysis are included in the Output section.*

In Screen	Do This	Step 5b
8.1	**gender** → *upper* ▸ → **ethnicity** → *middle* ▸ → **Cells**	
8.2	**Expected** → **Unstandardized** → **Continue**	
8.1	**Statistics**	
8.3	**Chi-square** → **Phi and Cramér's V** → **Continue**	
8.1	**OK**	Back1

Often we may wish to conduct a cross tabulation and chi-square analysis on a *subset* of a certain variable. For instance, in the **gender** by **ethnicity** crosstabulation described earlier, we may wish to delete the "Native" category from the analysis since there are only 5 of them and earlier analyses have indicated that there is a problem with low count cells. This means creating a 2 (levels of gender) × 4 (levels of ethnicity after excluding the first level) analysis. After you have selected the variables for crosstabulation, have chosen cell values and desired statistics, then click on the **Data** command in the Menu Bar at the top of the screen. In the menu of options that opens below, click on **Select Cases**. At this point, a new window will open (Screen 4.8 from Chapter 4). In this window, click the small circle to the right of **If condition is satisfied** (so a black dot fills it), then click the **If** button just below.

A new dialog box again opens (see Screen 4.3, also from Chapter 4) titled **Select Cases: If**. This window provides access to a wide variety of operations described in Chapter 4. For the present we are only concerned with how to select levels 2, 3, 4, and 5 of the **ethnicity** variable. First step is to select **ethnicity** from the variable list to the left, then click the ▸ to paste the variable into the "active" box, then click the >= (on the small keyboard below the active box), then click the 2. You have now indicated that you wish to select all levels of **ethnicity** greater or equal to 2. Then click **Continue**, Screen 4.8 will reappear, click **OK**, the original screen will appear (Screen 8.1) click the **OK** and your analysis will be completed with only four levels of ethnicity. The step-by-step sequence follows.

The starting point for this sequence is again Screen 8.1 It may be necessary to press **Reset**.

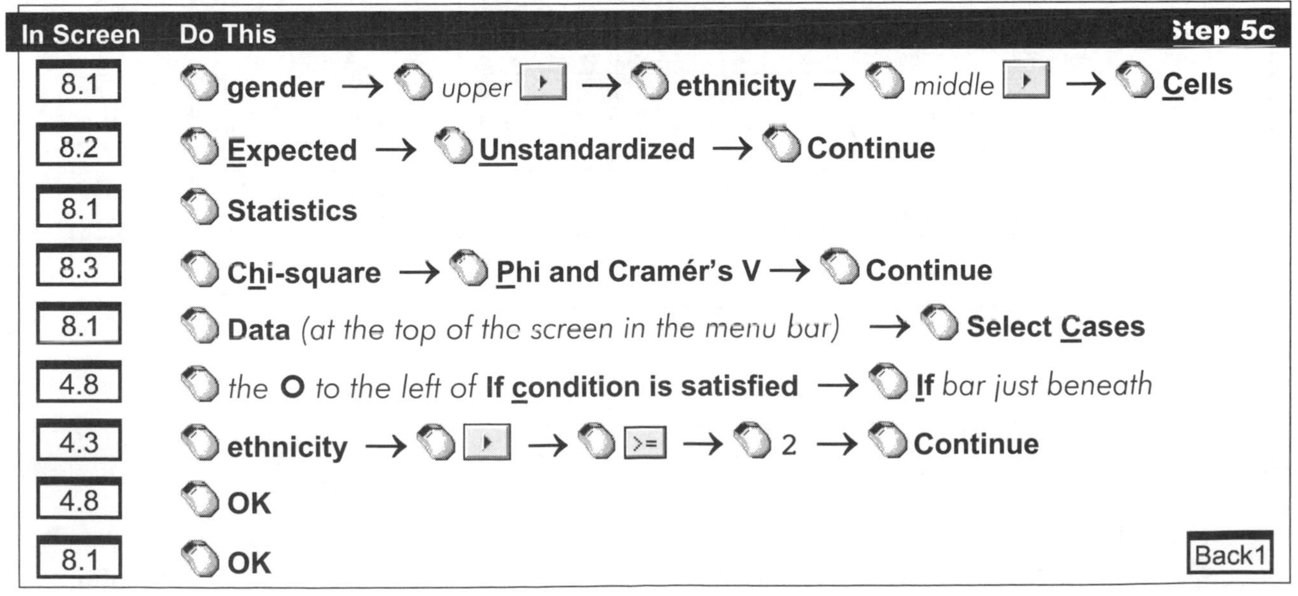

In Screen	Do This	Step 5c
8.1	**gender** → *upper* ▸ → **ethnicity** → *middle* ▸ → **Cells**	
8.2	**Expected** → **Unstandardized** → **Continue**	
8.1	**Statistics**	
8.3	**Chi-square** → **Phi and Cramér's V** → **Continue**	
8.1	**Data** *(at the top of the screen in the menu bar)* → **Select Cases**	
4.8	*the* **O** *to the left of* **If condition is satisfied** → **If** *bar just beneath*	
4.3	**ethnicity** → ▸ → >= → 2 → **Continue**	
4.8	**OK**	
8.1	**OK**	Back1

Upon completion of step 5, 5a, 5b, or 5c, the output screen will appear (Screen 1, inside back cover). All results from the just-completed analysis are included in the Output Navigator. Make use of the scroll bar arrows (▲▼▶◀) to view the results. Even when viewing output, the standard menu of commands is still listed across the top of the window. Further analyses may be conducted without returning to the data screen.

PRINTING RESULTS

Results of the analysis (or analyses) that have just been conducted requires a window that displays the standard commands (**File Edit Data Transform Analyze** . . .) across the top. A typical print procedure is shown below beginning with the standard output screen (Screen 1, inside back cover).

To print results, from the Output screen perform the following sequence of steps:

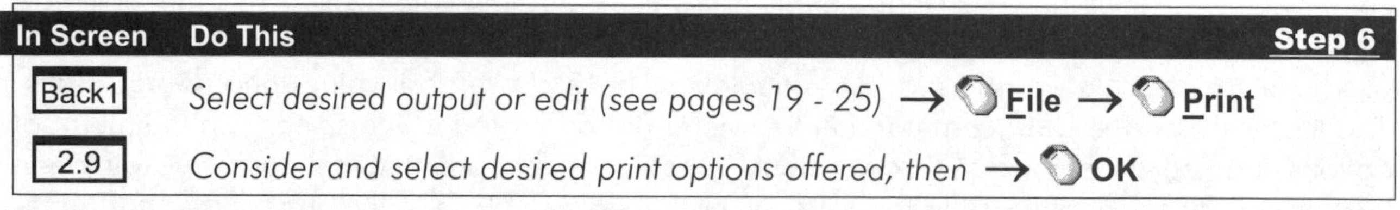

In Screen	Do This	Step 6
Back1	*Select desired output or edit (see pages 19 - 25)* → **File** → **Print**	
2.9	*Consider and select desired print options offered, then* → **OK**	

To exit you may begin from any screen that shows the **File** *command at the top.*

In Screen	Do This	Step 7
Menu	**File** → **Exit**	2.1

Note: After clicking **Exit**, there will frequently be small windows that appear asking if you wish to save or change anything. Simply click each appropriate response.

OUTPUT

Crosstabulation and Chi-Square (χ^2) Analyses

What follows is partial output from sequence step 5b, page 111.

			ethnicity					
			Native	Asian	Black	White	Hispanic	Total
gender	Female	Count	4	13	14	26	7	64
		Expected Count	3.0	12.2	14.6	27.4	6.7	64.0
		Residual	1.0	.8	-.6	-1.4	.3	
	Male	Count	1	7	10	19	4	41
		Expected Count	2.0	7.8	9.4	17.6	4.3	41.0
		Residual	-1.0	-.8	.6	1.4	-.3	
Total		Count	5	20	24	45	11	105
		Expected Count	5.0	20.0	24.0	45.0	11.0	105.0

	Value	D of Freedom	Significance
Pearson Chi-Square	1.193[a]	4	.879
Likelihood Ratio	1.268	4	.867
Linear-by-Linear Association	.453	1	.501
Phi	.107	--	.879
Cramér's V	.107	--	.879
N of valid cases	105		

a. 3 cells (30%) have expected count less than 5. The minimum expected count is 1.95

The first step in interpreting a crosstabulation or chi-square analysis is to observe the actual values and the expected values within each cell. Preliminary observation indicates that observed values and expected values are quite similar. The greatest discrepancy is for Whites (females, 26 actual count, 27.4 expected; and males 19 actual count, 17.6 expected). Note also that the residual value (the number beneath the other two) is simply the observed value minus the expected value. Even without looking at the chi-square statistics you would anticipate that observed values and expected values do not differ significantly (that is, gender and ethnicity in this sample are independent of each other). The results support this observation with a low Chi-square value (1.193) and a significance greater than 0.8 (.879). Notice that measures of association are also small and do not approach significance. As suggested in the Step by Step section, low-count cells is a problem. Three of 10 have an expected value less than 5. The usual response would be to redo the analysis after deleting the "Native" category. To further assist understanding, definitions of output terms follow.

Term	Definition/Description
COUNT	The top number in each of the 10 cells (4, 13, 14, . . .), indicates the number of subjects in each category.
EXPECTED COUNT	The second number in each of the 10 cells (3.0, 12.2, 14.6, . . .), indicates the number that would appear there if the two variables were perfectly independent of each other.
RESIDUAL	The observed value minus the expected value.
ROW TOTAL	The total number of subjects for each row (64 females, 41 males).
COLUMN TOTAL	The total number of subjects in each category for each column (5 Natives, 20 Asians, 24 Blacks, 45 Whites, 11 Hispanics).
CHI SQUARE: PEARSON and **LIKELIHOOD RATIO**	Two different methods for computing chi-square statistics. With a large *N*, these two values will be close to equal. The formula for the Pearson chi-square is: $\chi^2 = \Sigma[(f_o - f_e)^2/f_e]$
VALUE	For **PEARSON** and **MAXIMUM LIKELIHOOD** methods, as the test-statistic value gets larger the likelihood that the two variables are *not* independent (e.g., *are* dependent) also increases. The values close to 1 (1.193, 1.268) suggest that gender balance is not dependent on which ethnic groups are involved.
DEGREES OF FREEDOM	Degrees of freedom is the number of levels in the first variable minus 1 (2 - 1 = 1) times the number of levels in the second variable minus 1 (5 - 1 = 4); 1 × 4 = 4.
SIGNIFICANCE	The likelihood that these results could happen by chance. The large *p* value here indicates that observed values do *not* differ significantly from expected values.
LINEAR-BY-LINEAR ASSOCIATION	This statistic tests whether the two variables correlate with each other. This measure is often meaningless because there is no logical or numeric relation to the order of the variables. For instance, there is no logical order (from a low value to a high value) of ethnicity. Therefore, the correlation between gender and ethnicity is meaningless. If, however, the second variable were income, ordered from low to high, a valid correlation could result.
MINIMUM EXPECTED COUNT	The minimum expected count is for the first cell of the second row (male Natives). The expected value there is rounded off to the nearest tenth (2.0). The value accurate to two decimals is 1.95.
PHI	A measure of the strength of association between two categorical variables. A value of .107 represents a very weak association between gender and ethnicity. The equation: $\phi = \sqrt{\chi^2 / N}$
CELLS WITH EXPECTED COUNT < 5	Three of the 10 cells have an expected frequency less than 5. If you have many low-count cells (more than 25% is one accepted criteria), the overall chi-square value is less likely to be valid.

Term	Definition/Description
CRAMÉR'S V	A measure of the strength of association between two categorical variables. It differs from phi in that Cramér's V varies strictly between 0 and 1, while in certain cases phi may be greater than 1. The equation follows: (Note: k is the smaller of the number of rows and columns) $$V = \sqrt{\chi^2 / [N(k-1)]}$$
APPROXIMATE SIGNIFICANCE	This is the same as the significance for the Pearson chi-square. The high value (.879) indicates very weak association.

EXERCISES

Answers to selected exercises are available for download at **www.ablongman.com/george6e**.

For each of the chi-square analyses computed below:

1. Circle the observed (actual) values.
2. Box the expected values.
3. Put an * next to the unstandardized residuals.
4. Underline the significance value that shows whether observed and expected values differ significantly.
5. Make a statement about independence of the variables involved.
6. STATE THE NATURE OF THE RELATIONSHIP (in normal English, not statistical jargon).
7. Is there a significant linear association?
8. Does linear association make sense for these variables?
9. Is there a problem with low-count cells?
10. What would you do about it if there is a problem?

1. File: **grades.sav**. Variables: **gender** by **ethnicity**. Select: **observed count**, **expected count**, **unstandarized residuals**; Compute: **Chi-square**, **Phi and Cramer's V**; edit to fit on one page; print out; perform the 10 operations above.

2. File: **grades.sav**. Variables: **gender** by **ethnicity**. Prior to analysis, complete the procedure shown in Step 5c (page 111) to eliminate the "Native" category (due to low-count cells). Select: **observed count**, **expected count**, **unstandarized residuals**; Compute: **Chi-square**, **Phi and Cramer's V**; edit to fit on one page; print out; perform the 10 operations.

3. File: **helping3.sav**. Variables: **gender** by **problem**. Select: **observed count**, **expected count**, **unstandarized residuals**; Compute: **Chi-square**, **Phi and Cramer's V**; edit to fit on one page; print out; perform the 10 operations above.

4. File: **helping3.sav**. Variables: **school** by **occupat**. Prior to analysis, select cases: "**school** > 1 & **occupat** < 6". Select: **observed count**, **expected count**, **unstandarized residuals**; Compute: **Chi-square**, **Phi and Cramer's V**; edit to fit on one page; print out; perform the 10 operations above.

5. File: **helping3.sav**. Variables: **marital** by **problem**. Select: **observed count**, **expected count**, **unstandarized residuals**; Compute: **Chi-square**, **Phi and Cramer's V**; edit to fit on one page; print out; perform the 10 operations listed above.

The MEANS Procedure

WHILE THE **Crosstabs** procedure allows you to identify the frequency of certain types of categorical data (Chapter 8), the **Means** command allows you to explore certain characteristics of continuous variables within certain categories. By way of comparison, a crosstabulation of **ethnicity** by **gender** would indicate that there were 13 White females, 22 White males, 8 Hispanic females, 6 Hispanic males, and so forth. The **Means** command allows you to view certain characteristics of continuous variables (such as total points, GPAs, percents) by groups. Thus if you computed **total** (number of points) for **ethnicity** by **gender**, you would find that there were 13 White females who scored an average (mean) of 113.12 points, 22 White males who scored a mean of 115.34 points, 8 Hispanic females who scored a mean of 116.79, 6 Hispanic males who scored a mean of 113.45, and so forth. This information is, of course, presented in tabular format for ease of reading and interpretation. The utility of the **Means** command for data such as our sample file is several-fold. For a class with more than one section, we might like to see mean scores for each section, or to compare the scores of males with females, or the performance of upper-division with lower-division students.

The **Means** command is one of SPSS's simplest procedures. For the selected groups it will list the mean for each group, the standard deviation, and the number of subjects for each category. There is an additional **Options** subcommand with which you may conduct a one-way analysis of variance (ANOVA) based on the means and standard deviations you have just produced. We will include that option in this chapter but will save a detailed explanation of analysis of variance for the **One-Way ANOVA** and **General Linear Models** chapters (Chapters 12-14).

We again make use of the **grades.sav** file (*N* = 105) and the variables **total**, **percent**, **gpa**, **section**, **lowup**, and **gender** to illustrate the **Means** procedure.

STEP BY STEP

Describing Subpopulation Differences

*To enter SPSS, a click on **Start** in the taskbar (bottom of screen) activates the start menu:*

After clicking the SPSS program icon, Screen 1 (inside front cover) appears on the monitor.

Step 2
Create and name a data file or edit (if necessary) an already existing file (see Chapter 3).

*Screens 1 and 2 (inside front cover) allow you to access the data file used in conducting the analysis of interest. The following sequence accesses the **grades.sav** file for further analyses:*

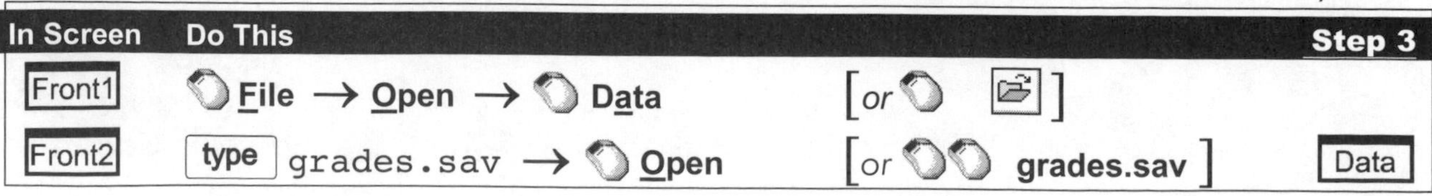

Whether first entering SPSS or returning from earlier operations the standard menu of commands across the top is required (shown below). As long as it is visible you may perform any analyses. It is not necessary for the data window to be visible.

This menu of commands disappears or modifies when using pivot tables or editing graphs. To uncover the standard menu of commands simply click on the or the icon.

After completion of Step 3 a screen with the desired menu bar appears. When you click a command (from the menu bar), a series of options will appear (usually) below the selected command. With each new set of options, click the desired item. The sequence to access **Means** *begins at any screen with the menu of commands visible:*

At this point a new window opens (Screen 9.1, below) that deals with designating for which variables you wish to compare means. At the top of the window is the **Dependent List** box. This is where the *continuous* variables that you wish to analyze will be placed. For instance, you may wish to compare mean scores for previous GPA, for final exam scores, for total points, or for percentage of total points. You may list several variables in this box, and SPSS will calculate a separate set of means for each variable.

The Means Window

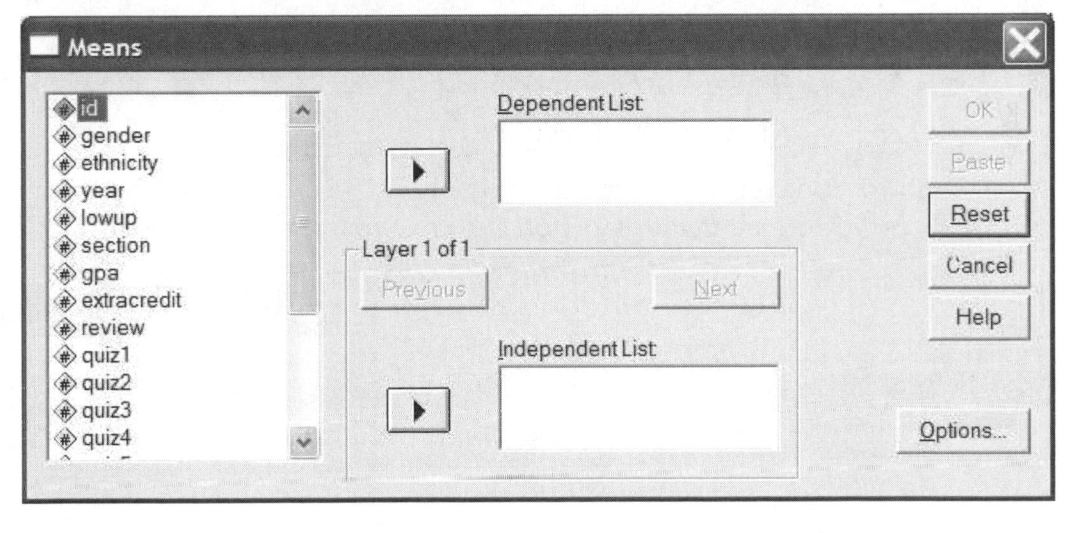

The lower **Independent List** box is where you identify the categorical variables, such as gender, section, grade, or ethnicity. If you include only one variable (such as **gender**) in the lower **Independent List** box, and a single variable in the upper **Dependent List** box (say, **total** points), the **Means** command will indicate the average (or mean) number of total points earned by women and the mean number of total points earned by men. For both men and women the *N* and standard deviations are also included by default.

More frequently the researcher will desire more than just two categories. A more common operation would be a **gender** by **section** analysis that would give mean total scores for males and females in each of the three sections, or a **gender** by **year** analysis that would yield mean **total** scores for males and females for each year in college. To specify more than one categorical variable, make use of the **Previous** and **Next** to the left and right of **Layer 1 of 1** in the middle of the screen. As you observe the sequences of steps shown below, it will become clear how these options are used. Each sequence that follows begins from Screen 9.1. *To determine the mean number of total points (**total**) in each **section**, perform the following sequence of steps. Screen 9.1 is the starting point.*

In Screen	Do This	Step 5
9.1	**total** → upper ▸ → **section** → lower ▸ → **OK**	Back1

*If you wish to include an additional categorical variable in the analysis (**lowup**—lower or upper division student) in addition to **section**, the appropriate sequence of steps follows:*

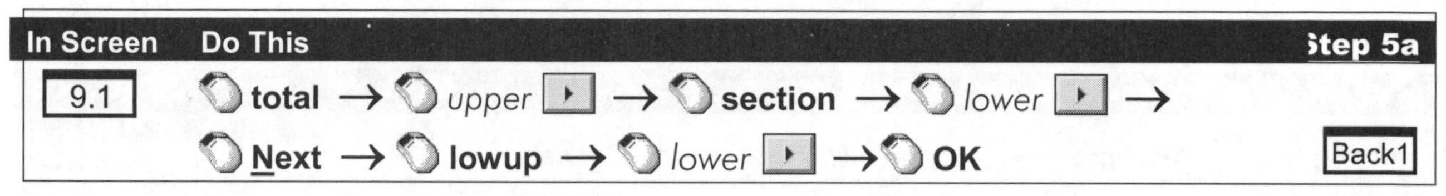

In Screen	Do This	Step 5a
9.1	**total** → upper ▸ → **section** → lower ▸ → **Next** → **lowup** → lower ▸ → **OK**	Back1

*You may list more than one dependent variable. SPSS will then produce as many columns in the table as there are dependent variables. In the procedure that follows, we compute means and standard deviations for **gpa**, **total**, and **percent** (3 dependent variables) for 6 categories (3 levels of **section** by 2 levels of **gender**) in the analysis.*

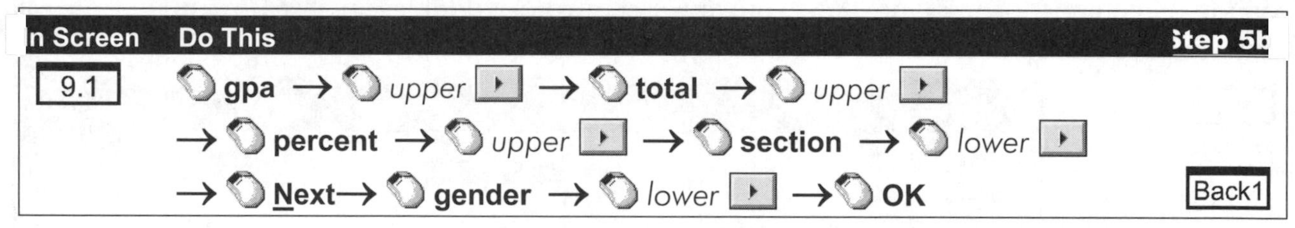

In Screen	Do This	Step 5b
9.1	**gpa** → upper ▸ → **total** → upper ▸ → **percent** → upper ▸ → **section** → lower ▸ → **Next** → **gender** → lower ▸ → **OK**	Back1

If you wish to conduct a one-way analysis of variance or to include additional output within each cell, a click on the **Options** button will open a new screen (Screen 9.2, shown below).

9.2

The Means: Options Window

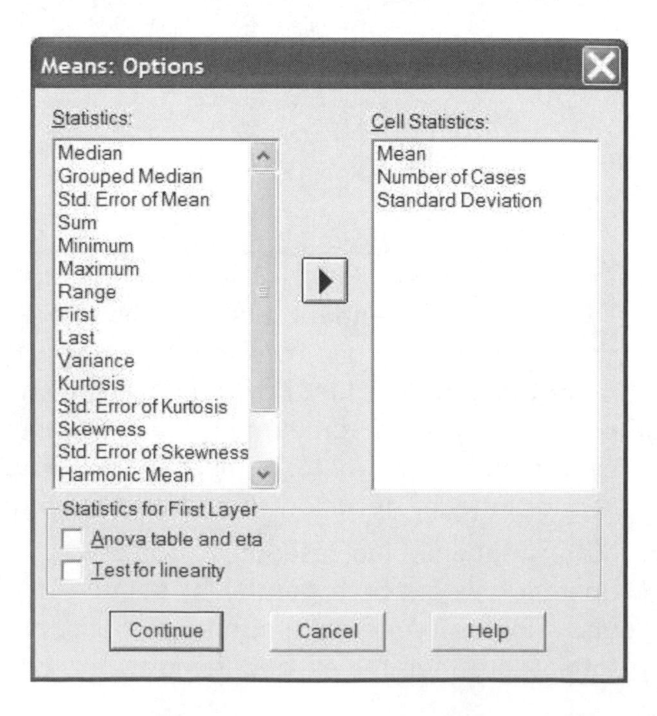

This screen allows the researcher to include additional output options for each analysis. For instance, in addition to the default (**Mean**, **Standard Deviation**, and **Number of Cases**), you may, by clicking over items from the box to the left, also include a number of other options. Using the **Means** command, it is possible to conduct a simple one-way analysis of variance (ANOVA). We will save detailed description of this procedure until Chapter 12, but will show how to access that option here. Under the **Statistics for the First Layer** box, click the **ANOVA table and eta** option. If you conduct this analysis with a "first layer" grouping variable of **section** and a dependent variable of **total**, then an analysis will be conducted that compares the mean total points for each of the three sections.

The following sequence of steps will produce means on **total** *points for the six categories of* **gender** *by* **section**. *It will also conduct a one-way analysis of variance for the first layer variable (***section***) on* **total** *points. We begin at Screen 9.1. Do whichever of steps 1-4 (pages 116-117) are necessary to arrive at this screen. You may need to click the* **Reset** *button.*

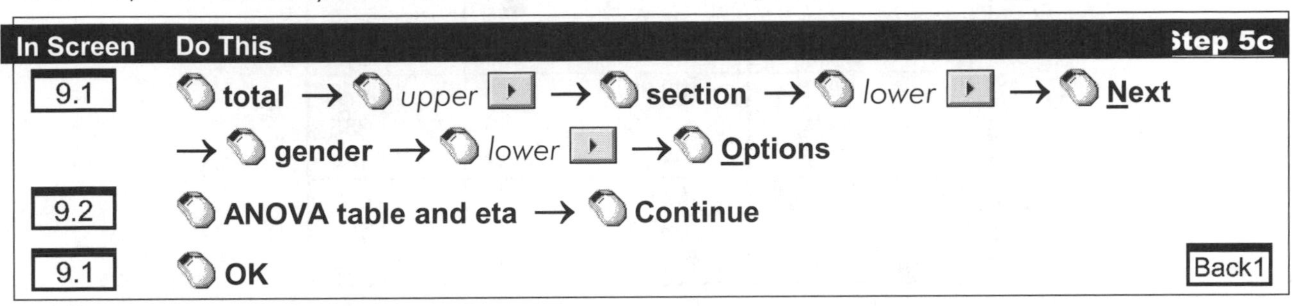

In Screen	Do This	Step 5c
9.1	**total** → *upper* ▸ → **section** → *lower* ▸ → **Next** → **gender** → *lower* ▸ → **Options**	
9.2	**ANOVA table and eta** → **Continue**	
9.1	**OK**	Back1

Upon completion of step 5, 5a, 5b, or 5c, the output screen will appear (Screen 1, inside back cover). All results from the just-completed analysis are included in the Output Navigator. Make use of the scroll bar arrows (▲▼▶◀) to view the results. Even when viewing output, the standard menu of commands is still listed across the top of the window. Further analyses may be conducted without returning to the data screen.

PRINTING RESULTS

Results of the analysis (or analyses) that have just been conducted requires a window that displays the standard commands (**File Edit Data Transform Analyze** . . .) across the top. A typical print procedure is shown below beginning with the standard output screen (Screen 1, inside back cover).

To print results, from the Output screen perform the following sequence of steps:

In Screen	Do This	Step 6
Back1	*Select desired output or edit (see pages 19 - 25)* → **File** → **Print**	
2.9	*Consider and select desired print options offered, then* → **OK**	

To exit you may begin from any screen that shows the **File** *command at the top.*

In Screen	Do This	Step 7
Menu	**File** → **Exit**	2.1

Note: After clicking **Exit**, there will frequently be small windows that appear asking if you wish to save or change anything. Simply click each appropriate response.

OUTPUT

Describing Subpopulation Differences

What follows is complete output from sequence step 5c, page 119.

SPSS for Windows: Means and One-Way ANOVA Results

total

section	gender	Mean	N	Std. Deviation
1	Female	103.95	20	18.135
	Male	106.85	13	13.005
	Total	105.09	33	16.148
2	Female	100.00	26	12.306
	Male	98.46	13	11.822
	Total	99.49	39	12.013
3	Female	102.83	18	10.678
	Male	90.73	15	21.235
	Total	97.33	33	17.184
Total	Female	102.03	64	13.896
	Male	98.29	41	17.196
	Total	100.57	105	15.299

ANOVA Table

		Sum of Squares	df	Mean Square	F	Sig.
TOTAL * SECTION	Between Groups (Combined)	1065.910	2	532.955	2.335	.102
	Within Groups	23277.804	102	228.214		
	Total	24343.714	104			

Measures of Association

	Eta	Eta Squared
TOTAL * SECTION	.209	.044

Note that the first portion of the output (table above) is the simple means and frequencies for the entire group and for each of the selected categories. SPSS lists the mean, standard deviation, sum of squares, and number of cases for the first of the two categorical variables (**section**) in the upper portion of the chart, and similar information for gender in the lower portion of the chart. Definitions of terms in the ANOVA output are listed below. With means of 105.1, 99.5, and 97.3, the test statistics yield a p value of .102. This finding suggests that the sections may have a trend toward a significant influence on the total scores. Pairwise comparisons are not possible with this very simple ANOVA procedure, but visual inspection reveals that the greatest difference is between Section 1 (M = 105.1) and Section 3 (M = 97.3). See Chapter 12 for a more complete description of one-way ANOVAs. We conclude with definitions of output terms and the exercises.

Term	Definition/Description
WITHIN-GROUP SUM OF SQUARES	The sum of the squared deviations between the mean for each group and the observed values of each subject within that group.
BETWEEN-GROUP SUM OF SQUARES	The sum of squared deviations between the grand mean and each group mean, weighted (multiplied) by the number of subjects in each group.
BETWEEN-GROUPS DEGREES OF FREEDOM	Number of groups minus one: (3 - 1 = 2)
WITHIN-GROUPS DEGREES OF FREEDOM	Number of subjects minus number of groups (105 - 3 = 102).
MEAN SQUARE	Sum of squares divided by degrees of freedom.
F RATIO	Between-groups mean square divided by within-groups mean square.
SIGNIFICANCE	The probability of the observed values happening by chance. The p value indicated here (p = .10) indicates that a marginally significant difference between means may exist among at least one of the three pairings of the three sections.
ETA	A measure of correlation between two variables when one of the variables is discrete.
ETA SQUARED	The proportion of the variance in the dependent variable accounted for by the independent variable. For instance, an eta squared of .044 indicates that 4.4% of the variance in the **total** scores is due to membership in one of the three **sections**.

EXERCISES

Answers to selected exercises are available for download at **www.ablongman.com/george6e**.

1. Using the **grades.sav** file use the Means procedure to explore the influence of **ethnicity** and **section** on **total**. Print output, fit on one page, in general terms describe what the value in each cell means.

2. Using the **grades.sav** file use the Means procedure to explore the influence of **year** and **section** on **final**. Print output, fit on one page, in general terms describe what the value in each cell means.

3. Using the **divorce.sav** file use the Means procedure to explore the influence of gender (**sex**) and marital status (**status**) on **spiritua** (spirituality—high score is spiritual). Print output and, in general terms, describe what the value in each cell means.

10

Bivariate CORRELATION

CORRELATIONS MAY be computed by making use of the SPSS command **Correlate**. Correlations are designated by the lower case letter *r*, and range in value from −1 to +1. A correlation is often called a *bivariate correlation* to designate a simple correlation between two variables, as opposed to relationships among more than two variables, as frequently observed in multiple regression analyses or structural equation modeling. A correlation is also frequently called the Pearson product-moment correlation or the Pearson *r*. Karl S. Pearson is credited with the formula from which these correlations are computed. Although the Pearson *r* is predicated on the assumption that the two variables involved are approximately normally distributed, the formula often performs well even when assumptions of normality are violated or when one of the variables is discrete. Ideally, when variables are not normally distributed, the Spearman correlation (a value based on the rank order of values) is more appropriate. Both Pearson and Spearman correlations are available using the **Correlate** command. There are other formulas from which correlations are derived that reflect characteristics of different types of data, but a discussion of these goes beyond the scope of this book. See the *SPSS for Windows Base System User's Guide* for additional information.

WHAT IS A CORRELATION?

Perfect positive ($r = 1$) correlation: A correlation of +1 designates a perfect, positive correlation. *Perfect* indicates that one variable is precisely predictable from the other variable. *Positive* means that as one variable increases in value, the other variable also increases in value (or conversely, as one variable decreases, the other variable also decreases).

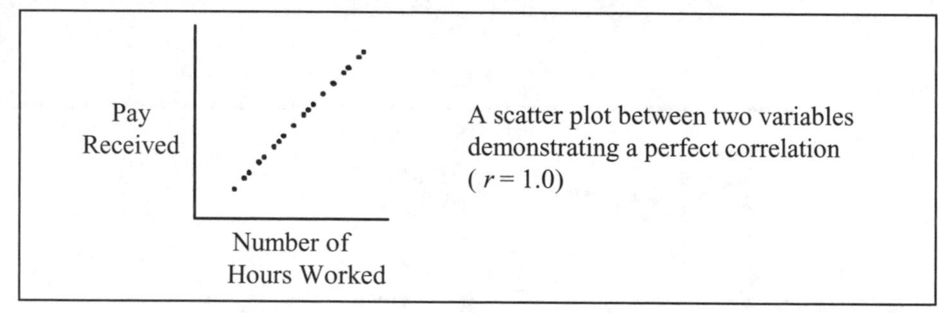

A scatter plot between two variables demonstrating a perfect correlation ($r = 1.0$)

Perfect correlations are essentially never found in the social sciences and exist only in mathematical formulas and direct physical or numerical relations. An example would be the relationship between the number of hours worked and the amount of pay received. As one number increases, so does the other. Given one of the values, it is possible to precisely determine the other value.

Positive ($0 < r < 1$) correlation: A positive (but not perfect) correlation indicates that as the value of one variable increases, the value of the other variable also *tends* to increase. The closer the correlation value is to 1, the stronger is that tendency; and the closer the correlation value is to 0, the weaker is that tendency.

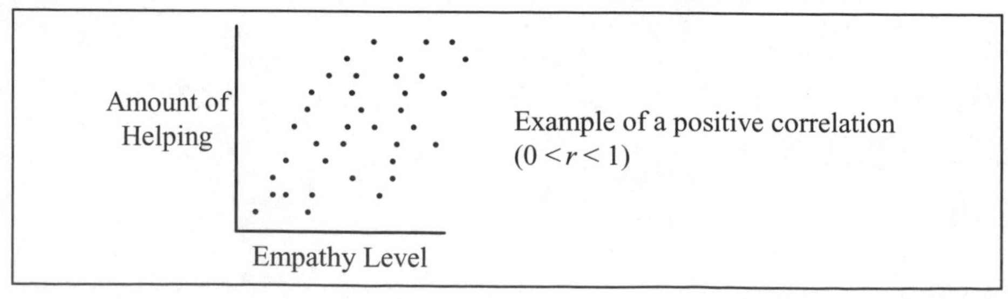

Example of a positive correlation ($0 < r < 1$)

An example of a strong positive correlation is the relation between height and weight in adult humans (r = .83). Tall people are usually heavier than short people. An example of a weak positive correlation is the relation between a measure of empathic tendency and amount of help given to a needy person (r = .12). Persons with higher empathic tendency scores give more help than persons who score lower, but the relationship is weak.

No (r = 0) correlation: A correlation of 0 indicates no relation between the two variables. For example, we would not expect IQ and height in inches to be correlated.

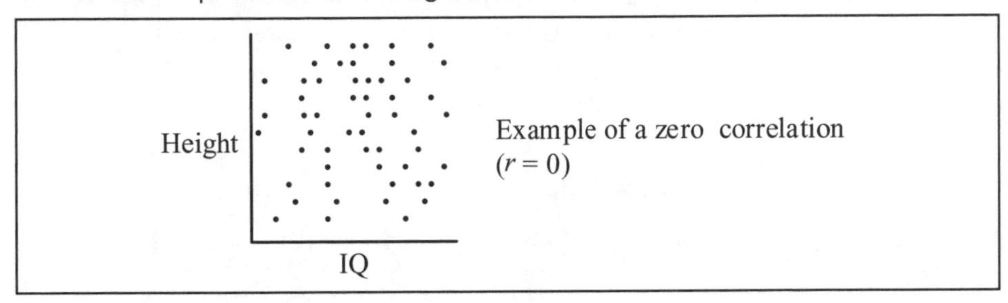

Example of a zero correlation ($r = 0$)

Negative (–1 < r < 0) correlation: A negative (but not perfect) correlation indicates a relation in which as one variable *increases* the other variable has a tendency to *decrease*. The closer the correlation value is to –1, the stronger is that tendency. The closer the correlation value is to 0, the weaker is that tendency.

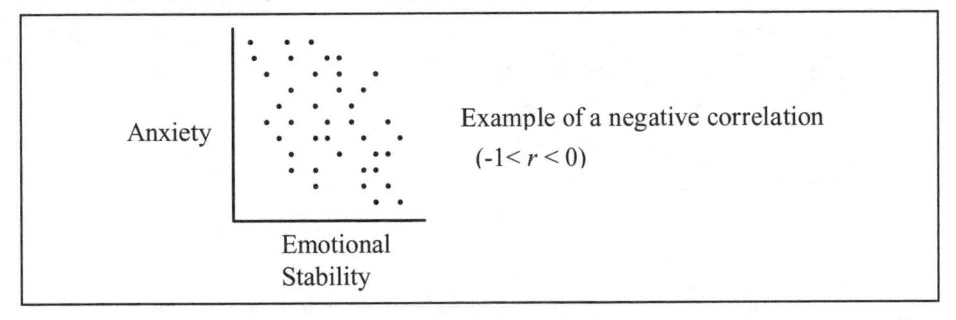

Example of a negative correlation ($-1 < r < 0$)

An example of a strong negative correlation is the relation between anxiety and emotional stability (r = –.73). Persons who score higher in anxiety tend to score lower in emotional stability. Persons who score lower in anxiety tend to score higher in emotional stability. A weak negative correlation is demonstrated in the relation between a person's anger toward a friend suffering a problem and the quality of help given to that friend (r = –.13). If a person's anger is *less* the quality of help given is *more*, but the relationship is weak.

Perfect negative (r = –1) correlation: Once again, perfect correlations (positive or negative) exist only in mathematical formulas and direct physical or numerical relations. An example of a perfect negative correlation is based on the formula **distance = rate × time**. When driving from point A to point B, if you drive *twice* as fast you will take *half* as long.

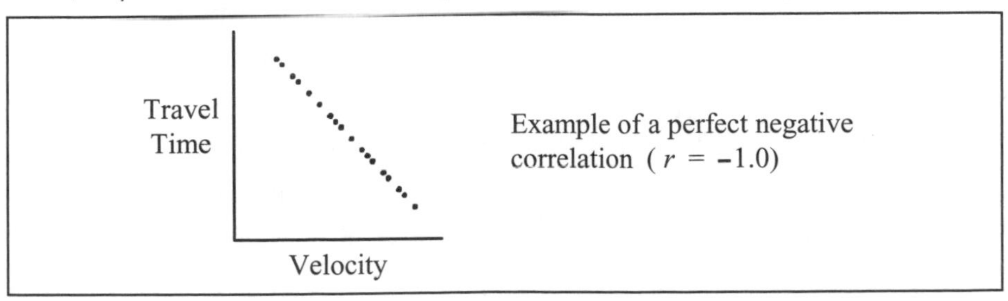

Example of a perfect negative correlation ($r = -1.0$)

ADDITIONAL CONSIDERATIONS

Linear versus Curvilinear

It is important to understand that the **Correlate** command measures only *linear* relationships. There are many relations that are not linear. Take the example of nervousness before a major exam. Too much or too little nervousness generally hurts performance while a moderate amount of nervousness typically aids performance. The relation on a scatter plot would look like an inverted U, but computing a Pearson correlation would yield no relation or a weak relation. The chapters on simple regression and multiple regression analysis (Chapters 15 and 16) will consider curvilinear relationships in some detail. It is often a good idea to create a scatter plot of your data before computing correlations, to see if the relationship between two variables is linear. If it is linear, the scatter plot will more or less resemble a straight line. While a scatter plot can aid in detecting linear or curvilinear relationships, it is often true that significant correlations may exist even though they can not be detected by visual analysis alone.

Significance

As with most other statistical procedures, a significance or probability is computed to determine the likelihood that a particular correlation could occur by chance. The significance (or *p* value) represents the *degree of rarity* of a certain result. A significance less than .05 ($p < .05$) means that there is less than a 5% chance that this relationship occurred by chance. SPSS has two different significance measures, one-tailed significance and two-tailed significance. To determine which to use, the rule of thumb generally followed is to use two-tailed significance when you compute a table of correlations in which you have little idea as to the *direction* of the correlations. If, however, you have prior expectations about the direction of correlations (positive or negative), then the statistic for one-tailed significance is generally used.

Causality

Correlation does not necessarily indicate causation. Sometimes causation is clear. If height and weight are correlated, it is clear that additional height causes additional weight. Gaining weight is not known to increase one's height. Also the relationship between gender and empathy shows that women tend to be more empathic than men. If a man becomes more empathic this is unlikely to change his gender. Once again, the *direction* of causality is clear: gender influences empathy, not the other way around.

There are other settings where direction of causality is likely but open to question. For instance self-efficacy (the belief of ability) is strongly correlated with helping. It would generally be thought that belief of ability will influence how much one helps, but one could argue that one who helps more may increase their belief of self-efficacy as a result of their actions. The former answer seems more likely but both may be partially valid.

Thirdly, sometimes it is difficult to have any idea of which causes which. Emotional stability and anxiety are strongly related (more emotionally stable people are less anxious). Does greater emotional stability result in less anxiety, or does greater anxiety result in lower emotional stability? The answer, of course, is yes. They both influence each other.

Finally there is the third variable issue. It is reliably shown that ice cream sales and homicides in New York City are positively correlated. Does eating ice cream cause one to become homicidal? Does committing murders give one a craving for ice cream? The answer is neither.

Both ice cream sales and murders are correlated with heat. When the weather is hot more murders occur and more ice cream is sold. The same issue is at play with the reliable finding that across many cities the number of churches is positively correlated with the number of bars. No, it's not that church going drives one to drink, nor is it that heavy drinking gives one an urge to attend church. There is again the third variable: population. Larger cities have more bars *and* churches while smaller cities have fewer of both.

Partial Correlation

We mention this issue because partial correlation is included as an option within the context of the **Correlate** command. We mention it only briefly here because it is covered in some detail in Chapter 14 in the discussion about covariance. Please refer to that chapter for a more detailed description of partial correlation. Partial correlation is the process of finding the correlation between two variables *after* the influence of other variables has been controlled for. If, for instance, we computed a correlation between GPA and total points earned in a class, we could include year as a covariate. We would anticipate that fourth year students would generally do better than first year students. By computing the partial correlation, that "partials out" the influence of year, we mathematically eliminate the influence of years of schooling on the correlation between total points and GPA. With the partial correlation option, you may include more than one variable as a covariate if there is reason to do so.

The file we use to illustrate the **Correlate** command is our example introduced in the first chapter. The file is called **grades.sav** and has an $N = 105$. This analysis computes correlations between five variables in the file: **gender**, previous GPA (**gpa**), the first and fifth quizzes (**quiz1**, **quiz5**), and the final exam (**final**).

STEP BY STEP

Correlations

To enter SPSS, a click on **Start** *in the taskbar (bottom of screen) activates the start menu:*

After clicking the SPSS program icon, Screen 1 (inside front cover) appears on the monitor.

Step 2
Create and name a data file or edit (if necessary) an already existing file (see Chapter 3).

Screens 1 and 2 (inside front cover) allow you to access the data file used in conducting the analysis of interest. The following sequence accesses the **grades.sav** *file for further analyses:*

Whether first entering SPSS or returning from earlier operations the standard menu of commands across the top is required (shown below). As long as it is visible you may perform any analyses. It is not necessary for the data window to be visible.

This menu of commands disappears or modifies when using pivot tables or editing graphs. To uncover the standard menu of commands simply click on the ▪ or the ▣ icon.

After completion of Step 3 a screen with the desired menu bar appears. When you click a command (from the menu bar), a series of options will appear (usually) below the selected command. With each new set of options, click the desired item. The sequence to access Correlations begins at any screen with the menu of commands visible:

After clicking **Bivariate**, a new window opens (Screen 10.1, below) that specifies a number of options available with the correlation procedure. First, the box to the left lists all the *numeric* variables in the file (note the absence of **firstname**, **lastname**, and **grade**—all *non*numeric). Moving variables from the list to the **Variables** box is similar to the procedures used in previous chapters. Click the desired variable in the list, click ▸, and that variable is pasted into the **Variables** box. This process is repeated for each desired variable. Also, if there are a number of consecutive variables in the variables list, you may "click & drag" from the first to the last desired variable to select them all. Then a single click of ▸ will paste all highlighted variables into the active box.

10.1

The Bivariate Correlations Window

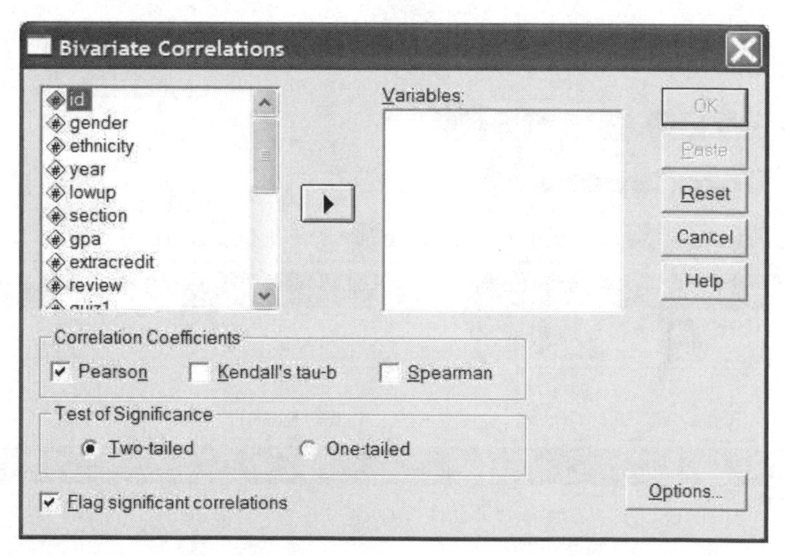

In the next box labeled **Correlation Coefficients**, the **Pearson** *r* is selected by default. If your data are *not* normally distributed then select **Spearman**. You may select both options and see how the values compare.

Under **Test of Significance**, **Two-tailed** is selected by default. Click on **One-tailed** if you have clear knowledge of the *direction* (positive or negative) of your correlations.

Flag significant correlations is selected by default and places an asterisk (*) or double asterisk (**) next to correlations that attain a particular level of significance (usually .05 and .01). Whether or not significant values are flagged, the correlation, the significance accurate to three decimals and the number of subjects involved in each correlation will be included.

For analyses demonstrated in this chapter we will stick with the **Pearson** correlation, the **Two-tailed** test of significance, and also keep **Flag significant correlations**. If you wish other options, simply click the desired procedure to select or deselect before clicking the final **OK**. For sequences that follow, the starting point is always Screen 10.1. If necessary, perform whichever of steps 1-4 (pp. 127-128) are required to arrive at that screen. A click of **Reset** may be necessary to clear former variables.

To produce a correlation matrix of **gender**, **gpa**, **quiz1**, **quiz5**, *and* **final**, *perform the following sequence of steps:*

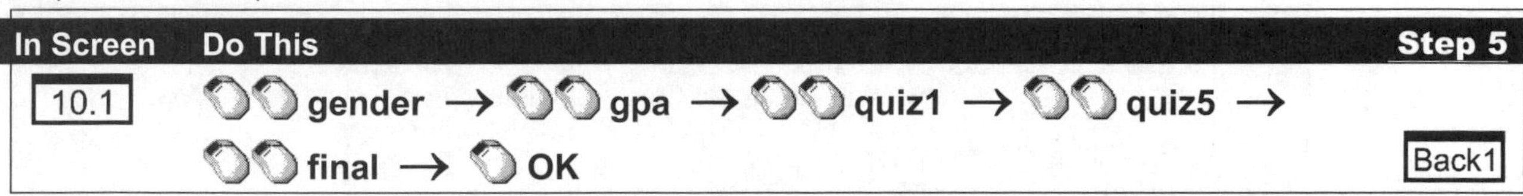

In Screen	Do This	Step 5
10.1	**gender** → **gpa** → **quiz1** → **quiz5** → **final** → **OK**	Back1

Additional procedures are available if you click the **Options** button in the lower right corner of Screen 10.1. This window (Screen 10.2, below) allows you to select additional statistics to be printed and to deal with missing values in two different ways. **Means and standard deviations** may be included by clicking the appropriate option, as may **Cross-product deviations and covariances**.

10.2

The Bivariate Correlations: Options Window

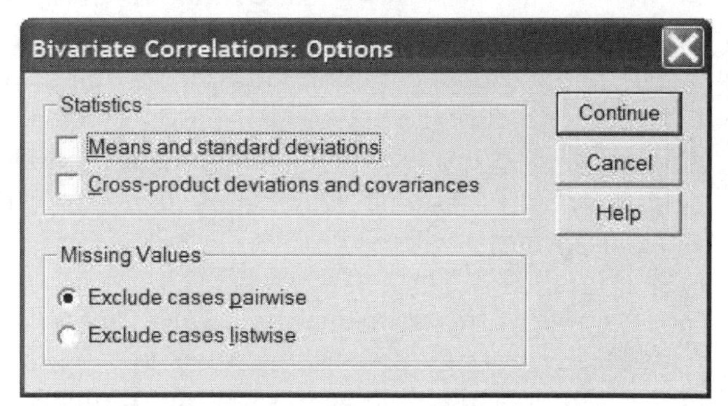

To **Exclude cases pairwise** means that for a particular correlation in the matrix, if a subject has one or two missing values for that comparison, then that subject's influence will not be included in that particular correlation. Thus correlations within a matrix may have different numbers of subjects determining each correlation. To **Exclude cases listwise**, means that if a subject has *any* missing values, all data from that subject will be eliminated from any analyses. Missing values is a thorny problem in data analysis and should be dealt with before you get to the analysis stage. See Chapter 4 for a more complete discussion of this issue.

The following procedure, in addition to producing a correlation matrix similar to that created in sequence Step 5, will print means, standard deviations, cross-product deviations, and covariances in two tables prior to printing the correlation matrix:

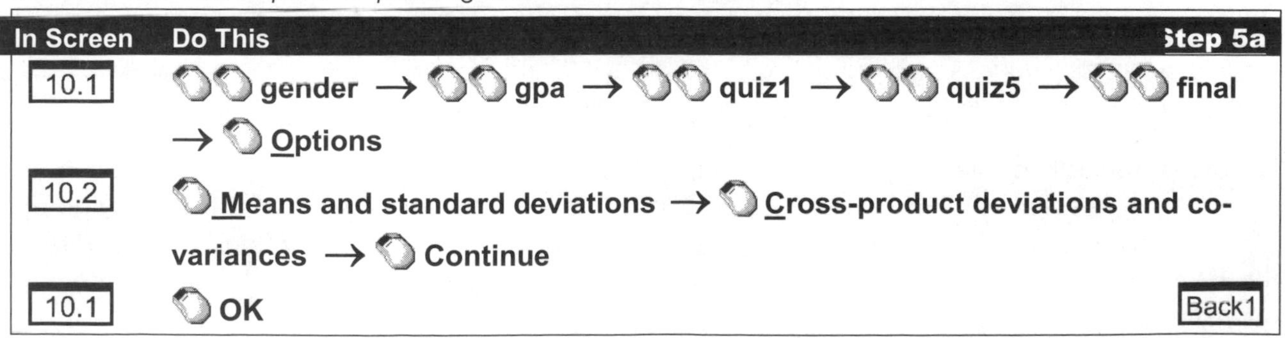

In Screen	Do This	Step 5a
10.1	**gender** → **gpa** → **quiz1** → **quiz5** → **final** → **Options**	
10.2	**Means and standard deviations** → **Cross-product deviations and covariances** → **Continue**	
10.1	**OK**	Back1

What we have illustrated thus far is the creation of a correlation matrix in which there are equal number of rows and columns. Often a researcher wishes to create correlations between one set of variables and another set of variables. For instance she may have created a 12 × 12 correlation matrix but wishes to compute correlations between 2 new variables and the original 12. The windows format does not allow this option and it is necessary to create a "command file", something familiar to users of the PC or mainframe versions of SPSS. If you attempt anything more complex than the sequence shown below, you will probably need to use the *SPSS Base System Syntax Reference Guide*.

10.3

The SPSS Syntax Editor Window

By clicking **File**, then **New**, then **Syntax**, a new screen opens that allows you to type in a command file to accomplish the procedure described above. Type in the words exactly as shown in sequence 5b (exchanging only the variable names you desire). Then highlight the text you have just typed and click the ▸ icon at the top of the screen to run the program. Note that the text is already typed in the Screen 10.3 display. All elements of the command line are necessary for it to run. If you omit equal signs, periods, or misspell words, an error message will flash and your program will not run.

You may begin this sequence from any screen that shows the standard menu of commands across the top of the screen. Below we create a 2 by 5 matrix of Pearson correlations comparing **gender** *and* **total** *with* **year**, **gpa**, **quiz1**, **quiz5**, *and* **final**.

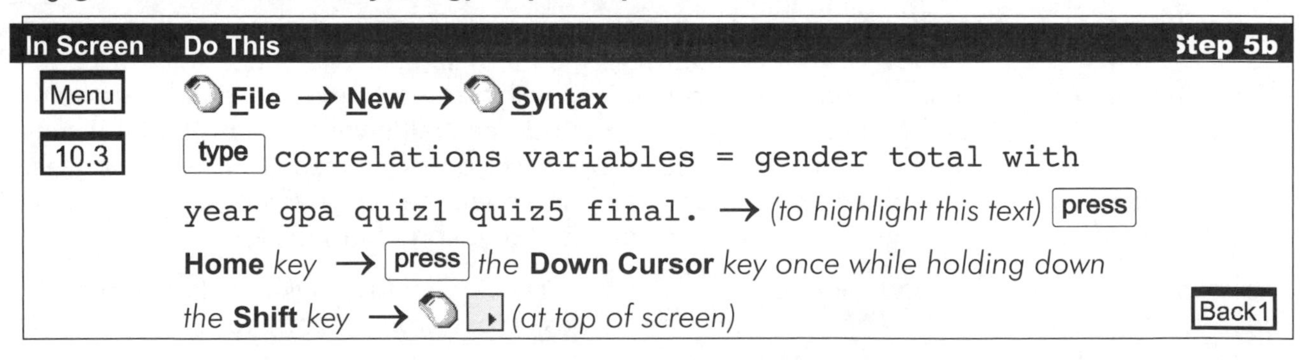

In Screen	Do This	Step 5b
Menu	**File** → **New** → **Syntax**	
10.3	type `correlations variables = gender total with year gpa quiz1 quiz5 final.` → *(to highlight this text)* press **Home** *key* → press *the* **Down Cursor** *key once while holding down the* **Shift** *key* → ▸ *(at top of screen)*	Back1

When you use this format, simply replace the variables shown here by the variables you desire. You may have as many variables as you desire both before and after the "`with`". If you get an error message, you probably spelled a variable name wrong or left off the period.

Upon completion of step 5, 5a, or 5b, the output screen will appear (Screen 1, inside back cover). All results from the just-completed analysis are included in the Output Navigator. Make use of the scroll bar arrows (▲▼▶◀) to view the results. Even when viewing output, the standard menu of commands is still listed across the top of the window. Further analyses may be conducted without returning to the data screen.

PRINTING RESULTS

Results of the analysis (or analyses) that have just been conducted requires a window that displays the standard commands (**File Edit Data Transform Analyze** . . .) across the top. A typical print procedure is shown below beginning with the standard output screen (Screen 1, inside back cover).

To print results, from the Output screen perform the following sequence of steps:

In Screen	Do This	Step 6
Back1	*Select desired output or edit (see pages 19 - 25)* → **File** → **Print**	
2.9	*Consider and select desired print options offered, then* → **OK**	

To exit you may begin from any screen that shows the **File** *command at the top.*

In Screen	Do This	Step 7
Menu	**File** → **Exit**	2.1

Note: After clicking **Exit**, there will frequently be small windows that appear asking if you wish to save or change anything. Simply click each appropriate response.

OUTPUT

Correlations

This output is from sequence step 5 (page 129) with two-tailed significance selected and significant correlations flagged.

		gender	gpa	quiz1	quiz5	final
gender	Pearson Correlation	1	-.194*	-.128	-.006	-.140
	Sig. (2-tailed)		.048	.195	.952	.156
	N	105	105	105	105	105
gpa	Pearson Correlation	-.194*	1	.246*	.262**	.498**
	Sig. (2-tailed)	.048		.011	.007	.000
	N	105	105	105	105	105
quiz1	Pearson Correlation	-.128	.246*	1	.504**	.535**
	Sig. (2-tailed)	.195	.011		.000	.000
	N	105	105	105	105	105
quiz5	Pearson Correlation	-.006	.262**	.504**	1	.472**
	Sig. (2-tailed)	.952	.007	.000		.000
	N	105	105	105	105	105
final	Pearson Correlation	-.140	.498**	.535**	.472**	1
	Sig. (2-tailed)	.156	.000	.000	.000	
	N	105	105	105	105	105

*. Correlation is significant at the 0.05 level (2-tailed).

**. Correlation is significant at the 0.01 level (2-tailed).

Notice first of all the structure of the output. The upper portion of each cell identifies the correlations between variables accurate to three decimals. The middle portion indicates the significance of each corresponding correlation. The lower portion records the number of subjects involved in each correlation. Only if there are missing values is it possible that the number of subjects involved in one correlation may differ from the number of subjects involved in others. The notes below the table identify the meaning of the asterisks and indicates whether the significance levels are one-tailed or two-tailed.

The diagonal of 1.000s shows that a variable is perfectly correlated with itself. Since the computation of correlations is identical regardless of which variable comes first, the half of the table above the diagonal of 1.000s has identical values to the half of the table below the diagonal. Note the strong positive relationship between **final** and **quiz5** ($r = .475$, $p < .001$). As described in the introduction of this chapter, these values indicate a strong positive relationship between the score on the fifth quiz and the score on the final. Those who scored higher on the fifth quiz tended to score higher on the final as well.

EXERCISES

Answers to selected exercises are available for download at **www.ablongman.com/george6e**.

1. Using the **grades.sav** file create a correlation matrix of the following variables; **id**, **ethnicity**, **gender**, **year**, **section**, **gpa**, **quiz1**, **quiz2**, **quiz3**, **quiz4**, **quiz5**, **final**, **total**; select one-tailed significance; flag significant correlations.
    - Draw a single line through the columns and rows where the correlations are meaningless.
    - Draw a double line through the correlations where there is linear dependency.
    - Circle 3 legitimate NEGATIVE correlations where the significance is $p < .05$ and explain what they mean.
    - Box 3 legitimate POSITIVE correlations where the significance is $p < .05$ and explain what they mean.
    - Create a scatterplot of **gpa** by **total** and include the regression line. (see Chapter 5 for instructions).

2. Using the **divorce.sav** file create a correlation matrix of the following variables; **sex**, **age**, **sep**, **mar**, **status**, **eth**, **school**, **income**, **avoicop**, **iq**, **close**, **locus**, **asq**, **socsupp**, **spiritua**, **trauma**, **lsatisfy**; select one-tailed significance; flag significant correlations. Note: Please make use of the **Data Files** descriptions starting on page 365 for meaning of all variables.
    - Draw a single line through the columns and rows where the correlations are meaningless.
    - Draw a double line through the correlations where there is linear dependency
    - Circle 3 legitimate NEGATIVE correlations where the significance is $p < .05$ and explain what they mean.
    - Box 3 legitimate POSITIVE correlations where the significance is $p < .05$ and explain what they mean.
    - Create a scatterplot of **close** by **lsatisfy** and include the regression line. (see Chapter 5 for instructions).
    - Create a scatterplot of **avoicop** by **trauma** and include the regression line.

11

The T TEST Procedure

A *T* TEST is a procedure used for comparing sample means to see if there is sufficient evidence to infer that the means of the corresponding population distributions also differ. More specifically, for an independent-samples *t* test, a sample is taken from two populations. The two samples are measured on some variable of interest. A *t* test will determine if the means of the two sample distributions differ significantly from each other. *T* tests may be used to explore issues such as: Does treatment A yield a higher rate of recovery than treatment B? Does one advertising technique produce higher sales than another technique? Do men or women score higher on a measure of empathic tendency? Does one training method produce faster times on the track than another training method? The key word is *two*: *t* tests always compare *two* different means or values.

In its chapter on *t* tests, the *SPSS for Windows Base System User's Guide* spends several pages talking about null hypotheses, populations, random samples, normal distributions, and a variety of research concerns. All its comments are germane and of considerable importance for conducting meaningful research. However, discussion of these issues goes beyond the scope of this book. The topic of this chapter is *t* tests: what they do, how to access them in SPSS, and how to interpret the results. We direct you to the manual if you wish to review some of those additional concerns.

INDEPENDENT-SAMPLES t TESTS

SPSS provides three different types of *t* tests. The first type, the independent-samples *t* test, compares the means of two different samples. The two samples share some variable of interest in common, but there is no overlap between membership of the two groups. Examples include: the difference between males and females on an exam score, the difference of performance on pull-ups by Americans and Europeans, the difference on achievement test scores for a sample of upper-class first graders and a sample of lower-class first graders, or the difference of life-satisfaction scores between those who are married and those who are unmarried. Note again that there is no overlap of membership between the two groups.

PAIRED-SAMPLES t TESTS

The second type of *t* test, the paired-samples *t* test, is usually based on groups of individuals who experience both conditions of the variables of interest. Examples include: students' scores on the first quiz versus the same students' scores on the second quiz; subjects' depression scores after treatment A as compared to the same subjects' scores after receiving treatment B; a set of students' scores on the SAT compared to the same students' scores on the GRE several years later; elementary school students' achievement test percentiles after one year at a low-SES school as compared to their achievement test percentiles after one year at a high-SES school. Note here that the same group experiences both levels of the variable.

ONE-SAMPLE t TESTS

The third type of test is a one-sample *t* test. It is designed to test whether the mean of a distribution differs significantly from some preset value. An example: Does a course offered to college seniors result in a GRE score greater than or equal to 1200? Did the performance of a particular class differ significantly from the professor's goal of an 82% average? During the previous season, the mean time for the cross country athletes' best seasonal performances was

18 minutes. The coach set a goal of 17 minutes for the current season. Did the athletes' times differ significantly from the 17-minute goal set by the coach? In this procedure, the sample mean is compared to a single fixed value.

TESTS OF SIGNIFICANCE

When using *t* tests to determine if two distributions differ significantly from each other, the test that measures the probability associated with the difference between the groups may be either a one-tailed or a two-tailed test of significance. The two-tailed test examines whether the mean of one distribution differs significantly from the mean of the other distribution, regardless of the direction (positive or negative) of the difference. The one-tailed test measures only whether the second distribution differs in a particular direction from the first. For instance, at a weight-loss clinic, the concern is only about the amount of weight *loss*. Any amount of weight gain is considered failure. Likewise, an advertising campaign concerns itself only with sales *increases*.

Usually the context of the research will make clear which type of test is appropriate. The only computational difference between the two is that the *p*-value of one is twice as much as the *p*-value of the other. If SPSS output produces a two-tailed significance value (this is the default), simply divide that number by two to give you the probability for a one-tailed test.

For demonstration of these commands we once more make use of the **grades.sav** file with $N = 105$. Variables of interest for this chapter include: **gender**, **total** (total points in the class), **year** (first, second, third or fourth year in college), the quizzes **quiz1** to **quiz5**, and **percent** (the final class percent).

STEP BY STEP

Computing t Tests

To enter SPSS, a click on **Start** *in the taskbar (bottom of screen) activates the start menu:*

After clicking the SPSS program icon, Screen 1 (inside front cover) appears on the monitor.

Step 2

Create and name a data file or edit (if necessary) an already existing file (see Chapter 3).

Screens 1 and 2 (inside front cover) allow you to access the data file used in conducting the analysis of interest. The following sequence accesses the **grades.sav** *file for further analyses:*

Whether first entering SPSS or returning from earlier operations the standard menu of commands across the top is required (shown below). As long as it is visible you may perform any analyses. It is not necessary for the data window to be visible.

This menu of commands disappears or modifies when using pivot tables or editing graphs. To uncover the standard menu of commands simply click on the ▬ or the ▣ icon.

INDEPENDENT-SAMPLES t TESTS

From any screen that shows the menu of commands, select the following three options:

In Screen	Do This	Step 4
Menu	**Analyze → Compare Means → Independent-Samples T Test**	11.1

At this point a new window opens (Screen 11.1, below) that allows you to conduct an independent-samples *t* test. Note the structure of this screen. To the left is the list of variables; to the right is a box to indicate the **Test Variable(s)**. Test variables are the *continuous* variables (such as total points, final grade, or others) for which we wish to make comparisons between two independent groups. A single variable or any number of variables may be included in this box. Below is the **Grouping Variable** box where the single variable that identifies the two groups will be indicated. This variable is usually discrete, that is, there are exactly two levels of the variable (such as gender or a pass/fail grade). It is possible, however, to use a variable with more than two levels (such as ethnicity—5 levels, or grade—5 levels) by indicating how you wish to divide the variable into exactly two groups. For instance, for grade, you might compare As and Bs (as one group) with Cs, Ds, and Fs (as the other group). A continuous variable could even be included here if you indicate the number at which to divide subjects into exactly two groups.

11.1

The Independent-Samples T Test Window

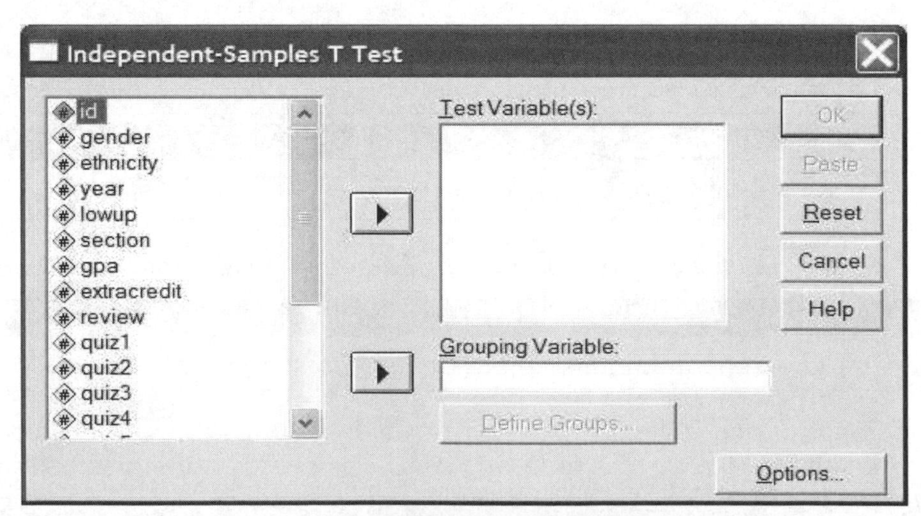

Once you have identified the grouping variable, you next click on the **Define Groups** button. Even for a variable that has exactly two levels, it is necessary to identify the two levels of the grouping variable. At this point a new window opens (Screen 11.2, below) that, next to **Group 1**, allows you to designate the single number that identifies the first level of the variable (e.g., female = 1), then, next to **Group 2**, the second level of the variable (e.g., male = 2). The **Cut point** option allows you to select a single dividing point for a variable that has more than two levels.

11.2

The Define Groups Window

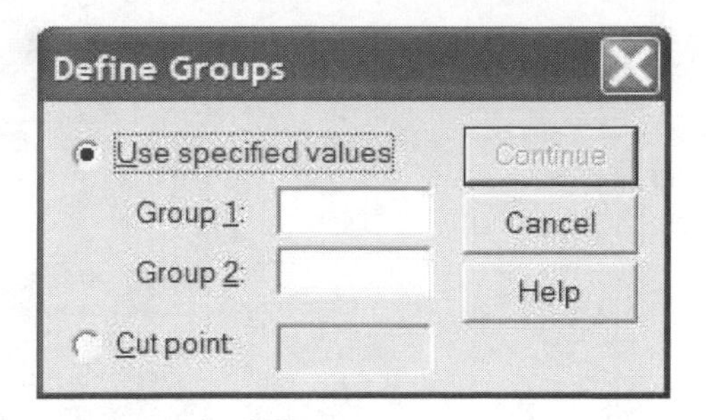

To test whether there is a difference between males and females on **total** *points earned:*

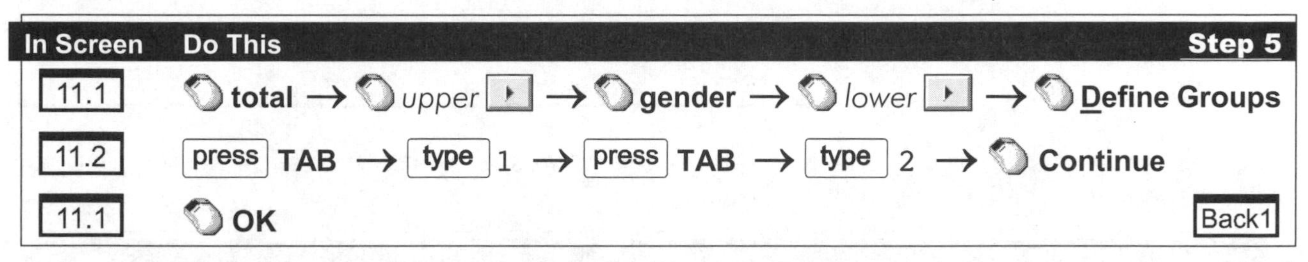

In Screen	Do This	Step 5
11.1	**total** → *upper* ▸ → **gender** → *lower* ▸ → **Define Groups**	
11.2	press TAB → type 1 → press TAB → type 2 → **Continue**	
11.1	**OK**	Back1

The **year** *variable has four levels and the use of the* **Cut point** *option is necessary to divide this variable into exactly two groups. The number selected (3 in this case) divides the group into the selected value and larger (3 and 4) for the first group, and all values smaller than the selected value (1 and 2) for the second group. The starting point for this procedure is Screen 11.1. Do whichever of steps 1-4 (pages 135-136) are necessary to arrive at this screen.*

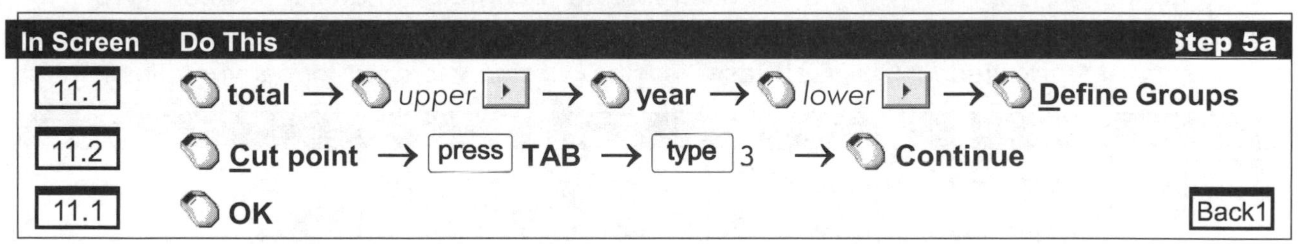

In Screen	Do This	Step 5a
11.1	**total** → *upper* ▸ → **year** → *lower* ▸ → **Define Groups**	
11.2	**Cut point** → press TAB → type 3 → **Continue**	
11.1	**OK**	Back1

PAIRED-SAMPLES t TESTS

From any screen that shows the menu of commands, select the following three options:

In Screen	Do This	Step 4a
Menu	**Analyze** → **Compare Means** → **Paired-Samples T Test**	11.3

The procedure for paired-samples *t* tests is actually simpler than for independent samples. Only one window is involved and you do not need to designate levels of a particular variable. Upon clicking the **Paired-Samples T Test** option, a new screen appears (Screen 11.3, below). To the left is the now familiar list of variable names, and, since you will be comparing all subjects on two different variables (**quiz1** and **quiz2** in the first example) you need to designate both variables before clicking the ▸ in the middle of the screen. You may select as many pairs of variables as you would like to paste into the **Paired Variables** box before conducting the analysis. There are, however, no automatic functions or click and drag options if you wish to conduct many comparisons. You must paste them, a pair at a time, into the **Paired Variables** box.

11.3

The Paired-Samples T Test Window

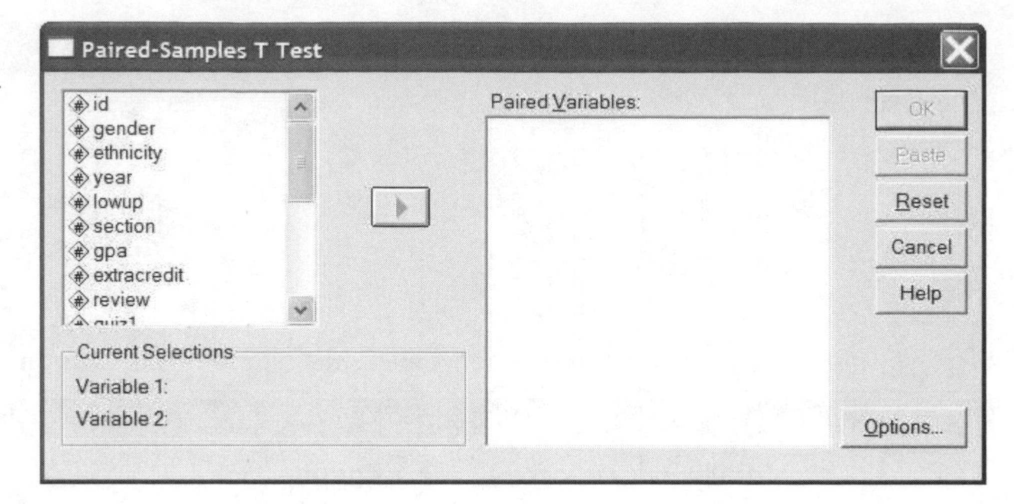

To compare the distribution of scores on **quiz1** *with scores on* **quiz2**, *the following sequence of steps is required.*

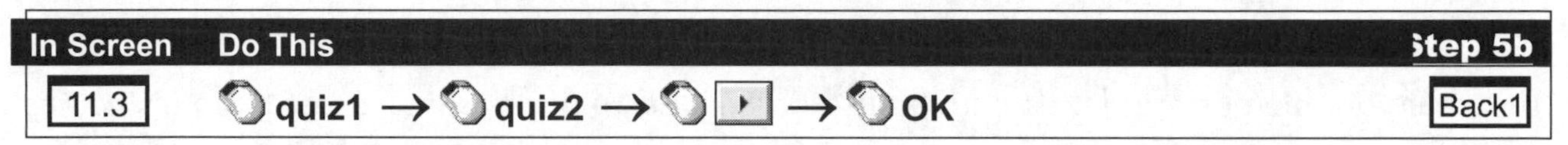

If you wish to compute several t tests in the same setting, you will paste all desired pairs of variables into the **Paired Variables** *box. In the sequence that follows, the scores on* **quiz1** *are compared with the scores on each of the other four quizzes (***quiz2** *to* **quiz5***). Screen 11.3 is the starting point for this sequence. Perform whichever of steps 1-4 (pages 135, 137) are necessary to arrive at this screen. A click of* **Reset** *may be necessary to clear former variables.*

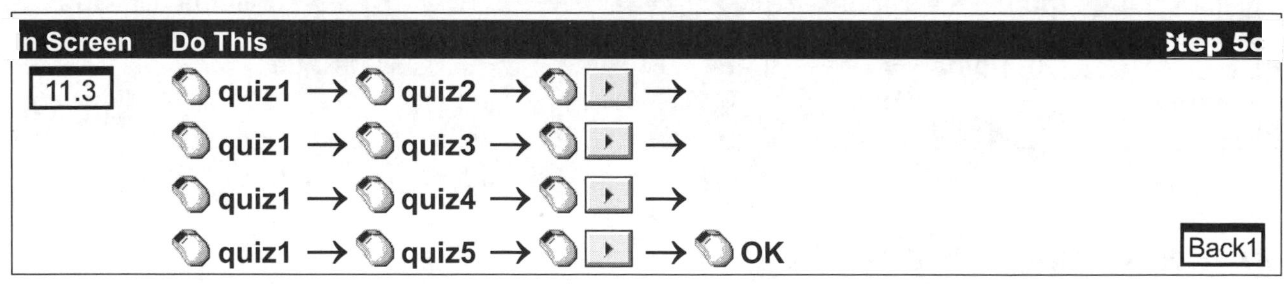

ONE-SAMPLE t TESTS

From any screen that shows the menu of commands, select the following three options.

It is often desirable to compare the mean of a distribution with some objective standard. With the **grades.sav** file, the instructor may have taught the class a number of times and has determined what he feels is an acceptable mean value for a successful class. If the desired value for final **percent** is 85, he may wish to compare the present class against that standard. Does this class differ significantly from what he considers to be an acceptable class performance?

11.4

The One-Sample T Test Window

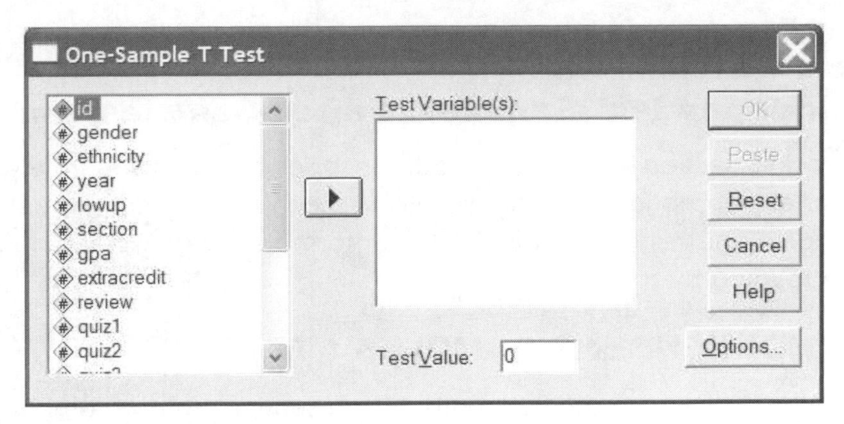

Upon clicking the **One-Sample T Test** option, the screen that allows one-sample tests to be conducted appears (Screen 11.4, above). The very simple procedure requires pasting variable(s) from the list of variables on the left into the **Test Variable(s)** box, typing the desired value into the box next to the **Test Value** label, then clicking the **OK**. For each variable selected this procedure will compare it to the designated value. Be sure, therefore, if you select several variables, that you want to compare them all to the same number. Otherwise, conduct separate analyses.

To determine if the **percent** *values for the entire class differed significantly from 85, conduct the following sequence of steps.*

In Screen	Do This	Step 5d
11.4	percent → ▸ → press TAB → type 85 → OK	Back1

Upon completion of step 5, 5a, 5b, 5c, or 5d, the output screen will appear (Screen 1, inside back cover). All results from the just-completed analysis are included in the Output Navigator. Make use of the scroll bar arrows (▲▼▶◀) to view the results. Even when viewing output, the standard menu of commands is still listed across the top of the window. Further analyses may be conducted without returning to the data screen.

PRINTING RESULTS

Results of the analysis (or analyses) that have just been conducted requires a window that displays the standard commands (**File Edit Data Transform Analyze** . . .) across the top. A typical print procedure is shown below beginning with the standard output screen (Screen 1, inside back cover).

To print results, from the Output screen perform the following sequence of steps:

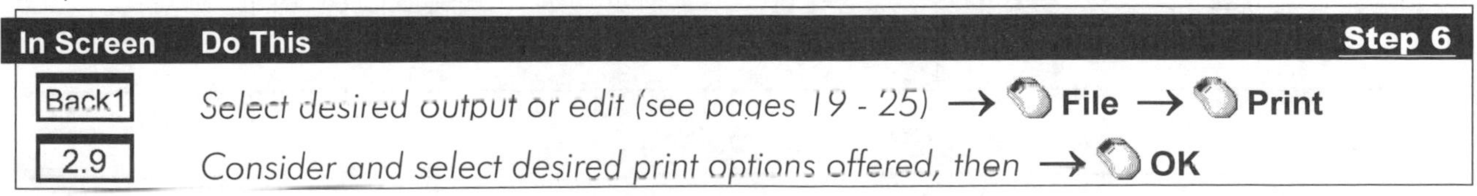

In Screen	Do This	Step 6
Back1	*Select desired output or edit (see pages 19 - 25)* → **File** → **Print**	
2.9	*Consider and select desired print options offered, then* → **OK**	

To exit you may begin from any screen that shows the **File** *command at the top.*

In Screen	Do This	Step 7
Menu	**File** → **Exit**	2.1

Note: After clicking **Exit**, there will frequently be small windows that appear asking if you wish to save or change anything. Simply click each appropriate response.

OUTPUT

Independent-Samples, Paired-Samples, and One-Sample t Tests

In this section we present the three types of *t* tests separately, each under its own heading. A brief description of findings follows each output section; definition of terms (for all three sections) finishes the chapter. Output format is slightly different (more space conserving) from that provided by SPSS.

INDEPENDENT-SAMPLES t TEST:

What follows is complete output from sequence step 5 on page 137.

Dependent Variable	gender	N	Mean	Std. Deviation	S.E. of Mean
total	Female	64	102.03	13.90	1.74
	Male	41	98.29	17.20	2.69

Independent-Samples t test

Variances	Levene's test for equality of variances		t-values	df	Sig. (2-tailed)	Mean Difference	St. Error Difference	95% Confidence Interval of the difference	
	F	Sig.							
Equal	2.019	.158	1.22	103	.224	3.74	3.05	-2.32	9.79
Unequal			1.17	72.42	.246	3.74	3.20	-2.64	10.11

This independent-samples *t* test analysis indicates that the 64 females had a mean of 102.03 total points in the class, the 41 males had a mean of 98.29 total points in the class, and the means did not differ significantly at the $p < .05$ level (note: $p = .224$). Levene's test for Equality of Variances indicates variances for males and females do not differ significantly from each other (note: $p = .158$). This result allows you to use the slightly more powerful equal-variance *t* test. If Levene's test did show significant differences, then it would be necessary to use the unequal variance test. Definitions of other terms are on pages 141 and 142.

PAIRED-SAMPLES t TEST:

On the following page is complete output from sequence step 5b on page 138.

Paired-Samples Statistics

Variables	N	Correlation	2-tail sig.	Mean	Std. Deviation	SE of Mean
quiz1	105	.673	.000	7.47	2.481	.242
quiz2				7.98	1.623	.158

Paired-Samples Test

Mean	Std. Deviation	SE or Mean	t-value	df	2-tailed sig.	95% Conf. Interval	
-.514	1.835	.179	-2.87	104	.005	-.869	-.159

This paired-samples *t* test analysis indicates that for the 105 subjects, the mean score on the second quiz ($M = 7.98$) was significantly greater at the $p < .01$ level (note: $p = .005$) than the mean score on the first quiz ($M = 7.47$). These results also indicate that a significant correlation exists between these two variables ($r = .673$, $p < .001$) indicating that those who score high on one of the quizzes tend to score high on the other.

ONE-SAMPLE t TEST:

What follows is complete output from sequence step 5d on page 139.

One-Sample Statistics

Variable	N	Mean	Std. Deviation	Std. Error of Mean
Percent	105	80.34	12.135	1.184

One-Sample Test

	Test Value = 85					
	t	Df	Sig. (2-tailed)	Mean difference	95% Confidence Interval	
Percent	-3.432	104	.000	-4.657	-7.01	-2.31

This one-sample *t* test analysis indicates that the mean percent for this class of 105 students ($M = 80.34$) was significantly lower at the $p < .001$ level than the instructor's goal (Test Value) of 85%. The Mean Difference is simply the observed mean (80.34) minus the Test Value (85.0).

DEFINITIONS OF TERMS

Term	Definition/Description
STD. ERROR	The standard deviation divided by the square root of *N*. This is a measure of stability or sampling error of the sample means.
F-VALUE	This value is used to determine if the variances of the two distributions differ significantly from each other. This is called a test of heteroschedasticity, a wonderful word with which to impress your friends.
P = (for Levene's test)	If variances do not differ significantly, then the *equal-variance estimate* may be used instead of the *unequal-variance estimate*. The *p*-value here, .158, indicates that the two variances do not differ significantly; so the statistically stronger equal-variance estimate may be used.
t-VALUES	Based on either the equal-variance estimate equation or the unequal-variance estimate equation. Conceptually, both formulas compare the within-group deviations from the mean with the between-group deviations from the mean. The slightly larger (absolute values) equal-variance estimate may be used here because variances do not differ significantly. The actual *t* value is the difference between means divided by the standard error.
df (degrees of freedom)	For the equal-variance estimate, number of subjects minus number of groups (105 – 2 = 103). The fractional degrees of freedom (72.42) for the unequal-variance estimate is a formula-derived value. For the paired-samples and one-sample tests, the value is number of subjects minus 1 (105 – 1 = 104).
2-TAIL SIG	(associated with the *t*-values) The probability that the difference in means could happen by chance.
MEAN DIFFERENCE	The difference between the two means.
STD. DEVIATION	This is the standard deviation of the difference, and it is used to calculate the *t*-value for the paired *t* test. For each subject in a paired *t* test, there is a difference between the two quizzes (sometimes 0, of course). This particular statistic is the standard deviation of this distribution of mean-difference scores.
CORRELATION	Measures to what extent one variable varies systematically with another variable. The statistic presented here is the Pearson product-moment correlation, designated with an *r*. See Chapter 10 for a description of correlations.
2-TAIL SIG (of the correlation)	The probability that such a pattern could happen by chance. In the paired-samples test, the $r = .67$ and $p < .001$ indicate a substantial and significant correlation between **quiz1** and **quiz2**.

Term	Definition/Description
95% CI (CONFIDENCE INTERVAL)	With *t* tests, the confidence interval deals with the difference-between-means value. If a large number of samples were drawn from a population, 95% of the mean differences will fall between the lower and upper values indicated.

EXERCISES

Answers to selected exercises are available for download at **www.ablongman.com/george6e**.

For questions 1- 7, perform the following operations:

a) Circle the two mean values that are being compared.

b) Circle the appropriate significance value (be sure to consider equal or unequal variance).

c) If the results are statistically significant, describe what the results mean.

1. Using the **grades.sav** file, compare men with women (**gender**) for **quiz1**, **quiz2**, **quiz3**, **quiz4**, **quiz5**, **final**, **total**.
2. Using the **grades.sav** file, determine whether the following pairings produce significant differences: **quiz1** with **quiz2**, **quiz1** with **quiz3**, **quiz1** with **quiz4**, **quiz1** with **quiz5**.
3. Using the **grades.sav** file, compare the GPA variable (**gpa**) with the mean GPA of the university of 2.89.
4. Using the **divorce.sav** file, compare men with women (**sex**) for **lsatisfy**, **trauma**, **age**, **school**, **cog-cope**, **behcope**, **avoicop**, **iq**, **close**, **locus**, **asq**, **socsupp**, **spiritua**.
5. Using the **helping3.sav** file, compare men with women (**gender**) for **age**, **school**, **income**, **tclose**, **hcontrot**, **sympathi**, **angert**, **hcopet**, **hseveret**, **empathyt**, **effict**, **thelplnz**, **tqualitz**, **tothelp**. Please see the **Data Files** section (page 365) for meaning of each variable.
6. Using the **helping3.sav** file, determine whether the following pairings produce significant differences: **sympathi** with **angert**, **sympathi** with **empathyt**, **empahelp** with **insthelp**, **empahelp** with **infhelp**, **insthelp** with **infhelp**.
7. Using the **helping3.sav** file, compare the age variable (**age**) with the mean age for North Americans (33.0).
8. In an experiment, 10 participants were given a test of mental performance in stressful situations. Their scores were 2, 2, 4, 1, 4, 3, 0, 2, 7, and 5. Ten other participants were given the same test after they had been trained in stress-reducing techniques. Their scores were 4, 4, 6, 0, 6, 5, 2, 3, 6, and 4. Do the appropriate t test to determine if the group that had been trained had different mental performance scores than the group that had not been trained in stress reduction techniques. What do these results mean?
9. In a similar experiment, ten participants were given a test of mental performance in stressful situations at the start of the study, were then trained in stress reduction techniques, and were finally given the same test again at the end of the study. In an amazing coincidence, the participants received the same scores as the participants in question 8: The first two people in the study received a score of 2 on the pretest, and a score of 4 on the posttest; the third person received a score of 4 on the pretest and 6 on the posttest; and so on. Do the appropriate t test to determine if there was a significant difference between the pretest and posttest scores. What do these results mean? How was this similar and how was this different than the results in question 1? Why?
10. You happen to know that the population mean for the test of mental performance in stressful situations is exactly three. Do a t test to determine whether the post-test scores in #9 above (the same numbers as the training group scores in #8) is significantly different than three. What do these results mean? How was this similar and how was this different than the results in question 2? Why?

12

The One-Way ANOVA Procedure

ONE-WAY analysis of variance is obtained through the SPSS **One-Way ANOVA** command. While a one-way analysis of variance could also be accomplished using the **General Linear Models** command (Chapters 13 and 14), the **One-Way ANOVA** command has certain options not available in **General Linear Models**, including post tests, such as Tukey and Scheffé, and planned comparisons of different groups or composites of groups.

INTRODUCTION TO ONE-WAY ANALYSIS OF VARIANCE

Analysis of variance is a procedure used for comparing sample means to see if there is sufficient evidence to infer that the means of the corresponding population distributions also differ. One-way analysis of variance is most easily explained by contrasting it with *t* tests (Chapter 11). Whereas *t* tests compare only two distributions, analysis of variance is able to compare many. If, for instance, a sample of students takes a 10-point quiz and we wish to see whether women or men scored higher on this quiz, a *t test* would be appropriate. There is a distribution of womens' scores and a distribution of mens' scores, and a t test will tell if the means of these two distributions differ significantly from each other. If, however, we wished to see if any of five different ethnic groups' scores differed significantly from each other on the same quiz, it would require one-way analysis of variance to accomplish this. If we were to run such a test, **One-Way ANOVA** could tell us if there are significant differences within any of the comparisons of the five groups in our sample. Further tests (such as the Scheffé test, which will be described in this chapter) are necessary to determine between *which* groups significant differences occur.

The previous paragraph briefly describes analysis of variance. What does the "one-way" part mean? Using the **One-Way ANOVA** command, you may have exactly *one* dependent variable (always continuous) and exactly *one* independent variable (always categorical). The independent variable illustrated above (ethnicity) is one variable, but it has several levels. In our example it has five: Native, Asian, Black, White, and Hispanic. **ANOVA Models** (next two chapters) may also have a maximum of *one* dependent variable, but it may have two or more independent variables. In MANOVA, multivariate analysis of variance (Chapters 23 and 24), there may be multiple dependent variables *and* multiple independent variables.

The explanation that follows gives a conceptual feel for what one-way analysis of variance is attempting to accomplish. The mean (average) quiz scores for each of the ethnic groups are compared with each other: Natives with Asians, Natives with Blacks, Natives with Whites, Natives with Hispanics, Asians with Blacks, Asians with Whites, Asians with Hispanics, Blacks with Whites, Blacks with Hispanics, and Whites with Hispanics. **One-Way ANOVA** will generate a significance value indicating whether there are significant differences within the comparisons being made. This significance value does not indicate where the difference is or what the differences are, but a Scheffé test can identify which groups differ significantly from each other. Be aware that there are other tests besides Scheffé that are able to identify pairwise differences. Along with the Scheffé procedure, Tukey's honestly significant difference (HSD), least-significant difference (LSD), and Bonferroni are also popular tests of bivariate comparisons.

The file we use to illustrate **One-Way ANOVA** is our familiar example. The file is called **grades.sav** and has an $N = 105$. This analysis contrasts scores on **quiz4** (the dependent variable) with five levels of **ethnicity** (the independent variable)—Native, Asian, Black, White, and Hispanic.

STEP BY STEP

One-Way Analysis of Variance

To enter SPSS, a click on **Start** *in the taskbar (bottom of screen) activates the start menu:*

After clicking the SPSS program icon, Screen 1 (inside front cover) appears on the monitor.

Step 2

Create and name a data file or edit (if necessary) an already existing file (see Chapter 3).

Screens 1 and 2 (inside front cover) allow you to access the data file used in conducting the analysis of interest. The following sequence accesses the **grades.sav** *file for further analyses:*

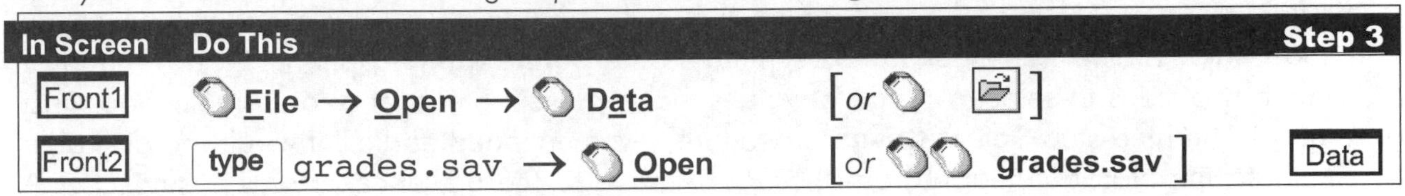

Whether first entering SPSS or returning from earlier operations the standard menu of commands across the top is required (shown below). As long as it is visible you may perform any analyses. It is not necessary for the data window to be visible.

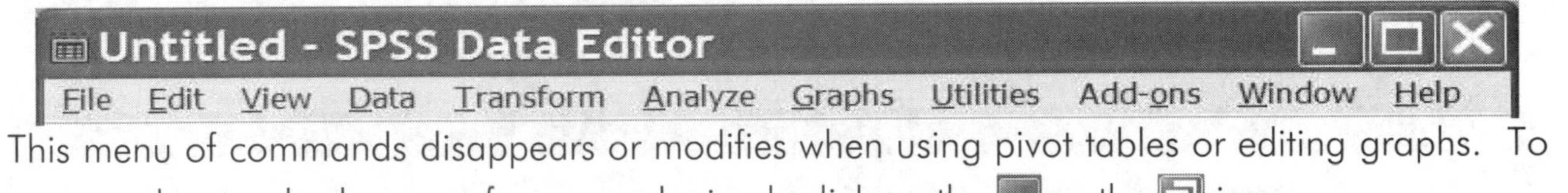

This menu of commands disappears or modifies when using pivot tables or editing graphs. To uncover the standard menu of commands simply click on the or the icon.

After completion of Step 3 a screen with the desired menu bar appears. When you click a command (from the menu bar), a series of options will appear (usually) below the selected command. With each new set of options, click the desired item. The sequence to access One-way analysis of variance begins at any screen with the menu of commands visible:

The initial screen that appears after clicking on **One-Way ANOVA** gives a clear idea of the structure of the command (Screen 12.1, following page). Often all three of the options (**Options, Post Hoc**, and **Contrasts**) will be employed in conducting a one-way analysis of variance. First note the familiar list of variables to the left. Then note the large box near the top of the screen (titled **Dependent List**). In this box will go a single continuous variable (**quiz4** in this example), or several continuous variables. SPSS will print out separate analyses of variance results for each dependent variable included. Beneath is the **Factor** box. Here a single categorical variable will be placed (**ethnicity** in our example). This analysis will compare the quiz scores for each of four ethnic groups. There are, of course, five ethnic groups listed under the **ethnicity** variable, but because there were only five subjects in the Native category, we will include only the other four groups. See pages 58-60 to refresh your memory on selecting cases.

12.1

The One-Way ANOVA Window

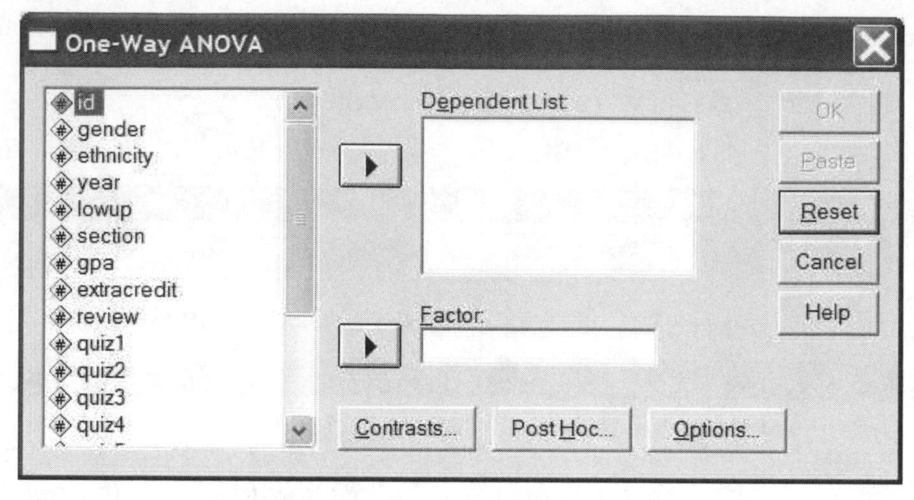

In a retrogressive move the magnitude of the demise of the rumble seat, the death of the drive-in theater, or the passing of the gas station that sells automotive products rather than candy and soda, the 12.0 version of SPSS has eliminated the **Define Range** option. Now, instead of the simple two steps to select a range of values that represent a subset of a variable, you must go through the nine-step **Select Cases** procedure. As mentioned earlier, it would be desirable to eliminate the Native category from the analysis due to the presence of only 5 subjects. A sequence step 4a is thus provided to remind you of that procedure (covered in Chapter 4). For the three Step-5 options that follow, if you wish to conduct the analysis for only four (rather than five) levels of ethnicity, then step 4a must precede them.

To select only the four levels of ethnicity described in the previous paragraph, perform the following sequence of steps. Screen 12.1 is an acceptable starting point.

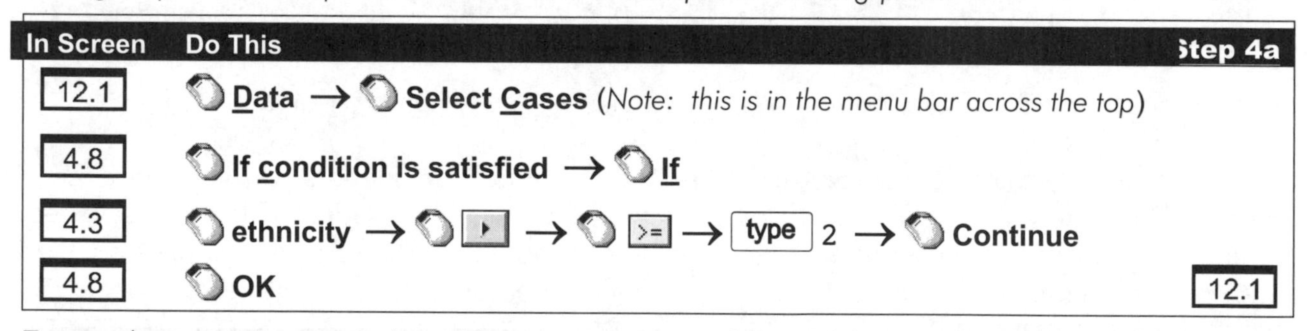

In Screen	Do This	Step 4a
12.1	**Data** → **Select Cases** (*Note: this is in the menu bar across the top*)	
4.8	**If condition is satisfied** → **If**	
4.3	**ethnicity** → ▸ → >= → type 2 → **Continue**	
4.8	**OK**	12.1

To conduct a one-way analysis of variance to see if any of four ethnic groups differ on their **quiz4** *scores, perform the following sequence:*

In Screen	Do This	Step 5
12.1	**quiz4** → *upper* ▸ → **ethnicity** → *lower* ▸ → **OK**	Back1

Two additional options now present themselves as desirable. Although the prior analysis (step 5, above) will tell you if a significant difference exists among the comparisons being made, it tells you little else. You don't know the mean values for each group, you have no information about the psychometric validity of your variables, and it is not possible to tell which groups differ from each other. The first omission is solved by clicking the **Options** button. Screen 12.2 opens allowing two important options. The **Descriptives** option provides means for each level, standard deviations, standard errors, 95% confidence limits, minimums, and maximums. The **Homogeneity-of-variance** selection also provides important psychometric information about the suitability of your variables for analysis. The **Means plot**, if selected, will produce a line graph that displays the mean of each category (each ethnic group in this case) in a plot-graph display. See Chapter 4 (pp. 48-50) for a perspective on missing values.

12.2

The One-Way ANOVA: Options Window

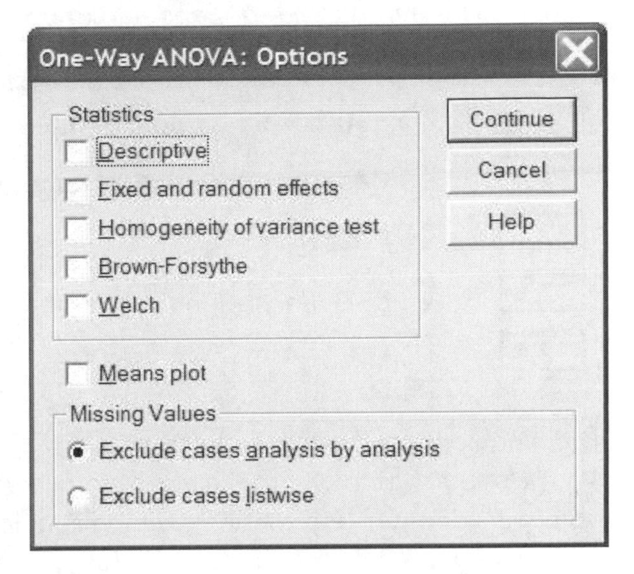

A second important issue considers pairwise comparisons—that is, comparisons of each possible pair of levels of the categorical variable. For instance in our ethnic comparisons, we are interested whether one group scores significantly higher than another: scores for Whites versus scores for Asians, Blacks versus Hispanics, and so forth. The title of the window (Screen 12.3, below) is **Post Hoc Multiple Comparisons**. *"Post Hoc"* means after the fact. "Multiple comparisons" means that all possible pairs of factors are compared. There are 14 options if equal variances for levels of the variable are assumed, and an additional 4 if equal variances are not assumed. The number of test options is beyond bewildering. We have never met anyone who knew them all, or even wanted to know them all. The **LSD** (least significant difference) is the most liberal of the tests (that is, you are most likely to show significant differences in comparisons) because it is simply a series of *t* tests. **Scheffe** and **Bonferroni** are probably the most conservative of the set. **Tukey** (HSD—honestly significant difference) is another popular option.

12.3

The One-Way ANOVA: Post Hoc Multiple Comparisons Window

One-Way ANOVA: Post Hoc Multiple Comparisons

Equal Variances Assumed

LSD
Bonferroni
Sidak
Scheffe
R-E-G-W F
R-E-G-W Q
S-N-K
Tukey
Tukey's-b
Duncan
Hochberg's GT2
Gabriel
Waller-Duncan
Type I/Type II Error Ratio: 100
Dunnett
Control Category: Last
Test
2-sided < Control > Control

Equal Variances Not Assumed

Tamhane's T2
Dunnett's T3
Games-Howell
Dunnett's C

Significance level: .05

Continue Cancel Help

*The procedure that follows will conduct a one-way ANOVA, request descriptive statistics, measures of heteroschedasticity (differences between variances), and select the least significant difference (**LSD**) method of testing for pairwise comparisons. As before, the categorical variable will be **ethnicity**, and the dependent variable, **quiz4**.*

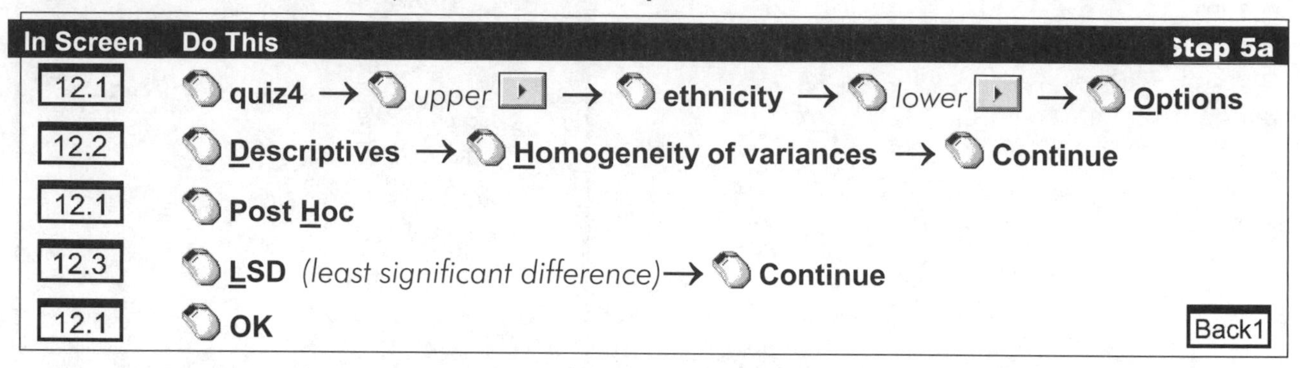

In Screen	Do This	Step 5a
12.1	quiz4 → *upper* ▸ → ethnicity → *lower* ▸ → Options	
12.2	Descriptives → Homogeneity of variances → Continue	
12.1	Post Hoc	
12.3	LSD *(least significant difference)* → Continue	
12.1	OK	Back1

The final option on the initial screen involves **Contrasts**. Upon clicking this button, Screen 12.4 appears (below). This procedure allows you to compare one level of the categorical variable with another (e.g., Asians with Hispanics), one level of the variable with the composite of others (e.g., Whites with non-Whites), or one composite with another composite (e.g., a White/Hispanic group with a Black/Asian group). Once the groups are identified, **Contrasts** computes a *t* test between the two groups. In this procedure, levels of a variable are coded by their value label. Present coding is Asians = 2, Blacks = 3, Whites = 4, and Hispanics = 5. In the Coefficients boxes, you need to insert numbers that contrast positive numbers in one group with negative numbers in another group. It is always required that these coefficient numbers sum to zero. For instance, a Hispanic-Asian comparison could be coded (-1 0 0 1); A Whites with non-Whites contrast (1 1 -3 1); and levels 2 and 3 contrasted with groups 4 and 5 (-1 -1 1 1). Note that each of these sequences sums to zero. The comparison shown in the screen below contrasts Hispanics (the 5th group) with non-Hispanics (groups 2, 3, and 4).

12.4

The One-Way ANOVA: Contrasts Window

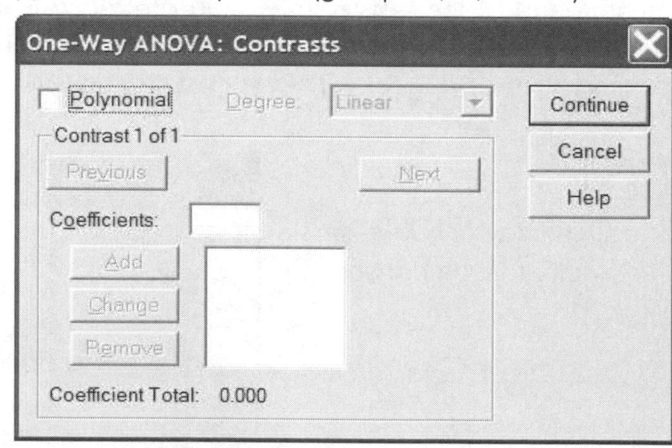

The procedure includes typing the number into the **Coefficients** box that represents the first level of the categorical variable, clicking the **Add**, then typing the number that represents the second level of the variable and clicking the **Add** button and so on until all levels of the variable have been designated. Then, if you wish another contrast, click the **Next** button to the right of **Contrast 1 of 1** (shown as **2 of 2** on the screen above) and repeat the procedure. The sequence that follows includes all the selections in step 5a, but also adds contrasts of Whites with Asians, and Hispanics with non-Hispanics on **quiz4** scores.

The contrasts selected below compare scores on the 4th quiz of 1) Whites and Asians with Hispanics and Blacks, and 2) Non-Hispanics with Hispanics. The starting point is Screen 12.1.

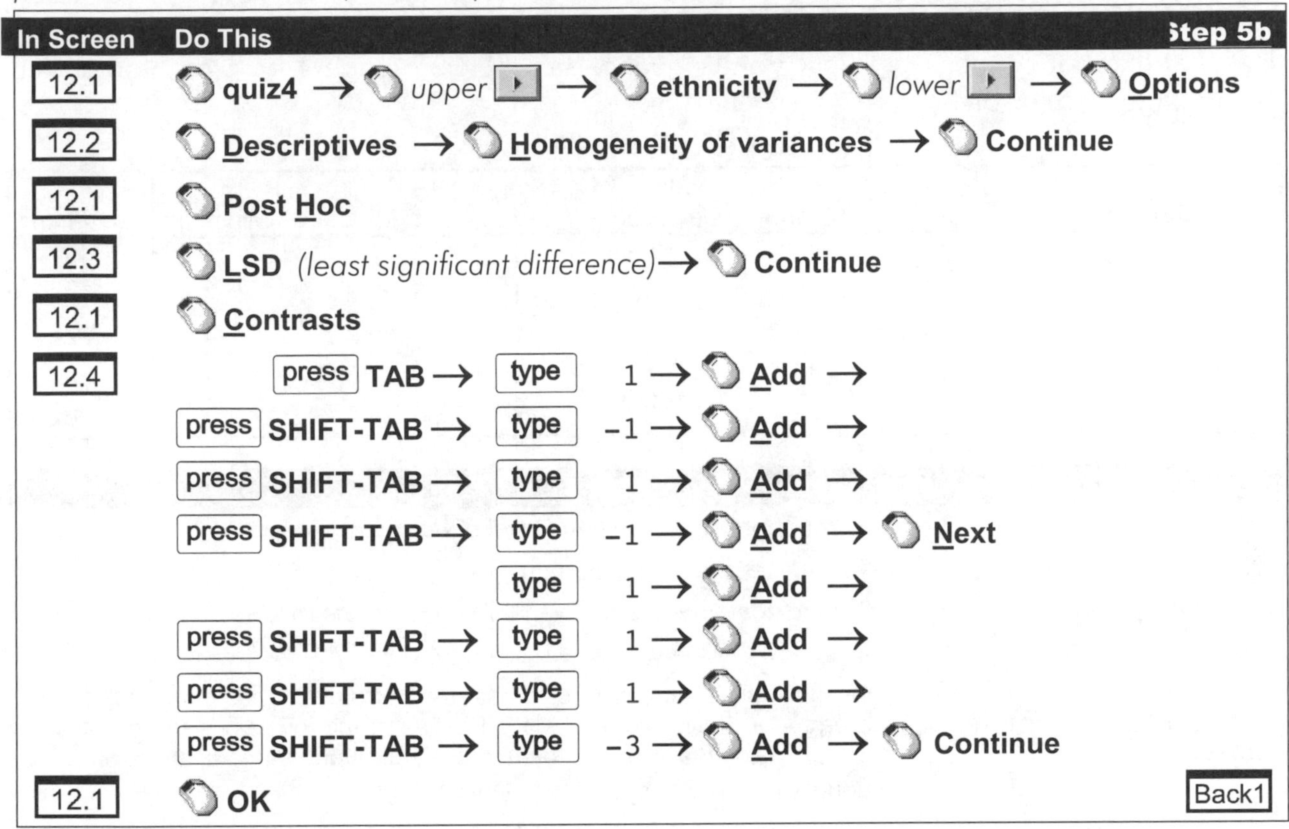

In Screen	Do This	Step 5b
12.1	quiz4 → upper ▸ → ethnicity → lower ▸ → **Options**	
12.2	**Descriptives** → **Homogeneity of variances** → **Continue**	
12.1	**Post Hoc**	
12.3	**LSD** *(least significant difference)* → **Continue**	
12.1	**Contrasts**	
12.4	press TAB → type 1 → **Add** →	
	press SHIFT-TAB → type −1 → **Add** →	
	press SHIFT-TAB → type 1 → **Add** →	
	press SHIFT-TAB → type −1 → **Add** → **Next**	
	type 1 → **Add** →	
	press SHIFT-TAB → type 1 → **Add** →	
	press SHIFT-TAB → type 1 → **Add** →	
	press SHIFT-TAB → type −3 → **Add** → **Continue**	
12.1	**OK**	Back1

Upon completion of step 5, 5a, or 5b, the output screen will appear (Screen 1, inside back cover). All results from the just-completed analysis are included in the Output Navigator. Make use of the scroll bar arrows (▲▼▶◀) to view the results. Even when viewing output, the standard menu of commands is still listed across the top of the window. Further analyses may be conducted without returning to the data screen.

PRINTING RESULTS

Results of the analysis (or analyses) that have just been conducted requires a window that displays the standard commands (**File Edit Data Transform Analyze** . . .) across the top. A typical print procedure is shown below beginning with the standard output screen (Screen 1, inside back cover).

To print results, from the Output screen perform the following sequence of steps:

In Screen	Do This	Step 6
Back1	*Select desired output or edit (see pages 19 - 25)* → **File** → **Print**	
2.9	*Consider and select desired print options offered, then* → **OK**	

To exit you may begin from any screen that shows the **File** *command at the top.*

In Screen	Do This	Step 7
Menu	**File** → **Exit**	2.1

Note: After clicking **Exit**, there will frequently be small windows that appear asking if you wish to save or change anything. Simply click each appropriate response.

OUTPUT

One-Way Analysis of Variance

This is somewhat condensed output from sequence step 5b from page 149. Don't forget step 4a on page 146, also.)

Descriptives

quiz4 Ethnicity	N	Mean	Std. Deviation	Std. Error	95% CI for Mean	Minimum	Maximum
Asian	20	8.35	1.53	.34	7.63 to 9.07	6	10
Black	24	7.75	2.13	.44	6.85 to 8.65	4	10
White	45	8.04	2.26	.34	7.73 to 8.72	2	10
Hispanic	11	6.27	3.32	1.00	4.04 to 8.50	2	10
Total	100	7.84	2.29	.23	7.39 to 8.29	2	10

Term	Definition/Description
N	Number of subjects in each level of **ethnic**.
MEAN	Average score for each group.
STANDARD DEVIATION	The standard measure of variability around the mean.
STANDARD ERROR	Standard deviation divided by square root of *N*.
95% CI (CONFIDENCE INTERVAL) FOR MEAN	Given a large number of samples drawn from a population, 95% of the means for these samples will fall between the lower and upper values. These values are based on the *t* distribution and are approximately equal to the mean $\pm$ 2 $\times$ the standard error.
MINIMUM/MAXIMUM	Smallest and largest observed values for that group.

Test of Homogeneity of Variance

	Levene Statistic	df1 between groups	df2 within groups	Significance
quiz4	5.517	3	96	.002

Levene's test for homogeneity of variance with a significance value of .002 indicates that variances for **quiz4** scores for each of the ethnic groups do indeed differ significantly. Note that these values vary between a narrow variance for Asians of 1.53^2 (= 2.34), to a much wider variance for Hispanics of 3.32^2 (= 11.02). Most researchers, upon seeing this result, would check the distributions for measures of normality (skewness and kurtosis), and if they found nothing unusual, would probably ignore these results and accept the ANOVA analysis as valid. These measures of homogeneity of variance act more as a warning than as a disqualifier. However, in the contrast coefficient matrices (below), you will use the slightly less powerful estimate based on assumption of unequal variances.

ANOVA

quiz4	Sum of Squares	df	Mean Square	F	Sig
Between groups	34.297	3	11.432	2.272	.085
Within Groups	483.143	96	5.033		
Total	517.440	99			

The key interpretive element of interest in the original ANOVA table is that, based on a p = .085, a marginally significant difference (or differences) exists within comparisons of **quiz4** scores among the four different ethnic groups. Definitions of terms in the ANOVA table follow.

Term	Definition/Description
WITHIN-GROUPS SUM OF SQUARES	The sum of squared deviations between the mean for each group and the observed values of each subject within that group.
BETWEEN-GROUPS SUM OF SQUARES	The sum of squared deviations between the grand mean and each group mean weighted (multiplied) by the number of subjects in each group.
BETWEEN-GROUPS DF	Number of groups minus one.
WITHIN-GROUPS DF	Number of subjects minus number of groups minus one.
MEAN SQUARE	Sum of squares divided by degrees of freedom.
F RATIO	Between-groups mean square divided by within-groups mean square.
SIGNIFICANCE	The probability of the observed value happening by chance. The result here indicates that there is/are marginally significant difference(s) between means of the four groups as noted by a probability value of .085.

Contrast Coefficients

Contrast	ETHNIC			
	Asian	Black	White	Hispanic
1	1	-1	1	-1
2	1	1	1	-3

Contrast Tests

		Contrast	Value of Contrast	Std. Error	t	df	Significance 2-tailed
quiz4	Assume equal variances	1	2.37	1.015	2.336	96	.022
		2	5.33	2.166	2.459	96	.016
	Do not assume equal variances	1	2.37	1.192	1.989	19.631	.061
		2	5.33	3.072	1.734	10.949	.111

The first chart simply restates the contrasts typed in earlier. Two types of *t* comparisons are made: Under the equal-variance estimates, both contrasts are significant: between Asians/Whites and Hispanics/Blacks, ($p = .022$); between Hispanics and non-Hispanics ($p = .016$). For the unequal-variance estimate, however, neither contrast achieves significance. Since variances *do* differ significantly, we should accept the unequal-variance estimate as valid, resulting in no significant differences. Terms in this portion of the analysis are defined below.

Term	Definition/Description
VALUE	Of little interest because it is a weighted number.
STANDARD ERROR	Standard deviation divided by square root of *N*.
T-VALUES	For either the equal- or unequal-variance estimate, *t* is determined by the **VALUE** divided by the standard error.
STANDARD ERROR	Standard deviation divided by square root of *N*.
DF (DEGREES OF FREEDOM)	Number of subjects minus number of groups for the equal-variance estimate. It is a little-known formula that computes the fractional degrees of freedom value for the unequal-variance estimate.
T PROBABILITY	The likelihood that these values would happen by chance. The results indicate that, for scores on **quiz4**, for the unequal-variance estimates, neither contrast achieves significance.

Post Hoc Tests: **Multiple Comparisons**

Ethnicity (I)	vs. Ethnicity (J)	Mean (I – J) Difference	Std. Error	Significance	95% Confidence Interval
Asian	Black	.600	.679	.379	-.75 to 1.95
	White	.306	.603	.613	-.89 to 1.50
	Hispanic	2.077*	.842	.015	.41 to 3.75
Black	White	-.294	.567	.605	-1.42 to .83
	Hispanic	1.477	.817	.074	-.14 to 3.10
White	Hispanic	1.772*	.755	.021	.27 to 3.27

The mean value (average score for **quiz4**) for each of the four groups is listed in the previous chart. The asterisks (*) indicate there are two pairs of groups whose means differ significantly (at the $p < .05$ level) from each other: According to these fictional data, Asians (M = 8.35) and Whites (M = 8.04) both scored significantly higher on **quiz4** than did Hispanics (M = 6.27). Note the associated significance values of .015 and .021. The fact that the overall ANOVA results showed only marginal significance ($p = .085$) and that pairwise comparisons yield two differences that are statistically significant is because the overall ANOVA compares all values simultaneously (thus weakening statistical power) while the LSD procedure is just a series of independent *t* tests (increasing experimentwise error, the chance that some of the significant differences are only due to chance).

EXERCISES

Answers to selected exercises are available for download at **www.ablongman.com/george6e**.

Perform one-way ANOVAs with the specifications listed below. If there are significant findings write them up in APA format (or in the professional format associated with your discipline). Examples of correct APA format are shown on the web site. Further, notice that the final five problems make use of the **helping3.sav** data file. This data set (and all data files used in this book) is also available for download at the website listed above. To assist in understanding the meaning and specification of each of the variables, make generous use of **Data Files** section of this book beginning on page 365.

1. File: **grades.sav**; dependent variable: **quiz4**; factor: **ethnicity** (2,5); use **LSD** procedure for post hoc comparisons, compute two planned comparisons. This problem asks you to reproduce the output on pages 150-152. Note that you will need to perform a select-cases procedure (see page 146) to delete the "1 = Native" category.

2. File: **helping3.sav**; dependent variable: **tothelp**; factor: **ethnicity** (1,4); use **LSD** procedure for post hoc comparisons, compute two planned comparisons.

3. File: **helping3.sav**; dependent variable: **tothelp**; factor: **problem** (1,4); use **LSD** procedure for post hoc comparisons, compute two planned comparisons.

4. File: **helping3.sav**; dependent variable: **angert**; factor: **occupat** (1,6); use **LSD** procedure for post hoc comparisons, compute two planned comparisons.

5. File: **helping3.sav**; dependent variable: **sympathi**; factor: **occupat** (1,6); use **LSD** procedure for post hoc comparisons, compute two planned comparisons.

6. File: **helping3.sav**; dependent variable: **effict**; factor: **ethnicity** (1,4); use **LSD** procedure for post hoc comparisons, compute two planned comparisons.

13

General Linear Models: Two-Way ANOVA

THIS CHAPTER describes a simple two-way (or two factor) analysis of variance (ANOVA). A two-way ANOVA is a procedure that designates a single dependent variable (always continuous) and utilizes exactly two independent variables (always categorical) to gain an understanding of how the independent variables influence the dependent variable. This operation requires the use of the **General Linear Models—Univariate** command because the **One-Way ANOVA** command (Chapter 12) is capable of conducting only one-way analyses (that is, analyses with one independent variable).

The three ANOVA chapters in this book (Chapters 12, 13, and 14) should be read sequentially. A thorough grasp of one-way ANOVA (Chapter 12) is necessary before two-way ANOVA (this chapter) can be understood. This paves the way for Chapter 14 that considers three-way ANOVA and the influence of covariates. The present chapter is (intentionally) the simplest of the three; it includes a simple two-factor hierarchical design and nothing more. Yet understanding this chapter will provide the foundation necessary to comprehend the considerably more complex Chapter 14. ANOVA is an involved process, and even in Chapter 14 we do not consider all the options that SPSS provides. Please refer to the *SPSS for Windows Base System User's Guide* for a more complete description of additional choices.

One "problem" when conducting ANOVA is that it is easy to get a statistical package such as SPSS to conduct a two-way, three-way, or even eight-way analysis of variance. The windows format makes it as simple as clicking the right buttons; the computer does all the arithmetic, and you end up with reams of impressive-looking output of which you understand very little. The ease of calculating analysis of variance on a computer has a tendency to mask the fact that a successful study requires many hours of careful planning. Also, although a one-way ANOVA is straightforward and simple to interpret, a two-way ANOVA requires some training and frequently involves a thorough examination of tables and charts before interpretation is clear. Understanding a three-way ANOVA usually requires an experienced researcher, and comprehending a four-way ANOVA extends beyond the abilities of most humans.

The present chapter focuses on the fundamentals of how to conduct a hierarchical analysis of variance with a single dependent variable (**total**) and two independent variables or factors (**gender** and **section**), then explains how to graph the cell means and interpret the results.

STATISTICAL POWER

One statistical concept needs to be introduced here, as it is not available for the previous procedures but is often useful for ANOVA: **Power**. Power is simply the probability of finding a significant difference in a sample, given a difference (between groups) of a particular size and a specific sample size. Often power is calculated before conducting a study. If you know the expected magnitude of a difference between groups, you can (for example) calculate how large a sample you need to be 80% sure of finding a significant difference between groups.

SPSS does not calculate power before you start a study, but it can calculate **observed power**. Observed power is the likelihood of finding a significant difference between groups in any particular sample with the sample size in your study, assuming that the differences between groups you find in your sample is the same size as the differences between groups in the population. (Feel free to read that sentence again.) This can be helpful in understanding what it means when a difference between groups does not reach statistical significance. For example, if you do not find a significant difference between groups and the observed power is .25, then there would only be a 25% chance of finding a significant difference with your sample size. You need to think about whether the difference between groups that you found is big or small, depending on what other researchers have found and what is theoretically interesting.

TWO-WAY ANALYSIS OF VARIANCE

As described in the last chapter, analysis of variance attempts to find significant differences between groups (or populations) by comparing the means of those groups on some variable of interest. To assist in understanding two-way ANOVA, we'll briefly summarize a one-way analysis. In the one-way analysis of variance described in the previous chapter, we sought to discover if four ethnic groups differed from each other on their performances for **quiz4**. The *one-way* part of the term indicates that there is exactly one independent variable (**ethnicity** in the Chapter 12 example), and the ANOVA part (as opposed to MANOVA) indicates that there is exactly one dependent variable as well (**quiz4**). For conducting a one-way analysis of variance, the **One-Way ANOVA** command is often preferable to the **General Linear Models** command. While **General Linear Models** is able to conduct two-way, three-way, or higher-order analyses, the **One-Way ANOVA** command is simpler.

The sample we will use to demonstrate two-way analysis of variance will again be the **grades.sav** file. The **total** variable (total points in the class) will serve as the dependent variable. The independent variables will be **gender** and **section**. We will be attempting to discover if **gender**, or **section**, or a **gender** by **section** interaction has an effect on performance in the class. The typical research questions of interest are:

1. Is there a *main effect for* **gender**? That is, do females and males differ significantly in number of points earned, and which group was higher than the other?
2. Is there a *main effect for* **section**? Do students in the three sections differ significantly in points scored in the class, and which section scored higher than the other?
3. Is there a **gender** *by* **section** *interaction*? Is the influence of the two variables idiosyncratic, such that in this case, gender has one effect in a particular section but a different effect in a different section? For example: Perhaps females score significantly higher in Section 1 than males, but males score significantly higher than females in Section 3. Interactions are often tricky to explain, and producing a plot displaying the cell means (shown in the output section) often helps to clarify.

SPSS can print cell means for all combinations of variables, and compute *F* values and associated significance values. These values will indicate if there are significant main effects and/or if there are significant interactions between variables. It is quite possible to have significant main effects but not to have a significant interaction. The reverse is also possible. The output section of this chapter clarifies some of these issues.

For additional material, two excellent books on analysis of variance are those written by Schulman (1998) and Lindman (1992). View the **Reference** section for more detailed information.

STEP BY STEP

Two-Way Analysis of Variance

To enter SPSS, a click on **Start** *in the taskbar (bottom of screen) activates the start menu:*

After clicking the SPSS program icon, Screen 1 (inside front cover) appears on the monitor.

Step 2

Create and name a data file or edit (if necessary) an already existing file (see Chapter 3).

Screens 1 and 2 (inside front cover) allow you to access the data file used in conducting the analysis of interest. The following sequence accesses the **grades.sav** *file for further analyses:*

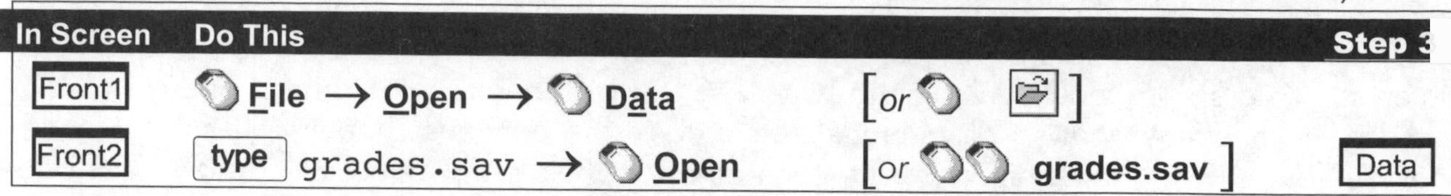

Whether first entering SPSS or returning from earlier operations the standard menu of commands across the top is required (shown below). As long as it is visible you may perform any analyses. It is not necessary for the data window to be visible.

After completion of Step 3 a screen with the desired menu bar appears. The sequence to access two-way analysis of variance begins at any screen with the menu of commands visible:

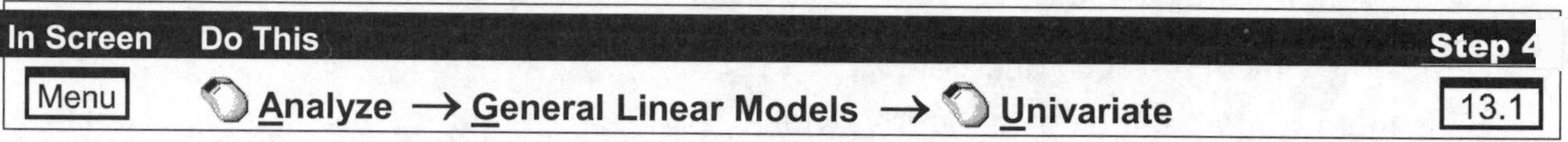

The first window that opens upon clicking **Univariate** (Screen 13.1, below) provides the structure for conducting a 2-way analysis of variance. To the left is the usual list of variables. To the right of the variables are five empty boxes. We will use the **Dependent Variable** and **Fixed Factor(s)** boxes in this chapter, and discuss the **Covariates** box in Chapter 14. The upper (**Dependent Variable**) box designates the single continuous dependent variable (**total** in this case), and the next (**Fixed Factor(s)**) box indicates independent variables or factors (**gender** and **section** in our example). **Random Factor(s)** and **WLS Weight** options are rarely used, and will not be described. Concerning the buttons to the right, only the final button **Options** will be discussed in this chapter. However, **Contrasts** and **Post Hoc** are nearly identical to similar options presented in the discussion of One-way ANOVA (Chapter 12). Please refer there for reference.

13.1

The Univariate General Linear Model Window

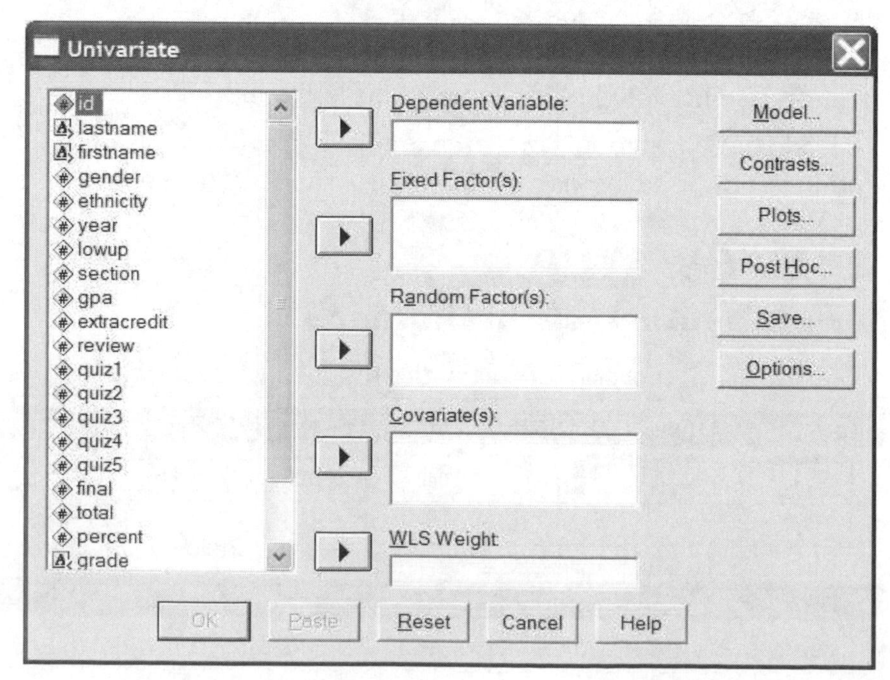

After both independent variables and dependent variables are selected, the next step is to click the **Options** button. A new window opens (Screen 13.2, below) that allows you to request that SPSS **Display** important statistics (such as descriptive statistics and effect sizes) to complete the analysis. In addition to cell means, **Descriptive statistics** produces standard deviations and sample sizes for each cell. Other frequently used options include, **Estimates of effect size**, which calculates the eta squared (η^2) effect size for each main effect and the interaction (indicating how large each effect is); and **Observed power**, which calculates the power given your effect size and sample size.

13.2

The Univariate: Options Window

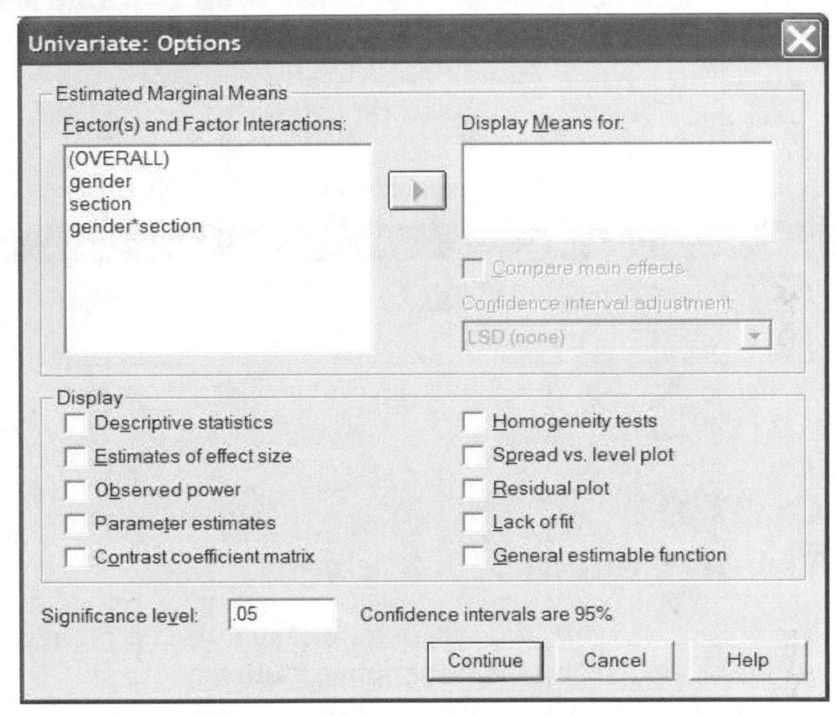

The step-by-step sequence below includes each of the options described above.

The following sequence begins with Screen 13.1. Do whichever of steps 1-4 (pp. 155-156) are necessary to arrive at this screen. Click the **Reset** *button if necessary to clear prior variables.*

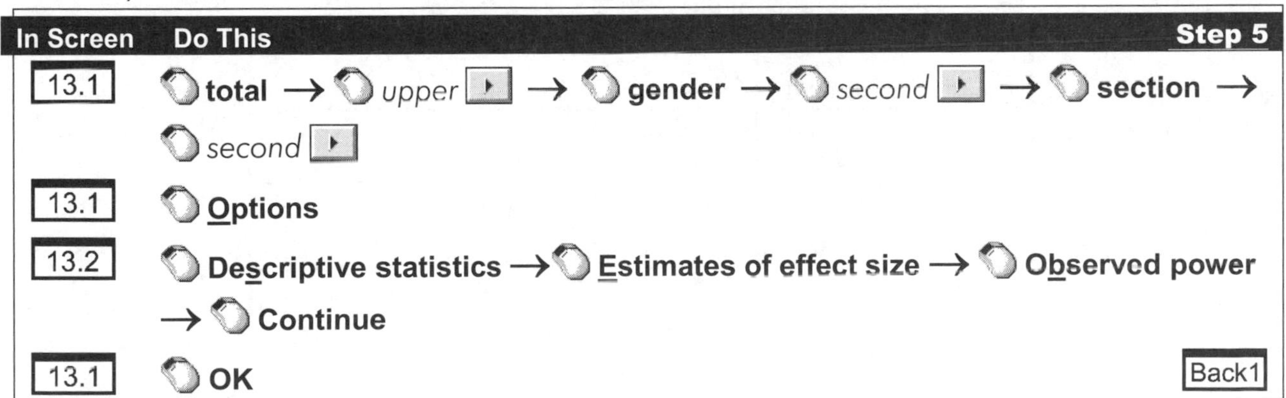

In Screen	Do This	Step 5
13.1	**total** → *upper* ▸ → **gender** → *second* ▸ → **section** → *second* ▸	
13.1	**Options**	
13.2	**Descriptive statistics** → **Estimates of effect size** → **Observed power** → **Continue**	
13.1	**OK**	Back1

Upon completion of step 5, the output screen will appear (Screen 1, inside back cover). All results from the just-completed analysis are included in the Output Navigator. Make use of the scroll bar arrows (▲ ▼ ▸ ◂) to view the results. Even when viewing output, the standard menu of commands is still listed across the top of the window. Further analyses may be conducted without returning to the data screen.

PRINTING RESULTS

Results of the analysis (or analyses) that have just been conducted requires a window that displays the standard commands (**File Edit Data Transform Analyze** . . .) across the top. A typical print procedure is shown below beginning with the standard output screen (Screen 1, inside back cover).

To print results, from the Output screen perform the following sequence of steps:

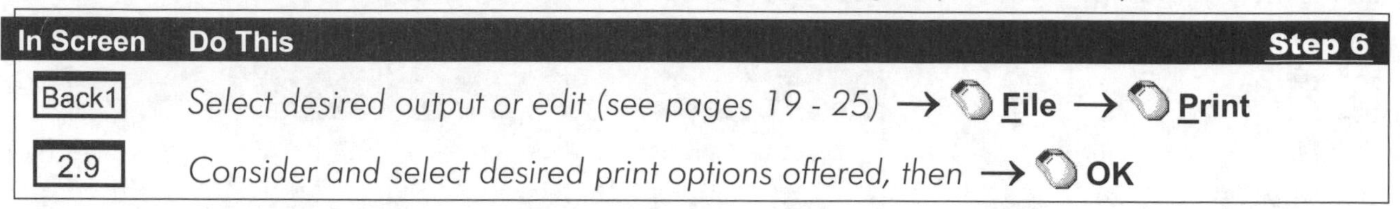

In Screen	Do This	Step 6
Back1	*Select desired output or edit (see pages 19 - 25)* → **File** → **Print**	
2.9	*Consider and select desired print options offered, then* → **OK**	

To exit you may begin from any screen that shows the **File** *command at the top.*

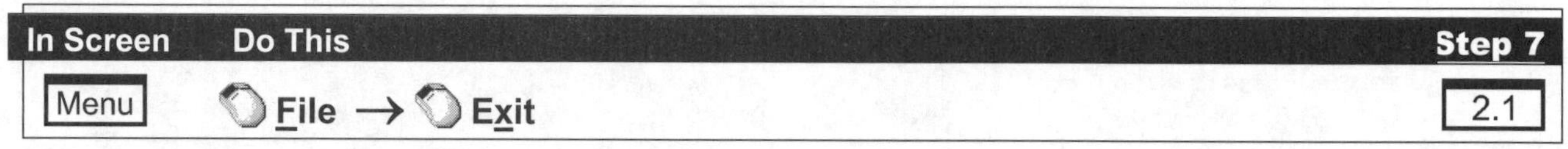

In Screen	Do This	Step 7
Menu	**File** → **Exit**	2.1

Note: After clicking **Exit**, there will frequently be small windows that appear asking if you wish to save or change anything. Simply click each appropriate response.

OUTPUT

Two-Way Analysis of Variance

This is slightly condensed output from sequence step 5 (previous page). We begin with the cell means produced with the **Descriptive statistics** option. A graphical display of the cell means and the ANOVA table (labeled "Tests of Between-Subject Effects") follow this. Commentary and definitions complete the chapter.

Descriptive Statistics

Dependent Variable: total

gender	section	Mean	Std. Deviation	N
Female	1	103.95	18.135	20
	2	100.00	12.306	26
	3	102.83	10.678	18
	Total	102.03	13.896	64
Male	1	106.85	13.005	13
	2	98.46	11.822	13
	3	90.73	21.235	15
	Total	98.29	17.196	41
Total	1	105.09	16.148	33
	2	99.49	12.013	39
	3	97.33	17.184	33
	Total	100.57	15.299	105

Graph of Gender by Section Interaction

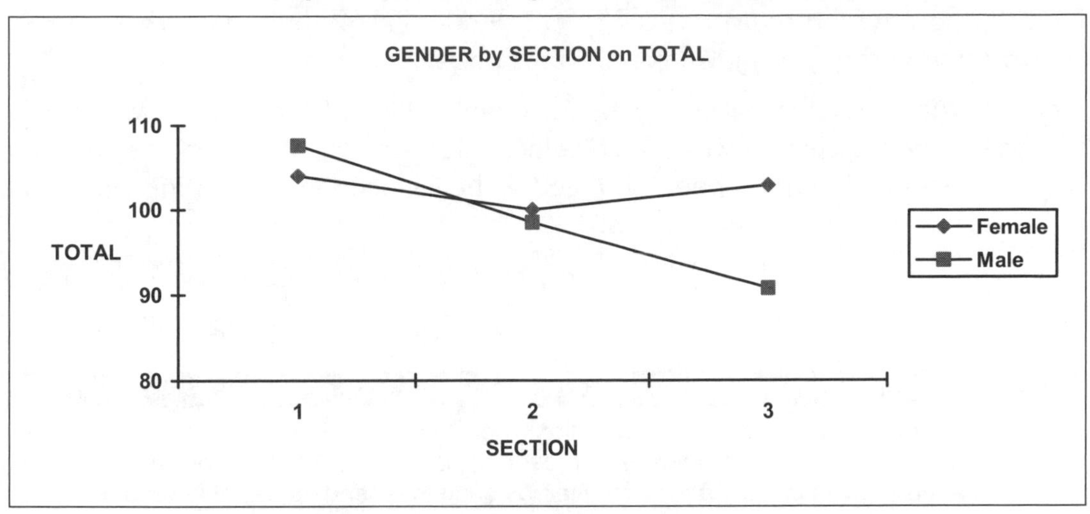

ANOVA (Tests of Between-Subjects Effects)

Dependent Variable: total

Source	Type III Sum of Squares	df	Mean Square	F	Sig.	Partial Eta Squared	Noncent. Parameter	Observed Power[a]
Corrected Mode	2350.408[b]	5	470.082	2.116	.070	.097	10.580	.678
Intercept	996892.692	1	996892.692	4487.382	.000	.978	4487.382	1.000
gender	316.564	1	316.564	1.425	.235	.014	1.425	.219
section	1262.728	2	631.364	2.842	.063	.054	5.684	.546
gender * section	960.444	2	480.222	2.162	.121	.042	4.323	.433
Error	21993.306	99	222.155					
Total	086378.000	105						
Corrected Total	24343.714	104						

Main effect for section

Main effect for gender

Two-way section by gender interaction

The first display (Descriptive Statistics) provides the mean total points for males, for females, for each of three sections, and for the six cells of the gender by section crosstabulation. The graph above displays the six gender by section means. You can easily plot this graph by hand, but if you want SPSS to produce this graph for you, you may follow the instructions at the end of Chapter 14 (see page 174). (Note that the **Plots** option in the **General Linear Models Univariate** procedure plots *estimated* means based on the model, rather than actual means of the data; this can be a problem in some models.) The tests of Between-Subjects Effects table (in this case, an ANOVA table) provides the answers to the three experimental questions:

1. There is no significant main effect for **gender**: Females (M = 102.0) did not score significantly higher than males (M = 98.3), $F(1, 99) = 1.425$, $p = .235$.
2. There is a marginally significant main effect for **section**: Results show that those in Section 1 (M = 105.09) scored higher than students in Sections 2 (M = 99.49) or 3 (M = 97.33), $F(2, 99) = 2.842$, $p = .063$.
3. There is no significant **gender** by **section** interaction: Despite the lack of statistical significance, we might note that while scores between males and females did not differ

much in the first two sections, in the third section females scored much higher (M = 102.83) than did males (M = 90.73), $F(2, 99) = 2.162$, $p = .12$. These data are displayed on the graph shown on the previous page.

Note that the F values for the main effects are slightly different than in old versions of SPSS. This is because there are several ways of calculating sums of squares, and the default method of calculating these numbers has been changed to better account for designs that don't have the same number of participants in each cell.

A more complete discussion of ANOVA interactions will take place in the following chapter. Definitions of related terms follow:

Term	Definition/Description
SUM OF SQUARES	This is the sum of squared deviations from the mean. In a broader sense, the corrected model sum of squares represents the amount of variation in the dependent variable (**total** in this case) explained by each independent variable or interaction, and corrected total sum of squares represents the total amount of variance present in the data. Error sum of squares (sometimes called residual sum of squares) represents the portion of the variance *not* accounted for by the independent variables or their interaction(s). In this example, only about 10% (2350.408/24343.714) of the variation in **total** is explained by **gender**, **section,** and the **gender** × **section** interaction. The remaining 90% remains unexplained by these two variables. Total and intercept values relate to the internal model that SPSS uses to calculate two-way ANOVA; it is not necessary to understand them to understand two-way ANOVA.
DEGREES OF FREEDOM (DF)	SECTION: (Levels of **section** – 1) = 3 – 1 = 2. GENDER: (Levels of **gender** – 1) = 2 – 1 = 1. **2-WAY INTERACTION**: (Levels of **gender** – 1)(Levels of **gender** – 1) = (2 – 1)(3 – 1) = 2. CORRECTED MODEL: (DF of Main effects + DF of Interaction) = 3 + 2 = 5. ERROR: (N – Explained DF – 1) = 105 – 5 – 1 = 99. CORRECTED TOTAL: (N – 1) = 105 – 1 = 104.
MEAN SQUARE	Sums of squares divided by degrees of freedom for each category.
F	The mean square for **gender** divided by the residual mean square; the mean square for **section** divided by the residual mean square; and the mean square for the **gender** × **section** interaction divided by the residual mean square.
SIG	Likelihood that each result could happen by chance.
PARTIAL ETA SQUARED	The effect-size measure. This indicates how much of the total variance is explained by each independent variable, the interaction, or (for the corrected model row) the effect of all main effects and interactions.
OBSERVED POWER	The probability of finding a significant effect (at the .05 level) with the sample size of the data being analyzed, assuming that the effect size in the population is the same as the effect size in the current sample. The probability of not making a Type II error.

EXERCISES

The exercises for 2-way ANOVAs are combined with the exercises for 3-way ANOVAs at the end of Chapter 14.

14

General Linear Models: Three-Way ANOVA

The previous two chapters—Chapter 12 (one-way ANOVA) and Chapter 13 (two-way ANOVA)—have overviewed the goals of analysis of variance. A three-way analysis of variance asks similar questions, and we are not going to reiterate here what we have already presented. We will address in this chapter those things that are unique to three-way analysis of variance (as contrasted with one-way or two-way ANOVAs), and also describe what a covariate is and how it can be used with the **General Linear Models** command.

This chapter will be organized in the same manner as all the analysis chapters (Chapters 6-27), but the emphasis will be somewhat different. This introductory portion will deal with the experimental questions that are usually asked in a three-way ANOVA and then explain what a covariate is and how it is typically used. The Step by Step section will include two different types of analyses. A three-way ANOVA will be described based on the **grades.sav** file, in which total points earned (**total**) will serve as the dependent variable, and **gender**, **section,** and **lowup** (lower- or upper-division student) will be designated as independent variables. The second analysis will be identical except that it will include one covariate, **gpa**, to "partial out" variance based on each student's previous GPA. The output from an analysis of variance is divided into two major sections: (1) The *descriptive statistics* portion, which lists the mean and frequency for each category created by the ANOVA; and (2) the ANOVA table, which indicates sums of squares, *F*-values, significance values, and other statistical information. The first portion (the descriptive statistics section) of the analysis will be identical whether or not a covariate is used, but the second portion (the ANOVA table) will often be substantially different.

The major difference between this chapter and previous chapters is that most of the explanation will occur in the Output section. A three-way or higher-order ANOVA is complex by any standards, and a very thorough and careful explanation of the analysis done here will hopefully assist you in untangling any three-way or higher-order ANOVAs that you might conduct. An almost universally-practiced procedure for helping to clarify ANOVA results is to graph the cell means for interactions. We will create and examine such graphs in this chapter (instructions for producing these graphs in SPSS are included at the end of the chapter). In fact, even *non*significant interactions will be graphed and related to the appropriate lines of output so that you can understand visually why there *is* a significant effect (when one occurs) and why there *isn't* (when a significant effect does *not* occur). A graph is not normally scrutinized for nonsignificant effects, but we hope that inclusion of them here will assist you toward a clearer understanding of the entire output.

THREE-WAY ANALYSIS OF VARIANCE

Restating briefly, the dependent variable for this analysis will be **total**, with independent variables of **gender**, **section**, and **lowup**. For this analysis (and for any three-way ANOVA) there will be seven possible experimental questions of interest. The first three questions deal with main effects, the next three questions deal with two-way interactions, and the final experimental question concerns whether there is a significant three-way interaction. The seven questions examined in this example are:

1. Is there a main effect for **gender** (i.e., do males score significantly differently from females on **total** points)?
2. Is there a main effect for **section** (i.e., do mean **total** points earned in one section differ significantly from the other sections)?
3. Is there a main effect for **lowup** (i.e., do upper-division students score significantly differently than lower-division students)?

4. Is there a significant interaction between **gender** and **section** on **total** points?
5. Is there a significant interaction between **gender** and **lowup** on **total** points?
6. Is there a significant interaction between **section** and **lowup** on **total** points?
7. Is there a significant interaction between **gender, section**, and **lowup** on **total**?

SPSS will print cell means for all combinations of variables and compute *F* values and associated significance values. These values will indicate if there are significant main effects and/or if there are significant interactions between variables. As stated in the previous chapter, it is quite possible to have significant main effects but *not* to have significant interactions, just as it is possible to have significant interactions without significant main effects. How to interpret interactions is discussed in some detail in the Output section.

THE INFLUENCE OF COVARIATES

The purpose of covariates is to partition out the influence of one or more variables before conducting the analysis of interest. A *covariate* could best be defined as a variable that has a substantial correlation with the dependent variable and is included in the experiment to adjust the results for differences existing among subjects before the start of the experiment. For example, a 1987 study explored the influence of personality traits on the competitive success (a standardized score of fastest racing times) of a sample of runners who ranged in ability from slow fitness runners to world-class athletes. In addition to a measure of 16 personality traits, each participant's weekly running mileage, weight/height ratio, and age were acquired. The influence of these physiological measures was well known prior to analysis (runners who run more miles, are skinnier, and who are younger, run faster). In the analysis, all three measures were used as covariates. The purpose was to exclude variance in the dependent variable (fast racing times) that was determined by the weekly mileage, weight/height ratio, and age. This would allow the researcher to see more clearly the influence of the psychological measures without the additional physiological factors included.

In our sample study **gpa** will be used as a covariate. In many different educational settings it has been demonstrated that students who have a higher overall GPA tend to do better in any given class. Thus if we wish to see the influence of **gender**, **section**, and **lowup** (lower- or upper-division student) *independently* of the influence of the student's previous GPA, then **gpa** can be included as a covariate. The inclusion of a covariate or covariates does not influence the cell means in the initial output, but it often has substantial influence (usually reduction) on the sum of squares, but can sometimes increase the *F* values (along with a corresponding decrease of *p*-values) because the error is reduced. Note that an ANOVA with a covariate is often abbreviated as ANCOVA. These issues will be further claritied in the Output section.

For additional material, two excellent books on analysis of variance are those written by Schulman (1998) and Lindman (1992). See the reference section for more detailed information.

STEP BY STEP

Three-Way Analysis of Variance

To enter SPSS, a click on **Start** *in the taskbar (bottom of screen) activates the start menu:*

After clicking the SPSS program icon, Screen 1 (inside front cover) appears on the monitor.

Step 2
Create and name a data file or edit (if necessary) an already existing file (see Chapter 3).

Screens 1 and 2 (inside front cover) allow you to access the data file used in conducting the analysis of interest. The following sequence accesses the **grades.sav** *file for further analyses:*

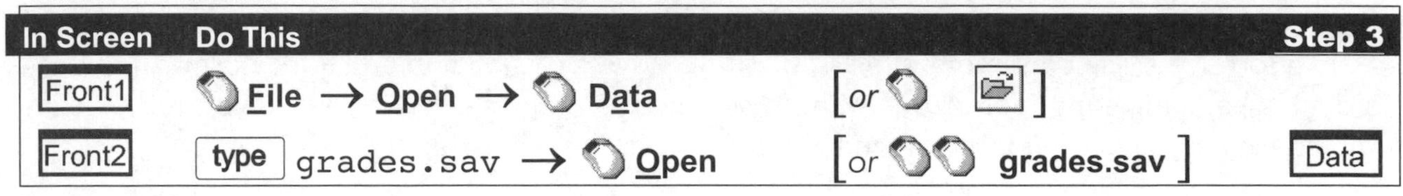

In Screen	Do This		Step 3
Front1	**File** → **Open** → **Data**	[or]	
Front2	type grades.sav → **Open**	[or **grades.sav**]	Data

Whether first entering SPSS or returning from earlier operations the standard menu of commands across the top is required (shown below). As long as it is visible you may perform any analyses. It is not necessary for the data window to be visible.

This menu of commands disappears or modifies when using pivot tables or editing graphs. To uncover the standard menu of commands simply click on the or the icon.

After completion of Step 3 a screen with the desired menu bar appears. When you click a command (from the menu bar), a series of options will appear (usually) below the selected command. With each new set of options, click the desired item. The sequence to access General Linear Models begins at any screen with the menu of commands visible:

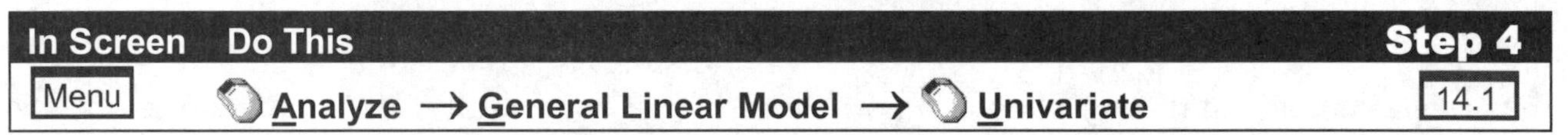

In Screen	Do This	Step 4
Menu	**Analyze** → **General Linear Model** → **Univariate**	14.1

First, a word concerning the four options that appear upon clicking the **General Linear Model** option. In addition to the **Univariate** choice that has been selected in the previous chapter (and in this one), there are three other options: **Multivariate, Repeated Measures**, and **Variance Components**. The **Multivariate** option includes operations more frequently identified by the acronyms MANOVA (multivariate analysis of variance) and MANCOVA (multivariate analysis of covariance) and is covered in Chapter 23. **Repeated Measures** is introduced in Chapter 24 and may be applied in either an ANOVA or MANOVA context. The **Variance Components** option allows for greater flexibility, but its complexity takes it beyond the scope of this book.

The first screen that appears upon clicking **Univariate** (Screen 14.1, following page) provides the structure for conducting a three-way analysis of variance. To the left is the usual list of variables. To the right of the variables are five boxes, three of which will interest us. The upper (**Dependent Variable**) box is used to designate the single continuous dependent variable, the variable **total** in this example. The next (**Fixed Factor(s)**) box is used to specify the categorical independent variables or factors (exactly three in this study), and a lower (**Covariate(s)**) box designates covariates, described in the introduction of this chapter. After selecting the dependent variable, **total**, and pasting it into the top box, each independent variable is pasted into the **Fixed Factor(s)** box one at a time.

14.1

The Univariate ANOVA Window

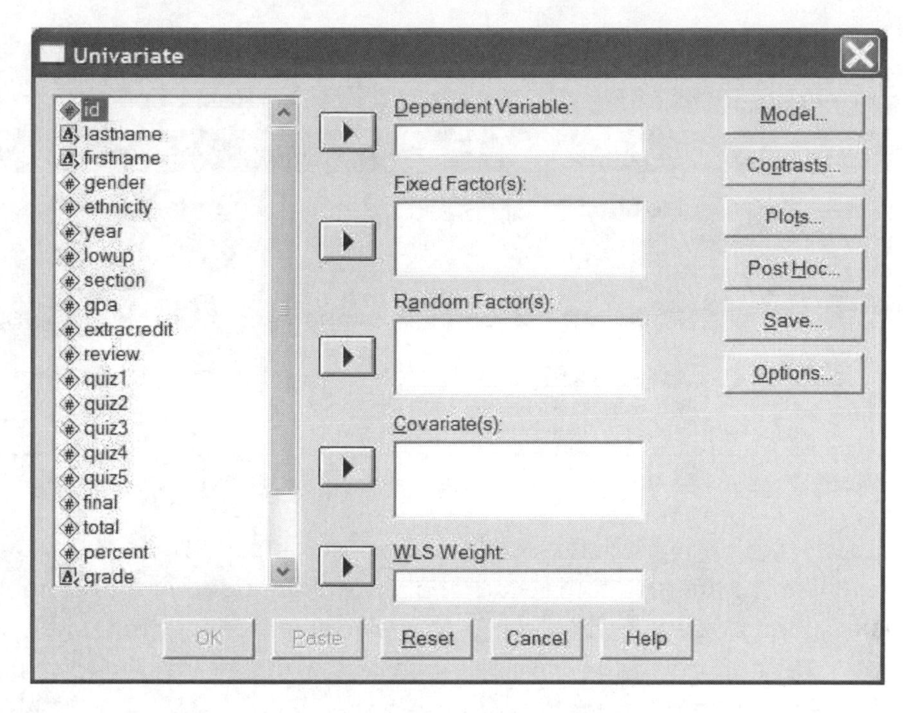

After the dependent variable (**total**) has been designated, all three independent variables (**gender**, **section**, **lowup**) have been selected and the covariate (**gpa**) chosen, the next step is to click the **Options** button. A new window opens (Screen 14.2, below) that allows you to **Display** important statistics to complete the analysis. **Descriptive statistics** produces means, standard deviations, and sample sizes for each cell. One other frequently used option, **Estimates of effect size**, calculates eta squared (η^2) effect size for each main effect and the interaction, indicating how large each effect is.

One technique that is often used to aid in the interpretation of main effects, the use of *post hoc* tests, will be discussed in Chapter 23, page 296. Although that chapter deals with a somewhat different procedure (**Multivariate** analysis of variance), the **Post Hoc** dialog box is identical to one used for **Univariate Post Hoc** tests.

14.2

The Univariate: Options Window

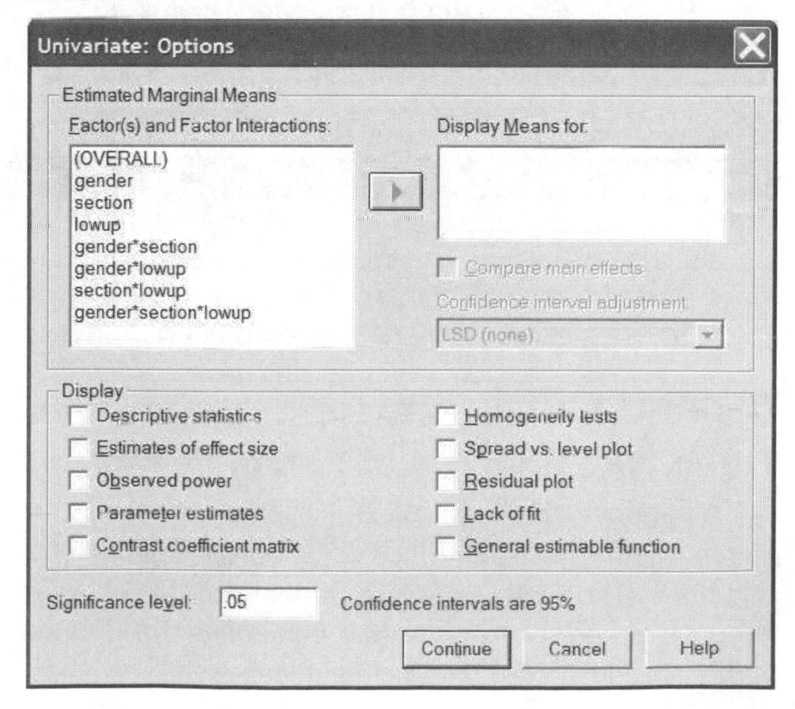

The following sequence begins with Screen 14.1. Do whichever of steps 1-4 (page 163-164) are necessary to arrive at this screen. Click the **Reset** *button if necessary.*

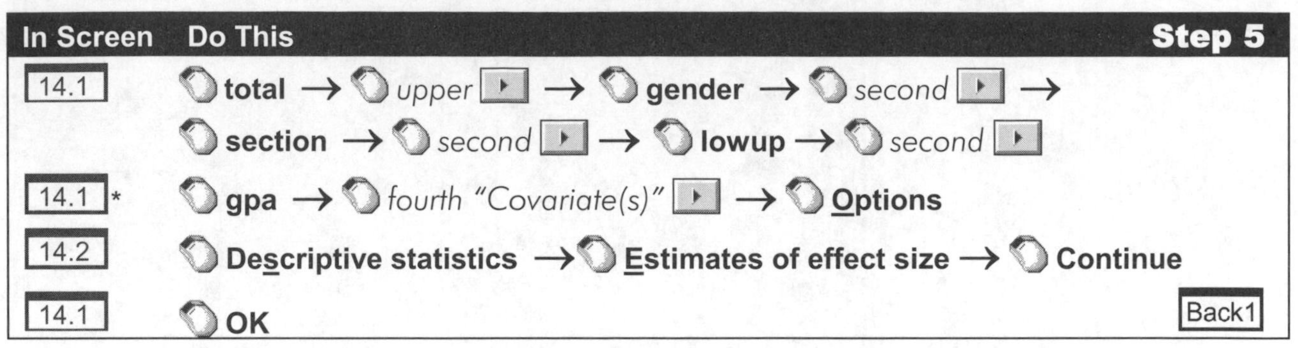

In Screen	Do This	Step 5
14.1	**total** → *upper* ▸ → **gender** → *second* ▸ → **section** → *second* ▸ → **lowup** → *second* ▸	
14.1 *	**gpa** → *fourth "Covariate(s)"* ▸ → **Options**	
14.2	**Descriptive statistics** → **Estimates of effect size** → **Continue**	
14.1	**OK**	Back1

*Note: If you wish to conduct the analysis without the covariate, simply omit this line selecting **gpa**.

Upon completion of step 5, the output screen will appear (Inside back cover, Screen 3). All results from the just-completed analysis are included in the Output Navigator. Make use of the scroll bar arrows (▲ ▼ ▶ ◀) to view the results. Even when viewing output, the standard menu of commands is still listed across the top of the window. Further analyses may be conducted without returning to the data screen.

PRINTING RESULTS

Results of the analysis (or analyses) that have just been conducted requires a window that displays the standard commands (**File Edit Data Transform Analyze** . . .) across the top. A typical print procedure is shown below beginning with the standard output screen (Screen 1, inside back cover).

To print results, from the Output screen perform the following sequence of steps:

In Screen	Do This	Step 6
Back1	*Select desired output or edit (see pages 19 - 25)* → **File** → **Print**	
2.9	*Consider and select desired print options offered, then* → **OK**	

To exit you may begin from any screen that shows the **File** *command at the top.*

In Screen	Do This	Step 7
Menu	**File** → **Exit**	2.1

Note: After clicking **Exit**, there will frequently be small windows that appear asking if you wish to save or change anything. Simply click each appropriate response.

OUTPUT

Three-Way Analysis of Variance and Analysis of Covariance

The format in this Output section will differ from most other output sections in this book. The output will be divided into two separate presentations. In the first of these, we show the results of a three-way analysis of variance that does *not* include a covariate. For this first analysis, instead of reproducing the output generated by SPSS we will integrate the descriptive statistics portion of the output (the first segment) with the ANOVA table portion (which follows). After each of the nine sections of the cell-means portion, the ANOVA line that relates to that portion will follow immediately. (For the real SPSS output, you'll need to hunt around to find the appropriate statistics.) For the five tables that involve interactions, a graph of the output will be included to help clarify the relation among the table, the graph, and the ANOVA output. These graphs may be produced by hand, or SPSS can produce these graphs for you; instructions for producing these graphs are included at the end of the chapter (pages 174-175). Explanation or clarification will follow after each table/graph/line presentation, rather than at the end of the Output section.

For the second presentation, the output from a three-way analysis of variance *that includes a covariate* will be shown. The tables of descriptive statistics and associated graphs (the first presentation) are identical whether or not a covariate is included; so these will not be produced a second time. What we will show is the complete ANOVA table output in similar format as that produced by SPSS. To demonstrate how mean square, *F*-, and *p*-values differ when a covariate is included, the ANOVA results *without* the covariate will *also* be included in the same table. The output that does *not* include the covariate will be shown *italicized* so as not to interfere in the interpretation of the original output. Explanation will then follow.

The following output is produced from sequence step 5 on page 166 (this initial portion are results without the covariate). Instructions for producing the graphs are on pages 174-175. For each result the cell means portion of the output is presented first followed by the associated ANOVA line.

Total Population

gender	section	lowup	Mean	Std. Deviation	N
Total	Total	Total	100.57	15.299	105

Source	Type III Sum of Squares	df	Mean Square	F	Sig.	Eta Squared
Corrected Total	24343.714	104				

This cell identifies the overall mean for the variable **total** for the entire sample ($N = 105$). The sum of squares is the total of squared deviations from the grand mean for all subjects. The degrees of freedom is $N - 1$ ($105 - 1 = 104$). There are no mean square, *F, p, or eta squared* statistics generated for a single value.

Main Effect for GENDER

gender	section	lowup	Mean	Std. Deviation	N
Female	Total	Total	102.03	13.896	64
Male			98.29	17.196	41

Source	Type III Sum of Squares	df	Mean Square	F	Sig.	Eta Squared
gender	454.062	1	454.062	2.166	.144	.023

The table indicates that 64 females scored an average of 102.03 **total** points while 41 males scored an average of 98.29. A visual display (e.g., graph) is rarely needed for main effects. It is clear that females scored higher than males. The number of levels of the variable minus one determines the degrees of freedom for main effects. Thus degrees of freedom for **gender** is:

2 – 1 = 1; for **section** is: 3 – 1 = 2; and for **lowup** is: 2 – 1 = 1. The *F*-value is determined by dividing the mean square for the variable of interest (**gender** in this case) by the residual mean square. The $F = 2.166$ and $p = .144$ verify that there is no significant main effect for gender; that is, the scores for females and males do not differ significantly. The Eta Squared value indicates the proportion of the variance that is accounted for by this variable; in this case, 2.3% of the variance in **total** is accounted for by **gender**.

Main Effect for SECTION

gender	section	lowup	Mean	Std. Deviation	N
Total	1	Total	105.09	16.148	33
	2		99.49	12.013	39
	3		97.33	17.184	33

Source	Type III Sum of Squares	df	Mean Square	F	Sig.	Eta Squared
section	1181.397	2	590.698	2.818	.065	.057

This table documents that 33 students from the first section scored a mean of 105.09 points, 39 students from the second section scored a mean of 99.49 points, and 33 students from the third section scored a mean of 97.33 points. An *F*-value of 2.82 ($p = .065$) indicates that there is only a marginally significant main effect for **section**—not enough to rely upon, but perhaps worthy of future study. Visual inspection verifies that the difference in scores between the first and the third sections is greatest.

Main Effect for LOWUP

SEX	SECTION	LOWUP	Mean	Std. Deviation	N
Total	Total	lower division	99.55	14.66	22
		upper division	100.84	15.54	83

Source	Type III Sum of Squares	df	Mean Square	F	Sig.	Eta Squared
LOWUP	14.507	1	14.507	.069	.793	.001

This result indicates that 22 lower-division students scored a mean **total** of 99.55, while 83 upper-division students scored a mean **total** of 100.84. A graph, once again, is unnecessary. An *F*-value of .40 and *p* of .529 (along with an eta squared of .001, indicating that only 0.1% of the variance of **total** is accounted for by **lowup**) indicate no significant differences for scores of upper-division and lower-division students.

Two-Way Interaction, GENDER by SECTION

gender	section	lowup	Mean	Std. Deviation	N
Female	1	Total	103.95	18.135	20
	2		100.00	12.306	26
	3		102.83	10.678	18
Male	1		106.85	13.005	13
	2		98.46	11.822	13
	3		90.73	21.235	15

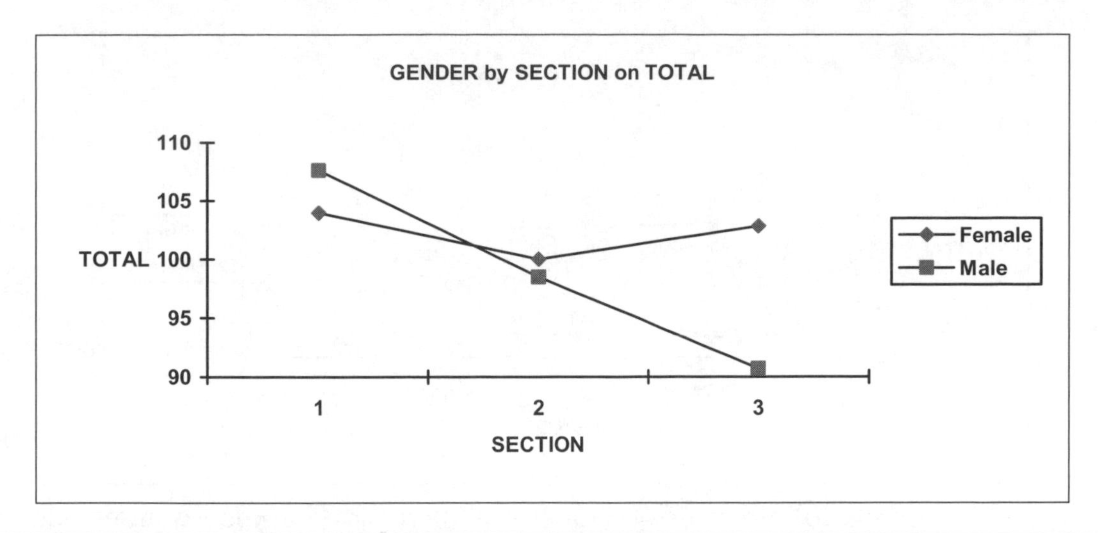

Source	Type III Sum of Squares	df	Mean Square	F	Sig.	Eta Squared
gender * section	118.155	2	59.078	.282	.755	.006

These results indicate no significant **gender** by **section** interaction ($F = .282$, $p = .755$). The top table identifies the mean **total** score and the number of subjects within each of six categories. One way to visually identify an interaction is to see whether two or more lines of the graph are parallel. If the lines are parallel, this indicates that there is *no* interaction. There is no *significant* interaction when the lines do not differ significantly from parallel. There is a significant interaction when the two lines *do* differ significantly from parallel. Be careful! Interactions cannot be determined by viewing a graph alone. The vertical and horizontal scales of a graph can be manipulated to indicate a greater effect than actually exists. The related output from the ANOVA table will indicate whether or not there is a significant effect. In the graph above, clearly the "action" is happening in the third section—but there is not enough "action" to be sure it is not just random (thus, it is not significant). These data could also be presented with the **gender** variable along the horizontal axis and the **section** variable coded to the right. The graphic configuration used is determined by which displays the relationship between variables more clearly.

Two-Way Interaction, GENDER by LOWUP

gender	section	lowup	Mean	Std. Deviation	N
Female	Total	lower division	102.50	14.482	16
		upper division	101.87	13.850	48
Male		lower division	91.67	13.095	6
		upper division	99.43	17.709	35

Source	Type III Sum of Squares	Df	Mean Square	F	Sig.	Eta Squared
gender * lowup	269.812	1	269.812	1.287	.260	.014

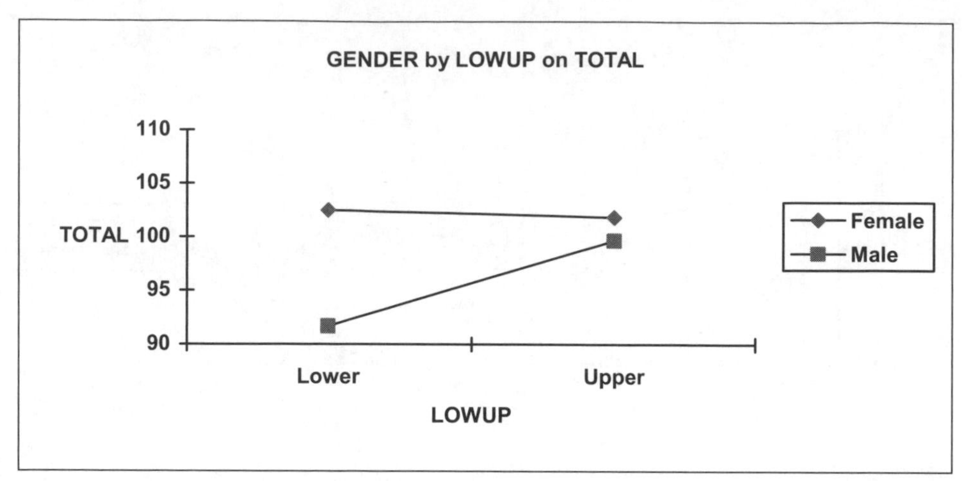

An *F*-value of 1.29 ($p = .26$) indicates no significant interaction between **gender** and **lowup**. Note that the two lines *appear* to deviate substantially from parallel, and yet there's no significance. Two factors influence this: (1) Values on the vertical axis vary only from 90 to 110, but **total** points vary from 48 to 125. We display only a narrow segment of the actual range, causing an exaggerated illusion of deviation from parallel. (2) The significance of a main effect or interaction is substantially influenced by the sample size. In the case above, two of the cells have only 6 and 16 subjects, decreasing the *power* of the analysis. Lower power means that a greater difference is required to show significance than if the sample size was larger.

Two-Way Interaction, SECTION by LOWUP

gender	section	lowup	Mean	Std. Deviation	N
Total	1	lower division	109.86	9.512	7
		upper division	103.81	17.436	26
	2	lower division	90.09	13.126	11
		upper division	103.18	9.444	28
	3	lower division	107.50	9.469	4
		upper division	95.93	17.637	29

Source	Type III Sum of Squares	df	Mean Square	F	Sig.	Eta Squared
section * lowup	1571.905	2	785.952	3.749	.027	.075

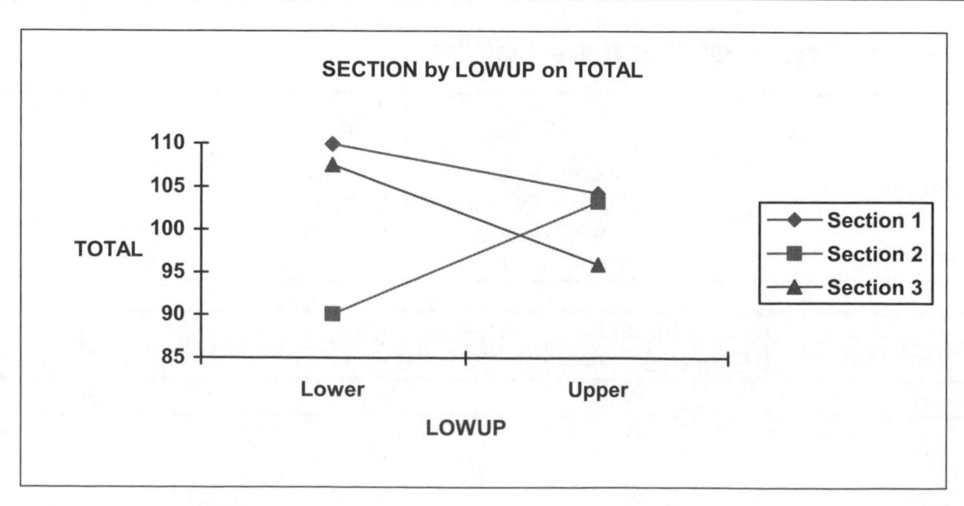

Both the ANOVA line (note the $F = 3.75$, $p = .027$) and the graph indicate a significant **section** by **lowup** interaction. Although the first and third sections' lines are relatively close to parallel, the second section shows a nearly opposite trend. A reasonable interpretation of the interaction might be: Although lower-division students tend to score higher than upper-division students in Sections 1 and 3, the reverse is true in Section 2. As indicated by the eta squared value, the **section** by **lowup** interaction accounts for 7.5% of the variance in **total**.

Three-Way Interaction, GENDER by SECTION by LOWUP

Three-way and higher-order interactions are often quite difficult to interpret. The researcher must have a strong conceptual grasp of the nature of the data set and what constitutes meaningful relationships. For the sample listed here, an $F = .409$ and $p = .665$ (table at the top of the following page) indicate that the three-way interaction does not even hint at significance. When there is no significant interaction, the researcher would usually not draw graphs. We have included them here to demonstrate one way that you might visually display a three-way interaction. Although one might note the almost identical pattern for females and males in the first chart, a more important thing to notice is the very low cell counts for all six cells of the lower division group (5, 8, 3, 2, 3, and 1, respectively). That is why it is so important to examine the F values in addition to the means and graphs. The degrees of freedom for a three-way interaction is the number of levels of the first variable minus one, times levels of the second variable minus one, times levels of the third variable minus one $[(2 - 1)\times(3 - 1)\times(2 - 1) = 2]$.

gender	section	Lowup	Mean	Std. Deviation	N
Female	1	lower division	113.20	6.458	5
		upper division	100.867	19.842	15
	2	lower division	93.000	13.743	8
		upper division	103.111	10.566	18
	3	lower division	110.000	9.849	3
		upper division	101.400	10.555	15
Male	1	lower division	101.500	13.435	2
		upper division	107.818	13.348	11
	2	lower division	82.333	8.737	3
		upper division	103.300	7.528	10
	3	lower division	100.000	.	1
		upper division	90.071	21.875	14

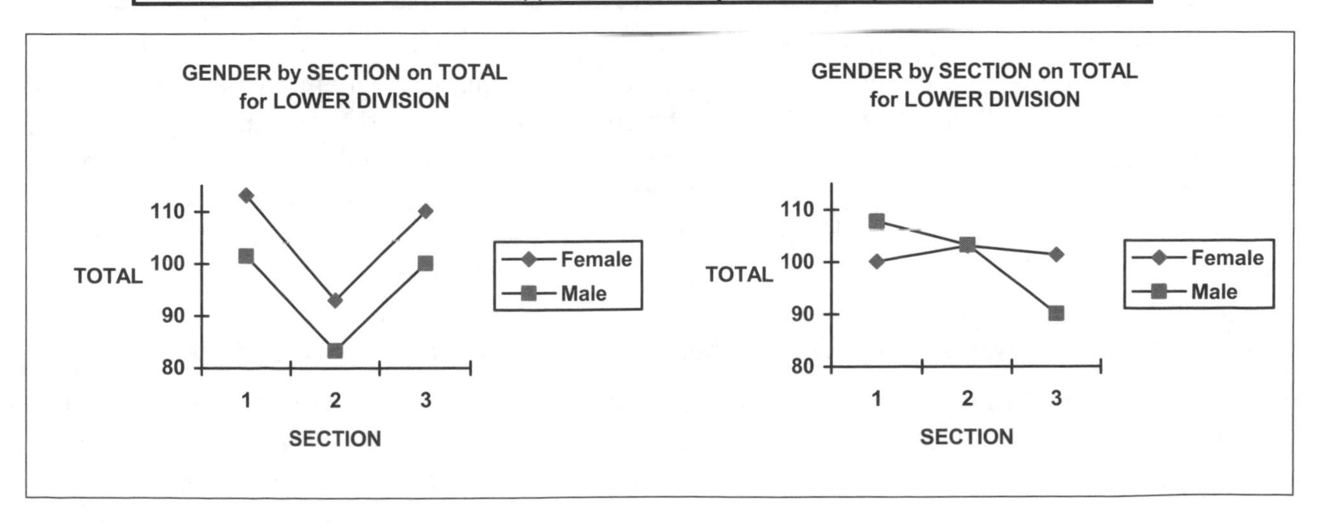

Source	Type III Sum of Squares	df	Mean Square	F	Sig.	Eta Squared
gender * section * lowup	171.572	2	85.786	.409	.665	.009

A THREE-WAY ANOVA THAT INCLUDES A COVARIATE

The following output is produced by sequence step 5 (page 166) and includes results both with and without a covariate. Values WITH a covariate are shown in **boldface**, *values WITHOUT a covariate are shown (italicized and in parentheses). Note that SPSS produces more significant digits than are presented in this table; this was done for space considerations. Commentary is on the following page.*

Source	Type III Sum of Squares	DF	Mean Square	F	Sig of F	Eta^2
Corrected Model	**8006.6** *(4846.0)*	**12** *(11)*	**667.2** *(440.5)*	**3.76** *(2.10)*	**.00** *(.03)*	**.33** *(.20)*
Intercept	**35588.6** *(494712.2)*	**1** *(1)*	**35588.6** *(494712.2)*	**200.4** *(2359.7)*	**.00** *(.00)*	**.69** *(.96)*
Covariate: **Gpa**	**3160.6** *(na)*	**1** *(na)*	**3160.6** *(na)*	**17.80** *(na)*	**.00** *(na)*	**.16** *(na)*
Main Effects						
Gender	**78.0** *(454.1)*	**1** *(1)*	**78.0** *(454.1)*	**0.44** *(2.17)*	**.51** *(.14)*	**.01** *(.02)*
Section	**896.7** *(1181.4)*	**2** *(2)*	**448.3** *(590.7)*	**2.53** *(2.82)*	**.09** *(.07)*	**.05** *(.06)*
Lowup	**53.6** *(14.5)*	**1** *(1)*	**53.6** *(14.5)*	**0.30** *(0.07)*	**.58** *(.79)*	**.00** *(.00)*
2-way Interactions						
gender * section	**38.1** *(118.2)*	**2** *(2)*	**19.1** *(59.1)*	**0.11** *(0.28)*	**.90** *(.76)*	**.00** *(.01)*
gender * lowup	**79.3** *(269.8)*	**1** *(1)*	**79.3** *(269.8)*	**0.45** *(1.29)*	**.51** *(.26)*	**.01** *(.01)*
section * lowup	**1510.0** *(1571.9)*	**2** *(2)*	**755.0** *(786.0)*	**4.25** *(3.75)*	**.02** *(.03)*	**.09** *(.08)*
3-way Interaction						
gender * section * lowup	**162.6** *(171.6)*	**2** *(2)*	**81.3** *(85.5)*	**0.46** *(0.41)*	**.63** *(.67)*	**.01** *(.01)*
Error	**16337.1** *(19497.7)*	**92** *(93)*	**177.6** *(209.7)*			
Total	**1086378.0** *(same)*	**105** *(105)*				
Corrected Total	**24343.7** *(same)*	**104** *(104)*				

In this final section of the output, we will show the ANOVA output in the form displayed by SPSS. The cell means and graphs (previous section) will be identical whether or not a covariate or covariates are included. What changes is the sum of squares and, correspondingly, the degrees of freedom (in some cases), the mean squares, *F* values, significance values, and eta squared. If the covariate has a substantial influence, the analysis will usually demonstrate lower mean square and *F* values for most of the main effects and interactions, and higher significance values. In the chart (previous page), the boldface material will duplicate the output (which includes **gpa** as a covariate) exactly as SPSS produces it. The results from the ANOVA table for the analysis *without* the covariate then follows, (*italicized and in parentheses*).

Some general observations concerning the output (previous page) are mentioned here, followed by the definition of terms. Note that the covariate **gpa** accounts for a substantial portion of the total variation in the dependent variable **total**. The eta squared value indicates the size of the effect; notice that **gpa** accounts for 16.2% of the variance in **total**. The *corrected model* variance (third line up from the bottom) is the sum of the sum of squares from all explained sources of variation—covariates, main effects, two-way interactions, and the three-way interaction.

Note also the comparisons of the components of the ANOVA table when comparing analyses with and without a covariate. Since the covariate "consumes" much of the variance, most of the *F* values are lower and corresponding *p* values (for main effects and interactions) are higher when the covariate is included. This is not strictly true, however. Notice that the main effect for **lowup** and the **section** × **lowup** interaction show higher *F* values and correspondingly lower *p* values when the covariate is included.

Term	Definition/Description
SUM OF SQUARES	This is the sum of squared deviations from the mean. In a broader sense, explained sum of squares represents the amount of variation in the dependent variable (**total** in this case) explained by each independent variable or interaction. Error sum of squares represents the portion of the variance *not* accounted for by the covariate(s), independent variables, or their interaction(s).
DEGREES OF FREEDOM	Covariates: 1 degree of freedom for each covariate. Main effects: Sum of the main effects degrees of freedom for each variable: 1 + 2 + 1 = 4. **gender**: (Levels of **gender** – 1): 2 – 1 = 1. **section**: (Levels of **section** – 1): 3 – 1 = 2. **lowup**: (Levels of **lowup** – 1): 2 – 1 = 1. Two-way interactions: Sum of degrees of freedom for the three interactions: 2 + 1 + 2 = 5. Three-way interactions: Degrees of freedom for the individual three-way interaction (2). Corrected Model: (DF covariates + DF main effects + DF of Interactions): 1 + 4 + 5 + 2 = 12. Error: (N – Explained DF – 1): 105 – 12 – 1 = 92. Total: (N – 1): 105 – 1 = 104.
MEAN SQUARE	Sums of squares divided by degrees of freedom for each category.
F	The mean square for each main effect or interaction divided by the residual mean square.
SIG.	Likelihood that each result could happen by chance.
ETA SQUARED	The effect-size measure. This indicates how much of the total variance is explained by each independent variable, the interaction, or (for the corrected model row) the effect of all main effects and interactions.

Graphing Interactions Using SPSS

The graph created here is identical to the first two-way interaction described in this chapter (and Chapter 13). This procedure works for any two-way interaction; if you want to produce three-way interactions, then you can simply produce several graphs for each level of a third independent variable. To select one level of an independent variable, use the **Select Cases** option (described in Chapter 4, pages 58-60).

To produce a graph of a two-way interaction, select the **Graphs** menu, then **Interactive**, followed by either the **Bar** option or the **Line** option. (Purists—and APA Style—say that you should use a bar graph for this kind of graph, because the X-axis is categorical instead of a scale. But tradition—and most statistics books—still use line graphs. Extra points for the first to figure out what your instructor likes.) Our example will produce a line chart, but there are only a few minor differences in creating a bar chart.

Once you have selected either a **Bar** or a **Line** graph, you will see Screen 5.1 (Chapter 5, page 71). This window allows you to specify what you want graphed. In this case, **total** will be on the Y-axis, **section** on the X-axis, and we will define **Styles** by **gender**.

To achieve a multiple line graph that represents an ANOVA interaction:

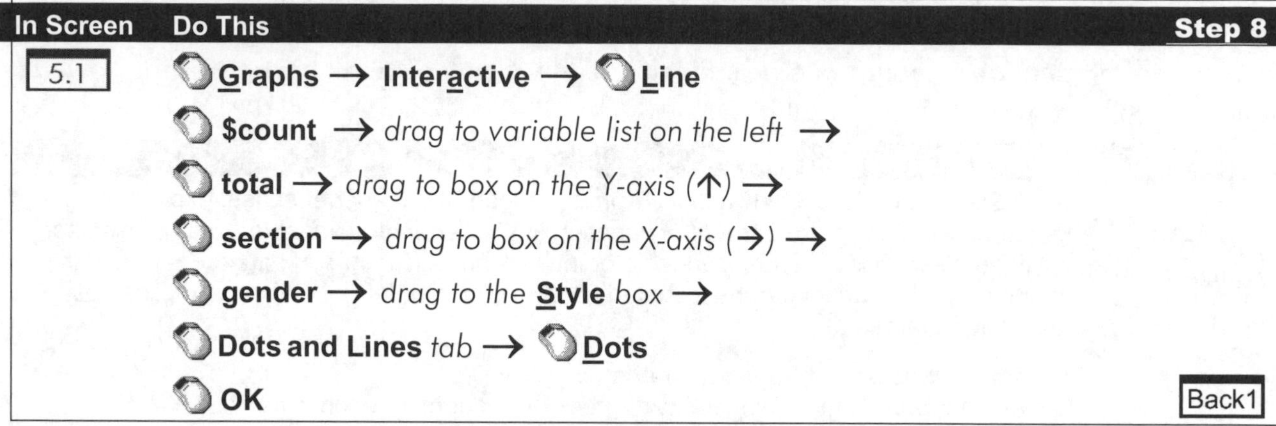

In Screen	Do This	Step 8
5.1	**Graphs** → **Interactive** → **Line**	
	$count → *drag to variable list on the left* →	
	total → *drag to box on the Y-axis (↑)* →	
	section → *drag to box on the X-axis (→)* →	
	gender → *drag to the* **Style** *box* →	
	Dots and Lines *tab* → **Dots**	
	OK	Back1

This procedure produces the following graph:

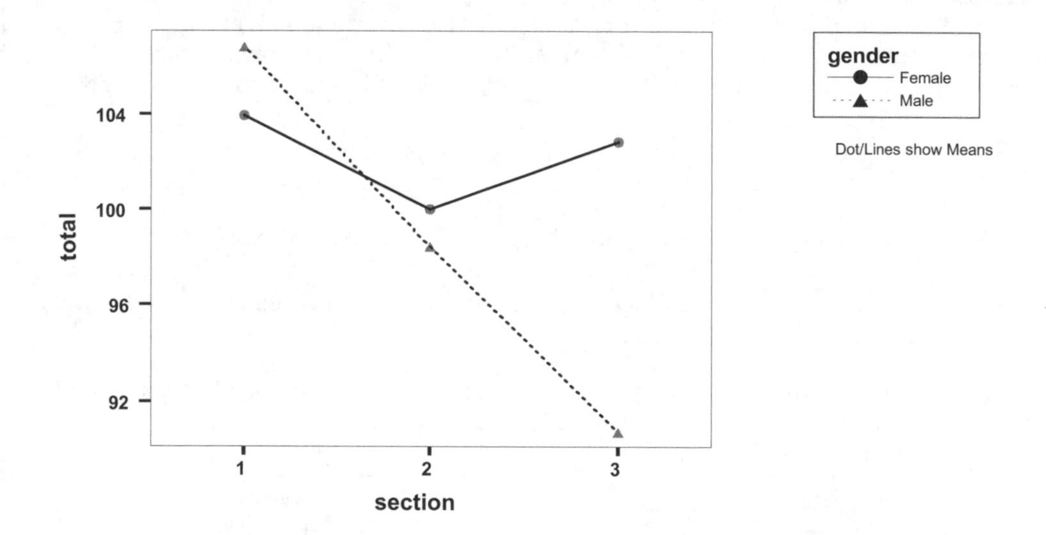

If you want a bar chart instead of a line graph, do this instead:

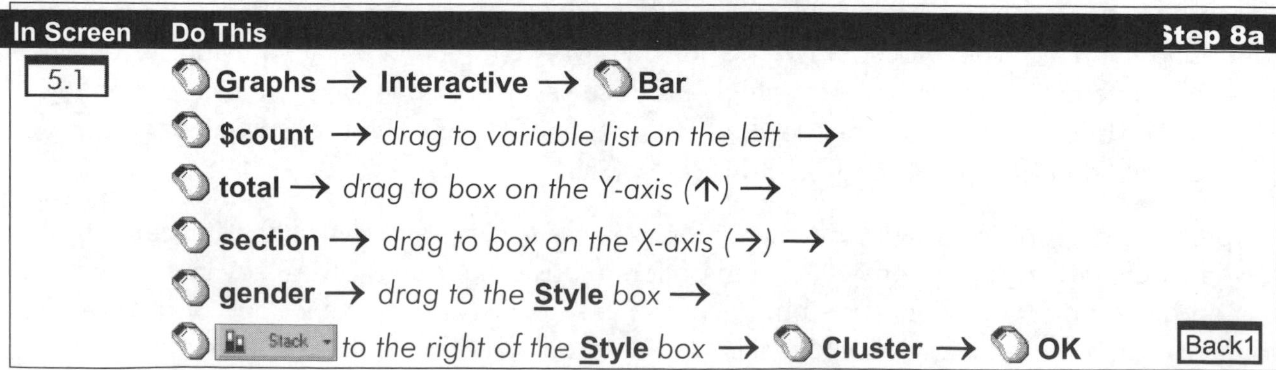

In Screen	Do This	Step 8a
5.1	**Graphs** → **Interactive** → **Bar**	
	$count → *drag to variable list on the left* →	
	total → *drag to box on the Y-axis (↑)* →	
	section → *drag to box on the X-axis (→)* →	
	gender → *drag to the* **Style** *box* →	
	Stack *to the right of the* **Style** *box* → **Cluster** → **OK**	Back1

This procedure produces the following graph. Note that the graph means the same thing as the line graph, but may be easier or more difficult to interpret, depending on your goals and what you've practiced before:

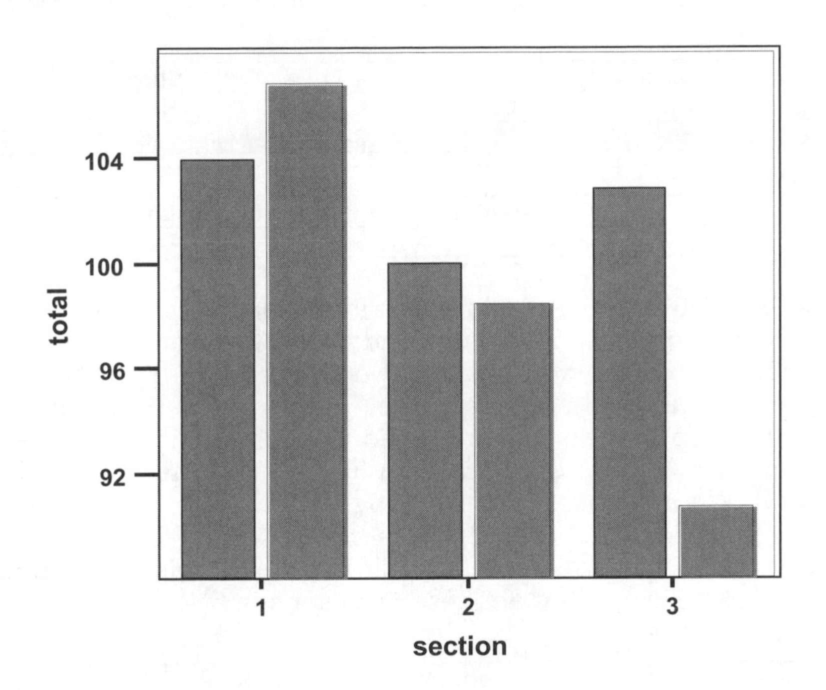

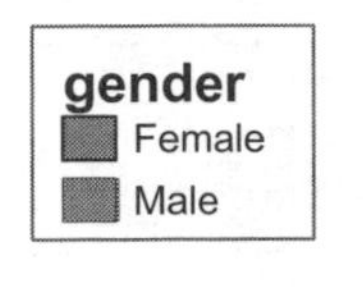

Bars show Means

EXERCISES

Answers to selected exercises are available for download at **www.ablongman.com/george6e**.

Notice that data files other than the **grades.sav** file are being used here. Please refer to the **Data Files** section starting on page 365 to acquire all necessary information about these files and the meaning of the variables. As a reminder, all data files are downloadable from the web address shown above.

For the first five problems below, perform the following:

- Print out the cell means portion of the output.
- Print out the ANOVA results (main effects, interactions, and so forth).
- Interpret and write up correctly (APA format) all main effects and interactions.
- Create multiple-line graphs (or clustered bar charts) for all significant interactions (see pp. 174-175).

1. File: **helping3.sav**; dependent variable: **tothelp**; independent variables: **gender**, **problem.**
2. File: **helping3.sav**; dependent variable: **tothelp**; independent variables: **gender**, **income.**
3. File: **helping3.sav**; dependent variable: **hseveret**; independent variables: **ethnic**, **problem.**
4. File: **helping3.sav**; dependent variable: **thelplnz**; independent variables: **gender**, **problem**; covariate: **tqualitz**.
5. File: **helping3.sav**; dependent variable: **thelplnz**; independent variables: **gender**, **income**, **marital.**

6. In an experiment, participants were given a test of mental performance in stressful situations. Some participants were given no stress-reduction training, some were given a short stress-reduction training session, and some were given a long stress-reduction training session. In addition, some participants who were tested had a low level of stress in their lives, and others had a high level of stress in their lives. Perform an ANOVA on these data (listed below). What do these results mean?

Training:	None										Short				
Life Stress:	High					Low					High				
Performance Score:	5	4	2	5	4	4	4	6	6	2	6	4	5	4	3

Training:	Short					Long									
Life Stress:	Low					High					Low				
Performance Score:	7	6	6	5	7	5	5	5	3	5	7	7	9	9	8

7. In an experiment, participants were given a test of mental performance in stressful situations. Some participants were given no stress-reduction training, and some were given a stress-reduction training session. In addition, some participants who were tested had a low level of stress in their lives, and others had a high level of stress in their lives. Finally, some participants were tested after a full night's sleep, and some were tested after an all-night study session on three-way ANOVA. Perform an ANOVA on these data (listed below question 8; ignore the "caffeine" column for now). What do these results mean?

8. In the experiment described in problem 7, data were also collected for caffeine levels. Perform an ANOVA on these data (listed below). What do these results mean? What is similar to and different than the results in question 1?

Training?	Stress Level	Sleep/Study	Performance	Caffeine
No	Low	Sleep	8	12
No	Low	Sleep	9	13
No	Low	Sleep	8	15
No	Low	Study	15	10
No	Low	Study	14	10
No	Low	Study	15	11
No	High	Sleep	10	14
No	High	Sleep	11	15
No	High	Sleep	11	16
No	High	Study	18	11
No	High	Study	19	10
No	High	Study	19	11
Yes	Low	Sleep	18	11
Yes	Low	Sleep	17	10
Yes	Low	Sleep	18	11
Yes	Low	Study	10	4
Yes	Low	Study	10	4
Yes	Low	Study	11	4
Yes	High	Sleep	22	14
Yes	High	Sleep	22	14
Yes	High	Sleep	23	14
Yes	High	Study	13	5
Yes	High	Study	13	5
Yes	High	Study	12	4

15

Simple Linear REGRESSION

THE **Regression** procedure is designed to perform either simple regression (Chapter 15) or multiple regression (Chapter 16). We split the command into two chapters largely for the sake of clarity. If the reader is unacquainted with *multiple* regression, this chapter, on simple regression, will serve as an introduction. Several things will be covered in the introductory portion of this chapter: (a) the concept of predicted values and the regression equation, (b) the relationship between bivariate correlation and simple regression, (c) the proportion of variance in one variable explained by another, and (d) a test for curvilinear relationships.

Several words of caution are appropriate here: First, a number of thick volumes have been written on regression analysis. We are in no way attempting in a few pages to duplicate those efforts. The introductions of these two chapters are designed primarily to give an overview and a conceptual feel for the regression procedure. Second, in this chapter (and Chapter 16), in addition to describing the standard linear relationships, we explain how to conduct regression that considers curvilinear tendencies in the data. We suggest that those less acquainted with regression should spend the time to thoroughly understand *linear* regression before attempting the much less frequently used tests for curvilinear trends.

PREDICTED VALUES AND THE REGRESSION EQUATION

There are many times when, given information about one characteristic of a particular phenomenon, we have some idea as to the nature of another characteristic. Consider the height and weight of adult humans. If we know that a person is 7 feet (214 cm) tall, we would suspect (with a fair degree of certainty) that this person probably weighs more than 200 pounds (91 kg). If a person is 4 feet 6 inches (137 cm) tall, we would suspect that such a person would weigh less than 100 pounds (45 kg). There is a wide variety of phenomena in which, given information about one variable, we have some clues about characteristics of another: IQ and academic success, oxygen uptake and ability to run a fast mile, percentage of fast-twitch muscle fibers and speed in a 100-meter race, type of automobile one owns and monetary net worth, average daily caloric intake and body weight, feelings of sympathy toward a needy person and likelihood of helping that person. Throughout a lifetime humans make thousands of such inferences (e.g., he's fat, he must eat a lot). Sometimes our inferences are correct, other times not. Simple regression is designed to help us come to more accurate inferences. It cannot guarantee that our inferences are correct, but it can determine the likelihood or *probability* that our inferences are sound; and given a value for one variable, it can predict the most likely value for the other variable based on available information.

To illustrate regression, we will introduce our example at this time. While it would be possible to use the **grades.sav** file to illustrate simple regression (e.g. the influence of previous GPA on final points), we have chosen to create a new data set that is able to demonstrate both linear regression and curvilinear regression. The new file is called **anxiety.sav** and consists of a data set in which 73 students are measured on their level of pre-exam anxiety on a *none* (1) to *severe* (10) scale, and then measured on a 100-point exam. The hypothesis for a linear relationship might be that those with very low anxiety will do poorly because they don't care much and that those with high anxiety will do better because they are motivated to spend more time in preparation. The dependent (criterion) variable is **exam**, and the independent (predictor) variable is **anxiety**. In other words we are attempting to predict the exam score from the anxiety score. Among other things that regression accomplishes, it is able to create a *regression equa-*

tion to calculate a person's *predicted* score on the exam based on his or her anxiety score. The regression equation follows the model of the general equation designed to predict a student's *true* or actual score. The equation for the student's true score follows:

$$\textbf{exam}_{(\text{true})} = \text{some } \textbf{constant} + \text{a coefficient} \times \textbf{anxiety} + \textbf{residual}$$

That is, the true **exam** score is equal to a **constant** plus some weighted number (coefficient) times the **anxiety** score plus the **residual**. The inclusion of the **residual** (also known as error) term is to acknowledge that *predicted* values in the social sciences are almost never exactly correct and that to acquire a *true* value requires the inclusion of a term that adjusts for the discrepancy between the predicted score and the actual score. This difference is called the *residual*. For instance, the equation based on our data set (with constant and coefficient generated by the regression procedure) follows:

$$\textbf{exam}_{(\text{true})} = 64.247 + 2.818(\textbf{anxiety}) + \textbf{residual}$$

To demonstrate the use of the equation, we will insert the anxiety value for student #24, who scored 6.5 on the anxiety scale.

$$\textbf{exam}_{(\text{true})} = 64.247 + 2.818(\textbf{6.5}) + \textbf{residual}$$

$$\textbf{exam}_{(\text{true})} = 82.56 + \textbf{residual}$$

The 82.56 is the student's *predicted* score based on his 6.5 anxiety score. We know that the actual exam score for subject 24 was 94. We can now determine the value of the **residual** (how far off our predicted value was), but we can do this only *after* we know the true value of the dependent variable (**exam** in this case). The residual is simply the true value minus the predicted value (94 – 82.56), or 11.44. The equation with all values inserted now looks like this:

94	=	82.56	+	11.44
true score	=	**predicted score**	+	**residual**

We have included a brief description of the residual term because you will see it so frequently in the study of statistics, but we now turn our attention to the issue of predicted values based on a calculated regression equation. A more extensive discussion of residuals takes place in Chapter 28. The regression equation for the predicted value of **exam** is:

$$\textbf{exam}_{(\text{predicted})} = 64.247 + 2.818(\textbf{anxiety})$$

To demonstrate computation, subjects 2, 43, and 72 scored 1.5, 7.0, and 9.0 anxiety points, respectively. Computation of the predicted scores for each of the three follows. Following the predicted value is the actual score achieved by the three subjects (in parentheses), to demonstrate how well (or poorly) the equation was able to predict the true scores:

Subject 2:	$\textbf{exam}_{(\text{predicted})} = 64.247 + 2.818(\textbf{1.5}) = \textbf{68.47}$	(actual score was 52)
Subject 43:	$\textbf{exam}_{(\text{predicted})} = 64.247 + 2.818(\textbf{7.0}) = \textbf{83.97}$	(actual score was 87)
Subject 72:	$\textbf{exam}_{(\text{predicted})} = 64.247 + 2.818(\textbf{9.0}) = \textbf{89.61}$	(actual score was 71)

We notice that for subject 2, the predicted value was much too high (68.47 vs. 52); for subject 43, the predicted value was quite close to the actual score (83.97 vs. 87); and for subject 72, the predicted value was also much too high (89.61 vs. 71). From this limited observation we might conclude that the ability of our equation to predict values is pretty good for midrange

anxiety scores, but much poorer at the extremes. Or, we may conclude that there are factors other than a measure of the subject's pre-exam anxiety that influence his or her exam score. The issue of *several* factors influencing a variable of interest is called *multiple* regression and will be addressed in the next chapter.

SIMPLE REGRESSION AND THE AMOUNT OF VARIANCE EXPLAINED

We are not left at the mercy of our intuition to determine whether or not a regression equation is able to do a good job of predicting scores. The output generated by the **Regression** command calculates four different values that are of particular interest to the researcher:

1. SPSS generates a score that measures the strength of relationship between the dependent variable (**exam**) and the independent variable (**anxiety**). This score is designated with a capital *R* and is simply our old friend, the bivariate correlation (*r*). The capital *R* is used (rather than a lower case *r*) because the **Regression** command is usually used to compute *multiple* correlations (that is, the strength of relationship between several independent variables and a single dependent variable). For a description of correlation, please refer to Chapter 10.
2. Along with the computation of *R*, SPSS prints out a probability value (*p*) associated with *R* to indicate the significance of that association. Once again, a $p < .05$ is generally interpreted as indicating a statistically significant correlation. If $p > .05$, the strength of association between the two variables is usually not considered statistically significant; or the relationship between the two constructs is considered weak or nonexistent.
3. *R* square (or R^2) is simply the square of *R*, but it has special significance. The R^2 value is the proportion of variance in one variable accounted for (or explained) by the other variable. For the relationship between **anxiety** and **exam**, SPSS calculated values of $R = .48794$ ($p < .0001$) and $R^2 = .23808$. The *R* square value indicates that 23.8% of the variance in the **exam** score is accounted for by pretest **anxiety**. The standard disclaimers must be inserted here: With a correlation, be cautious about inferring causation. In this case the *direction* of causation is safe to assume because an exam score cannot influence *pre*-exam anxiety.
4. SPSS calculates the constant and the coefficient (called *B*-values) for the regression equation. As already noted, the constant and coefficient computed for the regression equation identifying the relationship between **anxiety** and **exam** were 64.247 and 2.818, respectively.

TESTING FOR A CURVILINEAR RELATIONSHIP

Most knowledgeable people would consider it foolishness to think that higher pretest anxiety will produce higher scores on an exam. A widely held position is that very low level anxiety will result in poor scores (due to lack of motivation) and that as anxiety scores increase, motivation to do well increases and higher scores result. However, there comes a point when additional anxiety is detrimental to performance, and at the top end of the anxiety scale there would once again be a decrease of performance. Regression analysis (whether it be simple or multiple) measures a *linear* relationship between the independent variable(s) and the dependent variable. In the fictional data set presented earlier there is a fairly strong linear relationship between **anxiety** and the **exam** score ($R = .484$, $p < .0001$). But perhaps the regression equation would generate more accurate predicted values (yielding a better "fit" of the data) if a quadratic equation were employed that included an **anxiety**-squared (**anxiety**^{2}) term.

Usually, before one tests for a curvilinear trend, there needs to be evidence (theoretical or computational) that such a relationship exists. Frankly, in the social sciences, curvilinear trends happen in only a limited number of circumstances, but they can be critical to understanding the data when they do occur. To demonstrate, we produce the scatterplot between exam and anxiety.

The graph (below) shows the **exam** scores on the vertical axis (the scale ranges from 40 to 110), and **anxiety** on the horizontal axis with a range of 0 to 10. Initial visual inspection reveals what appears to be a curvilinear trend. For mid-range anxiety values the test scores are highest, and at the extremes they tend to be lower. One needs to be cautioned when attempting to read a scattergram. A relationship may appear to exist, but when tested is not statistically significant. More frequently, visual inspection alone does not reveal a curvilinear trend but a statistical test does. When exploring the relationship between two variables, **Regression** is able to reveal whether there is a significant linear trend, a significant curvilinear trend, significant linear *and* curvilinear trends, or neither.

15.1

Sample Scatter plot demonstrating a curvilinear trend

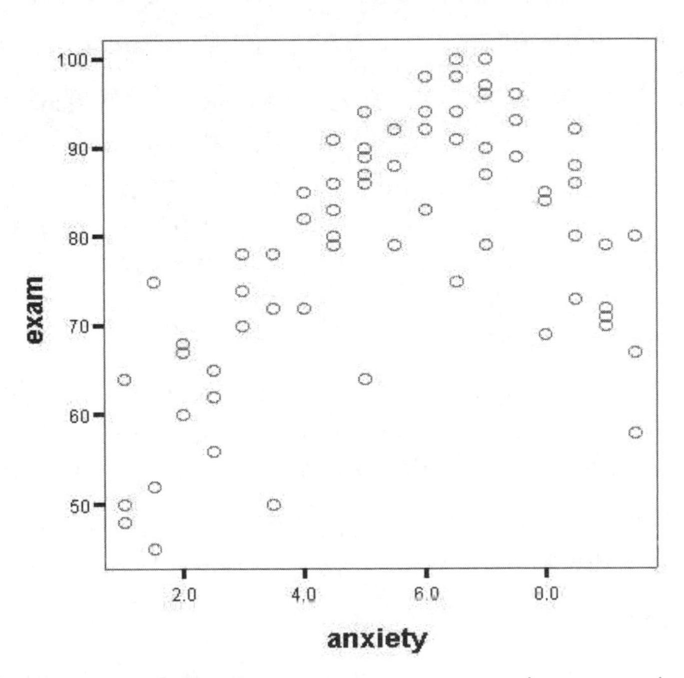

A simple procedure offered by SPSS within the context of the **Regression** command is a quick test to check for linear or curvilinear trends. You identify the dependent variable (**exam**), the independent variable (**anxiety**), then from the resulting dialog box, select **Linear** and **Quadratic.** An **OK** click will produce a two-line output (shown on the following page) that indicates if linear and/or curvilinear trends exist. The B-values are also included so that it is possible to write predicted-value equations for either linear or curvilinear relationships. This process also creates a chart showing the scattergram, the linear regression line (the straight one) and the curvilinear regression line (the curved one). Notice the similarity between the two charts in Screens 15.1 and 15.2. Also note that the constant and coefficients in the equations (following page) utilize values from the two lines of output.

15.2

Sample Scatter plot demonstrating a linear and a curvilinear trend

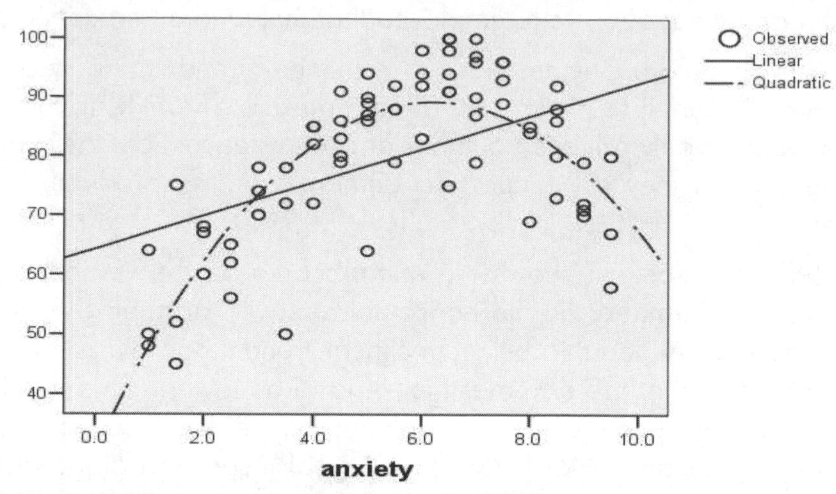

Dependent	Method	R-square	d.f.	F	Sig of F	b0	b1	b2
EXAM	linear	.238	71	22.19	.000	64.247	2.8178	
EXAM	quadratic	.641	70	62.52	.000	30.377	18.9256	-1.5212

The linear and curvilinear regression equations now follow:

Linear equation (the straight line): $\textbf{exam}_{(pred)} = 64.25 + 2.82(\textbf{anxiety})$

Quadratic equation (the curved line): $\textbf{exam}_{(pred)} = 30.38 + 18.93(\textbf{anxiety}) + -1.52(\textbf{anxiety})^2$

Observe that in the output (above), the R^2 value for the linear equation indicates that **anxiety** explains 23.8% of the **exam** performance, while the R^2 value for the quadratic equation (where both the linear and the curvilinear trend influences the outcome) indicates that 64.1% of the variance in **exam** is explained by **anxiety** and the square of **anxiety**. Under **Sig of F**, the .000 for both the linear and the curvilinear equation indicate that both trends are statistically significant ($p < .001$).

We would like to see if the quadratic equation is more successful at predicting actual scores than was the linear equation. To do so we substitute the **anxiety** values for the same subjects (numbers 2, 43, and 72) used to illustrate the linear equation:

Subject 2: $\textbf{exam}_{(pred)} = 30.38 + 18.93(\mathbf{1.5}) + -1.52(\mathbf{1.5})^2 = \mathbf{55.31}$ (actual score, 52)
Subject 43: $\textbf{exam}_{(pred)} = 30.38 + 18.93(\mathbf{7.0}) + -1.52(\mathbf{7.0})^2 = \mathbf{88.30}$ (actual score, 87)
Subject 72: $\textbf{exam}_{(pred)} = 30.38 + 18.93(\mathbf{9.0}) + -1.52(\mathbf{9.0})^2 = \mathbf{77.49}$ (actual score, 71)

A quick check of the results from the linear equation demonstrates the substantially superior predictive ability of the quadratic equation. Note the table:

Subject number	Actual score	Predicted linear score	Predicted quadratic score
2	52	68.47	55.31
43	87	83.97	88.30
72	71	89.61	77.49

A number of books are available that cover both simple, curvilinear, and multiple regression. Several that the authors feel are especially good include: Chatterjee and Price (1999); Gonick and Smith (1993); Schulman (1998); Sen and Srivastava (1997); and Weisberg (1985). Please see the reference section for more detailed information on these resources.

STEP BY STEP

Simple Linear and Curvilinear Regression

To enter SPSS, a click on **Start** *in the taskbar (bottom of screen) activates the start menu:*

After clicking the SPSS program icon, Screen 1 (inside front cover) appears on the monitor.

Step 2
Create and name a data file or edit (if necessary) an already existing file (see Chapter 3).

Screens 1 and 2 (inside front cover) allow you to access the data file used in conducting the analysis of interest. The following sequence accesses the **anxiety.sav** *file for further analyses:*

Whether first entering SPSS or returning from earlier operations the standard menu of commands across the top is required (shown below). As long as it is visible you may perform any analyses. It is not necessary for the data window to be visible.

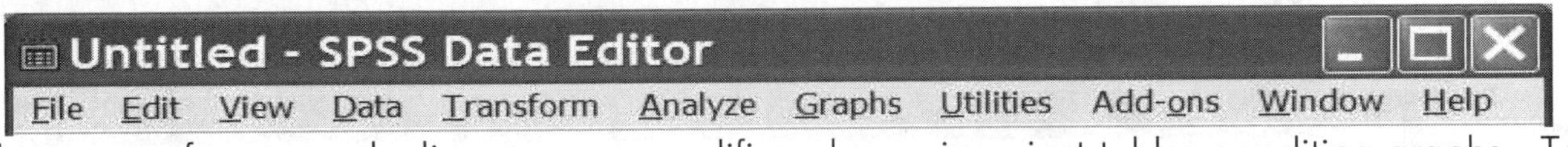

This menu of commands disappears or modifies when using pivot tables or editing graphs. To uncover the standard menu of commands simply click on the or the icon.

After completion of Step 3, a screen with the desired menu bar appears. When you click a command (from the menu bar), a series of options will appear (usually) below the selected command. With each new set of options, click the desired item. The sequence to access linear regression begins at any screen with the menu of commands visible:

At this point, a new dialog box opens (Screen 15.3, below) that allows you to conduct regression analysis. Because this box is much more frequently used to conduct *multiple* regression analysis than simple linear regression, there are many options available that we will not discuss until next chapter. A warning to parents with young children, however: Under no circumstances allow a child less than 13 years of age to click on the **Statistics** or **Plots** pushbuttons. Windows open with options so terrifying that some have never recovered from the trauma. The list to the left will initially contain only two variables, **anxiety** and **exam**. The procedure is to select **exam** and paste it into the **Dependent** box, select **anxiety** and paste it into the **Independent(s)** box, then click on the **OK** button. The program will then run yielding R and R^2 values, F values and tests of significance, the B values that provide constants and coefficients for the regression equation, and Beta (β) values to show the strength of association between the two variables. Some of the terms may be unfamiliar to you and will be explained in the output section. The step-by-step sequence follows Screen 15.3.

15.3

The Initial Linear Regression Window

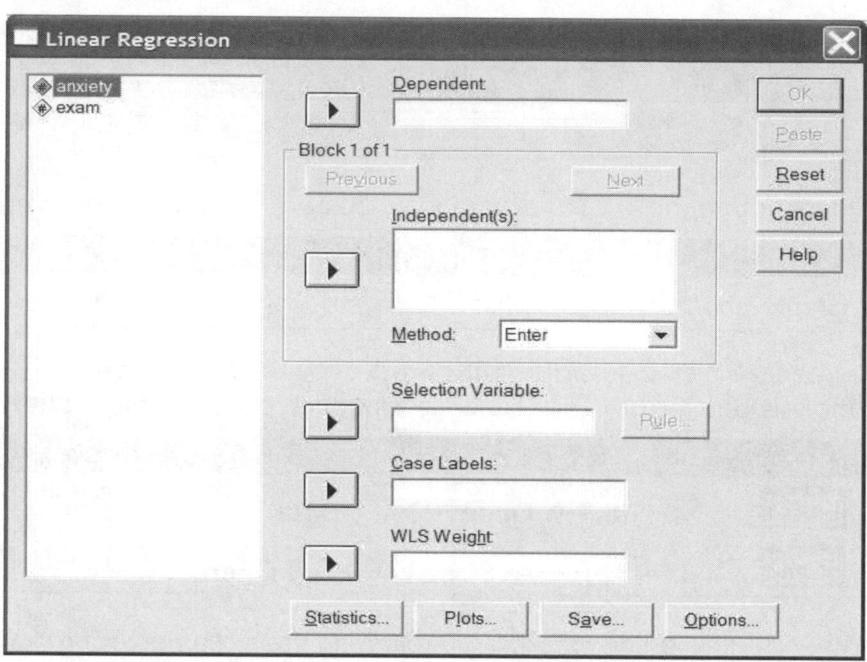

To conduct simple linear regression with a dependent variable of **exam** *and an independent variable of* **anxiety** *perform the following sequence of steps. Begin with Screen 15.3; results from this analysis are in the Output section.*

Curvilinear Trends

We will show two sequences that deal with curvilinear trends. The first deals with the simple two-line output reproduced in the introduction, and creation of the graph that displays linear and curvilinear trends (also displayed in the introduction). The second considers the more formal procedure of creating a quadratic variable that you may use in a number of different analyses. While much of what takes place in the second procedure could be produced by the first, we feel it is important that you understand what is really taking place when you test for a curvilinear trend. Furthermore, although we present curvilinear trends in the context of simple regression, the same principles (and access procedures) apply to the next chapter on multiple regression.

To access the **Curve Estimation** *chart requires a different step-4 sequence.*

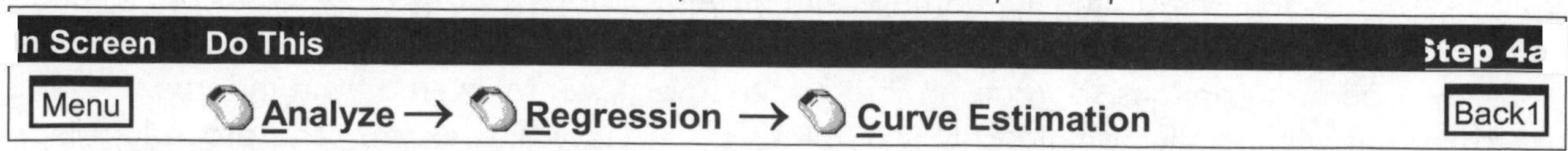

A new dialog box now opens (Screen 15.4, below) that allow a number of options for curve estimation. The early procedure is identical to that shown in the previous sequence of steps (sequence step 5): Select **exam** and paste it into the **Dependent** box, select **anxiety** and paste it into the **Independent** box. After the dependent and independent variables are selected, notice that three of the options in this dialog box are already selected by default: **Include constant in equation** provides the constant, necessary to create a regression equation. The **Plot models** option creates the graph that was illustrated in the introduction. The **Linear** model refers to testing for and then including as a line on the graph any linear trend in your data.

15.4

The Curve Estimation Window

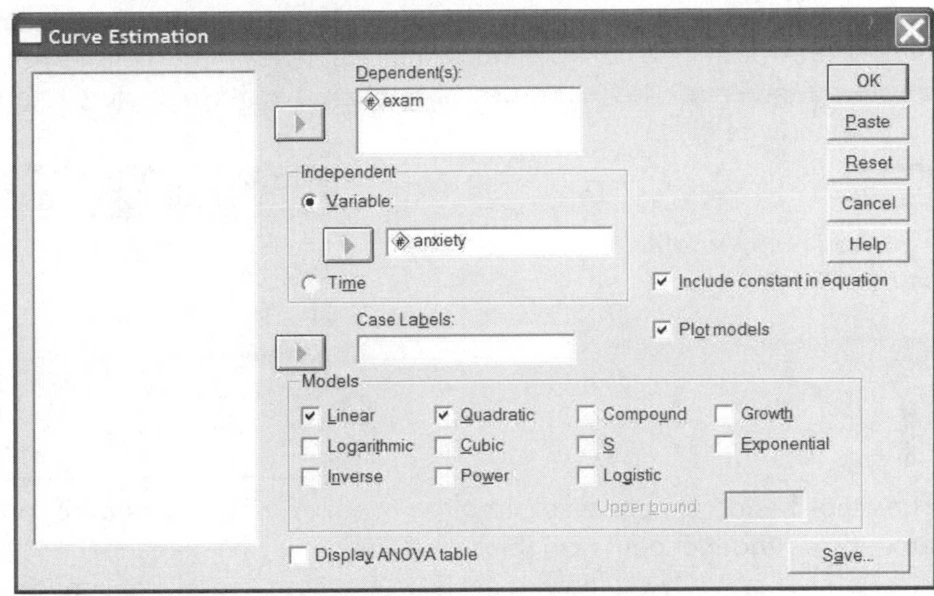

Within the **Models** box a variety of toys exist for the mathematical or statistical wizards. All of the curvilinear models included here have their unique applications, but none are used frequently, and only the **Quadratic** option will be addressed here. For the present analysis, keep the already-selected **Linear** option, and also select **Quadratic**. The steps now follow, beginning with Screen 15.4.

Beginning with a dependent variable of **exam** *and checking for linear and curvilinear trends in the independent variable (***anxiety***), perform the following sequence of steps. Perform the step 4a sequence (above) to arrive at Screen 15.4 results of this analysis are in the Introduction.*

In Screen	Do This	Step 5a
15.4	**exam** → *upper* ▸ → **anxiety** → *middle* ▸ → **Quadratic** → **OK**	Back1

What actually takes place when you formulate the regression equation that includes the influence of a quadratic term, is the creation of a new variable that is the square of the independent variable (**anxiety**), or **anxiety**^{2}. You all vividly remember from your Algebra I days that the inclusion of a variable squared produces a parabola, that is, a curve with a single change of direction that opens either upward (if the coefficient is positive) or downward (if the coefficient is negative). Notice in the equation on page 182 that the coefficient of the squared term is negative [-1.52(**anxiety**)^2], and that the parabola opens downward. Also note the influence of the linear trend (a tendency of data points to go from lower left to upper right) is also reflected in the curve. The left end of the curve is lower than the right end of the curve. Thus both linear and curvilinear (quadratic) trends are reflected in the graph. For the other models available in screen 15.4—exponential, cubic, etc.—pull out your high school algebra homework plots of these functions and see if your scatterplots look similar.

To go beyond this simple test for both linear and curvilinear trends it is necessary to create a new variable, the square of anxiety, assigned a variable name of **anxiety2**. Then when we begin the regression procedure two variables will be included as independent variables, **anxiety** and **anxiety2**. The method of creating new variables is described in Chapter 4, and we will step-by-step you through the sequence but will not reproduce the screen here (Screen 4.3 is on page 51 if you wish a visual reference). We do access one window that has not been displayed previously, that is the window that appears when you click on the **Type & Label** button displayed in Screen 4.3. The use of the box is self-explanatory and is reproduced (below) for visual reference only.

15.5

The Compute Variable: Type and Label Window

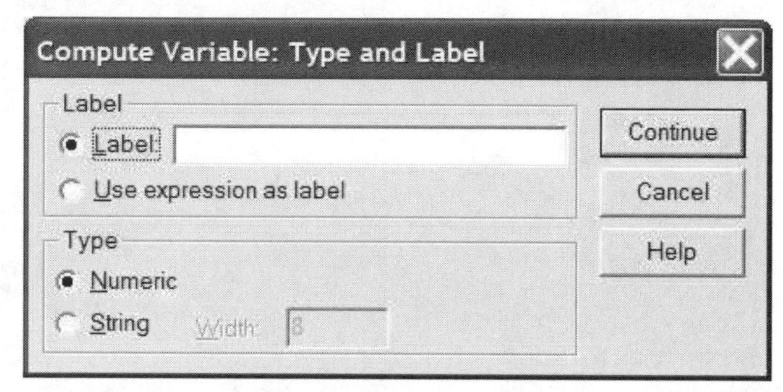

The step-by-step sequence of creating a new variable, **anxiety2**, and including it along with **anxiety** as independent variables that influence the dependent variable **exam** now follows. This is now actually *multiple* regression because more than one independent variable is included.

To run the regression procedure with a dependent variable of **exam** *and independent variables of* **anxiety** *and* (**anxiety**)^2 *perform the following sequence of steps. Begin at the Data Editor screen with the menu of commands showing. The procedure begins by creating the new variable* **anxiety2**. *Results from this analysis are included in the Output section.*

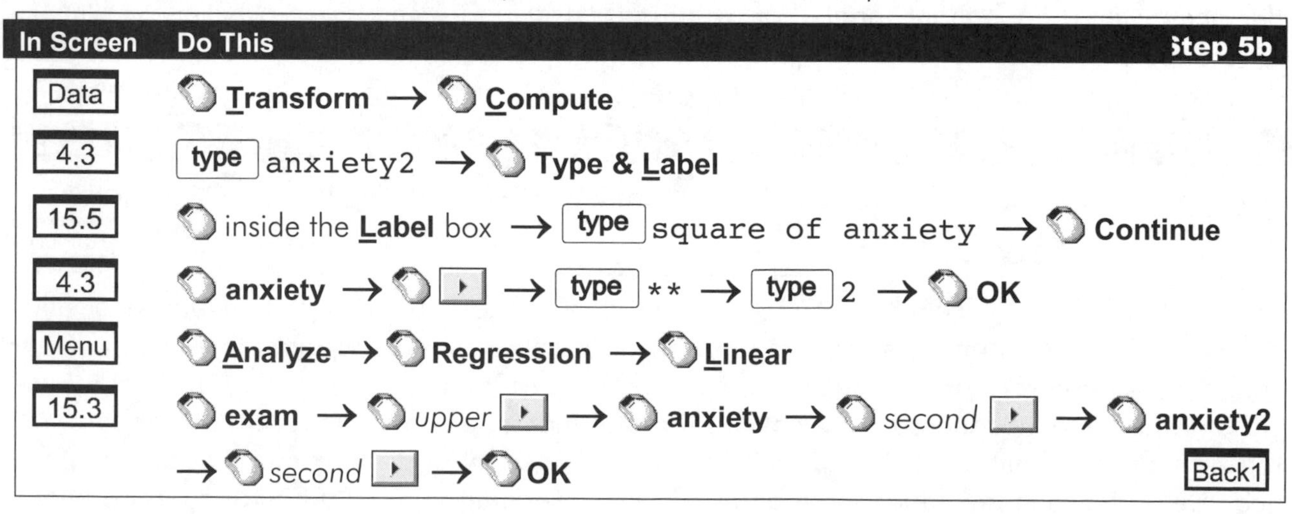

In Screen	Do This	Step 5b
Data	**Transform** → **Compute**	
4.3	type anxiety2 → **Type & Label**	
15.5	inside the **Label** box → type square of anxiety → **Continue**	
4.3	**anxiety** → ▸ → type ** → type 2 → **OK**	
Menu	**Analyze** → **Regression** → **Linear**	
15.3	**exam** → *upper* ▸ → **anxiety** → *second* ▸ → **anxiety2** → *second* ▸ → **OK**	Back1

Upon completion of step 5, 5a, or 5b, the output screen will appear (Screen 1, inside back cover). All results from the just-completed analysis are included in the Output Navigator. Make use of the scroll bar arrows (▲▼▶◀) to view the results. Even when viewing output, the standard menu of commands is still listed across the top of the window. Further analyses may be conducted without returning to the data screen.

PRINTING RESULTS

Results of the analysis (or analyses) that have just been conducted requires a window that displays the standard commands (**File Edit Data Transform Analyze** . . .) across the top. A typical print procedure is shown below beginning with the standard output screen (Screen 1, inside back cover).

To print results, from the Output screen perform the following sequence of steps:

In Screen	Do This	Step 6
Back1	*Select desired output or edit (see pages 19 - 25)* → **File** → **Print**	
2.9	*Consider and select desired print options offered, then* → **OK**	

To exit you may begin from any screen that shows the **File** *command at the top.*

In Screen	Do This	Step 7
Menu	**File** → **Exit**	2.1

Note: After clicking **Exit**, there will frequently be small windows that appear asking if you wish to save or change anything. Simply click each appropriate response.

OUTPUT

Simple Linear and Curvilinear Regression Analysis

What follows is output from sequence step 5 on page 184.

Variables Entered/Removed [b]

Model	Variables Entered	Variables Removed	Method
1	anxiety[a]	.	Enter

Model Summary

Model	R	R Square	Adjusted R Square	Std. Error of the Estimate
1	.488[a]	.238	.227	12.215

ANOVA

Source of Variation	df	Sum of Squares	Mean Square	F	Signif F
Regression	1	3310.476	3310.476	22.186	.000
Residual	71	10594.209	149.214		

Coefficients

	Unstandardized coefficients		Std Coefficients		
	B	Std. Error	Beta	t	Signif of t
(Constant)	64.247	3.602		17.834	.000
ANXIETY	2.818	.598	.488	4.710	.000

This output shows that there is a significant linear relationship between exam performance and anxiety such that a higher level of anxiety results in higher scores. The specific meaning of this output is summarized in the definitions of terms that follow.

Term	Definition/Description
R	Since there is only one independent variable, this number is the bivariate correlation (*r*) between **exam** and **anxiety**.
R SQUARE	The **R SQUARE** value identifies the proportion of variance in **exam** accounted for by **ANXIETY**. In this case 23.8% of the variance in **exam** is explained by **anxiety**.

ADJUSTED R SQUARE	**R SQUARE** is an accurate value for the sample drawn but is considered an optimistic estimate for the population value. The **ADJUSTED R SQUARE** is considered a better population estimate.
STANDARD ERROR	The standard deviation of the expected values for the dependent variable, **exam**.
REGRESSION	Statistics relating to the explained portion of the variance.
RESIDUAL	Statistics relating to the *un*explained portion of the variance.
DF	Degrees of freedom: For regression, the number of independent variables (1 in this case). For the residual, the number of subjects (73) minus the number of independent variables (1), minus 1: (73 – 1 – 1 = 71).
SUM OF SQUARES	For regression this is the between-groups sum of squares; for residual, the within-groups sum of squares. Note that in this case there is a larger portion of unexplained variance than there is of explained variance, a reality also reflected in the R^2 value.
MEAN SQUARE	Sum of squares divided by degrees of freedom.
F	Mean square regression divided by mean square residual.
SIGNIF F	Likelihood that this result could occur by chance.
B	Coefficient and constant for the linear regression equation: **exam$_{(pred)}$ = 64.247 + 2.818(anxiety)**.
STD ERROR	Standard error of *B*: A measure of the stability or sampling error of the *B*-values. It is the standard deviation of the *B*-values given a large number of samples drawn from the same population.
BETA	The standardized regression coefficients. This is the *B*-value for standardized scores (z-scores) of the variable **anxiety**. This value will always vary between ± 1.0 in linear relationships. For curvilinear relationships it will sometimes extend outside that range.
t	*B* divided by the standard error of *B*.
SIGNIF t	Likelihood that this result could occur by chance.

A REGRESSION ANALYSIS THAT TESTS FOR A CURVILINEAR TREND

What follows is output from sequence step 5b on page 186.

Variables Entered/Removed [b]

Model	Variables Entered	Variables Removed	Method
1	anxiety2, anxiety [a]	.	Enter

Model Summary

Model	R	R Square	Adjusted R Square	Std. Error of the Estimate
1	.801[a]	.641	.631	8.443

ANOVA

Source of Variation	df	Sum of Squares	Mean Square	F	Signif F
Regression	2	8194.538	4457.269	62.525	.0000
Residual	70	4990.146	71.288		

Coefficients

	Unstandardized coefficients		Std Coefficients		
	B	Std. Error	Beta	t	Signif of t
(Constant)	30.377	4.560		6.662	.000
ANXIETY	18.926	1.836	3.277	10.158	.000
ANXIETY2	-1.521	.172	-2.861	-8.866	.000

The terms described previously also apply to *this* output. Notice the high multiple *R*-value (.80), indicating a very strong relationship between the independent variable (**anxiety**) and its square (**anxiety2**) and the dependent variable, **exam**. An R^2 value of .641 indicates that 64.1% of the variance in **exam** is accounted for by **anxiety** and its square. The *F* and associated *p*-values (Signif F, Sig T) reflect the strength of the *overall* relationship between both independent variables and **exam** (F, and Signif F), and between *each individual* independent variable and **exam** (t and Signif t). In Chapter 16 we will address in some detail the influence of several variables (there are two here) on a dependent variable. Also note that while the betas always vary between ± 1 for simple linear equations, they extend beyond that range in this quadratic equation. Beneath the *B* in the lower table are the coefficients for the regression equation. Notice that the regression equation we tested (presented in the Introduction) had a constant of 30.377, an **anxiety** coefficient of 18.926, and an (**anxiety**)^2 coefficient of −1.521.

EXERCISES

Answers to selected exercises are available for download at **www.ablongman.com/george6e**.

1. Use the **anxiety.sav** file for exercises that follow (downloadable at the address above).

Perform the 4a - 5a sequences on pages 184 and 185.

- Include output in as compact a form as is reasonable
- Write the linear equation for the predicted exam score
- Write the quadratic equation for the predicted exam score

For subjects numbered 5, 13, 42, and 45

- Substitute values into the two equations and solve. Show work on a separate page.
- Then compare in a small table (shown below and similar to that on page 182)
  - → Linear equation results,
  - → Quadratic equation results, and
  - → Actual scores for sake of comparison.

subject #	anxiety score	predicted linear score	predicted quadratic score	actual score
5				
13				
42				
45				

2. Now using the **divorce.sav** file, test for linear and curvilinear relations between:

- physical closeness (**close**) and life satisfaction (**lsatisfy**)
- attributional style (**ASQ**) and life satisfaction (**lsatisfy**)

Print graphs and write linear and quadratic equations for both.

3. Now perform step 5b (p. 186) for the relationship between exam score, anxiety and anxiety squared (from the **anxiety.sav** file) and similar procedures for the two relationships shown in problem 2 (from the **divorce.sav** file).

   For each of the three analyses:

   - Box the Multiple R,
   - Circle the R Square,
   - Underline the two (2) B values, and
   - Double underline the two (2) Sig of T values.

   In a single sentence (just once, not for each of the 3 problems) identify the meaning of each of the four (4) bulleted items above.

4. A researcher is examining the relationship between stress levels and performance on a test of cognitive performance. She hypothesizes that stress levels lead to an increase in performance to a point, and then increased stress decreases performance. She tests ten participants, who have the following levels of stress: 10.94, 12.76, 7.62, 8.17, 7.83, 12.22, 9.23, 11.17, 11.88, and 8.18. When she tests their levels of mental performance, she finds the following cognitive performance scores (listed in the same participant order as above): 5.24, 4.64, 4.68, 5.04, 4.17, 6.20, 4.54, 6.55, 5.79, and 3.17. Perform a linear regression to examine the relationship between these variables. What do these results mean?

5. The same researcher tests ten more participants, who have the following levels of stress: 16, 20, 14, 21, 23, 19, 14, 20, 17, and 10. Their cognitive performance scores are (listed in the same participant order): 5.24, 4.64, 4.68, 5.04, 4.17, 6.20, 4.54, 6.55, 5.79, and 3.17. (Note that in an amazing coincidence, these participants have the same cognitive performance scores as the participants in question 4; this coincidence may save you some typing.) Perform a linear regression to examine the relationship between these variables. What do these results mean?

6. Create a scatterplot (see Chapter 5) of the variables in question 5. How do results suggest that linear regression might not be the best analysis to perform?

7. Perform curve estimation on the data from Question 5. What does this tell you about the data that you could not determine from the analysis in Question 5?

8. What is different about the data in Questions 4 and 5 that leads to different results?

16

MULTIPLE REGRESSION ANALYSIS

MULTIPLE REGRESSION is the natural extension of simple linear regression presented in Chapter 15. In simple regression, we measured the amount of influence one variable (the independent or predictor variable) had on a second variable (the dependent or criterion variable). We also computed the constant and coefficient for a *regression equation* designed to predict the values of the dependent variable, based on the values of the independent variable. While simple regression shows the influence of *one* variable on another, multiple regression analysis shows the influence of *two or more* variables on a designated dependent variable.

Another way to consider regression analysis (simple or multiple) is from the viewpoint of a slope-intercept form of an equation. When a simple correlation between two variables is computed, the intercept and slope of the *regression line* (or line of best fit) may be requested. This line is based on the regression equation mentioned above with the y-intercept determined by the constant value and the slope determined by the coefficient of the independent variable. We describe here a simple regression equation as a vehicle for introducing multiple regression analysis. To assist in this process we present a new example based on a file called **helping1.sav**. This data file is related to a study of helping behavior; it is real data derived from a sample of 81 subjects.

THE REGRESSION EQUATION

In this chapter we introduce the **helping1.sav** data mentioned above. Two of the variables used to illustrate simple regression are **zhelp** (a measure of total amount of time spent helping a friend with a problem, measured in *z* scores) and **sympathy** (the amount of sympathy felt by the helper in response to the friend's problem, measured on a *little* [1] to *much* [7] scale). Although correlation is often bidirectional, in this case **zhelp** is designated as the dependent variable (that is, one's sympathy influences how much help is given *rather than* the amount of help influencing one's sympathy). A significant correlation ($r = .46$, $p < .0001$) was computed, demonstrating a substantial relationship between the amount of sympathy one feels and the amount of help given. In addition, the intercept value (–1.892) and the slope (.498) were also calculated for a regression equation showing the relationship between the two variables. From these numbers we can create the formula to determine the *predicted value* of **zhelp**:

$$\textbf{zhelp}_{(\text{predicted})} = -1.892 + .498(\textbf{sympathy})$$

If a person measured 5.6 on the **sympathy** scale, the predicted value for that person for **zhelp** would be .897. In other words, if a person measured fairly high in sympathy (5.6 in this case), it is anticipated that he or she would give quite a lot of help (a *z* score of .897 indicates almost one standard deviation more than the average amount of help). The sequence below illustrates the computation of this number:

$$\textbf{zhelp}_{(\text{pred})} = -1.892 + .498(\textbf{5.6}) \longrightarrow \textbf{zhelp}_{(\text{pred})} = -1.892 + 2.789 \longrightarrow \textbf{zhelp}_{(\text{pred})} = \textbf{.897}$$

Multiple regression analysis is similar but allows *more than* one independent variable to have an influence on the dependent variable.

In this example, two other measures were also significantly correlated with **zhelp**: **anger** (angry or irritated emotions felt by the helper toward the needy friend, on a *none* (1) to *much* (7) scale), and **efficacy** (self-efficacy, or the helper's belief that he or she had the resources to be of help to the friend, on a *little* (1) to *much* (7) scale). The multiple regression analysis generated the following *B* values (The *B* can be roughly translated as the *slope* or *weighted constant* for the variable of interest.): $B_{(sympathy)}$ = .4941, $B_{(anger)}$ = .2836, and $B_{(efficacy)}$ = .4125, and the constant (intercept) = –4.3078. From these numbers a new equation may be generated to determine the predicted value for **zhelp**:

$$\textbf{zhelp}_{(predicted)} = -4.3078 + .4941(\textbf{sympathy}) + .2836(\textbf{anger}) + .4125(\textbf{efficacy})$$

Inserting numbers from an actual subject, subject 9 in this case:

$$\textbf{zhelp}_{(predicted)} = -4.3078 + .4941(\textbf{3.5}) + .2836(\textbf{1.0}) + .4125(\textbf{2.9}) \quad = \quad \textbf{-1.09}$$

This result suggests that this person who measured midrange on sympathy (3.5), low in anger (1.0), and low in self-efficacy (2.9) would be expected to give little help (a z score of –1.09 is more than one standard deviation below average). Subject 9's *actual* **zhelp** score was –.92. The prediction in this case was fairly close to accurate.

A *positive* value for one of the *B* coefficients indicates that a higher score on the associated variable will increase the value of the dependent variable (i.e., more sympathy yields more help). A *negative* coefficient on a predictor variable would *decrease* the value of the dependent variable (the equation above does not illustrate this; an example might be *more* cynicism yields *less* help). The greater the *B* value (absolute values), the greater the influence on the value of the dependent variable. The smaller the *B* value (absolute values) the less influence that variable has on the dependent variable.

However, *B* values often cannot be compared directly because different variables may be measured on different scales, or have different *metrics*. To resolve this difficulty, statisticians have generated a standardized score called *Beta* (β), which allows for direct comparison of the relative strengths of relationships between variables. β varies between ±1.0 and is a *partial correlation*. A partial correlation is the correlation between two variables in which the influence of all other variables in the equation have been partialed out. In the context of the present example, the *Beta* between **anger** and **zhelp** is the correlation between the two variables *after* **sympathy** and **efficacy** have been entered and the variability due to the subjects' sympathy and efficacy have already been calculated. Thus *Beta* is the *unique* contribution of one variable to explain another variable. The *Beta* weight, often called the *standardized regression coefficient*, is not only an important concept in regression analysis, but is the construct used in structural equation modeling to show the magnitude and direction of the relationships between all variables in a model. (Structural equation modeling is becoming increasingly popular in social science research but requires the purchase of an additional SPSS module.)

In the previous equation, we find, as expected, that higher **sympathy** and higher **efficacy** scores correlated with higher **zhelp** scores. Contrary to intuition, we find that more **anger** also correlated positively with **zhelp**. Why this is true is a matter of discussion and interpretation of the researcher. When an unexpected result occurs, the researcher would be well advised to recheck their data to ensure that variables were coded and entered correctly. SPSS gives no clue as to why analyses turn out the way they do.

REGRESSION AND R^2: THE AMOUNT OF VARIANCE EXPLAINED

In multiple regression analysis, any number of variables *may* be used as predictors, but many variables are not necessarily the ideal. It is important to find variables that *significantly* influence the dependent variable. SPSS has procedures by which only *significant* predictors will be entered into the regression equation. With the **Forward** entry method a dependent variable and any number of predictor (independent) variables are designated. **Regression** will first compute which predictor variable has the highest bivariate correlation with the dependent variable. SPSS will then create a regression equation with this one selected independent variable. This means that after the first step, a regression equation has been calculated that includes the designated dependent variable and *only one* independent variable. Then **Regression** will enter the second variable, which explains the greatest amount of *additional* variance. This second variable will be included only if it explains a *significant* amount of additional variation. After this second step, the regression equation has the same dependent variable but now has *two* predictor variables. Then, if there is a third variable that significantly explains more of the variance, it too will be included in the regression equation. This process will continue until no additional variables significantly explain additional variance. By default, **Regression** will cease to add new variables when *p* value associated with the inclusion of an additional variable increases above the .05 level of significance. The researcher, however, has the option to designate a different level of significance as a criterion for entry into the equation.

The measure of the strength of relationship between the independent variables (note, plural) and the dependent variable is designated with a capital *R* and is usually referred to as *multiple R*. This number squared (R^2) yields a value that represents the proportion of variation in the dependent variable that is explained by the independent variables. In the regression analysis that produced the regression equation shown above, the value of multiple *R* was .616, and the R^2 was .380. This indicates that 38% of the variance in **zhelp** was accounted for by **sympathy**, **anger**, and **efficacy**. See the Output section of this chapter for additional information about multiple *R*.

CURVILINEAR TRENDS, MODEL BUILDING, AND REFERENCES

Regression, like correlation, tests for *linear* relationships. In the previous chapter we described the procedure to test for a *curvilinear* relationship. The same process operates with multiple regression. If there is theoretical or statistical evidence that one or more of the predictor variables demonstrates a curvilinear association with the dependent variable, then a quadratic (the variable squared) factor may be added as a predictor. Please refer to the previous chapter for an explanation of this process and other possible non-linear (curved) relationships.

A final critical item concerns model building, that is, conducting a regression analysis that is conceptually sound. Certain fundamental criteria are necessary for creating a reliable model:

1. Your research must be thoughtfully crafted and carefully designed. The arithmetic of regression will not correct for either meaningless relationships or serious design flaws. Regrettably, many details of what constitutes research that is "thoughtfully crafted and carefully designed" extend well beyond the scope of this book.
2. The sample size should be large enough to create meaningful correlations. There are no hard rules concerning acceptable sample size, but as *N* drops below 50, the validity of your results become increasingly questionable. Also there is the consideration of the number of variables in the regression equation. The more variables involved, the larger the sample size needs to be to produce meaningful results.
3. Your data should be examined carefully for outliers or other abnormalities.
4. The predictor variables should be approximately normally distributed, ideally with skewness and kurtosis values between ± 1. However, good results can often be obtained

with an *occasional* deviation from normality among the predictor variables, or even the inclusion of a discrete variable (such as gender). A normal distribution for the dependent variable is also urged; but discriminant analysis (Chapter 22) uses a *discrete* measure as the criterion variable in a regression procedure.

5. Be keenly aware of the issue of linear dependency between the predictor variables. Never use two variables when one is partially or entirely dependent upon the other (such as points on the final and total points in the class). Also avoid variables that are conceptually very similar (such as worry and anxiety). To compute a matrix of correlations between potential predictor variables is always wise. Variables that correlate higher than $r = .5$ should be scrutinized carefully before both are included in a regression analysis. The power, stability, and interpretability of results are substantially compromised when variables that are linearly dependent are included in the analysis. This problem is sometimes called "collinearity" (or even "multicollinearity", if "collinearity" is too easy for you.)

Multiple regression analysis is not a simple procedure. Like analysis of variance, a number of thick volumes have been written on the topic, and we are in no way attempting to duplicate those efforts. It is suggested that, before you attempt to conduct multiple regression analysis, you take a course in the subject. A number of books are available that cover both simple and multiple regression. Several that the authors feel are especially good include: Chatterjee and Price (1999), Gonick and Smith (1993), Schulman (1998), Sen and Srivastava (1997), and Weisberg (1985). Please see the reference section for more detailed information.

The purpose of the previous four pages has been to remind you of the rationale behind regression if you have been away from it for a while. The purpose of the pages that follow is to explain step by step how to conduct multiple regression analysis with SPSS and how to interpret the output.

The data we use by way of example are from the study already discussed on pages 192-194. The file name is **helping1.sav,** $N = 81$. The following variables will be used in the description; all variables except **zhelp** are measured on a *little* (1) to *much* (7) scale.

- **zhelp**: The dependent variable. The standardized score (z score) for the amount of help given by a person to a friend in need on a –3 to +3 scale.
- **sympathy**: Feelings of sympathy aroused in the helper by the friend's need.
- **anger**: Feelings of anger or irritation aroused in the helper by the friend's need.
- **efficacy**: Self-efficacy of the helper in relation to the friend's need.
- **severity**: Helper's rating of how severe the friend's problem was.
- **empatend**: Empathic tendency of the helper as measured by a personality test.

STEP BY STEP

Multiple Regression Analysis

To enter SPSS, a click on **Start** *in the taskbar (bottom of screen) activates the start menu:*

After clicking the SPSS program icon, Screen 1 (inside front cover) appears on the monitor.

Step 2

Create and name a data file or edit (if necessary) an already existing file (see Chapter 3).

Screens 1 and 2 (inside front cover) allow you to access the data file used in conducting the analysis of interest. The following sequence accesses the **helping1.sav** *file for further analyses:*

In Screen	Do This		Step 3
Front1	File → Open → Data	[or]	
Front2	type helping1.sav → Open	[or helping1.sav]	Data

Whether first entering SPSS or returning from earlier operations the standard menu of commands across the top is required (shown below). As long as it is visible you may perform any analyses. It is not necessary for the data window to be visible.

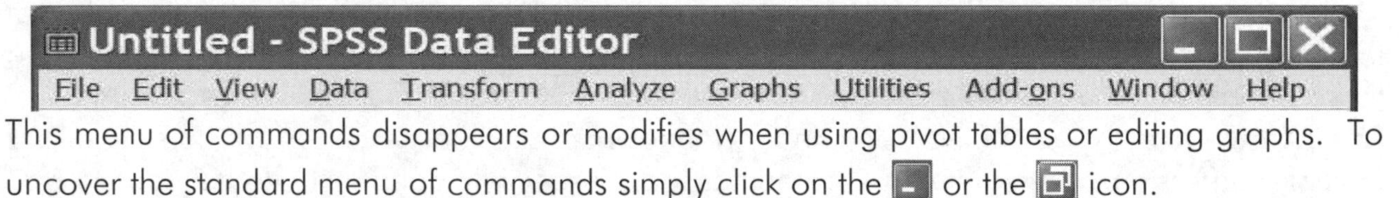

This menu of commands disappears or modifies when using pivot tables or editing graphs. To uncover the standard menu of commands simply click on the or the icon.

After completion of Step 3 a screen with the desired menu bar appears. When you click a command (from the menu bar), a series of options will appear (usually) below the selected command. With each new set of options, click the desired item. The sequence to access multiple regression analysis begins at any screen with the menu of commands visible:

In Screen	Do This	Step 4
Menu	Analyze → Regression → Linear	16.1

The screen that appears at this point (Screen 16.1, next page) is identical to the dialog box shown (Screen 15.1) from the previous chapter. We reproduce it here for sake of visual reference and to show the new list of variables in the **helping1.sav** file.

At the top of the window is the **Dependent** box that allows room for a single dependent variable. In the middle of the screen is the **Independent(s)** box where one or more carefully selected variables will be pasted. There are no SPSS restrictions to the *number* of independent variables you may enter; there are common sense restrictions described in the introduction. The **Block 1 of 1** option allows you to create and conduct more than one regression analysis in a single session. After you have chosen the dependent variable, independent variables, and selected all options that you desire related to an analysis, then click the **Next** pushbutton located below **Block 1 of 1**. The dependent variable will remain, all selected options and specifications will remain, but the independent variables will revert to the variable list. Then you may select specifications for a second regression analysis. You may specify as many different analyses in a session as you like; a click of the **OK** pushbutton will run them in the same order as they were created. Several additional options are described in the paragraphs that follow.

16.1

The Linear Regression Window

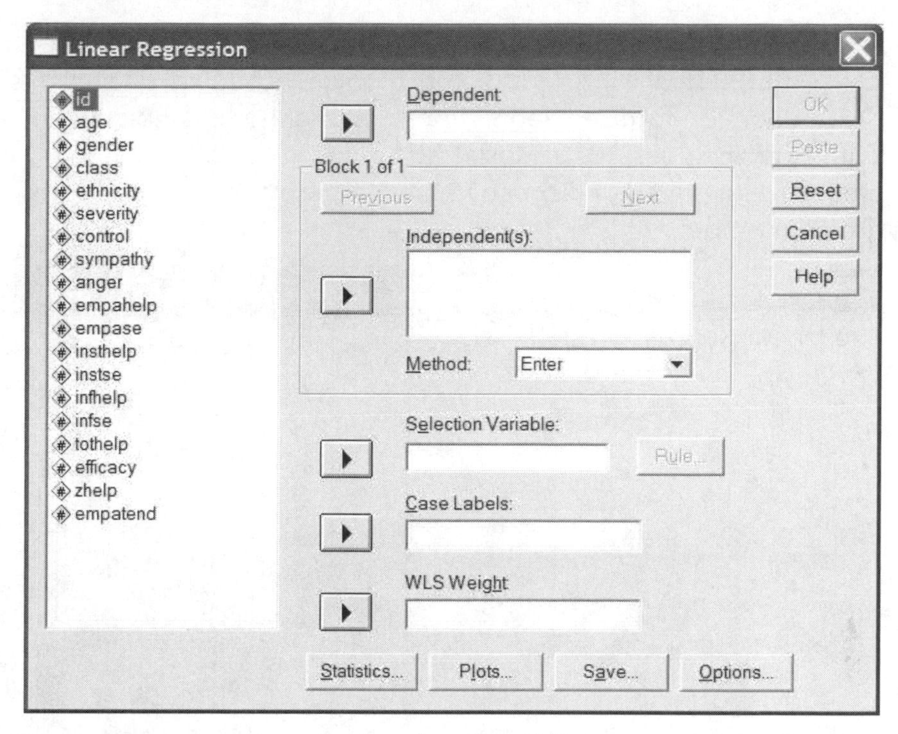

The menu box labeled **Method** is an important one. It identifies five different methods of entering variables into the regression equation. With a click of the ▾, the five options appear. **Enter** is the default method of entering variables. Each one is described below:

- **Enter**: This is the forced entry option. SPSS will enter at one time all specified variables regardless of significance levels.
- **Forward**: This method will enter variables one at a time, based on the designated significance value to enter. The process ceases when there are no additional variables that explain a significant portion of additional variance.
- **Backward**: This method enters *all* independent variables at one time and then removes variables one at a time based on a preset significance value to remove. The default value to remove a variable is $p \geq .10$. When there are no more variables that meet the requirement for removal, the process ceases.
- **Stepwise**: This method combines both **Forward** and **Backward** procedures. Due to the complexity of intercorrelations, the variance explained by certain variables will change when new variables enter the equation. Sometimes a variable that qualified to enter loses some of its predictive validity when other variables enter. If this takes place, the **Stepwise** method will remove the "weakened" variable. **Stepwise** is probably the most frequently used of the regression methods.
- **Remove**: This is the forced removal option. It requires an initial regression analysis using the **Enter** procedure. In the next block (**Block 2 of 2**) you may specify one or more variables to remove. SPSS will then remove the specified variables and run the analysis again. It is also possible to remove variables one at a time for several blocks.

The **WLS Weight** box (Weighted Least Squares) allows you to select a single variable (not already designated as a predictor variable) that weights the variables prior to analysis. This is an infrequently used option.

The **Plots** option deals with plots of residuals only and will be considered in the final chapter of the book (Chapter 28).

A click on **Statistics** opens a small dialog box (Screen 16.2). Two options are selected by default. **Estimates** will produce the *B* values (used as coefficients for the regression equation), *Betas* (the standardized regression coefficients) and associated standard errors, *t* values, and significance values. The **Model fit** produces the Multiple *R*, R^2, an ANOVA table, and associated *F* ratios and significance values. These two options represent the essence of multiple regression output.

16.2

The Linear Regression: Statistics Window

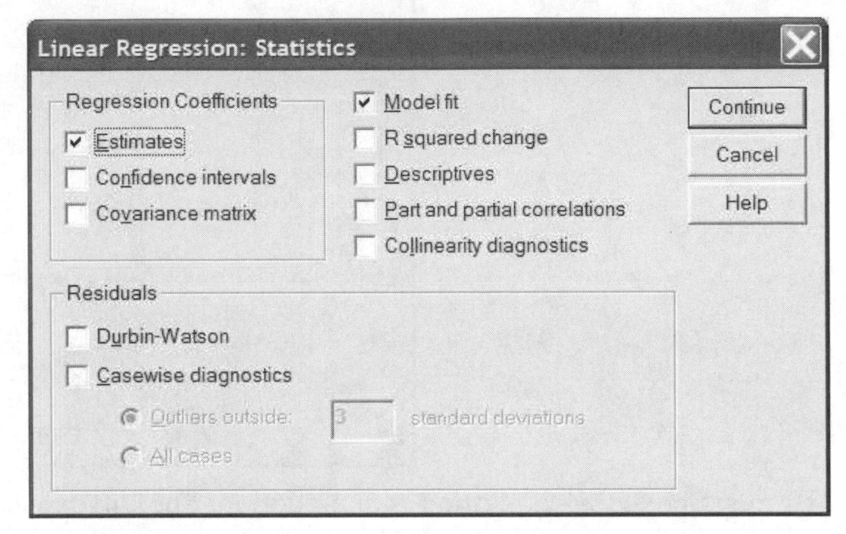

Other frequently used options include:

- **Confidence intervals**: Produces 95% confidence intervals for the *B* values.
- **Covariance matrix**: Produces a covariance-variance-correlation matrix with covariances below the diagonal, variances on the diagonal, and correlations above the diagonal. This is just one tool used in the regression procedure for identifying collinearity among variables.
- **R squared change**: In the **Forward** and **Stepwise** procedures documents the change in the R^2 value as each new variable is entered into the regression equation.
- **Descriptives**: The variable means, standard deviations, and a correlation matrix.
- **Collinearity diagnostics**: Assists in exploring whether collinearity exists among predictor and criterion variables.

A click on the **Save** button opens a dialog window (Screen 16.3, following page) with many terms unfamiliar even to the mathematically sophisticated. In general, selection of an item within this dialog box will save in the data file one or more new variable-values determined by a number of different procedures. Contents of the five boxes are briefly described below.

- **Residuals**: Some of the five options listed here are covered in Chapter 28.
- **Influence Statistics**: This box deals with the very useful function of what happens to a distribution or an analysis with the deletion of a particular subject or case. It may be necessary to delete even valid data from time to time. I once had a student in a class at UCLA who could run a 4:00 mile (Jim Robbins). If I were measuring physical fitness based on time to run a mile, the class average would probably be about 8:00 minutes. Despite the fact that Jim's result is valid, it is certainly abnormal when compared to the general population. Thus deletion of this value may have a significant influence on the results of an analysis. Unfortunately a more thorough discussion of these procedures extends well beyond the scope of this book.

- **Prediction Intervals**: These are 95% (or any desired percentage) confidence intervals for either the mean of a variable or of an individual case.
- **Distances**: Three different ways of measuring distances between cases, a concept introduced in the chapter on Cluster Analysis (Chapter 21).
- **Predicted Values:** This is the only portion of this entire window that is illustrated in one of the step-by-step sequences that follow. The Unstandardized predicted values are often useful to compare with actual values when considering the validity of the regression equation. Also it is one of the most sophisticated ways to replace missing values (see Chapter 4, pages 48-49). The Standardized predicted values are the predicted values for z scores of the selected variables.

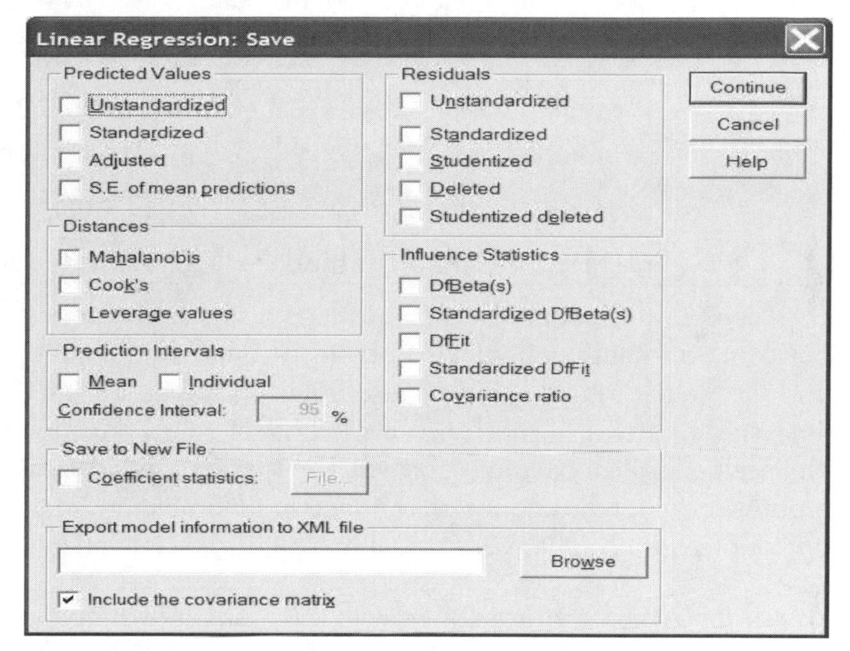

16.3

The Linear Regression: Save Window

Finally we consider the dialog box that appears when you click the **Options** button (Screen 16.4, below). The **Include constant in equation** is selected by default and should remain unless there is some specific reason for deselecting it. The **Stepping Method Criteria**, depending on your setting, may be used fairly frequently. This allows you to select the *p* value for a variable to enter the regression equation (the default is .05) for the **Forward** and **Stepwise** methods, or a *p* value to remove an already entered variable (the default is .10) for the **Stepwise** and **Backward** methods. If you wish you may instead choose *F* values as criteria for entry or removal from the regression equation. The issue of **Missing Values** was discussed in Chapter 4 and will not be considered here.

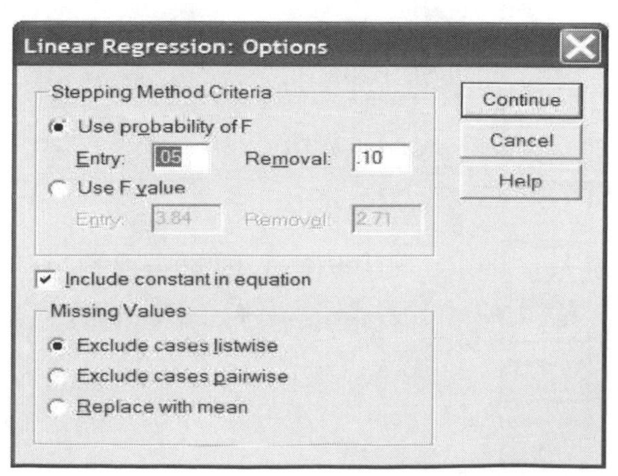

16.4

The Linear Regression: Options Window

Below we produce two step-by-step sequences. The first (step 5) is the simplest possible default regression analysis with **zhelp** as the dependent variable and **sympathy**, **severity**, **empatend**, **efficacy**, and **anger** as independent (or predictor) variables. We select the **Forward** method of entering variables. The second (step 5a) is an analysis that includes a number of the options discussed above. The italicized introduction to each box will specify what functions are being utilized in each procedure.

To run the regression procedure with a dependent variable of **zhelp** *and independent variables of* **sympathy**, **severity**, **empatend**, **efficacy**, *and* **anger**, *using the forward method of selection, perform the following sequence of steps. The starting point is screen 16.1. Perform the 4-step sequence (pages 195-196) to arrive at this screen.*

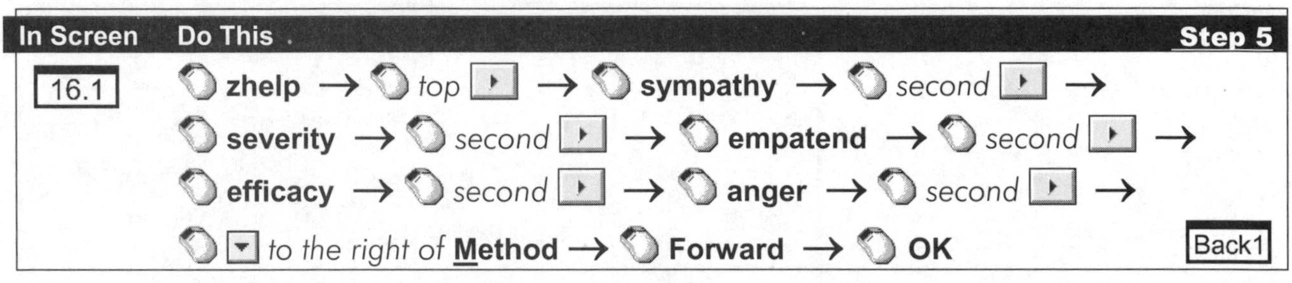

In Screen	Do This	Step 5
16.1	**zhelp** → *top* ▸ → **sympathy** → *second* ▸ → **severity** → *second* ▸ → **empatend** → *second* ▸ → **efficacy** → *second* ▸ → **anger** → *second* ▸ → ▾ *to the right of* **Method** → **Forward** → **OK**	Back1

The result of this sequence is to produce a regression analysis that identifies which of the independent variables (sympathy, problem severity, empathic tendency, self-efficacy of the helper, and the helper's anger) have the greatest influence on the dependent variable, the amount of time spent helping. The forward method of selection will first enter the independent variable having the highest bivariate correlation with help, then enter a second variable that explains the greatest additional amount of variance in time helping, then enter a third variable and so forth until there are no other variables that significantly influence the amount of help given.

To run the regression procedure with a dependent variable of **zhelp** *and independent variables of* **sympathy**, **severity**, **empatend**, **efficacy**, *and* **anger**, *using the* **Stepwise** *method of selection, including statistics of* **Estimates**, **Descriptives**, *and* **Model fit**, *selecting the* **Unstandardized predicted values** *as an additional saved variable, and establishing an* **Entry value** *of .10 and a* **Removal value** *of .20, perform the following sequence of steps. The starting point is Screen 16.1. (Refer to sequence step 4 on page 196.)*

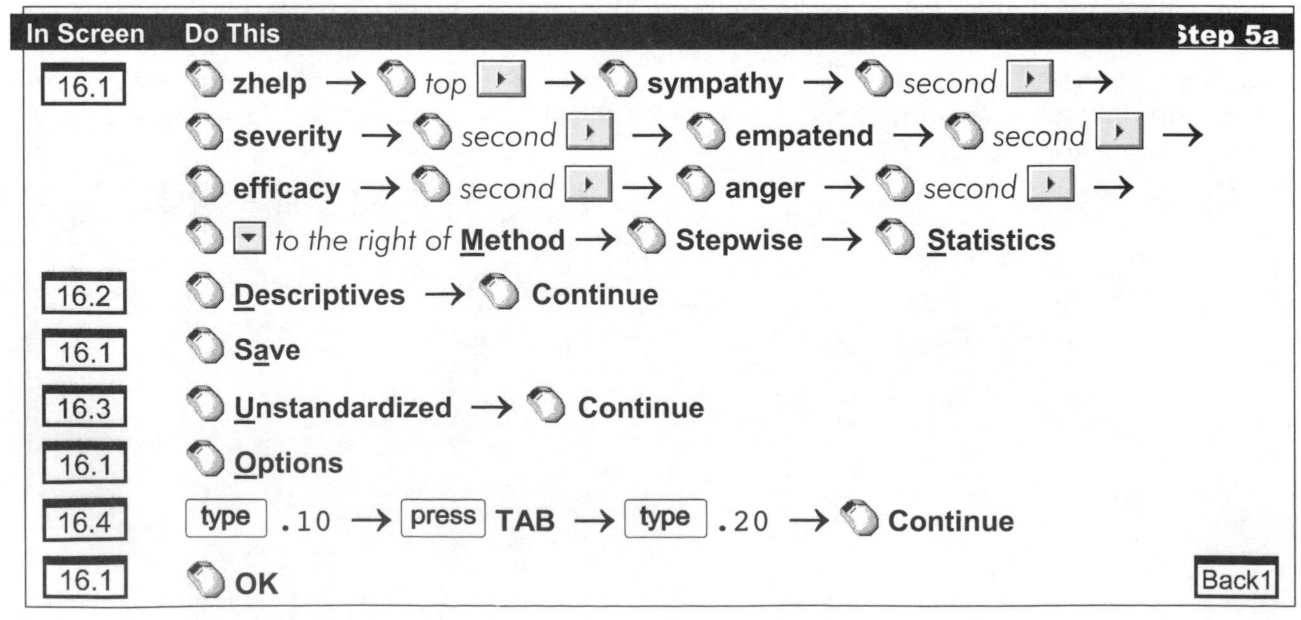

In Screen	Do This	Step 5a
16.1	**zhelp** → *top* ▸ → **sympathy** → *second* ▸ → **severity** → *second* ▸ → **empatend** → *second* ▸ → **efficacy** → *second* ▸ → **anger** → *second* ▸ → ▾ *to the right of* **Method** → **Stepwise** → **Statistics**	
16.2	**Descriptives** → **Continue**	
16.1	**Save**	
16.3	**Unstandardized** → **Continue**	
16.1	**Options**	
16.4	type .10 → press **TAB** → type .20 → **Continue**	
16.1	**OK**	Back1

The result of this sequence is to produce a regression analysis that identifies which of the independent variables (sympathy, problem severity, empathic tendency, self-efficacy of the helper, and the helper's anger) have the greatest influence on the dependent variable (amount of time spent helping). The stepwise method of selection will first enter the independent variable with the highest bivariate correlation with help, then enter the variable that explains the greatest additional amount of variance, then enter a third variable and so forth until no other variables significantly (significance is specified as $p \le .10$ for this analysis) influence the amount of help given. If the influence of any variables increases above a significance of .20 after entry into the regression analysis, it will be dropped from the regression equation. We request the saving of a new variable that will include the predicted amount of time helping for each subject. We also request the addition of descriptive statistics and the correlation matrix of all variables.

Upon completion of step 5, or 5a, the output screen will appear (Screen 1, inside back cover). All results from the just-completed analysis are included in the Output Navigator. Make use of the scroll bar arrows (▲▼▶◀) to view the results. Even when viewing output, the standard menu of commands is still listed across the top of the window. Further analyses may be conducted without returning to the data screen.

PRINTING RESULTS

Results of the analysis (or analyses) that have just been conducted requires a window that displays the standard commands (**File Edit Data Transform Analyze** . . .) across the top. A typical print procedure is shown below beginning with the standard output screen (Screen 1, inside back cover).

To print results, from the Output screen perform the following sequence of steps:

In Screen	Do This	Step 6
Back1	*Select desired output or edit (see pages 19 - 25)* → **File** → **Print**	
2.9	*Consider and select desired print options offered, then* → **OK**	

To exit you may begin from any screen that shows the **File** *command at the top.*

In Screen	Do This	Step 7
Menu	**File** → **Exit**	2.1

Note: After clicking **Exit**, there will frequently be small windows that appear asking if you wish to save or change anything. Simply click each appropriate response.

OUTPUT

Multiple Regression Analysis

The following output is produced by sequence step 5 on page 200. The format of the output below is fairly different (and much neater) than actual SPSS output, but all relevant data are included with identical terminology. As procedures become more complex (in chapters that follow) there is an increasing necessity to condense format to only the most relevant output. Hopefully that way when you get a book-length output from SPSS, you can see where the most essential parts of the output are.

Model Summary

R	.616
R square	.380
Adjusted R square	.355
Std Error of Estimate	1.0058

Model 3

Method: Forward (Criterion: Probability of F to enter <= .050)

Dependent Variable: ZHELP

Predictors: (Constant, SYMPATHY, ANGER, EFFICACY)

ANOVA

Source of Variation	df	Sum of Squares	Mean Square	F	Signif F
Regression	3	47.654	15.885	15.701	.000
Residual	77	77.899	1.012		

Coefficients

	Unstandardized coefficients		Std Coefficients		
	B	Std. Error	Beta	t	Signif of t
(Constant)	-4.308	.732		-5.885	.000
SYMPATHY	.494	.100	.451	4.938	.000
ANGER	.284	.083	.310	3.429	.001
EFFICACY	.412	.132	.284	3.134	.002

Excluded Variables

Model	Variable	Beta in	t	Signif of t	Partial Correla-	Collinearity Stats
						Tolerance
3	SEVERITY	.077	.787	.443	.090	.844
	EMPATEND	.119	1.269	.208	.144	.909

An initial look identifies key elements of the analysis: Three models were tested (only the third is shown here), with the three variables that met the entry requirement included in the final equation (**sympathy**, **anger**, and **efficacy**). Two variables did not meet the entry requirement (**severity** and **empatend**). The multiple *R* shows a substantial correlation between the three predictor variables and the dependent variable **zhelp** ($R = .616$). The *R*-square value indicates that about 38% of the variance in **zhelp** is explained by the three predictor variables. The β values indicate the relative influence of the entered variables, that is, sympathy has the greatest influence on help ($\beta = .45$), followed by anger ($\beta = .31$), and then efficacy ($\beta = .28$). The direction of influence for all three is positive.

CHANGE OF VALUES AS EACH NEW VARIABLE IS ADDED

What follows is partial output from sequence step 5a on page 200. Once again this output is a greatly condensed version of the SPSS output to demonstrate the changes in the variables from step to step as new variables are entered into the regression equation.

The purpose of displaying the output from the three steps in this format is to allow you to see how the computed values change as each new variable is added. Note for instance how the multiple *R*, the *R* square, and the adjusted *R* square *increase* in value with the addition of each new variable. Note also how the standard error and *R*-square change *decrease* in value with the addition of new variables. Similar patterns may be noted for degrees of freedom, sum of squares, and mean squares.

Dependent Variable ZHELP	Model #	Model 1	Model 2	Model 3
Statistics	Variables in Equation	SYMPATHY	SYMPATHY ANGER	SYMPATHY ANGER EFFICACY
Multiple R		.455	.548	.616
R Square		.207	.300	.380
Adjusted R square		.197	.282	.355
Standard Error		1.123	1.061	1.006
R Square change		.207	.093	.079

degrees of freedom	regression	1	2	3
	residual	79	78	77
SUM of SQUARES	regression	26.008	37.715	47.654
	residual	99.544	87.837	77.899
MEAN SQUARES	regression	26.008	18.858	15.884
	residual	1.260	1.126	1.012
Overall F		20.641	16.746	15.701
Significance of F		.0000	.0000	.0000

Additional comment is included in the definitions of terms that follow.

Term	Definition/Description
PROB OF F TO ENTER	Probability value to enter a variable into the regression equation. In this case p = .05, the default value.
PROB OF F TO RE-MOVE	For **stepwise** or **backward** procedures only. Probability value to remove an already entered variable from the regression equation.
R	The multiple correlation between the dependent variable **zhelp** and the three variables in the regression equation, **sympathy**, **anger**, and **efficacy**.
R SQUARE	The R^2 value identifies the portion of the variance accounted for by the independent variables; that is, approximately 38% of the variance in **zhelp** is accounted for by **sympathy**, **anger**, and **efficacy**.
ADJUSTED R SQUARE	R^2 is an accurate value for the sample drawn but is considered an optimistic estimate for the *population* value. The Adjusted R^2 is considered a better population estimate and is useful when comparing the R^2 values between models with different numbers of independent variables.
STANDARD ERROR	**STANDARD ERROR** is the standard deviation of the expected value of (in this case) **zhelp**. Note that this value shrinks as each new variable is added to the equation (see output from bottom box on previous page).
REGRESSION	Statistics relating to the explained portion of the variance.
RESIDUAL	Statistics relating to the *un*explained portion of the variance.

Term	Definition/Description
DF	For regression, the number of independent variables in the equation. For residual, *N,* minus the number of independent variables entered, minus 1: 81 – 3 – 1 = 77.
SUM OF SQUARES	The regression sum of squares corresponds to the between-group sum of squares in ANOVA, and the residual sum of squares corresponds to the within-group sum of squares in ANOVA. Note that in this case there is a larger portion of *un*explained variance than there is of explained variance, a reality also reflected in the **R SQUARE** value.
MEAN SQUARE	Sum of squares divided by degrees of freedom.
F	Mean square regression divided by mean square residual.
SIGNIF F	Probability that this *F*-value could occur by chance. The present results show that the likelihood of the given correlation occurring by chance is less than 1 in 10,000.
B	The coefficients and constant for the regression equation that measures predicted values for **ZHELP**: **zhelp**$_{(predicted)}$ = –4.308 + .494 (**sympathy**) + .284(**anger**) + .412(**efficacy**)
STD ERROR	Standard error of *B*: A measure of the stability or sampling error of the *B* values. It is the standard deviation of the *B* values given a large number of samples drawn from the same population.
BETA (β)	The standardized regression coefficients. This is the *B* value for standardized scores (*z* scores) of the independent variables. This value will always vary between ± 1.0 in linear relationships. For curvilinear relationships it will sometimes extend outside that range.
t	*B* divided by the standard error of *B*. This holds both for variables in the equation and those variables *not* in the equation.
SIGNIF t	Likelihood that these *t* values could occur by chance.
BETA IN	The beta values for the *excluded* variables if these variables were actually in the regression equation.
PARTIAL CORRELATION	The partial correlation coefficient with the dependent variable **zhelp**, adjusting for variables already in the regression equation. For example, the simple correlation (*r*) between **zhelp** and **severity** is .292. After **sympathy**, **anger**, and **efficacy** have "consumed" much of the variance, the *partial* correlation between **zhelp** and **severity** is only .089. **empatend**, on the other hand, shrinks only a little, from .157 to .144.
MINIMUM TOLERANCE	Tolerance is a commonly used measure of collinearity. It is defined as $1 - R_i^2$, where R_i is the *R*-value of variable *i* when variable *i* is predicted from all other independent variables. A low tolerance value (near 0) indicates extreme collinearity, that is, the given variable is almost a linear combination of the other independent variables. A high value (near 1) indicates that the variable is relatively independent of other variables. This measure deals with the issue of linear dependence that was discussed in the Introduction. When variables are included that are linearly dependent, they inflate the standard errors, thus weakening the power of the analysis. It also conceals other problems, such as curvilinear trends in the data.
R SQUARE CHANGE	This is simple subtraction of the R^2 value for the given line minus the R^2 value from the previous line. Note that the value for **anger** .093 = .300 – .207 (within rounding error). The number (.09324) indicates that the inclusion of the second variable (**anger**) explains an additional 9.3% of the variance.

EXERCISES

Answers to selected exercises are available for download at **www.ablongman.com/george6e**.

Use the **helping3.sav** file for the exercises that follow (downloadable at the address shown above).

Conduct the following THREE regression analyses:

Criterion variables:

1. **thelplnz**: Time spent helping
2. **tqualitz**: Quality of the help given
3. **tothelp**: A composite help measure that includes both time and quality

Predictors: (use the same predictors for each of the three dependent variables)
age: range from 17 to 89
angert: Amount of anger felt by the helper toward the needy friend
effict: Helper's feeling of self-efficacy (competence) in relation to the friend's problem
empathyt: Helper's empathic tendency as rated by a personality test
gender: 1 = female, 2 = male
hclose: Helper's rating of how close the relationship was
hcontrot: helper's rating of how controllable the cause of the problem was
hcopet: helper's rating of how well the friend was coping with his or her problem
hseveret: helper's rating of the severity of the problem
obligat: the feeling of obligation the helper felt toward the friend in need
school: coded from 1 to 7 with 1 being the lowest education, and 7 the highest (> 19 years)
sympathi: The extent to which the helper felt sympathy toward the friend
worry: amount the helper worried about the friend in need

- Use **entry value** of .06 and **removal value** of .11.
- Use **stepwise** method of entry.

Create a table (example below) showing for each of the three analyses Multiple R, R^2, then each of the variables that significantly influence the dependent variables. Following the R^2, List the name of each variable and then (in parentheses) list its β value. Rank order them from the most influential to least influential from left to right. Include only significant predictors.

Dependent Variable	Multiple R	R^2	1st var (β)	2nd var (β)	3rd var (β)	4th var (β)	5th var (β)	6th var (β)
Time helping								
Help quality								
Total help								

4. A researcher is examining the relationship between **stress** levels, **self-esteem**, **coping skills**, and **performance** on a test of cognitive performance (the dependent measure). His data are shown below. Perform multiple regression on these data, entering variables using the stepwise procedure. Interpret the results.

Stress	Self-esteem	Coping skills	Performance
6	10	19	21
5	10	14	21
5	8	14	22
3	7	13	15
7	14	16	22
4	9	11	17
6	9	15	28
5	9	10	19
5	11	20	16
5	10	17	18

17

NONPARAMETRIC Procedures

THIS CHAPTER deals with (as the title suggests) nonparametric tests. Before we become involved in *non*parametric tests, it might be well to consider first, what is a *parametric* test? A parametric test is one that is based on certain parameters. The critical parameter that most of the procedures described in this book are based on is that data from samples (and the populations from which they are drawn) are normally distributed (or something close to normal). Although some operations are based on other assumptions or parameters (e.g., binomial or Poisson distributions of data), the **Nonparametric Tests** procedure deals primarily with populations that are *not* normally distributed and considers how to conduct statistical tests if the assumption of normality is violated.

For data that are not normally distributed the nonparametric tests use other statistical techniques to test hypotheses. These techniques may include, among others, analyses based on:

- ranked values,
- summation of how many values in one distribution are larger (or smaller) than values in another distribution,
- use of weighted comparisons,
- tests to determine whether a distribution of values deviates from randomness or is binomially distributed,
- single-group tests of deviation from normality,
- comparisons of frequencies, and
- calculation of the frequency of values above or below a grand median to compare groups.

In addition, nonparametric tests may calculate statistics about one sample or make comparisons between two or more samples. Despite this seeming complexity, most nonparametric tests are quite understandable and easy to conduct. Further, general description is not attempted here because it is more profitable to explain the tests one at a time.

To demonstrate nonparametric tests, we will make use, once more, of the original data file first presented in Chapters 1 and 2: the **grades.sav** file. This is an excellent file to demonstrate nonparametric tests because the contents of the file are so easily understood. Tests will be conducted on the following variables within the file: **gender**, the five 10-point quizzes (**quiz1** to **quiz5**), **final** (the score on a 75-point final exam), **ethnicity** (the ethnic makeup of the sample), and **section** (membership of subjects in the three sections of the class). The *N* for this sample is 105.

This will be one of only two analysis chapters (Chapters 6-27) that deviates substantially from the traditional format. In this chapter, the Step by Step and the Output sections will be combined into one. We begin the Step by Step section with the (by now) familiar first three steps and end with the equally familiar final two steps. After the first three steps are presented, two screens will be presented and discussed that are used in several of the operations that follow. Then, nine different nonparametric procedures will be presented, one per page. There are more than nine nonparametric tests, but we present only the ones that are used most frequently. Each section will have its own introduction, followed by step 4 (or 4a, 4b, etc.), then the output, an interpretation of the output, and definitions of terms. The nine procedures covered include:

1. **Mann-Whitney rank-sum test**: A test of whether two groups differ from each other based on ranked scores.

2. **Sign test**: Tests whether two distributions differ based on a comparison of paired scores. That is, for how many of the paired scores is the value for group A larger than the value for group B (positive sign), or is the group B score larger than the group A score (negative sign)?
3. **Wilcoxon matched-pairs signed-ranks test**: The same as the sign test except the positive and negative signs are weighted by the mean rank of positive versus negative comparisons.
4. **Runs test**: Tests whether the elements of a single dichotomous group differ from a random distribution.
5. **Binomial test**: Tests whether the elements of a single dichotomous group differ from a binomial distribution (each outcome equally likely).
6. **Kolmogorov-Smirnov one-sample test**: Tests whether the distribution of the members of a single group differ significantly from a **normal** (or **uniform**, or **Poisson**) distribution.
7. **One-sample chi-square test**: Tests whether observed scores differ significantly from expected scores for levels of a single variable.
8. **Fridman one-way ANOVA**: Tests whether three or more groups differ significantly from each other, based on the average rank of groups rather than the distribution of values.
9. **K-sample median test**: Tests whether two or more groups differ on the number of instances (within each group) greater than the median value or less than the median value.

For additional information, the most comprehensive book on nonparametric statistics discovered by the authors is a text by Siegel and Castellan (1988). Please see the reference section for additional detail.

STEP BY STEP

Nonparametric Tests

To enter SPSS, a click on **Start** *in the taskbar (bottom of screen) activates the start menu:*

After clicking the SPSS program icon, Screen 1 (inside front cover) appears on the monitor.

Step 2

Create and name a data file or edit (if necessary) an already existing file (see Chapter 3).

Screens 1 and 2 (inside front cover) allow you to access the data file used in conducting the analysis of interest. The following sequence accesses the **grades.sav** *file for further analyses:*

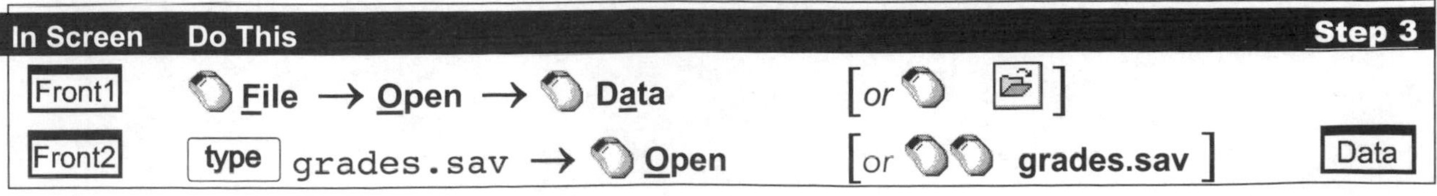

Whether first entering SPSS or returning from earlier operations the standard menu of commands across the top is required (shown below). As long as it is visible you may perform any analyses. It is not necessary for the data window to be visible.

This menu of commands disappears or modifies when using pivot tables or editing graphs. To uncover the standard menu of commands simply click on the or the icon.

Screens Used in Several of the Nonparametric Procedures

Some nonparametric procedures require that you identify the different levels of a variable. When this is true it is necessary to click the **Define Groups** button and a window (here titled Screen 17.0) will open. If the variable has two levels coded 1 and 2, when the box opens, type 1, press **TAB**, type 2, then click **Continue**. The variable will then have a defined range and you may continue with the analysis.

17.0

The Generic Define Groups Window

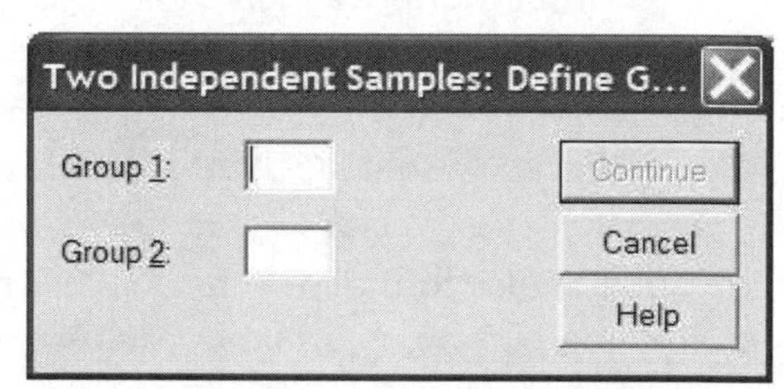

The second dialog box (at right) is available in many of the nonparametric procedures. With a click of the **Options** button in the main dialog window you may request basic **Descriptives** (mean, standard deviation, minimum, maximum, and *N*), or the **Quartiles**. You have already dealt with Missing Values during data entry, right?

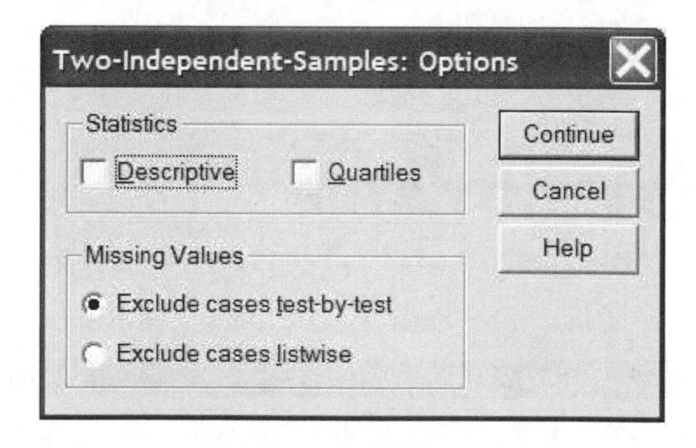

MANN-WHITNEY RANK-SUM TEST

The Mann-Whitney rank-sum *U* test accomplishes essentially what a *t* test does when the distributions of the two samples deviate significantly from normal. If the distributions do *not* differ significantly from normal, the *t* test should be used because it has greater power. In our example we consider whether females and males (the **gender** variable) differ significantly on their scores on the **final** exam. The Mann-Whitney procedure rank orders all 105 scores, determines the rank of each subject, and then computes the average rank for the two groups. Clearly the group with the higher average rank scored higher on the test. The *U* test determines whether that difference is significant.

17.1

The Mann-Whitney Rank-Sum Test Window

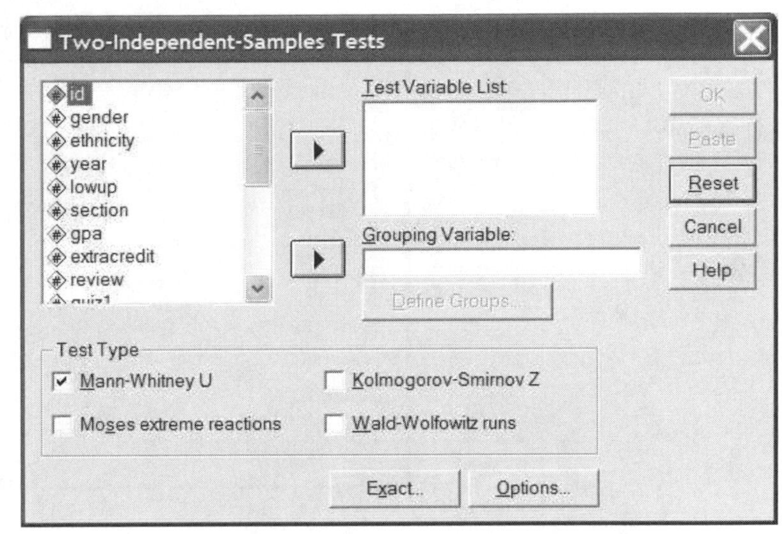

Step by Step

Using Screen 17.1 for visual reference, perform the following steps to determine whether women score higher than men on the final exam.

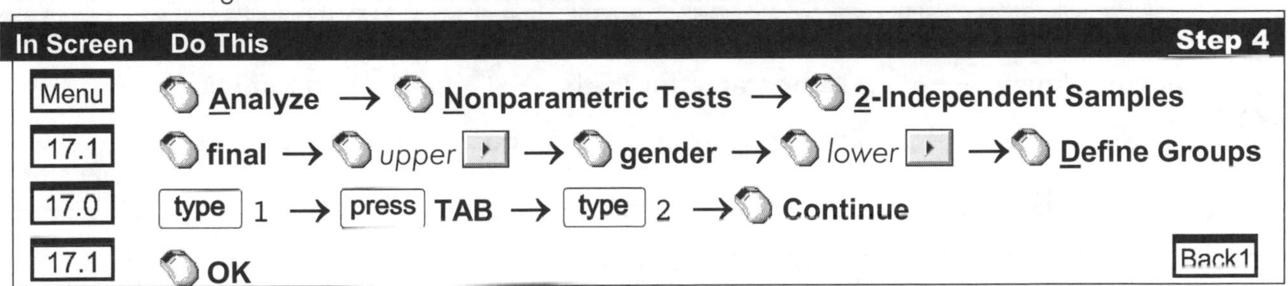

Output

Ranks

	gender	N	Mean Rank	Sum of Ranks
final	Female	64	55.81	3572.00
	Male	41	48.61	1993.00
	Total	105		

Test Statistics[a]

	final
Mann-Whitney U	1132.000
Wilcoxon W	1993.000
Z	-1.184
Asymp. Sig. (2-tailed)	.237

Note that the mean rank for females is higher (55.81) than the mean rank for males, (48.61), indicating that females scored higher than males. The *U* statistic is the number of times members of the lower-ranked group (males) precede members of the higher-ranked group (females). The *Z* is the standardized score associated with the significance value ($p = .237$). Since the *p* value is large, we conclude that women did not score significantly higher than men.

THE SIGN TEST

The sign test utilizes pairwise comparisons of two different distributions to identify which is larger than which, and then from this information it determines if the two distributions differ significantly from each other. To demonstrate, we compare the scores on **quiz1** with the scores on **quiz2**. By default, the sign test compares the second distribution with the first distribution. For the first subject, **quiz1** was 9 and **quiz2** was 7. This would rate as a negative (–) difference. For the second subject, **quiz1** was 6, **quiz2** was 7. This would rate as a positive (+) difference. The signs test sums all the positives, negatives, and ties and then computes a *z* score and a *p* value associated with the frequency of the positives and negatives.

17.2

The Sign Test Window

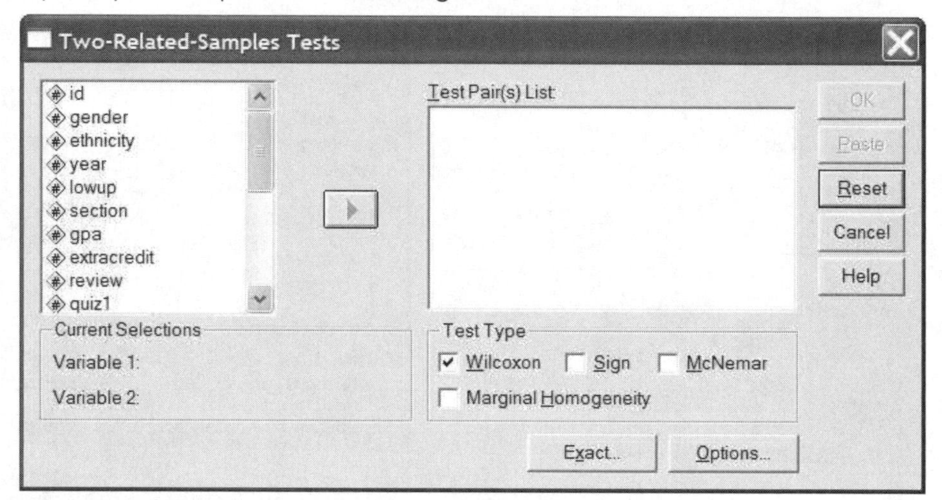

Step by Step

Using Screen 17.2 for visual reference, perform the following steps to determine whether the scores on **quiz 1** *were significantly higher than the scores on* **quiz 2**.

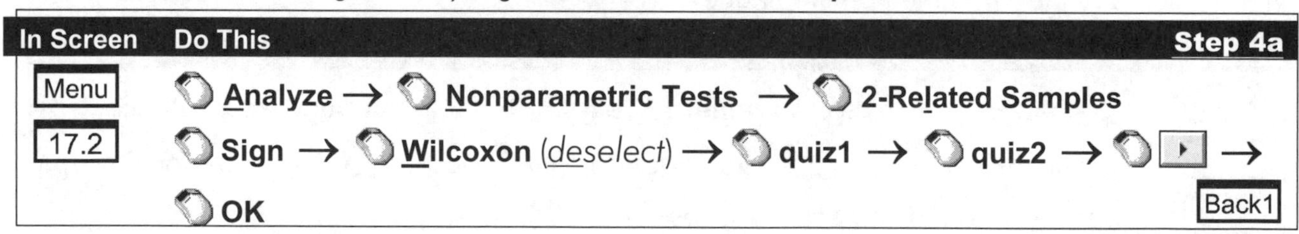

		N
quiz2 - quiz1	Negative Differences [a]	34
	Positive Differences [b]	50
	Ties [c]	21
	Total	105

a. quiz2 < quiz1

b. quiz2 > quiz1

c. quiz2 = quiz1

	quiz2 - quiz1
Z	-1.637
Asymp. Sig. (2-tailed)	.102

a. Sign Test

Output

Note that in 34 instances **quiz2** was *less* than **quiz1**, in 50 instances **quiz2** was *greater* than **quiz1**, and in 21 instances both quizzes had the same score. The *z* score determined from these values is –1.637, with an associated *p* value of .102. This suggests that the quizzes do not differ significantly from each other. It should be noted that, since the distributions of both quizzes *are* normal, a *t* test would be appropriate here. If a *t* test is conducted on these data, **quiz2** *is* significantly greater than **quiz1** (p = .005). The *t* test is significant, whereas the sign test is not because the *sign test has lower* statistical power.

WILCOXON MATCHED-PAIRS SIGNED-RANKS TEST

The difficulty with the sign test is that a difference between paired quizzes of 10 (10 on one, 0 on the other) and a difference of 1 (e.g., 6 on one, 5 on the other) will be coded identically (as a negative). The Wilcoxon matched-pairs signed-ranks test incorporates information about the magnitude of the differences between paired values. To compute this value, first the magnitude of the differences (ignoring the signs) are ranked from high to low. Then the ranks for the negative signs (**quiz2** < **quiz1**) are summed and averaged, and the ranks for the positive signs (when **quiz2** > **quiz1**) are summed and averaged. Finally, significance values are calculated based on z scores.

17.3

The Wilcoxon Matched-Pairs Signed-Ranks Test Window

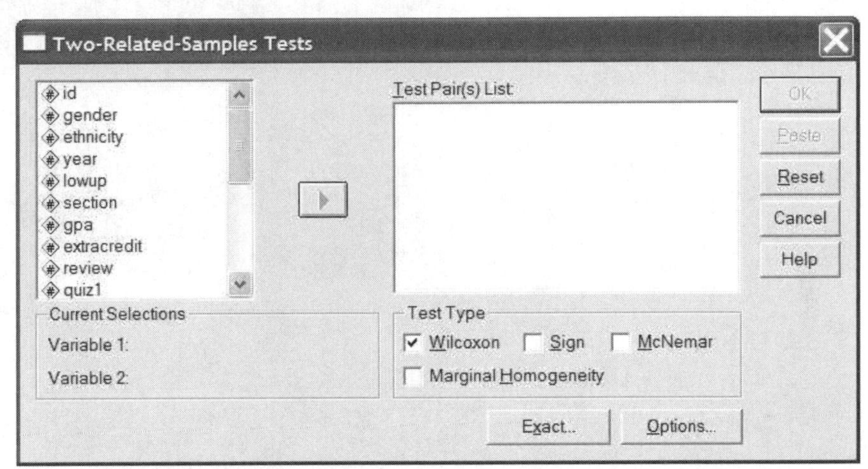

Step by Step

Using Screen 17.3 for visual reference, perform the following steps to make use of the Wilcoxon matched-pairs signed-ranks test to compare scores on **quiz 1** *with scores on* **quiz 2**.

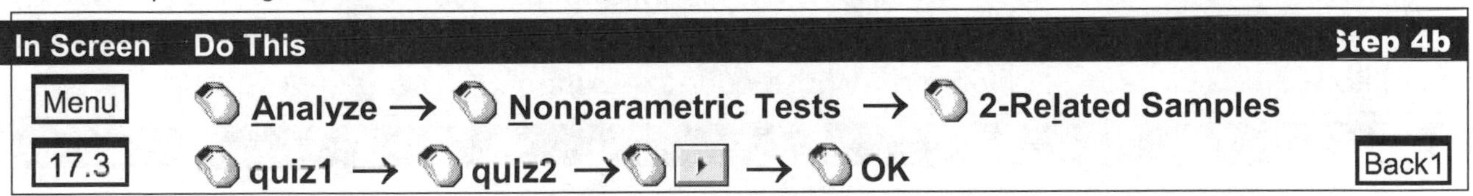

In Screen	Do This	Step 4b
Menu	Analyze → Nonparametric Tests → 2-Related Samples	
17.3	quiz1 → quiz2 → ▸ → OK	Back1

		N	Mean Rank	Sum of Ranks
quiz2 - quiz1	Negative Ranks	34[a]	35.62	1211.00
	Positive Ranks	50[b]	47.18	2359.00
	Ties	21[c]		
	Total	105		

a. quiz2 < quiz1
b. quiz2 > quiz1
c. quiz2 = quiz1

	quiz2 - quiz1
Z	-2.612[a]
Asymp. Sig. (2-tailed)	.009

a. Based on negative ranks.
b. Wilcoxon Signed Ranks Test

Output

Note the similarities to the sign test (previous page). The frequency of negative ranks, positive ranks, and ties are the same. Additional information includes the mean rank of each group based on the overall magnitude of differences. While visual inspection of output from the sign test shows that **quiz2** scores are higher than **quiz1** scores, the additional information of the magnitude of the differences now produces a much larger z score (–2.61) and a much smaller *p* value (.009). While the sign test did not reveal a significant difference between the two groups, the Wilcoxon test was able to do so. The test, while much improved over the sign test, is still not as powerful as the *t* test, which yields a *p* value of .005. If the distributions are normally distributed, use *t* tests rather than nonparametric tests.

THE RUNS TEST

The runs test is used to see if the elements of a particular data set are randomly distributed. If the sequence HHTHTTHTTHTHTTTTHTH resulted from flipping a coin, does this sequence differ significantly from randomness? In other words, are we flipping a biased coin? Unfortunately this procedure works only with dichotomous data (exactly two possible outcomes). It is not possible to test, for instance, if we are rolling a loaded die. Sticking fiercely by our determination to use the **grades.sav** file to demonstrate all procedures in this chapter, we will test whether the males and females in our file are distributed randomly throughout our data set.

17.4

The Runs Test Window

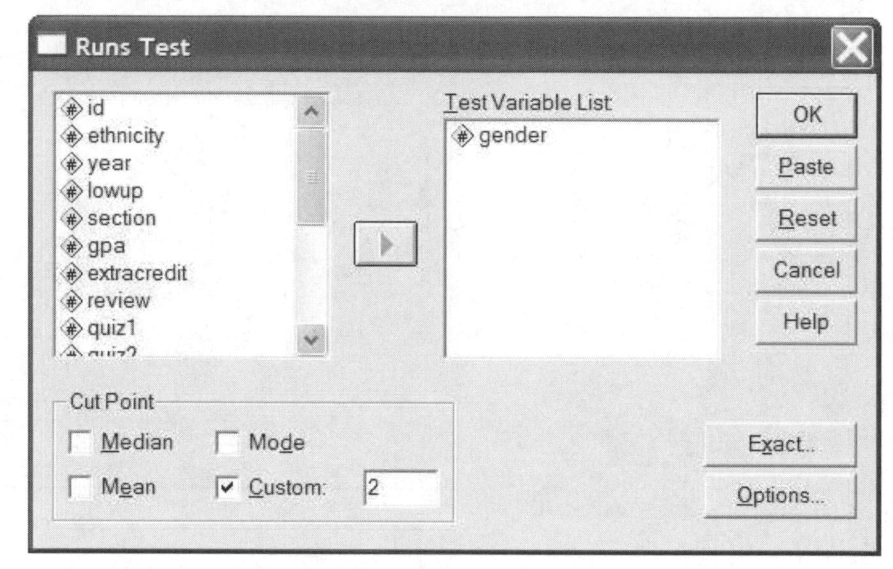

Step by Step

Using Screen 17.4 for visual reference, perform the following sequence of steps to see if males and females are distributed randomly. Note that the **Custom** *option divides the* **gender** *variable into subjects coded 2 or higher and those coded less than 2.*

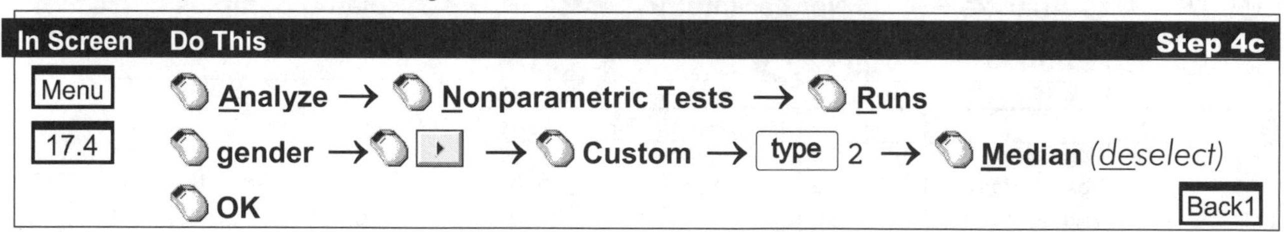

	gender
Test Value[a]	2.00
Total Cases	105
Number of Runs	50
Z	-.202
Asymp. Sig. (2-tailed)	.840

Output

This output indicates that there were 50 runs. This measures the number of times in the data set when there was a switch from one code to the other code. Thus a single value (that switches at the next case or subject) is considered a run of 1. Longer runs are of course also included in the 50 shown in the output. Test value is the number that discriminates between the two groups. The output utilizes 64 females (< 2) and 41 males (≥ 2). The z and p values are dependent on the total number of runs. This test converts the runs number into a z statistic from which the probability is determined. The significance indicated here ($p = .840$) suggests that the order of males and females on the roster does not deviate significantly from randomness.

THE BINOMIAL TEST

The binomial test measures whether a distribution of values is binomially distributed. Binomial distribution assumes that any outcome is equally likely ($p = .5$). If you tossed an unbiased coin 100 times you would expect approximately 50 heads and 50 tails. We will apply the binomial test to the distribution of males and females in our data set. We already know there are 41 men and 64 women, so the use of the binomial test is mainly to demonstrate how the procedure works. It is of legitimate interest, however, to see if this distribution differs significantly from 52.5 men and 52.5 women.

17.5

The Binomial Test Window

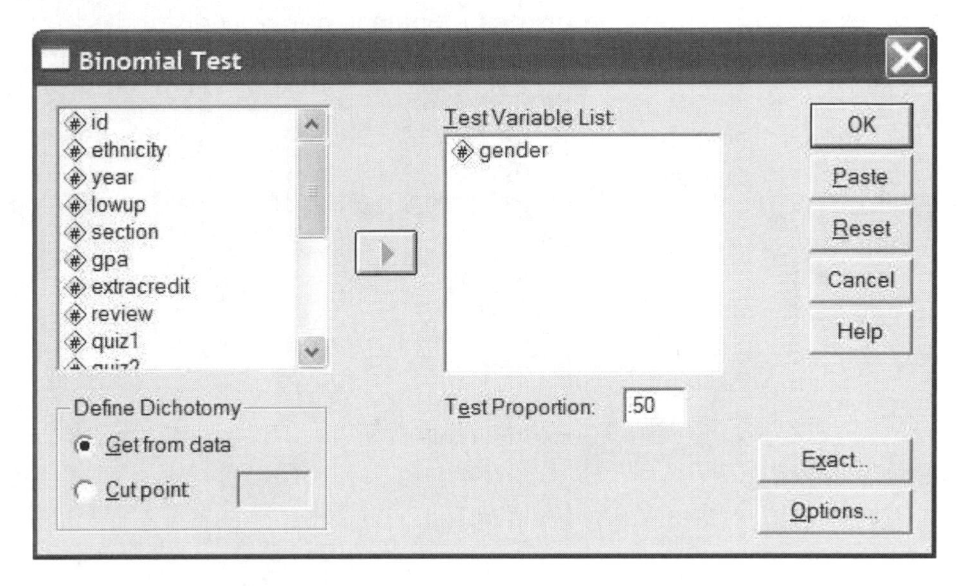

Step by Step

Using Screen 17.5 for visual reference, perform the following sequence of steps to utilize the binomial test to see if the number of men and women differ significantly from equal probability.

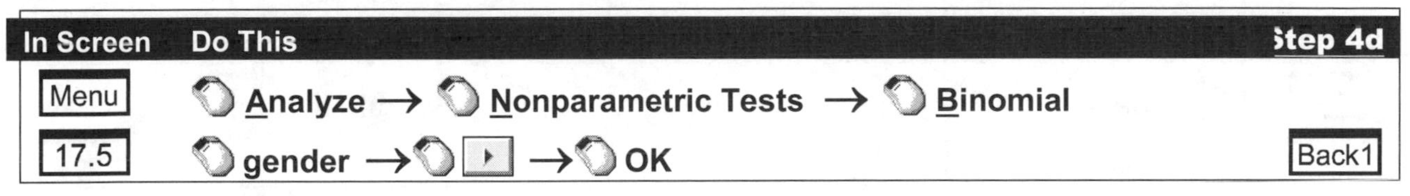

Output

gender	N	Observed Proportion	Test Proportion	Sig. (2-tailed)
(Group 1) Male	41	.39	.50	.031
(Group 2) Female	64	.61		
Total	105	1.00		

The test proportion of .50 is the expected proportion for a binomial distribution. The observed proportion is the larger of the two numbers (64 females) divided by the total number of observations (105). The p value associated with this comparison is .023, indicating that the number of men and women do differ significantly from the binomial assumption of equal probability of either.

THE KOLMOGOROV-SMIRNOV ONE-SAMPLE TEST

This test is designed to measure whether a particular distribution differs significantly from a **Normal** distribution (skewness and kurtosis of the distribution = 0), a **Uniform** distribution (values are distributed evenly, such as the numbers 1-100 consecutively), a **Poisson** distribution (the value λ equals the mean and the variance of the distribution; as λ becomes large, the distribution approximates normality), or an **Exponential** distribution. This procedure is based on a comparison of the sample cumulative distribution to the hypothesized (normal, uniform, or Poisson) cumulative distribution. To demonstrate this process we will see if the distribution of **final** exam scores from the **grades.sav** file is normally distributed.

17.6

The One-Sample Kolmogorov-Smirnov Test Window

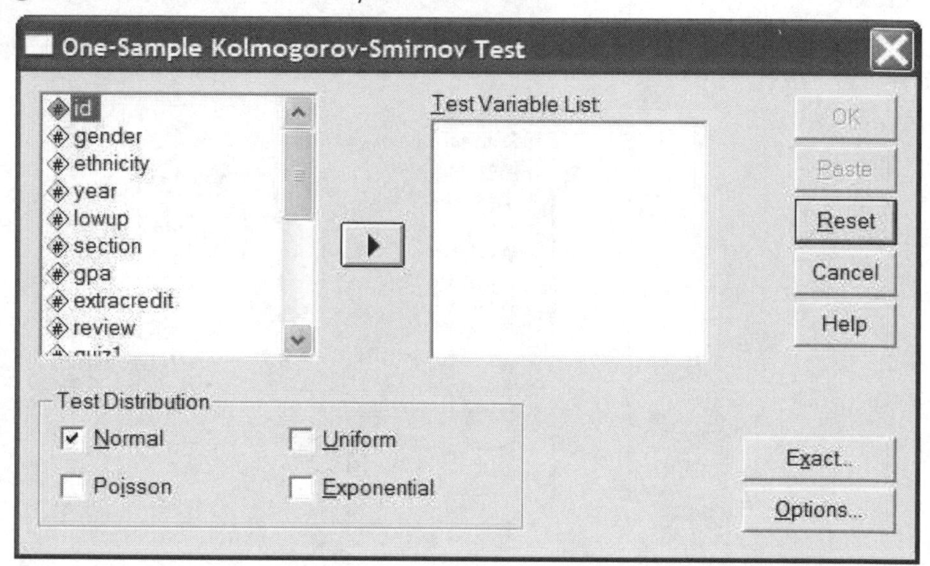

Step by Step

Using Screen 17.6 for visual reference, perform the following sequence of steps to test whether the **final** *variable deviates significantly from normal.*

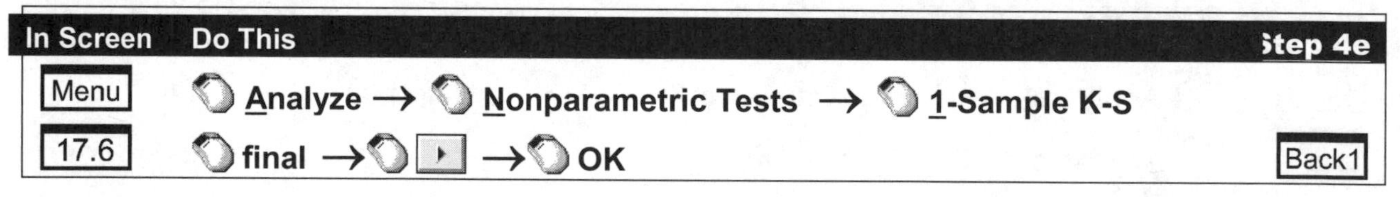

In Screen	Do This	Step 4e
Menu	**Analyze** → **Nonparametric Tests** → **1-Sample K-S**	
17.6	**final** → ▸ → **OK**	Back1

Output

FINAL	N	Normal Parameters		Most Extreme Differences			K-S Z	2-tail Sig.
		Mean	Std. Dev	Absolute	Positive	Negative		
	105	61.48	7.94	.064	.048	-.064	.660	.777

The mean, standard deviation, and *N* of the sample are first identified. The most extreme differences identify the greatest positive difference and the greatest negative difference between the sample and the hypothesized distributions (in z scores). The Kolmogorov-Smirnov z indicates a probability of .777. The large significance value indicates that the distribution of **final** points does not differ significantly from normal, so you can use other procedures than those described in this chapter to analyze this variable.

THE ONE-SAMPLE CHI-SQUARE TEST

This procedure conducts a one-sample chi-square test rather than the more traditional chi-square test of crosstabulated data. The expected values are simply the total number of cases divided by the number of levels of a variable. To demonstrate the procedure, we will conduct a chi-square analysis on the **ethnicity** variable from the **grades.sav** file. With five levels of **ethnicity** and an *N* = 105, the expected value for each cell will be 105/5 = 21. A nice feature of the dialog box is that if you don't expect the distribution to be equal, you may click the **Values** option and specify what proportions you expect. If you are drawing from a population that is 10% Native, 20% Asian, 20% Black, 40% White, and 10% Hispanic, you may test against that distribution by: type 1, click **Add**, type 2, click **Add**, type 2, click **Add**, type 4, click **Add**, type 1, click **Add**. Then the procedure will test your sample against those proportions.

17.7

The Chi-Square Test Window

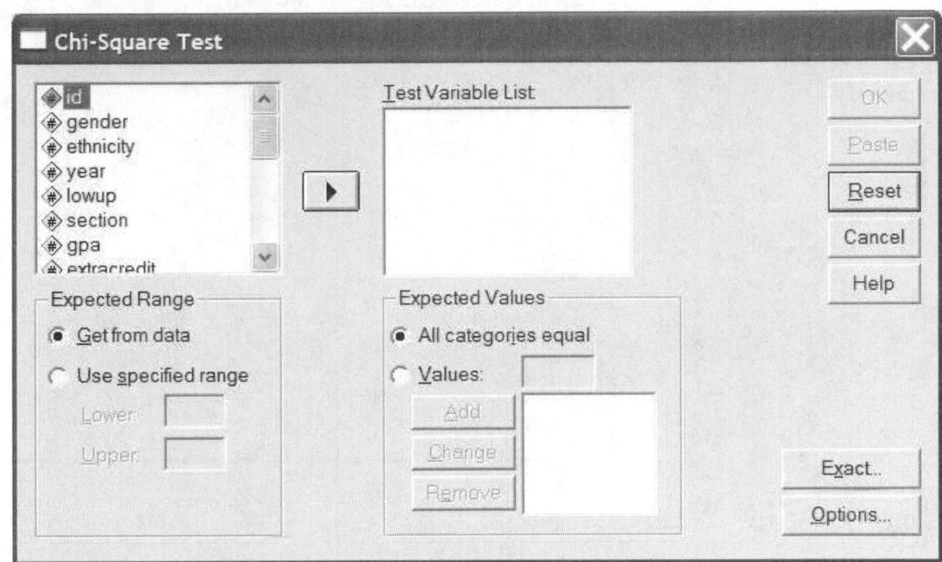

Step by Step

Using Screen 17.7 for visual reference, perform the following sequence of steps to test whether the number of subjects in each ethnic group differ significantly.

In Screen	Do This	Step 4f
Menu	Analyze → Nonparametric Tests → Chi-Square	
17.7	ethnicity → ▸ → OK	Back1

	Observed N	Expected N	Residual
Native	5	21.0	-16.0
Asian	20	21.0	-1.0
Black	24	21.0	3.0
White	45	21.0	24.0
Hispanic	11	21.0	-10.0
Total	105		

Output

Visual inspection of the differences between observed and expected values reveals wide discrepancies. The residuals are the observed values minus the corresponding expected values. The chi-square formula is described in Chapter 8. The degrees of freedom is the number of levels minus 1. The very small significance level demonstrates that the ethnic breakdown of the class deviates substantially from the expected values (equal frequency of each ethnic group).

	ethnicity
Chi-Square[a]	44.857
df	4
Asymp. Sig.	.000

THE FRIEDMAN ONE-WAY ANOVA

The Friedman one-way ANOVA is similar to traditional analysis of variance with two notable exceptions: (a) Comparisons in the Friedman procedure are based on mean rank of variables rather than on means and standard deviations of raw scores, and (b) rather than calculating an *F* ratio, Friedman compares ranked values with expected values in a chi-square analysis. The power of the Friedman operation is less than that of normal analysis of variance, but if your distributions deviate to far from normality, the Friedman one-way ANOVA should be used. This is a simple procedure that does not allow for post hoc tests such as Scheffé or Tukey, nor does it allow for planned contrasts. To demonstrate, we will see whether scores on the five quizzes (**quiz1** to **quiz5**) from the **grades.sav** file differ significantly from each other.

17.8

The Friedman One-Way ANOVA Window

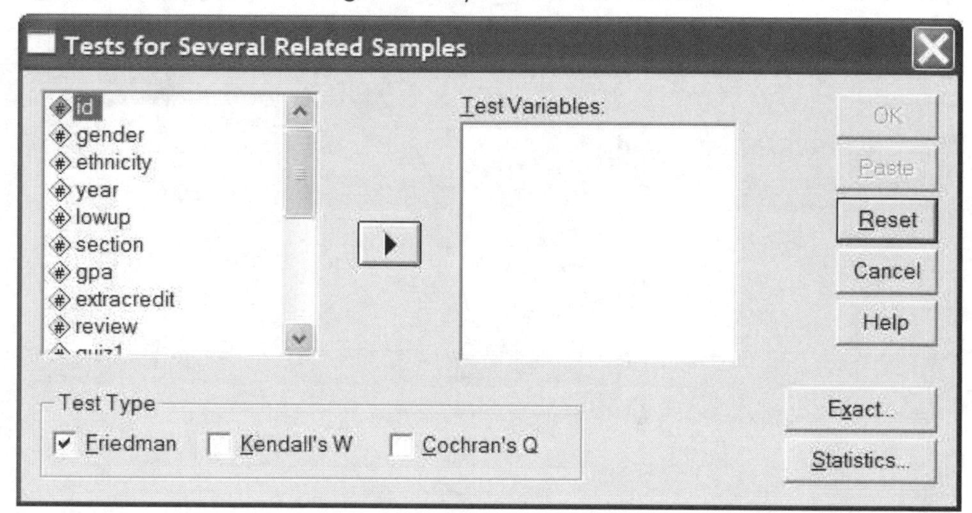

Step by Step

Using Screen 17.8 for visual reference, perform the following sequence of steps to conduct a one-way analysis of variance comparing the means of the five quizzes.

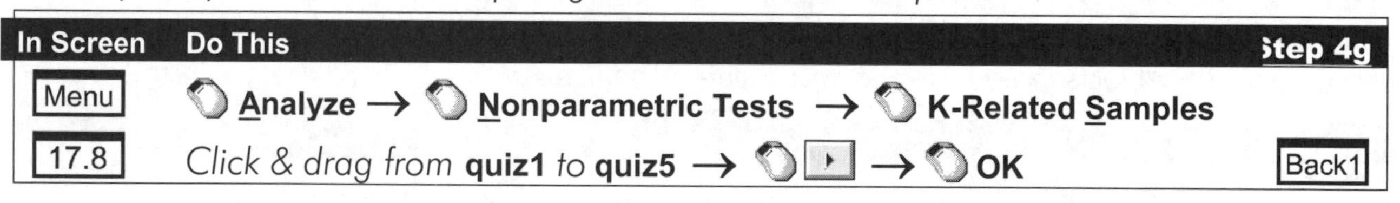

In Screen	Do This	Step 4g
Menu	Analyze → Nonparametric Tests → K-Related Samples	
17.8	*Click & drag from* **quiz1** *to* **quiz5** → ▶ → **OK**	Back1

Output

	quiz1	quiz2	quiz3	quiz4	quiz5	N	Chi-Square	df	Sig.
Mean rank	2.68	3.07	3.34	3.04	2.88	105	12.411	4	.015

The Mean Rank values are determined as follows: All 525 quiz scores (105 × 5) are ranked from high to low and numbered from 1 (for the lowest score) to 11 (for the highest). There are 11 possible scores (0 to 10), and there will be, of course, many frequencies for each level. The ranks for each of the five quizzes are summed and then divided by 105. The significance value associated with the chi-square analysis (p = .015) indicates that there is a significant difference between the five quizzes. This difference could be anywhere within the possible pairwise comparisons. Visual inspection would indicate that **quiz1** (M = 2.68) probably differs significantly from **quiz3** (M = 3.34).

THE K-SAMPLE MEDIAN TEST

The final procedure in this chapter involves computing the median of two or more distributions and then comparing whether the number of values *below* the grand median (median for all groups) differs from the number of values *above* the grand median for each group compared. A chi-square analysis is used to calculate significance levels. To demonstrate we will compare **final** scores in each of the three **section**s. The procedure is to rank order all scores from all three sections combined to determine the grand median. Then for each section the number of scores above this median and the number of scores below this median are calculated. If any section deviates from approximately equal number of scores above and below the grand median, this would indicate that some biasing factor may be present for that section.

17.9

The K-Sample Median Test Window

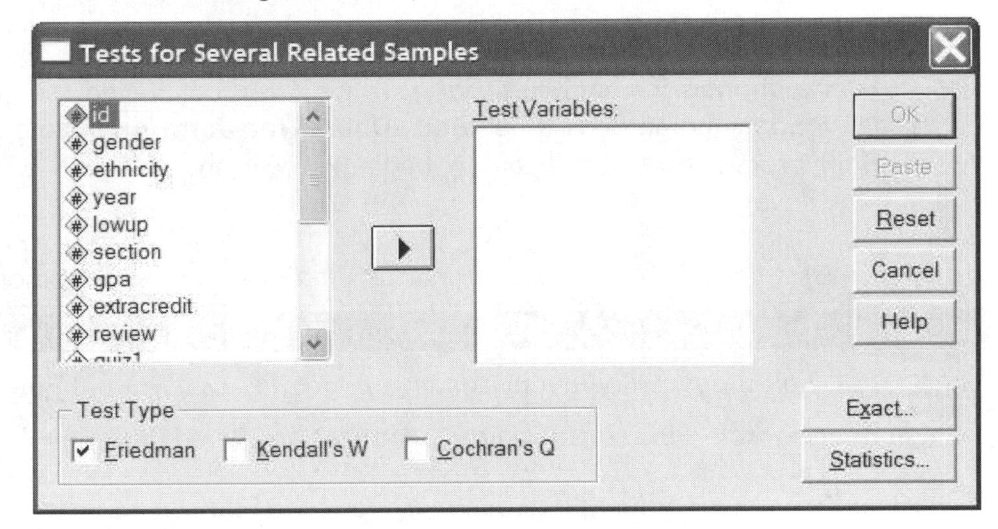

Step by Step

Using Screen 17.9 for visual reference, perform the following sequence of steps to see if the distribution of final exam values from the three sections differ from each other.

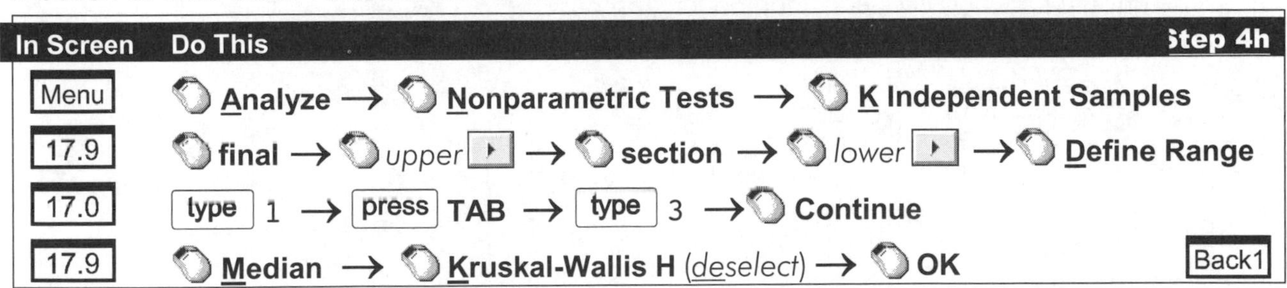

In Screen	Do This	Step 4h
Menu	Analyze → Nonparametric Tests → K Independent Samples	
17.9	final → upper ▸ → section → lower ▸ → Define Range	
17.0	type 1 → press TAB → type 3 → Continue	
17.9	Median → Kruskal-Wallis H (*deselect*) → OK	Back1

Output

final	SECTION 1	SECTION 2	SECTION 3	N	Median	Chi-Square	df	Signif.
> Median	17	16	15	105	62.0	.794	2	.672
<= Median	16	23	18					

The tables portion of the chart is straightforward, indicating the number of scores in each section greater than the median and the number of scores in each section less than the median. Below the chart the overall median value (62) is indicated, along with the chi-square value determined by comparing the observed values with the expected values. Degrees of freedom are the levels of the one variable minus one (3 – 1) times the levels of the other variable minus one (2 – 1). The significance value (p = .672) indicates that the distribution of scores in each section does not differ significantly from predicted values.

Upon completion of any of steps 4 through 4h the output screen will appear (Screen 1, inside back cover). All results from the just-completed analysis are included in the Output Navigator. Make use of the scroll bar arrows (▲▼▶◀) to view the results. Even when viewing output, the standard menu of commands is still listed across the top of the window. Further analyses may be conducted without returning to the data screen.

PRINTING RESULTS

Results of the analysis (or analyses) that have just been conducted requires a window that displays the standard commands (**File Edit Data Transform Analyze** . . .) across the top. A typical print procedure is shown below beginning with the standard output screen (Screen 1, inside back cover).

To print results, from the Output screen perform the following sequence of steps:

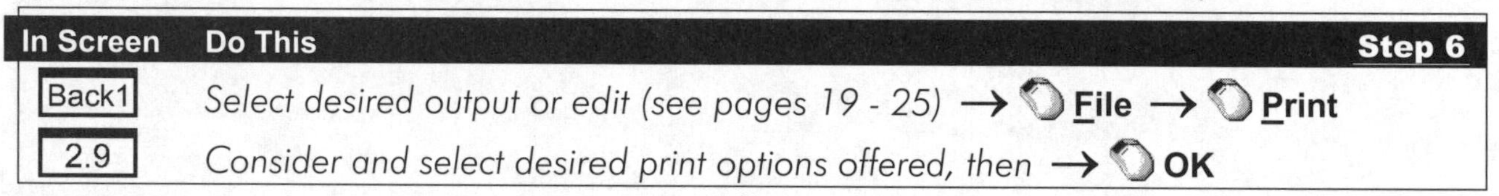

In Screen	Do This	Step 6
Back1	*Select desired output or edit (see pages 19 - 25)* → **File** → **Print**	
2.9	*Consider and select desired print options offered, then* → **OK**	

To exit you may begin from any screen that shows the **File** *command at the top.*

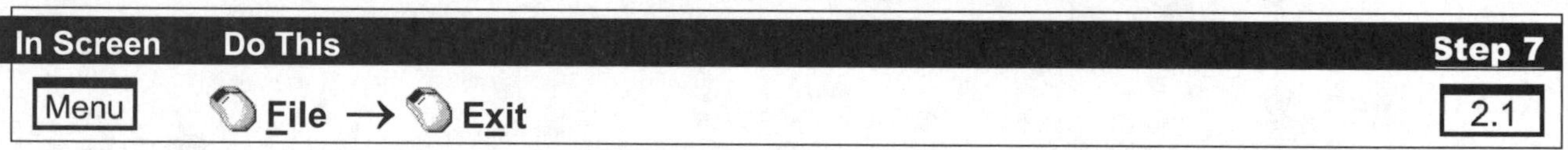

In Screen	Do This	Step 7
Menu	**File** → **Exit**	2.1

Note: After clicking **Exit**, there will frequently be small windows that appear asking if you wish to save or change anything. Simply click each appropriate response.

18

RELIABILITY ANALYSIS

MANY CONSTRUCTS are measured in which a subset of relevant items is selected, administered to subjects, and scored—and then inferences are made about the true population values. Examples abound: An introductory psychology course administers a final exam of 100 questions to test students' knowledge; performance on the final is designed to be representative of knowledge of the entire course content. The quantitative section of the GRE is designed to measure general mathematical ability. The 11 subscales of the Wechsler Intelligence Scale are designed to measure general intellectual aptitude—intelligence. Thirty-three questions on the *Mehrabian-Epstein Scale of Empathic Tendency* are designed to measure the subject's general empathic tendency. And this process continues outside of academia: When Reynaldo Nehemiah, former world record holder in the 110-meter hurdles, expressed interest in playing football with the San Francisco 49ers, they measured his 40-yard speed, his vertical leaping ability, and his strength in a number of lifts as indicators of his ability to play football successfully.

Of the thousands of measurement scales that have been constructed, two critical questions are asked of each: "Is it reliable?" and "Is it valid?" The question of *reliability* (the topic of this chapter) addresses the issue of whether this instrument will produce the same results each time it is administered to the same person in the same setting. Instruments used in the social sciences are generally considered reliable if they produce similar results regardless of who administers them and regardless of which forms are used. The tests given to Nehemiah were certainly reliable: The same tests given many times over several weeks would yield a 40-yard dash of about 4.2 seconds, a vertical leap of about 53 inches, and a bench press of about 355 pounds. But this raised the second question. The cognoscenti of football coined the phrase, "Yeah, but can he take a hit?" They acknowledged Nehemiah's exceptional physical skills, but he hadn't played football since high school. Were these measures of physical skill good predictors of his ability to play in the NFL? They were concerned about the *validity* of the measure used by the 49ers. In a general sense, validity asks the question, "Does it actually measure what it is trying to measure?" In the case of Nehemiah, they wondered if measures of 40 yards speed, leaping ability, and strength actually measured his ability to play professional football.

This chapter deals with the issue of *reliability*. We addressed the issue of *validity* briefly because the two words are so often linked together, and many budding researchers get the two mixed up. We do not have a chapter on validity, unfortunately, because validity is frequently determined by nonstatistical means. Construct validity is sometimes assessed with the aid of factor analysis, and discriminant analysis can assist in creating an instrument that measures well what the researcher is attempting to measure; but face validity is established by observational procedures, and construct validity (although factor analysis may be used to assist) is often theory based.

The two types of *reliability* discussed in this chapter are Chronbach's alpha (also referred to as *coefficient alpha* or α) and *split-half reliability*. There are other measures of reliability (these will be mentioned but not explained in the Step by Step section), but α is the measure that is most widely used. In this chapter, we will first explain the theoretical and statistical procedure on which coefficient alpha is based and then, more briefly, do the same for split-half reliability. We will then present the example that will be used to demonstrate coefficient alpha and split-half reliability.

COEFFICIENT ALPHA (α)

Chronbach's alpha is designed as a measure of internal consistency; that is, do all items within the instrument measure the same thing? Alpha is measured on the same scale as a Pearson *r* (correlation coefficient) and typically varies between 0 and 1. Although a negative value is possible, such a value indicates a scale in which some items measure the opposite of what other items measure. The closer the alpha is to 1.00, the greater the internal consistency of items in the instrument being assessed. At a more conceptual level, coefficient alpha may be thought of as the correlation between a test score and all other tests of equal length that are drawn randomly from the same population of interest. For instance, suppose 1,000 questions existed to test students' knowledge of course content. If ten 100-item tests were drawn randomly from this set, coefficient alpha would approximate the average correlation between all pairs of tests. The formula that determines alpha is fairly simple and makes use of the number of items in the scale (*k*) and the average correlation between pairs of items (*r*):

$$\alpha = \frac{kr}{1+(k-1)r}$$

As the number of items in the scale (*k*) increases, the value of α becomes larger. Also, if the intercorrelation between items is large, the corresponding α will also be large.

SPLIT-HALF RELIABILITY

Split-half reliability is most frequently used when the number of items is large and it is possible to create two halves of the test that are designed to measure the same thing. An example for illustration is the *16 Personality Factor Questionnaire* (16PF) of Raymond Cattell. This 187-item test measures 16 different personality traits. There are 10 to 14 questions that measure each trait. If you wished to do a split-half reliability measure with the 16PF, it would be foolish to compare the first half of the test with the second half, because questions in each half are designed to measure many different things. It would be better to compute reliabilities of a single trait. If you wished to check the reliability of **aggression** (assessed by 14 questions), the best procedure would be to select 7 questions randomly from the 14 and compare them to the other 7. Once item selection is completed, SPSS conducts split-half reliability by computing correlations between the two parts. The split-half procedure can also be utilized if two different forms of a test are taken, or if the same test is administered more than once.

THE EXAMPLE

We use real data from the **helping2.sav** file (*N* = 517) to demonstrate reliability analysis. One segment of this file contains 14 questions that measure the subjects' empathic tendency. These are questions from the *Mehrabian-Epstein Scale of Empathic Tendency* mentioned earlier. The variables are named **empathy1** to **empath14**.

STEP BY STEP

Reliability Analysis

To enter SPSS, a click on **Start** *in the taskbar (bottom of screen) activates the start menu:*

After clicking the SPSS program icon, Screen 1 (inside front cover) appears on the monitor.

Step 2
Create and name a data file or edit (if necessary) an already existing file (see Chapter 3).

Screens 1 and 2 (inside front cover) allow you to access the data file used in conducting the analysis of interest. The following sequence accesses the **helping2.sav** *file for further analyses:*

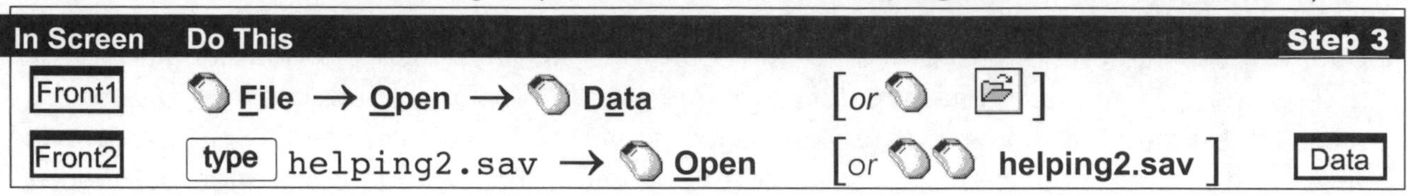

In Screen	Do This		Step 3
Front1	**File** → **Open** → **Data**	[*or*]	
Front2	type `helping2.sav` → **Open**	[or **helping2.sav**]	Data

Whether first entering SPSS or returning from earlier operations the standard menu of commands across the top is required (shown below). As long as it is visible you may perform any analyses. It is not necessary for the data window to be visible.

This menu of commands disappears or modifies when using pivot tables or editing graphs. To uncover the standard menu of commands simply click on the or the icon.

After completion of Step 3 a screen with the desired menu bar appears. When you click a command (from the menu bar), a series of options will appear (usually) below the selected command. With each new set of options, click the desired item. The sequence to access reliability analysis begins at any screen with the menu of commands visible:

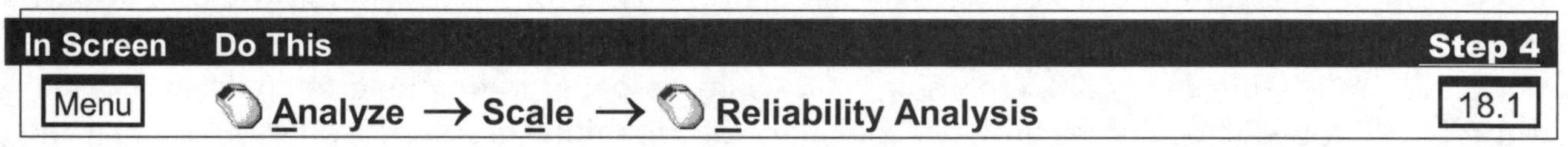

In Screen	Do This	Step 4
Menu	**Analyze** → **Scale** → **Reliability Analysis**	18.1

Reliability Analysis is a popular and frequently used procedure and the SPSS method of accessing reliability analysis is user friendly and largely intuitive. There are only two dialog boxes that provide different selections for the types of analyses offered: the main dialog box (Screen 18.1, below) and a variety of **Statistics** options (Screen 18.2, page 226).

18.1

The Reliability Analysis Window

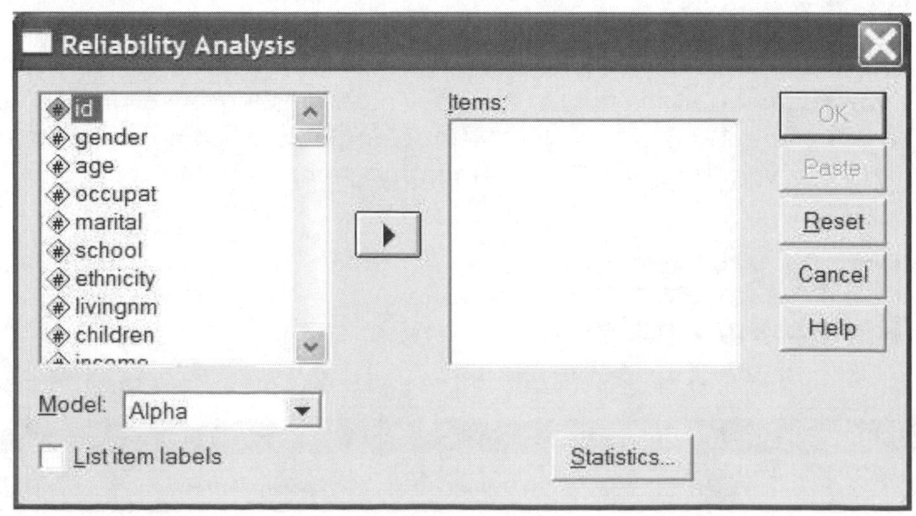

Within the main dialog window several options are available. Similar to other main dialog windows, the variables are listed to the left and the five standard-function pushbuttons to the right. A single **Items** box provides room to enter the variables you wish to include in reliability analysis. When you place variables into the active box (the **Items** box for reliability analysis), you may select them in any order. If they are already in consecutive order this makes your job that much easier. In the menu of items to the right of **Model**, the default (**Alpha**) appears in the small window, but a click on the ▾ will reveal several others.

- **Split-half**: From a single list of variables the split-half procedure will compare variables in the first half of the list (the first half is one larger if there are an odd number of items) with variables in the second half of the list.
- **Guttman**: This option calculates reliability based on a lower bounds procedure.
- **Parallel**: Computes reliability based on the assumption that all included items have the same variance.
- **Strict parallel**: Computes a reliability based on the assumption that all items have the same mean, the same true score variance, and the same error variance over multiple administrations of the instrument.

A click on the **Statistics** pushbutton opens a dialog box (Screen 18.2, following page) that presents the remaining options for this procedure. All items will be briefly described here except for items in the **ANOVA Table** box. The ANOVA and Chi-square options provided there are described in some detail elsewhere in this book, and, frankly, are rarely used in reliability analysis. First the **Descriptives for** items will be described, followed by **Summaries**, **Inter-Item**, **Hotelling's T-square**, and **Tukey's test of additivity**.

- **Descriptives for Item**: This provides simple means and standard deviations for each variable included in the analysis.
- **Descriptives for Scale**: This option provides the mean, variance, standard deviation and *N* for the *sum* of all variables in the scale. More specifically, for each subject the scores for the 14 empathy questions are summed. Descriptives are then provided for this single summed variable.
- **Descriptives for Scale if item deleted**: For each variable this option identifies the resulting alpha value if that item were deleted from the scale. Almost always this option should be selected.
- **Summaries** for **Means**: This option computes the mean for all subjects for each of the entered variables. In the present example, it computes the mean value for **empathy1**, the mean value for **empathy2**, the mean value for **empathy3**, and so forth up to the mean value of **empath14**. Then it lists the *mean* of these 14 means, the *minimum* of the 14 means, the *maximum* of the 14 means, the *range*, the *maximum divided by the minimum*, and the *variance* of the 14 means. Although a bit complex, this is a central construct of reliability analysis.
- **Summaries** for **Variances**: The same as the previous entry except for the variances.
- **Summaries** for **Covariances**: This procedure computes the covariance between each entered variable and the sum of all other variables. In the present example, a covariance would be computed between **empathy1** and the sum of the other 13, between **empathy2** and the sum of the other 13, and so forth for all fourteen variables. Then the final table would provide the mean of these 14 covariances, the minimum, maximum, range, minimum divided by maximum, and variance.

- **Summaries** for **Correlations**: The same as the procedure for covariances except that all values are computed for correlations instead. In the present example, the mean of these 14 correlations is the *r* in the alpha formula.
- **Inter-Item Correlations**: The simple correlation matrix of all entered variables.
- **Inter-Item Covariances**: The simple covariance matrix of all entered variables.
- **Hotelling's T-square**: This is a multivariate T-test that tests whether all means of the entered variables are equal.
- **Tukey's test of additivity**: This is essentially a test for linear dependency between entered variables.

18.2

The Reliability Analysis: Statistics Window

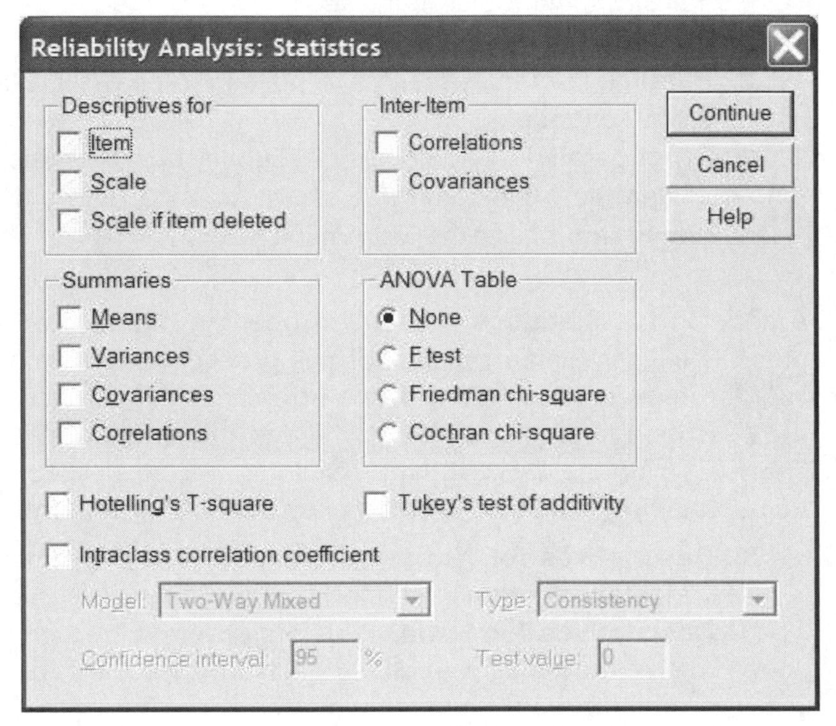

Three different step-by-step sequences will be presented: Step 5 will describe the simplest default procedure for conducting reliability analysis. In some settings a default procedure may be quite useful. Often you may want a quick check of how reliable some measure is without thought about further analyses. Step 5a will create a reliability analysis that includes a number of options discussed earlier. It turns out that when using all 14 of the empathy questions the reliability is fairly low. By deleting five of the items that hurt the internal consistency of the measure, nine items are left that make a reliable scale. This second analysis will include only those variables that contribute to the maximum alpha value. Finally, step 5b will demonstrate how to access a split-half reliability with the nine (rather than the 14) variables. The italicized text before each sequence box will identify the specifics of the analysis. All three analyses are included in the Output section.

*To conduct a reliability analysis with the fourteen empathic tendency questions (**empathy1** to **empath14**) with all the default options, perform the following sequence of steps. The starting point is Screen 18.1. Carry out the step-4 sequence (page 224) to arrive at this screen.*

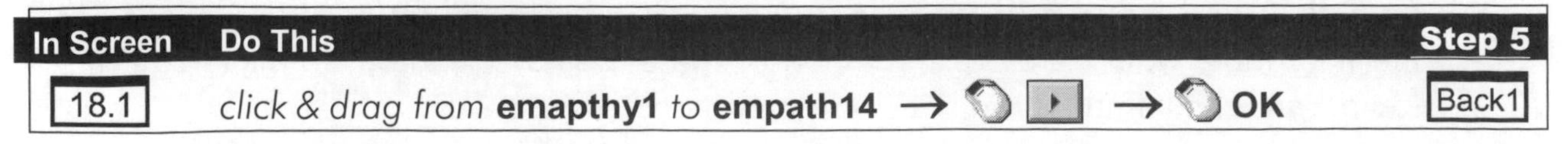

The click and drag operation will select all 14 questions in the original order. The output from this sequence includes the alpha value, the number of subjects and the number of variables.

*To conduct a reliability analysis based on coefficient alpha with the nine empathic tendency questions that yielded the highest alpha (***empathy1-3-4-5-6-7-9-11-12***), to request* **Scale**, **Scale if item deleted**, **Means**, **Variances**, **Correlations**, **Inter-Item correlations** *and a list of the value labels; perform the following sequence of steps. The starting point is Screen 18.1. Carry out the 4-step sequence (page 224) to arrive at this screen.*

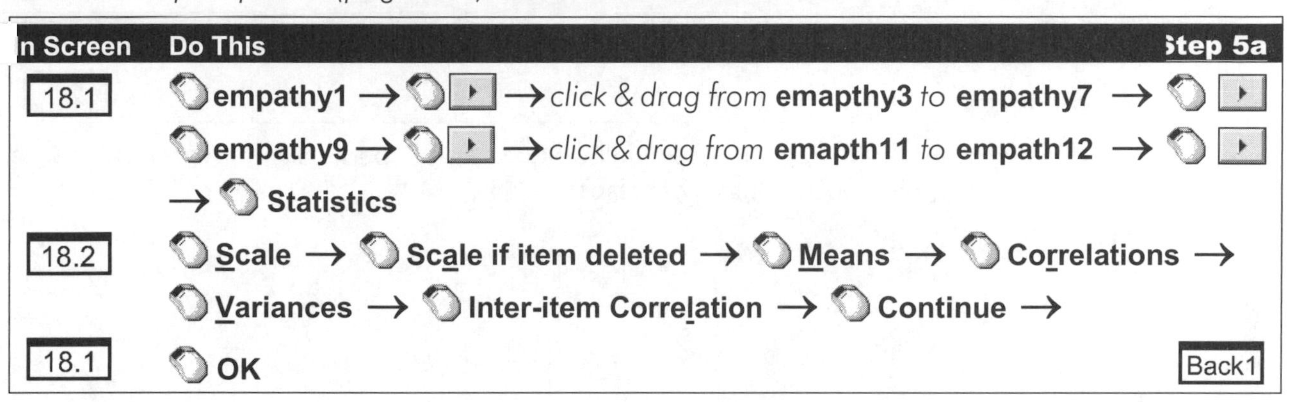

In Screen	Do This	Step 5a
18.1	**empathy1** → ▸ → *click & drag from* **emapthy3** *to* **empathy7** → ▸ **empathy9** → ▸ → *click & drag from* **emapth11** *to* **empath12** → ▸ → **Statistics**	
18.2	**Scale** → **Scale if item deleted** → **Means** → **Correlations** → **Variances** → **Inter-item Correlation** → **Continue** →	
18.1	**OK**	Back1

If you wish to perform an identical procedure to that demonstrated in sequence step 5a but conducting **Split-half** *reliability rather than using* **Alpha**, *perform the following sequence of steps. The starting point is Screen 18.1. Carry out the step-4 sequence (page 224) to arrive at this screen.*

In Screen	Do This	Step 5b
18.1	**empathy1** → ▸ → *click & drag from* **emapthy3** *to* **empathy7** → ▸ **empathy9** → ▸ → *click & drag from* **emapth11** *to* **empath12** → ▸ → ▾ *next to* **Model** → **Split-half** → **Statistics**	
18.2	**Scale** → **Scale if item deleted** → **Means** → **Correlations** → **Variances** → **Inter-item correlation** → **Continue** →	
18.1	**OK**	Back1

Upon completion of step 5, 5a, or 5b, the output screen will appear (Screen 1, inside back cover). All results from the just-completed analysis are included in the Output Navigator. Make use of the scroll bar arrows (▲▼▶◀) to view the results. Even when viewing output, the standard menu of commands is still listed across the top of the window. Further analyses may be conducted without returning to the data screen.

PRINTING RESULTS

Results of the analysis (or analyses) that have just been conducted requires a window that displays the standard commands (**File Edit Data Transform Analyze** . . .) across the top. A typical print procedure is shown below beginning with the standard output screen (Screen 1, inside back cover).

To print results, from the Output screen perform the following sequence of steps:

In Screen	Do This	Step 6
Back1	*Select desired output or edit (see pages 19 - 25)* → **File** → **Print**	
2.9	*Consider and select desired print options offered, then* → **OK**	

To exit you may begin from any screen that shows the **File** *command at the top.*

In Screen	Do This	Step 7
Menu	**File** → **Exit**	2.1

Note: After clicking **Exit**, there will frequently be small windows that appear asking if you wish to save or change anything. Simply click each appropriate response.

OUTPUT

Reliability Analysis

What follows is partial output from the three sequence steps (5, 5a, and 5b) presented on page 227. We have included an **Item-total statistics** for all 14 variables (even though it was not selected in the default procedure) for sake of illustration. A slightly condensed version of the output in the standard format is reproduced here. We begin with the list of variable names and labels. This is followed by the item-total statistics for all 14 variables, used to illustrate the criteria by which certain variables may be eliminated to improve the alpha. This is followed by scale statistics, inter-item correlations, then the item-total statistics for the nine remaining items after five have been removed from the analysis. The section concludes with output from a split-half analysis. Definitions and text are included between sections to aid in interpretation. Below are variable names and labels for your reference.

#	Variable	Label
1.	EMPATHY1	Sad to see lonely stranger
2.	EMPATHY2	Annoyed at sorry-for-self people (*neg)
3.	EMPATHY3	Emotionally involved with friends problem
4.	EMPATHY4	Disturbed when bring bad news
5.	EMPATHY5	Person crying upsetting
6.	EMPATHY6	Really involved in a book or movie
7.	EMPATHY7	Angry when someone ill-treated
8.	EMPATHY8	Amused at sniffling at movies (*neg)
9.	EMPATHY9	Do not feel ok when others depressed
10.	EMPATH10	Hard to see why others so upset (*neg)
11.	EMPATH11	Upset to see animal in pain
12.	EMPATH12	Upset to see helpless old people
13.	EMPATH13	Irritation rather than sympathy at tears (*neg)
14.	EMPATH14	Remain cool when others excited (*neg)

Item-Total Statistics

	Scale Mean if Item Deleted	Scale Variance if Item Deleted	Corrected Item-Total Correlation	Squared Multiple Correlation	Cronbach's Alpha if Item Deleted
empathy1	62.27	81.313	.403	.326	.654
empathy2	63.37	94.520	-.053	.120	.718
empathy3	62.57	81.323	.457	.350	.648
empathy4	62.17	81.702	.453	.415	.649
empathy5	62.40	79.322	.473	.372	.644
empathy6	62.31	81.852	.385	.232	.657
empathy7	61.22	84.746	.398	.240	.659
empathy8	62.13	84.895	.228	.272	.680
empathy9	62.83	85.395	.271	.211	.672
empath10	62.75	91.286	.048	.121	.704
empath11	61.60	82.256	.400	.325	.655
empath12	61.51	81.987	.436	.411	.651
empath13	61.45	84.628	.303	.250	.668
empath14	63.34	88.461	.164	.101	.686

Reliability Statistics

Cronbach's Alpha	Cronbach's Alpha Based on Standardized Items	N of Items
.685	.698	14

The first chart lists the variables for which a reliability check is being conducted. Note that five of the items are marked with a "(*neg)". These are questions that have been phrased negatively to control for response bias. The scoring on these items is reversed before beginning calculations. The second chart is designed to designate similarities and differences between each variable and composites of the other variables. Some terminology (defined below) will aid understanding.

Term	Definition/Description
SCALE MEAN IF ITEM DELETED	For each subject the 13 variables (excluding the variable to the left) are summed. The values shown are the means for the 13 variables across all 517 subjects.
SCALE VARIANCE IF ITEM DELETED	The variance of summed variables when the variable to the left is deleted.
CORRECTED ITEM-TOTAL CORRELATION	Correlation of the designated variable with the sum of the other 13.
ALPHA IF ITEM DELETED	The resulting alpha if the variable to the left is deleted.

Notice that for four of the variables, the correlations between each of them and the sum of all other variables is quite low; in the case of **empathy2**, it is even negative (-.053). Correspondingly, the **Alpha** value would increase (from the .685 shown on the lower chart) if these items were deleted from the scale. In the analyses that follow, all items that *reduce* the alpha value have been deleted. After removing the four designated variables, a second analysis was run, and it was found that a fifth variable's removal would also increase the value of **Alpha**. Five variables were eventually dropped from the scale: variables 2, 8, 10, 13, and 14. Please be vividly aware that variables are not *automatically* dropped just because a higher **Alpha** results. There are often theoretical or practical reasons for keeping such variables.

Note: Interpretation of the descriptive statistics and correlation matrix are straightforward and not shown. The output that follows involves analyses *after* the five variables were removed.

Scale Statistics

Mean	Variance	Std. Deviation	N of Items
44.75	71.131	8.434	9

Summary Item Statistics

	Mean	Minimum	Maximum	Range	Max /Min	Variance	N of Items
Item Means	4.973	4.242	5.849	1.607	1.379	.286	9
Item Variances	2.317	1.632	2.663	1.031	1.631	.110	9
Inter-Item Correlations	.303	.120	.521	.401	4.343	.009	9

These items were described and explained earlier; the definitions of terms follow.

Term	Definition/Description
SCALE STATISTICS	There are a total of nine variables being considered. This line lists descriptive information about the *sum* of the nine variables for the entire sample of 517 subjects.
ITEM MEANS	This is descriptive information about the 517 subjects' means for the nine variables. In other words, the $N = 9$ for this list of descriptive statistics. The "mean of means" = 4.973. The minimum of the nine means = 4.242, and so forth.
ITEM VARIANCES	A similar construct as that used in the line above (**ITEM MEANS**). The first number is the mean of the nine variances, the second is the lowest of the nine variances, and so forth.
INTER-ITEM CORRELATIONS	This is descriptive information about the correlation of each variable with the sum of the other eight. There are nine correlations computed: the correlation between the first variable and the sum of the other eight variables, the correlation between the second variable and the sum of the other eight, and so forth. The first number listed is the mean of these nine correlations (.303), the second is the lowest of the nine (.120), and so forth. The mean of the inter-item correlations (.303) is the r in the $\alpha = rk/[1 + (k - 1)r]$ formula.

The table that follows is the same as that presented earlier except that values shown relate to 9 rather than 14 variables, and an additional column has been added: Squared Multiple Correlation. Further the **Alpha** has increased to .795 (from .685). Also notice that the overall alpha value cannot be improved by deleting another variable. Four of the columns have been described on the previous page. Terms new to this chart are defined after the following output.

Item-Total Statistics

	Scale Mean if Item Deleted	Scale Variance if Item Deleted	Corrected Item-Total Correlation	Squared Multiple Correlation	Cronbach's Alpha if Item Deleted
empathy1	39.95	55.847	.522	.316	.770
empathy3	40.26	57.477	.509	.337	.772
empathy4	39.85	56.141	.590	.400	.762
empathy5	40.08	54.658	.572	.368	.763
empathy6	39.99	58.630	.399	.221	.787
empathy7	38.91	61.109	.420	.209	.783
empathy9	40.51	59.014	.399	.185	.787
empath11	39.28	58.630	.432	.314	.782
empath12	39.20	57.263	.527	.403	.770

Reliability Statistics

Cronbach's Alpha	Cronbach's Alpha Based on Standardized Items	N of Items
.795	.796	9

Term	Definition/Description
SQUARED MULTIPLE CORRELATION	These values are determined by creating a multiple regression equation to generate the *predicted correlation* based on the correlations for the other eight variables. The numbers listed are these predicted correlations.
ALPHA	Based on the formula: $\alpha = rk/[1 + (k - 1)r]$, where k is the number of variables considered (9 in this case) and r is the mean of the inter-item correlations (.303, from previous page). The alpha value is inflated by a larger number of variables; so there is no set interpretation as to what is an acceptable alpha value. A rule of thumb that applies to most situations is: $\alpha > .9$—excellent; $\alpha > .8$—good; $\alpha > .7$—acceptable; $\alpha > .6$—questionable; $\alpha > .5$—poor; $\alpha < .5$—unacceptable
ALPHA BASED ON STANDARDIZED ITEMS	This is the alpha produced if the composite items (in the chart on the previous page, *not* the scores for the 517 subjects) are changed to *z*-scores before computing the **Alpha**. The almost identical values produced here (.795 vs. .796) indicate that the means and variances in the original scales do not differ much, and thus standardization does not make a great difference in the **Alpha**.

Split-Half Reliability

The output from a split-half reliability follows. Recall that we are conducting a split-half analysis of the same nine empathy questions. This would typically be considered too small a number of items upon which to conduct this type of analysis. However, it serves as an adequate demonstration of the procedure and does not require description of a different data set. Note that much information in the **Split-half** output is identical to that for the **Alpha** output. Only output that is different is included here.

Scale Statistics

	Mean	Variance	Std. Deviation	N of Items
Part 1	23.64	31.112	5.578	5
Part 2	21.12	17.108	4.136	4
Both Parts	44.75	71.131	8.434	9

Summary Item Statistics

		Mean	Minimum	Maximum	Range	Max / Min	Variance	N of Items
Item Means	Part 1	4.727	4.499	4.901	.402	1.089	.023	5
	Part 2	5.279	4.242	5.849	1.607	1.379	.505	4
	Both Parts	4.973	4.242	5.849	1.607	1.379	.286	9
Item Variances	Part 1	2.446	2.120	2.663	.543	1.256	.069	5
	Part 2	2.155	1.632	2.474	.842	1.516	.138	4
	Both Parts	2.317	1.632	2.663	1.031	1.631	.110	9
Inter-Item Correlations	Part 1	.389	.260	.504	.244	1.940	.005	5
	Part 2	.332	.226	.521	.295	2.306	.010	4
	Both Parts	.303	.120	.521	.401	4.343	.009	9

Reliability Analysis—Split-half

Alpha For Part 1 (5 Items)	.759
Alpha For Part 2 (4 Items)	.662
Correlation Between Forms	.497
Equal-Length Spearman-Brown	.664
Guttman Split-Half	.644
Unequal Length Spearman-Brown	.666

The Statistics for Item Means, Item Variances, and Inter-Item Correlations sections are identical to those shown on page 230 and it has equivalent numbers for the two halves of the scale as well as for the entire scale. The Scale Statistics values are also identical to those on page 230. Please refer there for an explanation. Notice that the alpha values (last line of output) for each half are lower than for the whole scale. This reflects the reality that a scale with a fewer number of items produces a deflated alpha value. Definitions of new terms follow.

Term	Definition/Description
CORRELATION BETWEEN FORMS	An estimate of the reliability of the measure if it had five items.
EQUAL-LENGTH SPEARMAN-BROWN	The reliability of a 10-item test (.6636) if it was made up of equal parts that each had a five-item reliability of .4965.
GUTTMAN SPLIT-HALF	Another measure of the reliability of the overall test, based on a lower-bounds procedure.
UNEQUAL-LENGTH SPEARMAN-BROWN	The reliability acknowledging that this test has 9 and not 10 items.

EXERCISES

Answers to selected exercises are available for download at **www.ablongman.com/george6e**.

Use the **helping3.sav** file for the exercises that follow (downloadable at the address shown above). Measure the internal consistency (coefficient alpha) of the following sets of variables. An "h" in front of a variable name, refers to assessment by the help giver; an "r" in front of a variable name refers to assessment by the help recipient.

Compute Coefficient alpha for the following sets of variables, then delete variables until you achieve the highest possible alpha value. Print out relevant results.

1. **hsevere1, hsevere2, rsevere1, rsevere2** measure of problem severity
2. **sympath1, sympath2, sympath3, sympath4** measure of helper's sympathy
3. **anger1, anger2, anger3, anger4** measure of helper's anger
4. **hcompe1, hcompe2, hcope3, rcope1, rcope2, rcope3** how well the recipient is coping
5. **hhelp1-hhelp15** helper rating of time spent helping
6. **rhelp1-rhelp15** recipient's rating of time helping
7. **empathy1-empath14** helper's rating of empathy
8. **hqualit1, hqualit2, hqualit3, rqualit1, rqualit2, rqualit3** quality of help
9. **effic1-effic15** helper's belief of self efficacy
10. **hcontro1, hcontro2, rcontro1, rcontro2** controllability of the cause of the problem

From the **divorce.sav** file:

11. **drelat-dadjust** (16 items) factors disruptive to divorce recovery
12. **arelat-amain2** (13 items) factors assisting recovery from divorce
13. **sp8-sp57** (18 items) spirituality measures

19

MULTIDIMENSIONAL SCALING

MULTIDIMENSIONAL SCALING is useful because a picture is often easier to interpret than a table of numbers. In multidimensional scaling, a matrix made up of dissimilarity data (for example, ratings of how different products are, or distances between cities) is converted into a one, two, or three-dimensional graphical representation of those distances. (Although it is possible to have more than three dimensions in multidimensional scaling, that is rather rare).

A particular advantage of multidimensional scaling graphs is that the distances between two points can be interpreted intuitively. If two points on a multidimensional scaling graph are far apart from each other, then they are quite dissimilar in the original dissimilarity matrix; if two points are close together on the graph, then they are quite similar in the original matrix.

Multidimensional scaling has some similarities with factor analysis (see Chapter 20). Like factor analysis, multidimensional scaling produces a number of dimensions along which the data points are plotted; unlike factor analysis, in multidimensional scaling the meaning of the dimensions or factors is not central to the analysis. Like factor analysis, multidimensional scaling reduces a matrix of data to a simpler matrix of data; whereas factor analysis typically uses correlational data, however, multidimensional scaling typically uses dissimilarity ratings. Finally, factor analysis produces plots in which the angles between data points are the most important aspect to interpret; in multidimensional scaling, the location in space of the data points (primarily the distance between the points) is the key element of interpretation.

There are variations on the multidimensional scaling theme that use similarity data instead of dissimilarity data. These are included in the SPSS Categories option, which is not covered in this book.

There are also a number of similarities between multidimensional scaling and cluster analysis (see Chapter 21). In both procedures, the proximities or distances between cases or variables may be examined. In cluster analysis, however, a qualitative grouping of the cases into clusters is typical; in multidimensional scaling, rather than ending with clusters we end with a graph in which we can both visually and quantitatively examine the distances between each case.

The SPSS multidimensional scaling procedure (historically known as ALSCAL) is in fact a collection of related procedures and techniques rather than a single procedure. In this chapter, we will focus on three analyses that exemplify several key types of data that may be analyzed, and attempt to briefly yet clearly define the key terms involved. If this is your first experience with multidimensional scaling, we would suggest that you refer to Davidson (1983) or Young (1987).

In the first example, a sociogram is computed for a selection of students in a class; in this case, ratings of how much they dislike each other are converted into a graph of how far apart they want to be from each other. In the second example, we will let SPSS compute dissimilarity scores for students' quiz scores, and produce a graph of how far apart the students are in their patterns of quiz scores. Finally, in the third example a small group of student' ratings of dissimilarities between television shows are analyzed.

SQUARE ASYMMETRICAL MATRIXES (THE SOCIOGRAM EXAMPLE)

Imagine that an instructor wanted to create the perfect seating chart of a class. In order to do this, she asked each student to rate each other student, asking them, "On a scale of 1 (not at all) to 5 (very much), how much do you dislike _____?" She asked each of the 20 students in her class to rate the other 19 students (they did not rate themselves). The raw data for this fictitious example is included in the **grades-mds.sav** file. The teacher would like a visual representation of how far apart each student wants to sit from each other student, so she performs a multidimensional scaling analysis.

In this case, the teacher is starting from a 20 × 20 matrix of dissimilarities. She would like to convert this matrix to a two-dimensional picture indicating liking or disliking. To do this, multidimensional scaling will develop a much simpler matrix (20 students × 2 dimensions) that will, as much as possible, represent the actual distances between the students. This is a square asymmetrical dissimilarity matrix. To translate:

- Square matrix: The rows and columns of the matrix represent the same thing. In this case, the rows represent raters, and the columns represent the people being rated. This is a square matrix because the list of students doing the rating is the same as the list of people being rated…they are rating each other. The matrix would be called rectangular if the raters were rating something else other than themselves. The analysis and interpretation of rectangular matrices is beyond the scope of this book.
- Asymmetrical matrix: The matrix is called asymmetrical because it is quite possible that, for example, Shearer dislikes Springer less than Springer dislikes Shearer. If the numbers below the diagonal in the matrix can be different than the numbers above the diagonal, as in this example, then the matrix is asymmetrical; if the numbers below the diagonal in the matrix are always a mirror image of the numbers above the diagonal in the matrix, then the matrix is symmetric. Correlation tables are a good example of a matrix that is always symmetric.
- Dissimilarity matrix: In the data, higher numbers mean more dissimilarity; in this case, higher numbers mean more disliking. If a similarity matrix is used (for example, if students rated how much they liked each other), students that were farthest apart would be the ones that liked each other most (that's why you shouldn't use multidimensional scaling to analyze similarity matrixes).

SQUARE SYMMETRICAL MATRIXES WITH CREATED DISTANCES FROM DATA (THE QUIZ SCORE EXAMPLE)

After successfully using her seating chart for one term, our fictional instructor decides that she wants to use multidimensional scaling to create a seating chart again; this time, she decides to seat students according to their quiz scores on the first five quizzes. In this case, because the quiz scores are ***not*** dissimilarity data, SPSS will be used to calculate dissimilarities. In this case, we are starting from a data matrix containing 20 students × 5 quizzes, computing a data matrix containing 20 students × 20 students dissimilarities, and then using multidimensional scaling to produce a 20 students × 2 dimensions matrix, represented visually on a graph, from which we can derive our seating chart. We will again be using data included in the **grades-mds.sav** file, and the analysis will involve a square symmetrical matrix:

- Square matrix: Although our original matrix is not square (because we have cases that are students, and variables that are quizzes), we are asking SPSS to compute dissimilarities. The dissimilarity matrix that SPSS computes will have students in rows and in columns; because the rows and columns represent the same thing, the matrix is square.
- Symmetrical matrix: The matrix is symmetrical because the dissimilarity between, for example, between Liam and Cha is identical to the dissimilarity between Cha and Liam.

INDIVIDUAL DIFFERENCE MODELS (THE TV SHOW EXAMPLE)

In each of the previous examples, there was one dissimilarity matrix that was used to compute distances and create a multidimensional mapping of the data. In some cases, however, it is possible to have several dissimilarity matrixes used in a single analysis. This can happen, for example, if you want to replicate your experiment in three different samples; in this case, you would have three dissimilarity matrixes (this is known as replicated multidimensional scaling). You may also have several dissimilarity matrixes if you have several people rating the dissimilarity between things. In our example, we will focus on the later situation (known as individual difference models, or weighted multidimensional scaling models).

Imagine that, because our fictitious teacher has worn herself out developing seating charts, she decides to have her students watch television for an hour each day. After watching several episodes each of *ER*, *60 Minutes*, *The Simpsons*, and *Seinfeld*, she has five of her students rate how different each of the programs are from each of the other programs. These five symmetrical dissimilarity matrixes will be examined to determine how each of the television shows is perceived.

Because SPSS is rather exacting as to the format of the data when using individual difference models, the data for this analysis is included in the a separate **grades-mds2.sav** file. In particular, with individual difference data, you must have several matrixes of the same size in your data file; in our example, because we have five students rating dissimilarity between four television shows, our data file needs to have (5) 4 × 4 matrixes. The first four rows of the file will be the dissimilarity matrix for the first student, the second four rows of the file will be the dissimilarity matrix for the third student, and so forth, for a total of twenty rows in the file.

In the Step by Step section, different versions of Step 5 will be used to describe how to complete each of the three sample analyses.

STEP BY STEP

Multidimensional Scaling

To enter SPSS, a click on **Start** *in the taskbar (bottom of screen) activates the start menu:*

After clicking the SPSS program icon, Screen 1 (inside front cover) appears on the monitor.

Step 2

Create and name a data file or edit (if necessary) an already existing file (see Chapter 3).

Screens 1 and 2 (displayed on the inside front cover) allow you to access the data file used in conducting the analysis of interest. The following sequence accesses the **grades-mds.sav** *file for further analyses; if you wish to access the* **grades-mds2.sav** *file or a different file, simply substitute the appropriate file name.*

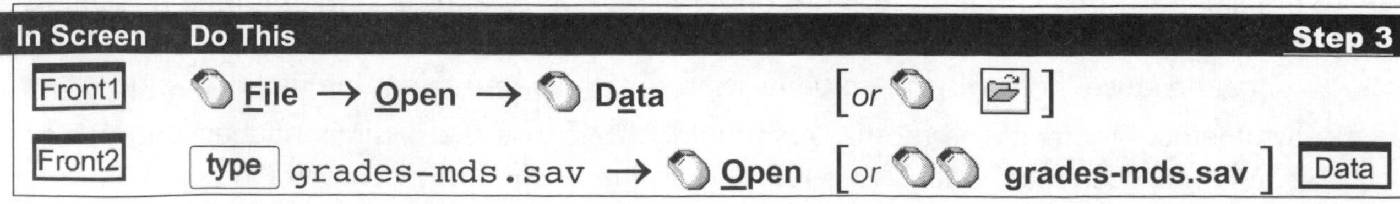

Whether first entering SPSS or returning from earlier operations the standard menu of commands across the top is required (shown below). As long as it is visible you may perform any analyses. It is not necessary for the data window to be visible.

This menu of commands disappears or modifies when using pivot tables or editing graphs. To uncover the standard menu of commands simply click on the ▬ or the ▣ icon.

After completion of Step 3 a screen with the desired menu bar appears. When you click a command (from the menu bar), a series of options will appear (usually) below the selected command. With each new set of options, click the desired item. The sequence to access multidimensional scaling begins at any screen with the menu of commands visible:

In Screen	Do This	Step 4
Menu	**Analyze → Scale → Multidimensional Scaling (ALSCAL)**	19.1

After this sequence step is performed, a window opens (Screen 19.1, below) in which you select the variables that you want to analyze. There are two different ways of doing a multidimensional scaling analysis in SPSS: You can either provide a dissimilarity matrix, or you can ask SPSS to create one.

19.1

Multidimensional Scaling Dialog

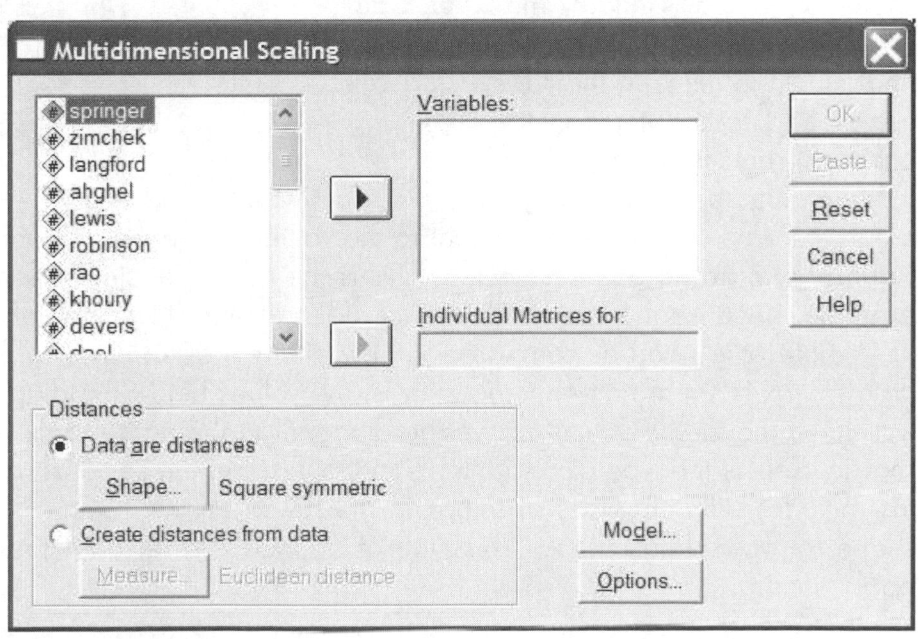

If you are providing a dissimilarity matrix, then here you select which columns in you data matrix you want to analyze. It is possible to have more columns in your data file than you want to include in your analysis, but it is not possible to include more rows in the data file than you want to analyze (unless you use the select command; see page 56). If you are analyzing a square matrix (in which the variables correspond to the cases in your data file, as in our sociogram example), and your data file contains dissimilarity data (rather than having SPSS compute the dissimilarities), then the number of variables that you select should be equal to the number of rows in your data file.

If you want SPSS to calculate a dissimilarity matrix from other data in your file, then you select which variables you want SPSS to use in calculating distances between cases (or, you may have SPSS calculate distances between variables you select using all of the cases).

In either case, once you have selected the variables from the list at the left that you want to analyze, click the ▶ button to move them to the **Variables** box.

If you want SPSS to compute distances from your data, and you are also planning on using a replication model or an individual differences (weighted) model, then you can select a variable for the **Individual Matrices for** box. This is a fairly complex procedure, so we do not further discuss this option here.

Once you have selected which variables you want to include in the analysis, it is time to tell SPSS exactly what the structure of your data file is. This is done through the options and dialog buttons in the lower left of Screen 19.1, labeled **Distances**. If your data file contains dissimilarity data (that is, it contains data indicating how different or how far apart cases and variables are) then you should keep the default setting, **Data are distances**. If your data matrix is square symmetric (that is, the dissimilarity values above the diagonal are equal to the data values below the diagonal), then you do not need to further specify what kind of data you have. If you are using square asymmetric or rectangular data, then click on the **Shape...** button and Screen 19.2 will pop up.

19.2

Multidimensional Scaling: Shape of Data Window

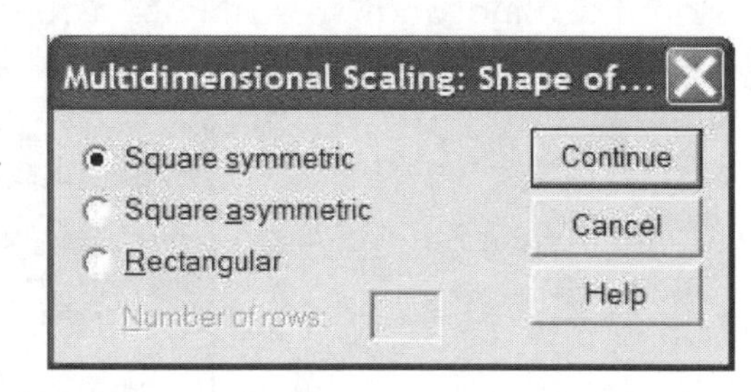

In this screen, you select the shape of your dissimilarity matrix. If your matrix is square (that is, if the items in the rows correspond to the items in the columns) and symmetric (that is, the values below the diagonal in the data matrix are equivalent to the values above the diagonal), then you may leave the **Square symmetric** option selected. This is the type of dissimilarity matrix used in the TV show example, because there is no difference, for example, in ratings of how different *Seinfeld* and *60 Minutes* are rated as compared to how different *60 Minutes* and *Seinfeld* are rated. If your data matrix is square asymmetric (that is, the values below the diagonal in the data matrix are not the same as the values above the diagonal in the data matrix), then select **Square asymmetric**. This is the type of dissimilarity matrix used in the sociogram example. If your data matrix includes different types of things in the rows and the columns, then select a **Rectangular** shape for your data matrix. This type of analysis is rare, and goes beyond the scope of this book.

If your data file is not a dissimilarity matrix, but instead you want to calculate dissimilarities from other variables in your data file, then in Screen 19.1 you would select **Create distances from data**. SPSS will automatically create a dissimilarity matrix from the variables you have selected. If you want to use Euclidian distance between data points (think high school geometry and the Pythagorean theorem: $a^2 + b^2 = c^2$) and you want SPSS to calculate distances between variables, then you do not need to further define the **Distances**. If you want to use a different kind of geometry, or if you want to have SPSS calculate the distances between cases instead of between variables (as in our quiz score example), then click the **Measure** button and Screen 19.3 will appear.

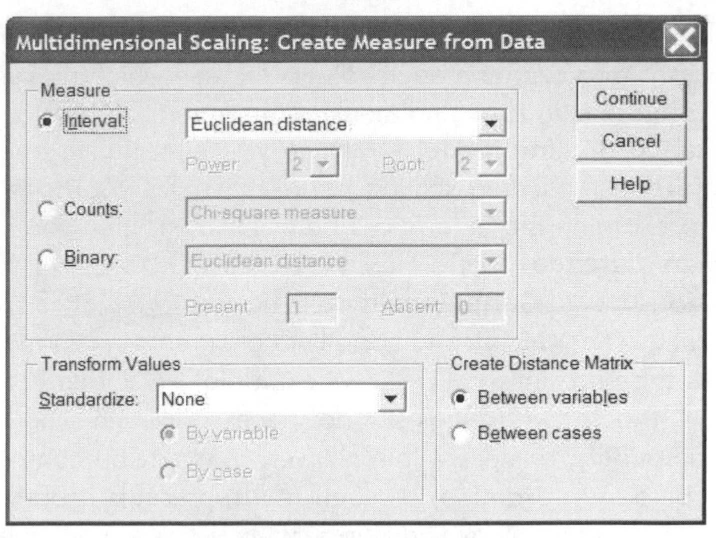

19.3

Multidimensional Scaling: Create Measures from Data Window

In this dialog window you select the **Measure**, whether to **Transform Values**, and whether to **Create Distance Matrix** for variables or cases. In addition to Euclidean distance (described above), other popular distance equations include Minkowski (similar to Euclidian distance, except that you specify the power; if you set power to 4, for example, then $a^4 + b^4 = c^4$) and block (think of a city block, in which you have to walk only on the streets and can't cut across within a block; this may be appropriate if your variables are measuring different constructs). If all of your variables are measured in the same scale (e.g., they are all using a 1-5 scale), then you do not need to **Transform Values**, and may leave the **Standardize** option at the default setting (None). If your variables are measured using different scales (e.g., some 1-5 scales, some 1-20 scales, and some distances in meters), then you may want to **Transform Values** (with a Z score transformation being the most common). Finally, by default SPSS calculates the distances **Between variables**. It is quite possible that you want the distances **Between cases**; if this is the case, select the **Between cases** option.

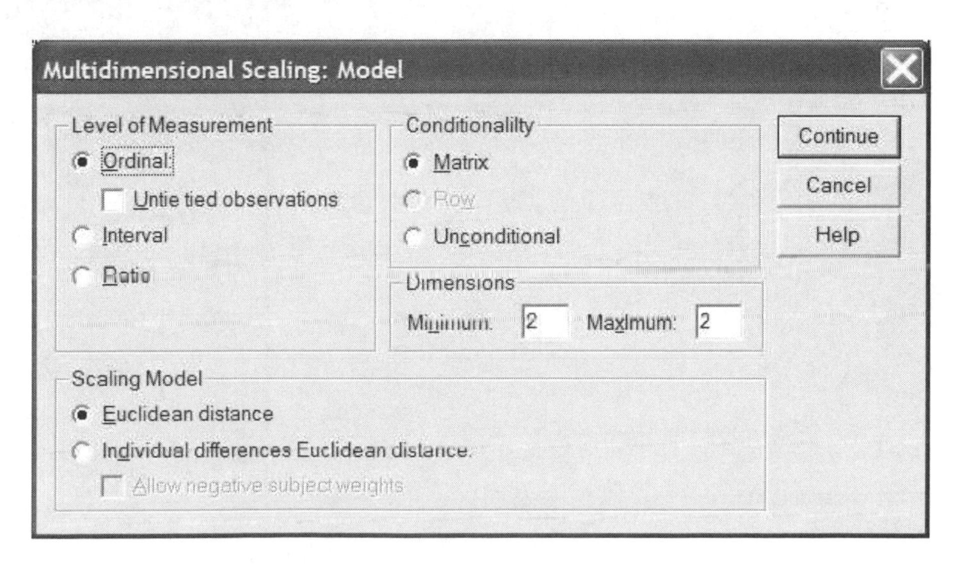

19.4

Multidimensional Scaling: Model Window

Once you have defined the **Distances**, you can now tell SPSS what type of **Model** you wish to use. Once you click the **Model** button, Screen 19.4 appears. Of key interest here are the **Dimensions** and the **Scaling Model**. Unlike factor analysis (see Chapter 20), multidimensional scaling cannot automatically determine the number of dimensions that you should use. As typically only 1, 2, or 3 dimensions are desired, this is not a major problem, as you can always do multidimensional scaling for a **Minimum** of 1 dimension through a **Maximum** of 3 dimensions and examine each solution to see which makes the most sense (here's where some of the art comes in to multidimensional scaling). In most cases, you will want to use a **Euclidian distance Scaling Model** (in which you either have one dissimilarity matrix, or in which you have several replicated scaling matrixes), but if you have ratings from several participants and wish to examine the differences between them then you should click **Individual differences Euclidian distance**. This is the case in the TV shows example, in which five students rated four television shows. **Conditionality** allows you to tell SPSS which comparisons are or are not meaningful: With **Matrix** conditionality, cells with a given matrix may be compared with each other (this is the case in our TV Shows example, as student's ratings may be assumed to be comparable with their own ratings but not necessarily with other students' ratings). In the case of **Row** conditionality, all cells within a given row may be compared with each other (this is the case in our sociogram example, as each row represents a particular rater; raters may use different scales, but presumably an individual rater is using the same scale throughout). In **Unconditional** conditionality, SPSS may compare any cell with any other cell. For **Level of Measurement**, you can tell SPSS how to treat your dissimilarity data; there are not often major differences between the models produced when treating data as **Ordinal**, **Interval**, or **Ratio**.

Once the **Model** is specified and you return to Screen 19.1, a click on the **Options** button will bring up Screen 19.5. In this window, you may request **Group plots** (you will almost always want this, as without selecting this option you will have to get out the graph paper to draw the picture yourself!), the **Data matrix** (this shows your raw data, and is primarily useful if you are having SPSS compute the dissimilarity matrix for you and you want to take a peek), and **Model and options summary** (this is useful to remind yourself later what type of model you ran). Other options in this screen are rarely needed or advised.

19.5

Multidimensional Scaling: Options Window

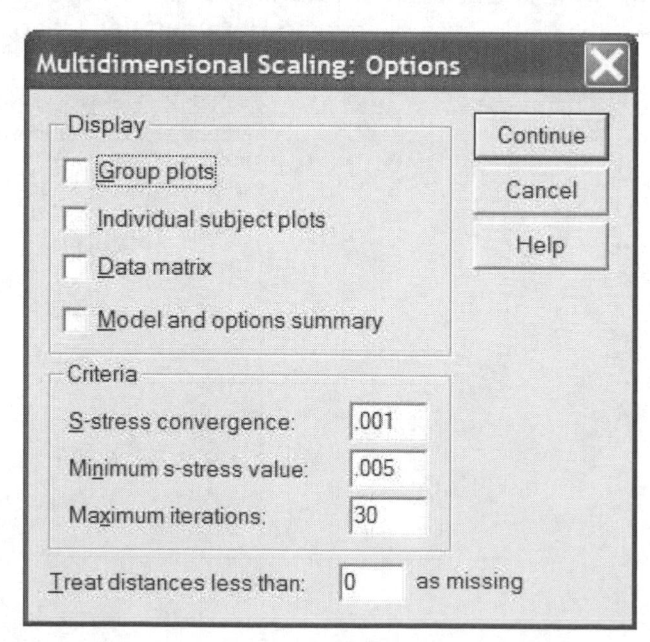

To perform a multidimensional scaling analysis with two dimensions for the 20 students' disliking ratings (the sociogram example), perform the following sequence of steps. This analysis is for a square asymmetric dissimilarity matrix of ordinal data. The starting point is Screen 19.1. Please perform whichever of steps 1-4 (pages 236-237) are necessary to arrive at this screen.

In Screen	Do This	Step 5
19.1	**springer** → hold shift and **shearer** → *top* ▶ → **Shape**	
19.2	**Square asymmetric** → **Continue**	
19.1	**Model**	
19.4	**Row** → **Continue**	
19.1	**Options**	
19.5	**Group plots** → **Continue**	
19.1	**OK**	Back1

If you wish to perform an analysis with two dimensions using a square symmetric matrix than SPSS computes using other variables in your data file, perform the following sequence of steps. This analysis has SPSS calculate distances based on quiz scores (the quiz score example), and uses that square symmetrical dissimilarity matrix (treated as interval data) to perform multidimensional scaling. The starting point is Screen 19.1. Please perform whichever of steps 1-4 (pages 236-237) are necessary to arrive at this screen.

In Screen	Do This	Step 5a
19.1	**quiz1** → hold shift and **quiz5** → *top* ▶ → **Create distances from data** → **Measure**	
19.3	**Between cases** → **Continue**	
19.1	**Model**	
19.4	**Interval** → **Continue**	
19.1	**Options**	
19.5	**Group plots** → **Continue**	
19.1	**OK**	Back1

If you wish to perform an analysis with two dimensions using a square symmetric matrix with an individual differences model, perform the following sequence of steps. Please note that this example uses the **grades-mds2.sav** *file; if you wish to try this procedure, please open this file in Step 3, above. In this analysis, 5 participants' ratings of dissimilarity between four television shows are analyzed; the matrixes are square symmetric, and the data is assumed to be ratio data. The starting point is Screen 19.1. Please perform whichever of steps 1-4 (pages 236-237) are necessary to arrive at this screen.*

In Screen	Do This	Step 5b
19.1	**er** → hold shift and **seinfeld** → *top* ▶ → **Model**	
19.4	**Ratio** → **Individual differences Euclidian distance** → **Continue**	
19.1	**Options**	
19.5	**Group plots** → **Continue**	
19.1	**OK**	Back1

Upon completion of step 5, the output screen will appear (Screen 1, inside back cover). All results from the just-completed analysis are included in the Output Navigator. Make use of the scroll bar arrows (▲ ▼ ▶ ◀) to view the results. Even when viewing output, the standard menu of commands is still listed across the top of the window. Further analyses may be conducted without returning to the data screen.

PRINTING RESULTS

Results of the analysis (or analyses) that have just been conducted requires a window that displays the standard commands (**File Edit Data Transform Analyze** . . .) across the top. A typical print procedure is shown below beginning with the standard output screen (Screen 1, inside back cover).

To print results, from the Output screen perform the following sequence of steps:

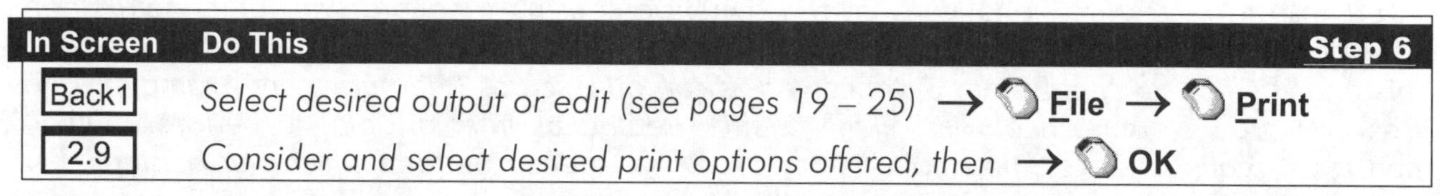

In Screen	Do This	Step 6
Back1	*Select desired output or edit (see pages 19 – 25)* → **File** → **Print**	
2.9	*Consider and select desired print options offered, then* → **OK**	

To exit you may begin from any screen that shows the **File** *command at the top.*

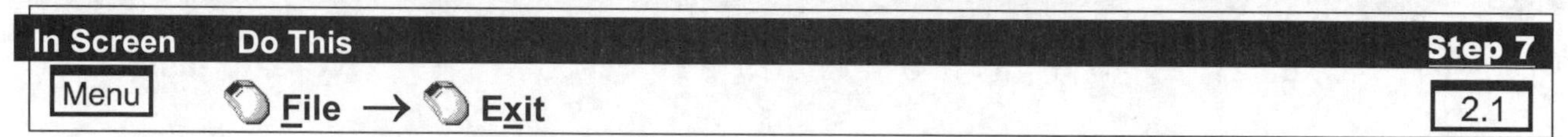

In Screen	Do This	Step 7
Menu	**File** → **Exit**	2.1

Note: After clicking **Exit**, there will frequently be small windows that appear asking if you wish to save or change anything. Simply click each appropriate response.

OUTPUT

Multidimensional Scaling

What follows is partial output from sequence step 5 (the sociogram example) on page 241. In several cases in which the output differs substantially with step 5a (the quiz score example) or step 5b (the TV shows example), selected output from those procedures is included.

Iteration History

```
Iteration history for the 2 dimensional solution (in squared distances)
                Iteration       S-stress       Improvement
                    1             .34159
                    2             .32376          .01783
                    .               .               .
                    .               .               .
                    .               .               .
                    8             .30867          .00120
                    9             .30779          .00088
Iterations stopped because S-stress improvement is less than   .001000
```

SPSS attempts a first model, and the S-Stress formula indicates how far off the model is from the original dissimilarity matrix (lower numbers mean less stress, and a better model). SPSS then tries to improve the model, and it keeps improving until the S-Stress doesn't improve very much. If you reach 30 iterations, that probably means that your data has some problems.

Stress and Squared Correlations in Distances

```
Stress and squared correlation (RSQ) in distances
             Stimulus     Stress      RSQ
                1           .328     .773
                2           .220     .892
                3           .216     .890
                :            :         :

             Averaged (rms) over stimuli
          Stress  =   .259       RSQ =  .764
```

For each of the rows in a square matrix that you have specified as row conditional, for each matrix in an individual differences model, and in all cases for the overall model, SPSS calculates the stress and R^2. The stress is similar in concept to the S-Stress output above, but uses a different equation that makes comparisons between different analyses with different computer programs easier. RSQ (R^2) is an indicator of how much of the variance in the original dissimilarity matrix is accounted for by the multidimensional scaling model (thus, higher numbers are good). If you are running the analysis with several different dimensions (for example, seeing how the graph looks in 1, 2, or 3 dimensions), then the stress and R^2 values can help you decide which model is most appropriate.

Stimulus Coordinates

```
Stimulus    Stimulus        Dimension
Number        Name            1        2

1       SPRINGER     .8770    -.0605
2       ZIMCHEK     1.1382     .1515
3       LANGFORD    1.0552     .2871
.          .           .         .
:          :           :         :
```

For each of the items on the multidimensional scaling graph, the coordinates on the graph are provided (in case you want to create your own graph, or if you want to further analyze the data). For a two dimensional graph, these are *x* and *y* coordinates on the graph.

Subject Weights (Individuals Difference or Weighted Models Only)

If you are performing an individual differences model or weighted multidimensional scaling, then SPSS will calculate weights indicating the importance of each dimension to each subject. In this TV show example (following page), each of the five subjects has weights on each of the two dimensions. The weights indicate how much each student's variance in his or her ratings was related to each of the two dimensions. An overall weighting of each dimension including all of the students is also included. "Weirdness" indicates the imbalance in the weights for a particular subject; in this case, subject number 4 (who seemed to base his or her ratings almost totally along dimension 1) has the highest possible weirdness score of 1.

Subject Weights (Individual Difference or Weighted Models Only)

Subject Number	Weird-ness	Dimension 1	Dimension 2
1	.1777	.8233	.4144
2	.0652	.6283	.2151
3	.1616	.8739	.4285
4	1.0000	.9915	.0000
5	.3044	.8190	.5110
Overall importance of each dimension:		.6980	.1325

Derived Stimulus Configuration

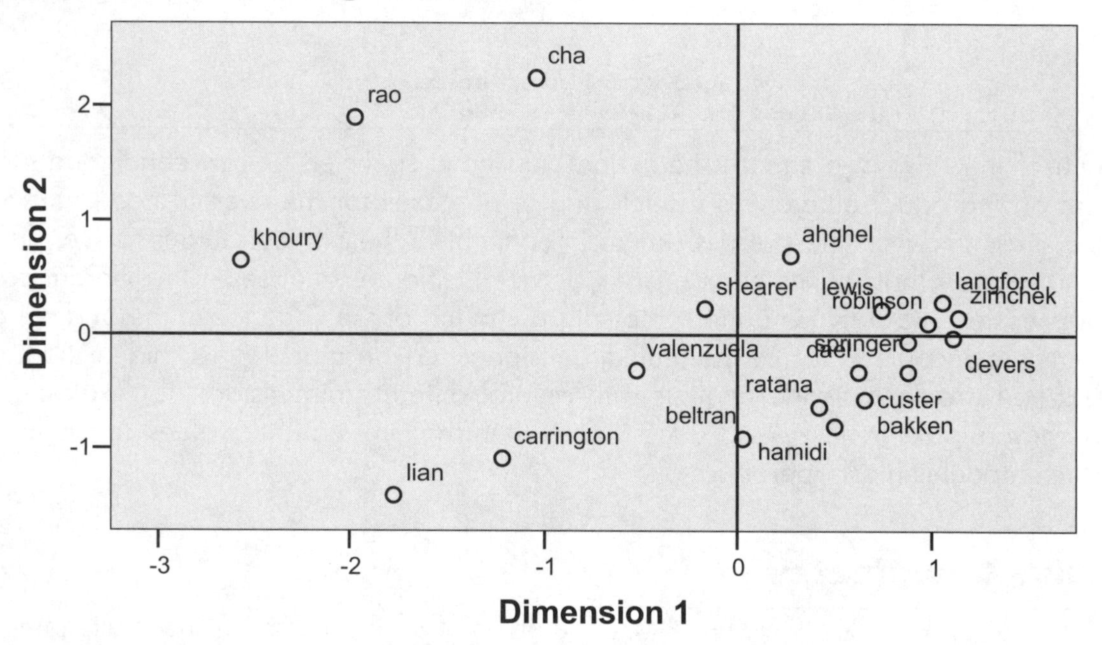

This graph displays the final multidimensional scaling model. For the sociogram example, this displays the 20 students in the class arranged so that the farther apart the cases are in the graph, the farther apart they are in the original dissimilarity matrix. In this case, we can see which students are most unpopular (Khoury, Rao, and Cha), and which students are most popular (those in the cluster in the middle right of the graph). Note that if you ask SPSS to compute distances for you (as in the quiz scores example), the points on the graph will be labeled "Case 1," "Case 2," and so forth; you will need to go back to your original data file to determine the meaning of each case. The two dimensions do not necessarily have an obvious meaning; the main thing of interest here is the distance between the points.

In this case, one group of students (the clique in the middle right of the graph) is so close together that it is difficult to read their names. In this case, you should either a) look back at the stimulus coordinates above, or b) double click on the graph to go into the output graph editor (see Chapter 5), and use the button to select which labels you want displayed.

20

FACTOR ANALYSIS

OVER THE past 30 or 40 years, factor analysis has gained increasing acceptance and popularity. Raymond B. Cattell drew attention to the procedure when he used factor analysis to reduce a list of more than 4,500 trait names to fewer than 200 questions that measured 16 different personality traits in his personality inventory called the *16 Personality Factor Questionnaire* (16PF). Cattell's use of factor analysis underlines its primary usefulness, that is, to take a large number of observable instances to measure an unobservable construct or constructs. For instance: "Attends loud parties," "talks a lot," "appears comfortable interacting with just about anyone," and "is usually seen with others" are four behaviors that can be observed that may measure an unobservable construct called "outgoing". Factor analysis is most frequently used to identify a small number of factors (e.g., outgoing) that may be used to represent relationships among sets of interrelated variables (e.g., the four descriptors).

Four basic steps are required to conduct a factor analysis:

1. Calculate a correlation matrix of all variables to be used in the analysis.
2. Extract factors.
3. Rotate factors to create a more understandable factor structure.
4. Interpret results.

The first three steps are covered in this introduction. For step 4, we give a conceptual feel of how to interpret results, but most of the explanation will take place in the Output section.

CREATE A CORRELATION MATRIX

Calculating a correlation matrix of all variables of interest is the starting point for factor analysis. This starting point provides some initial clues as to how factor analysis works. Even at this stage it is clear that factor analysis is derived from some combinations of intercorrelations among descriptor variables. It is not necessary to type in a correlation matrix for factor analysis to take place. If you are starting from raw data (as we have in all chapters up to this point), the **Factor** command will automatically create a correlation matrix as the first step. In some instances the researcher may not have the raw data, but only a correlation matrix. If this is the case, it is possible to conduct factor analysis by inserting the correlation matrix into an SPSS syntax command file. The process is complex, however, and a description of it extends beyond the scope of this book. Please consult *SPSS Base System Manual* for specifics.

FACTOR EXTRACTION

The purpose of the factor-extraction phase is to extract the factors. *Factors* are the underlying constructs that describe your set of variables. Mathematically, this procedure is similar to a forward run in multiple regression analysis. As you recall (Chapter 16), the first step in multiple regression is to select and enter the *independent* variable that significantly explains the greatest amount of variance observed in the *dependent* variable. When this is completed, the next step is to find and enter the independent variable that significantly explains the greatest *additional* amount of variation in the dependent variable. Then the procedure selects and enters the variable that significantly explains the *next* greatest additional amount of variance and so forth, until there are no variables that significantly explain further variance.

With factor analysis the procedure is similar, and a conceptual understanding of the factor-extraction phase could be gained by rewriting the previous paragraph and omitting "*dependent variable*", "*significantly*"; and changing *independent variable* to *variables* (plural). Factor analysis does not begin with a dependent variable. It starts with a measure of the total amount of variation observed (similar to total sum of squares) in all variables that have been designated for factor analysis. Please note that this "variation" is a bit difficult to understand conceptually (Whence all this variance floating about? How does one see it? smell it? grasp it? but I digress . . .), but it is quite precise mathematically. The first step in factor analysis is for the computer to select the combination of variables whose shared correlations explain the greatest amount of the total variance. This is called Factor 1. Factor analysis will then extract a second factor. This is the combination of variables that explains the greatest amount of the variance that remains, that is, variation *after* the first factor has been extracted. This is called Factor 2. This procedure continues for a third factor, fourth factor, fifth factor, and so on, until as many factors have been extracted as there are variables.

In the default SPSS procedure, each of the variables is initially assigned a *communality* value of 1.0. Communalities are designed to show the proportion of variance that the factors contribute to explaining a particular variable. These values range from 0 to 1 and may be interpreted in a matter similar to a Multiple *R*, with 0 indicating that common factors explain none of the variance in a particular variable, and 1 indicating that all the variance in that variable is explained by the common factors. However, for the default procedure at the initial extraction phase, each variable is assigned a communality of 1.0.

After the first factor is extracted, SPSS prints an eigenvalue to the right of the factor number (e.g., Factor number = 1; eigenvalue = 5.13). Eigenvalues are designed to show the proportion of variance accounted for by each factor (*not* each *variable* as do communalities). The first eigenvalue will always be largest (and always be greater than 1.0) because the first factor (by the definition of the procedure) always explains the greatest amount of total variance. It then lists the percent of the variance accounted for by this factor (the eigenvalue divided by the number of variables), and this is followed by a cumulative percent. For each successive factor, the eigenvalue printed will be smaller than the previous one, and the cumulative percent (of variance explained) will total 100% after the final factor has been calculated. The absence of the word *significantly* is demonstrated by the fact that the **Factor** command extracts as many factors as there are variables, regardless of whether or not subsequent factors explain a significant amount of additional variance.

FACTOR SELECTION and ROTATION

The factors extracted by SPSS are almost never *all* of interest to the researcher. If you have as many factors as there are variables, you have not accomplished what factor analysis was created to do. The goal is to explain the phenomena of interest with fewer than the original number of variables, usually substantially fewer. Remember Cattell? He started with 4,500 descriptors and ended up with 16 traits.

The first step is to decide *which* factors you wish to retain in the analysis. The common-sense criterion for retaining factors is that each retained factor must have some sort of face validity or theoretical validity; but prior to the rotation process, it is often impossible to interpret what each factor means. Therefore, the researcher usually selects a mathematical criterion for determining which factors to retain. The SPSS default is to keep any factor with an eigenvalue larger than 1.0. If a factor has an eigenvalue less than 1.0 it explains less variance than an original variable and is usually rejected. (Remember, SPSS will print out as many factors as there are variables, and usually for only a few of the factors will the eigenvalue be larger than 1.0.) There are other criteria for selection (such as the scree plot), or conceptual reasons (based on your knowledge of the data) that may be used. The procedure for selecting a number other than the default will be described in the Step by Step section.

Once factors have been selected, the next step is to rotate them. Rotation is needed because the original factor structure is mathematically correct but is difficult to interpret. The goal of rotation is to achieve what is called *simple structure*, that is, high *factor loadings* on one factor and low loadings on all others. Factor loadings vary between $\pm$ 1.0 and indicate the strength of relationship between a particular variable and a particular factor, in a way similar to a correlation. For instance, the phrase "enjoys loud parties" might have a high loading on an "outgoing" factor (perhaps $> .6$) and have a low loading on an "intelligence" factor (perhaps $< .1$). This is because an enjoys-loud-parties statement is thought to be related to outgoingness, but unrelated to intelligence. Ideally, simple structure would have variables load entirely on one factor and not at all on the others. On the second graph (below), this would be represented by all asterisks being *on* the rotated factor lines. In social science research, however, this never happens, and the goal is to rotate the axes to have data points as close as possible to the rotated axes. The following graphs are good representations of how an unrotated structure and a rotated structure might look. SPSS will print out graphs of your factor structure (after rotation) and include a table of coordinates for additional clarity.

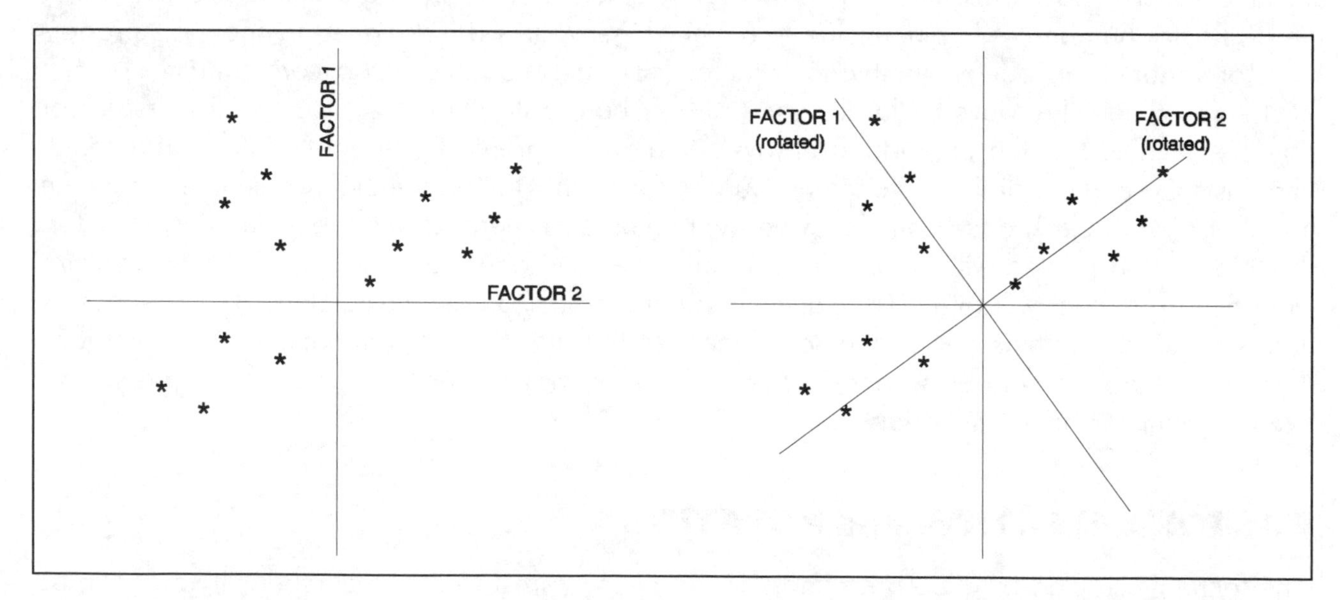

Rotation does not alter the mathematical accuracy of the factor structure, just as looking at a picture from the front rather than from the side does not alter the picture, and changing the measure of height in inches to height in centimeters does not alter how tall a person is. Rotation was originally done by hand, and the researcher would position the axes in the location that appeared to create the optimal factor structure. Hand rotation is not possible with SPSS, but there are several mathematical procedures available to rotate the axes to the best simple structure. Varimax is the default procedure used by SPSS, but there are several others (mentioned in the Step by Step section).

Oblique Rotations: Varimax rotations are called *orthogonal rotations* because the axes that are rotated remain at right angles to each other. Sometimes it is possible to achieve a better simple structure by diverging from perpendicular. The **Oblimin** and **Promax** procedures allow the researcher to deviate from orthogonal to achieve a better simple structure. Conceptually, this deviation means that factors are no longer uncorrelated with each other. This is not necessarily disturbing, because few factors in the social sciences *are* entirely uncorrelated. The use of oblique rotations can be quite tricky, and (here we insert our standard disclaimer) you should not attempt to use them unless you have a clear understanding of what you are doing. Let's expand: You should probably not attempt to conduct factor analysis *at all* unless you have had a course in it and/or have a clear conceptual understanding of the procedure. The technique for specifying **Oblimin** or **Promax** rotation is described in the Step by Step section. We do not demonstrate this procedure in the Output section because it requires more attention than we can accord it here.

INTERPRETATION

In an ideal world, each of the original variables will load highly (e.g., $> .5$) on one of the factors and low (e.g., $< .2$) on all others. Furthermore, the factors that have the high loadings will have excellent face validity and appear to be measuring some underlying construct. In the real world, this rarely happens. There will often be two or three irritating variables that end up loading on the "wrong" factor, and often a variable will load onto two or three different factors. The output of factor analysis requires considerable understanding of your data, and it is rare for the arithmetic of factor analysis alone to produce entirely clear results. In the Output section we will clarify with a real example. The example will be introduced in the following paragraphs.

We draw our example from real data; it is the same set of data used to demonstrate reliability analysis (Chapter 18), the **helping2.sav** file. It has *N* of 517 and measures many of the same variables as the **helping1.sav** and **helping3.sav** files used in earlier chapters. In the **helping2.sav** file, self-efficacy (belief that one has the ability to help effectively) was measured by 15 questions, each paired with an amount-of-help question that measured a particular type of helping. An example of one of the paired questions follows:

9a) Time spent expressing sympathy, empathy, or understanding.

none	0-15 minutes	15-30 minutes	30-60 minutes	1-2 hours	2-5 hours	___ hours

9b) Did you believe you were capable of expressing sympathy, empathy, or understanding to your friend?

1	2	3	4	5	6	7
not at all			some		very much so	

There were three categories of help represented in the 15 questions: Six questions were intended to measure empathic types of helping, four questions were intended to measure informational types of helping, four questions were intended to measure instrumental ("doing things") types of helping, and the fifteenth question was open-ended to allow any additional type of help given to be inserted. Factor analysis was conducted on the 15 self-efficacy questions to see if the results would yield the three categories of self-efficacy that were originally intended.

A final word: In the Step by Step section, there will be two step-5 variations. Step 5 will be the simplest default factor analysis that is possible. Step 5a will be the sequence of steps that includes many of the variations that we present in this chapter. Both will conduct a factor analysis on the 15 questions that measure self-efficacy in the **helping2.sav** file.

For additional information on factor analysis, the best textbooks on the topic known by the authors are by Comrey and Lee (1992) and Gorsuch (1983). Please see the reference section for additional information about this book.

STEP BY STEP

Factor Analysis

To enter SPSS, a click on **Start** *in the taskbar (bottom of screen) activates the start menu:*

After clicking the SPSS program icon, Screen 1 (inside front cover) appears on the monitor.

Step 2

Create and name a data file or edit (if necessary) an already existing file (see Chapter 3).

Screens 1 and 2 (inside front cover) allow you to access the data file used in conducting the analysis of interest. The following sequence accesses the **helping2.sav** *file for further analyses:*

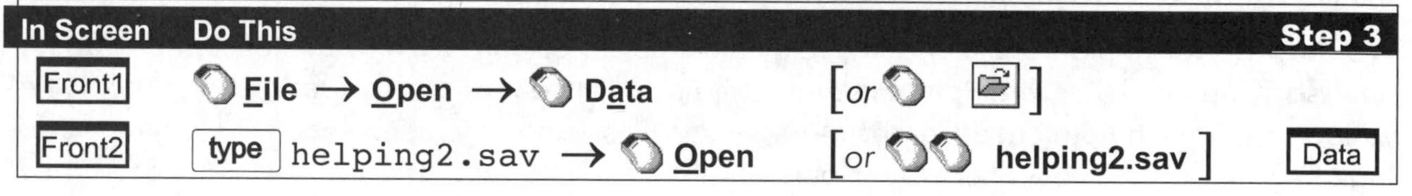

Whether first entering SPSS or returning from earlier operations the standard menu of commands across the top is required (shown below). As long as it is visible you may perform any analyses. It is not necessary for the data window to be visible.

This menu of commands disappears or modifies when using pivot tables or editing graphs. To uncover the standard menu of commands simply click on the ▬ or the ⧉ icon.

After completion of Step 3 a screen with the desired menu bar appears. When you click a command (from the menu bar), a series of options will appear (usually) below the selected command. With each new set of options, click the desired item. The sequence to access Factor Analysis begins at any screen with the menu of commands visible:

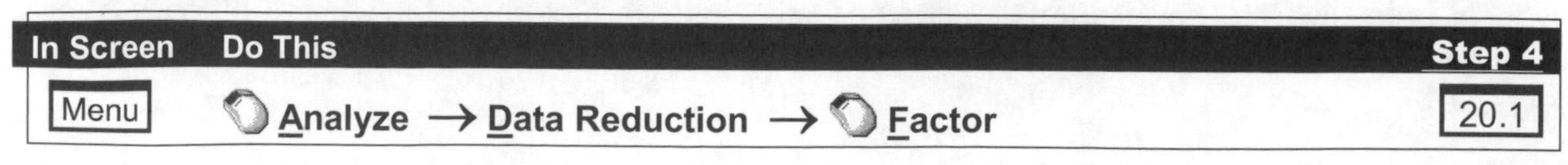

20.1

The Factor Analysis Opening Dialog Window

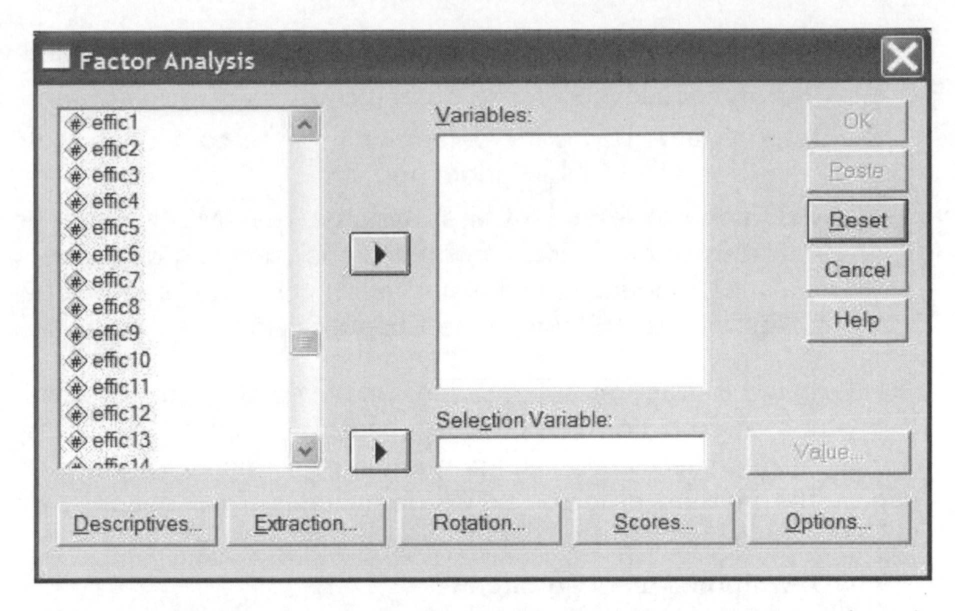

The click on **Factor** opens the main dialog window for factor analysis (Screen 20.1, above). The initial screen looks innocent enough: the box of available variables to the left, a single active (**Variables**) box to the right and five push buttons below representing different options. Although the arithmetic of factor analysis is certainly complex, a factor analysis may be conducted by simply pasting some variables into the active box, selecting **Varimax** rotations, and clicking **OK**. Indeed, *any* factor analysis will begin with the pasting of carefully-selected variables into the **Variables** box. For the present illustration we will select the 15 self-efficacy questions (**effic1** to **effic15**) for analysis. The efficacy questions are displayed in the variable list box in screen 20.1 above. The five pushbuttons (lower portion of the screen) represent many available options and may be processed in any order. We begin with **Descriptives**.

20.2

The Factor Analysis: Descriptives Window

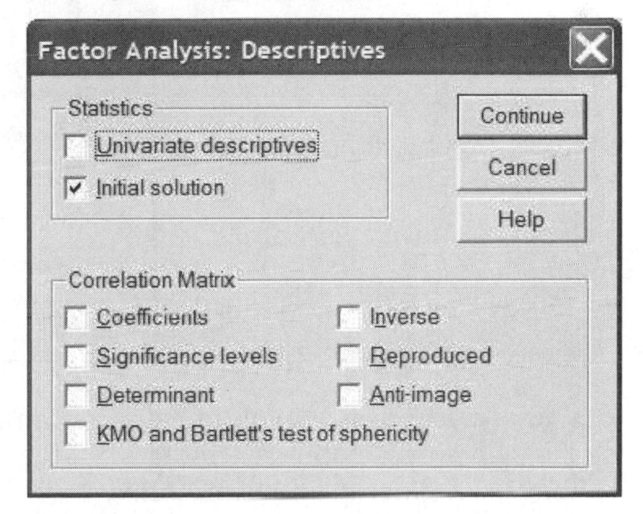

This option opens up Screen 20.2, shown above. For this and other windows we will describe only those selections that are most frequently used by researchers. The **Univariate descriptives** option is quite useful. It lists in four neat columns the variable names, the means, the standard deviations, and the always handy variable labels. This chart will be referred to frequently during the course of analysis. The **Initial solution** is selected by default and lists the variable names, the initial communalities (1.0 by default), the factors, the eigenvalues, and the percent and cumulative percent accounted for by each factor. The correlation matrix is the starting point of any factor analysis. We describe four of the frequently-used statistics related to the correlation matrix:

- **Coefficients**: This is simply the correlation matrix of included variables.
- **Significance levels**: These are the p-values associated with each correlation.
- **Determinant**: This is the determinant of the correlation matrix. It is used in computing values for tests of multivariate normality.
- **KMO and Bartlett's test of sphericity**: The KMO test and Bartlett's test of sphericity are both tests of multivariate normality and sampling adequacy (the adequacy of your variables for conducting factor analysis). This test is selected by default. This option is discussed in greater detail in the Output section.

A click on the **Extraction** button opens an new dialog box (Screen 20.3, below) that deals with the method of extraction, the criteria for factor selection, the display of output related to factor extraction, and specification of the number of iterations for the procedure to converge to a solution. The **Method** of factor extraction includes seven options. **Principal components** is the default procedure; a click of the ▾ reveals the six others:

- **Unweighted least squares**
- **Generalized least squares**
- **Maximum likelihood**
- **Principal-axis factoring**
- **Alpha factoring**
- **Image factoring**

The principal components method is, however, the most frequently used (with maximum likelihood the next most common), and space limitations prohibit discussion of the other options.

20.3

The Factor Analysis: Extraction Window

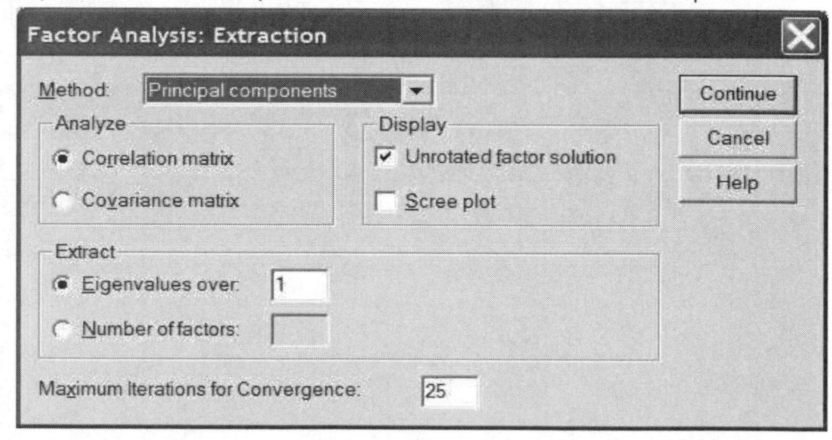

The **Analyze** box allows you to select either the **Correlation matrix** or the **Covariance matrix** as the starting point for your analysis. Under **Extract**, selection of factors for rotation is based either on the eigenvalue being greater than 1 (the default), selection of a different eigenvalue, or simple identification of how many factors you wish to select for rotation. Under **Display**, the unrotated factor solution is selected by default, but unless you are quite mathematically sophisticated, the unrotated factor solution rarely reveals much. Most researchers would choose to deselect this option. You may also request a **Scree plot** (illustrated in the Output section). Finally you may designate the number of iterations you wish for convergence. The default of 25 is almost always sufficient.

A click on **Rotation** moves on to the next step in factor analysis, rotating the factors to final solution. Screen 20.4 displays available options. There are three different *orthogonal* methods for rotation, **Varimax** (the most popular method, yes, but using it requires weathering the disdain of the factor analytic elite), **Equamax**, and **Quartimax**. The **Direct Oblimin** and **Promax** procedures allow for a nonorthogonal rotation of selected factors. Both oblique procedures' parameters (δ and κ) can usually be left at the default values. As suggested in the introduction, don't even think of attempting oblique rotations unless you've taken a course in factor analysis.

20.4

The Factor Analysis: Rotation Window

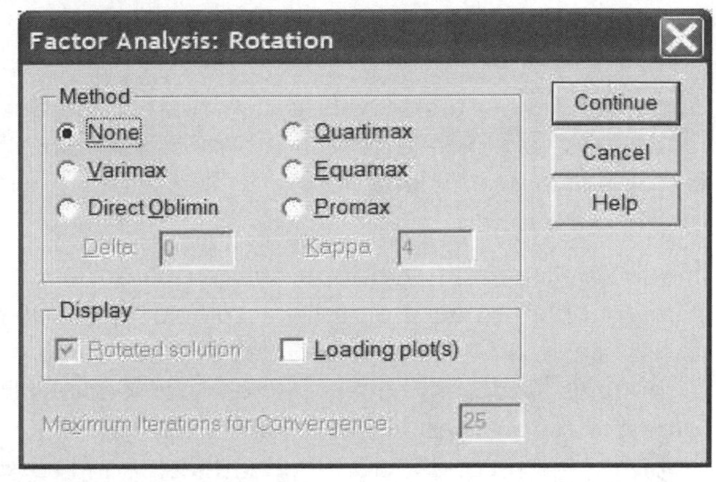

For display, the **Rotated solution** is selected by default and represents the essence of what factor analysis is designed to do. The Output section includes several paragraphs about interpretation of a rotated solution. If you wish a visual display of factor structure after rotation, click on the **Loading plot(s)** option. The plot(s) option provides by default a 3-D graph of the first three factors. The chart is clever; you can drop lines to the floor and view it from a variety of perspectives, but no matter what manipulations you enact, one constant remains: the graph's meaning is almost unintelligible. Two-dimensional graphs are much easier to interpret. SPSS Factor Analysis options do not include this possibility. If, however, you wish to convert the 3-D graph to a much simpler 2-D graph, just open the chart by double-clicking on it, and select **Edit → Properties**. Then, drag the name Component 3 to the Excluded box. Once you **Apply** the changes, you will have a much simpler graph (shown here in screen 20.5 with reference lines added; see chapter 5 for details on that procedure).

20.5

Factor Plot of Two Rotated Factors, Efficacy for Empathic Help with Efficacy for Instrumental Help

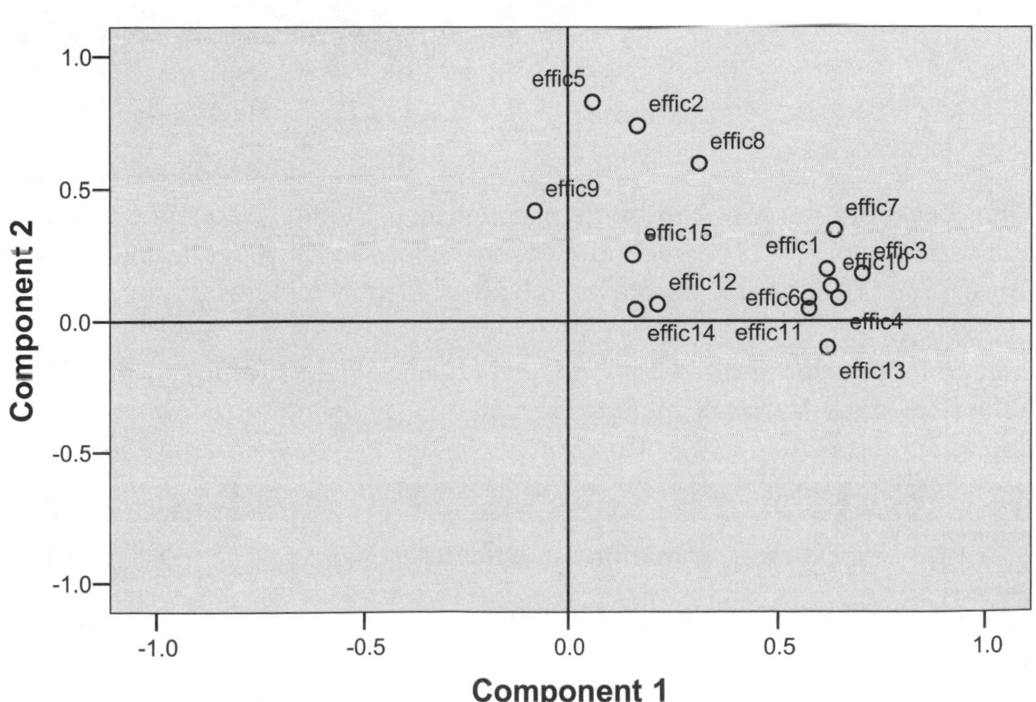

These 2-D charts are quite interpretable, particularly if you refer to the values listed in the rotated component (or factor) matrix. Notice that on the chart (previous page), the horizontal axis is Factor 1 (efficacy for empathic types of helping) and the vertical axis is Factor 2 (efficacy for informational types of helping). Note also that the eight data points at the extreme right feature high loadings (.6 to .8) on the efficacy for empathic-helping (Factor 1), and low loadings (–.1 to .4) on efficacy for informational helping (Factor 2).

Back to the main dialog window: A click on the **Scores** pushbutton opens a small dialog box that allows you to save certain scores as variables. This window is not shown but the **Display factor score coefficient matrix** (when selected) will display the component score coefficient matrix in the output.

Finally, a click on the **Options** button opens a dialog box (Screen 20.6, below) that allows two different options concerning the display of the rotated factor matrix. The **Sorted by size** selection is very handy. It sorts variables by the magnitude of their factor loadings by factor. Thus, if 6 variables load onto Factor 1, the factor loadings for those six variables will be listed from the largest to the smallest in the column titled Factor 1. The same will be true for variables loading highly on the second factor, the third factor and so forth. The Output section demonstrates this feature. Then you can suppress factor loadings that are less than a particular value (default is .10) if you feel that such loadings are insignificant. Actually, you can do it no matter how you feel. Anyone able to conduct factor analysis has already dealt with **Missing Values** and would be insulted by the implication that she or he would resort to an automatic procedure at this stage of the process.

20.6

The Factor Analysis: Options Window

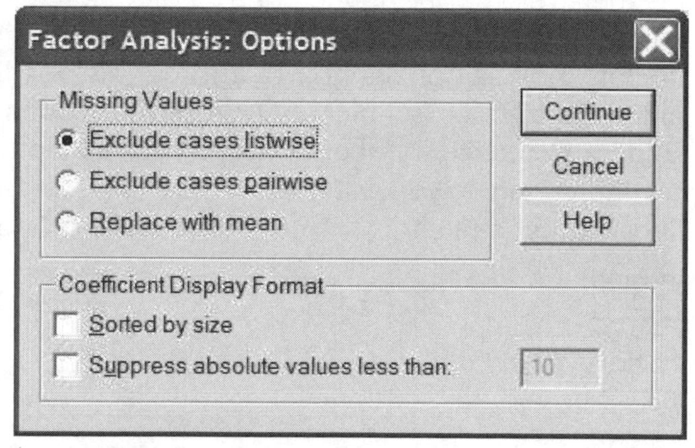

What follow are the two step-by-step sequences. The first (sequence step 5) is the simplest default factor analysis. The second (sequence step 5a) includes many of the options described above. The italicized text before each box will identify features of the procedure that follows.

*To conduct a factor analysis with the fifteen efficacy items (**effic1** to **effic15**) with all the default options (plus the **Varimax** method of rotation), perform the following sequence of steps. The starting point is Screen 20.1. Carry out the 4-step sequence (page 250) to arrive at this screen.*

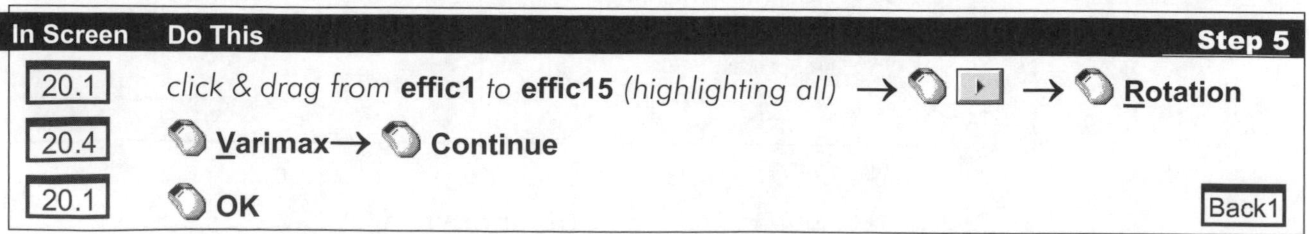

In Screen	Do This	Step 5
20.1	*click & drag from* **effic1** *to* **effic15** *(highlighting all)* → ▸ → **Rotation**	
20.4	**Varimax** → **Continue**	
20.1	**OK**	Back1

These five clicks perform a factor analysis that
1. calculates a correlation matrix of the 15 efficacy questions from the data file,
2. extracts 15 factors by the principal-components method,
3. selects as factors to be rotated all factors that have an eigenvalue greater than 1.0,
4. rotates the selected factors to a Varimax solution, and
5. prints out the factor transformation matrix.

*To conduct a factor analysis with the fifteen efficacy items (***effic1** *to* **effic15***), request* **Univariate descriptives** *for the 15 variables, correlation* **Coefficients** *and the standard measures of multivariate normality (***KMO and Bartlett's test of sphiricity***); request the principal components method of extraction (default) and the* **Scree plot***; access the* **Varimax** *method of rotation and the available* **Loading plots** *of factor scores; and requesting that factor loadings be sorted by factor number and* **Sorted by size***, perform the following sequence of steps. The starting point is Screen 20.1. Carry out the 4-step sequence (page 250) to arrive at this screen.*

In Screen	Do This	Step 5a
20.1	*click & drag from* **effic1** *to* **effic15** *(highlighting all)* → ▸ → **Descriptives**	
20.2	**Univariate descriptives** → **Coefficients** → **KMO & Bartlett's test of sphiricity** → **Continue**	
20.1	**Extraction**	
20.3	**Scree plot** → **Continue**	
20.1	**Rotation**	
20.4	**Varimax** → **Loading Plots** → **Continue**	
20.1	**Options**	
20.6	**Sorted by size** → **Continue**	
20.1	**OK**	Back1

OUTPUT

Factor Analysis

What follows is output from sequence step 5a (above). The output presented below is rendered in a fairly simple, intuitive format, with variable names, not labels. See page 26 for switching between using variable names or labels. As in most of the advanced procedures, output is condensed and slightly reformatted.

KMO and Bartlett's Test

Test	Test Statistic	df	Significance
Kaiser-Meyer-Olkin Measure of Sampling Adequacy	.871		
Bartlett's Test of Sphericity	1321.696	105	.000

Term	Definition/Description
KAISER-MAYER-OLKIN	A measure of whether your distribution of values is adequate for conducting factor analysis. Kaiser himself (enjoying the letter m) designates levels as follows: A measure > .9 is marvelous, > .8 is meritorious, > .7 is middling, > .6 is mediocre, > .5 is miserable, and < .5 is unacceptable. In this case .871 is meritorious, almost marvelous.
BARTLETT TEST OF SPHERICITY	This is a measure of the multivariate normality of your set of distributions. It also tests whether the correlation matrix is an identity matrix (factor analysis would be meaningless with an identity matrix). A significance value < .05 indicates that that these data do NOT produce an identity matrix (or "differ significantly from identity") and are thus approximately multivariate normal and acceptable for factor analysis.

Communalities / **Total Variance Explained**

			Initial Eigenvalues		
Variable	**Communality**	**Component**	**Total**	**% of Variance**	**Cumulative %**
effic1	1.00000	1	5.133	34.2	34.2
effic2	1.00000	2	1.682	11.2	45.4
effic3	1.00000	3	1.055	7.0	52.5
effic4	1.00000	4	1.028	6.9	59.3
effic5	1.00000	5	.885	5.9	65.2
effic6	1.00000	6	.759	5.1	70.3
effic7	1.00000	7	.628	4.2	74.5
effic8	1.00000	8	.624	4.2	78.6
effic9	1.00000	9	.589	3.9	82.5
effic10	1.00000	10	.530	3.5	86.1
effic11	1.00000	11	.494	3.3	89.4
effic12	1.00000	12	.458	3.1	92.5
effic13	1.00000	13	.429	2.9	95.3
effic14	1.00000	14	.398	2.7	98.0
effic15	1.00000	15	.300	2.0	100.0

Extraction method: Principle Component Analysis

The two columns to the left refer only to the variables and the communalities. The four columns to the right refer to the components or factors. Note that there are four factors with eigenvalues larger than 1.0 and they account for almost 60% of the total variance. The definitions below clarify others items shown in this output.

Term	Definition/Description
PRINCIPAL-COMPONENT ANALYSIS	The default method of factor extraction used by SPSS. Other options are available and are listed on page 252.
VARIABLE	All 15 efficacy variables used in the factor analysis are listed here.
COMMUNALITY	The default procedure assigns each variable a communality of 1.00. Different communalities may be requested. See the SPSS manuals for details.
COMPONENT	The number of each component (or factor) extracted. Note that the first two columns provide information about the *variables*, and the last four provide information about the *factors*.
EIGENVALUE	The proportion of variance explained by each factor.

Term	Definition/Description
% OF VARIANCE	The percent of variance explained by each factor, the eigenvalue divided by the sum of the communalities (15 in this case).
CUMULATIVE %	The sum of each step in the previous column.

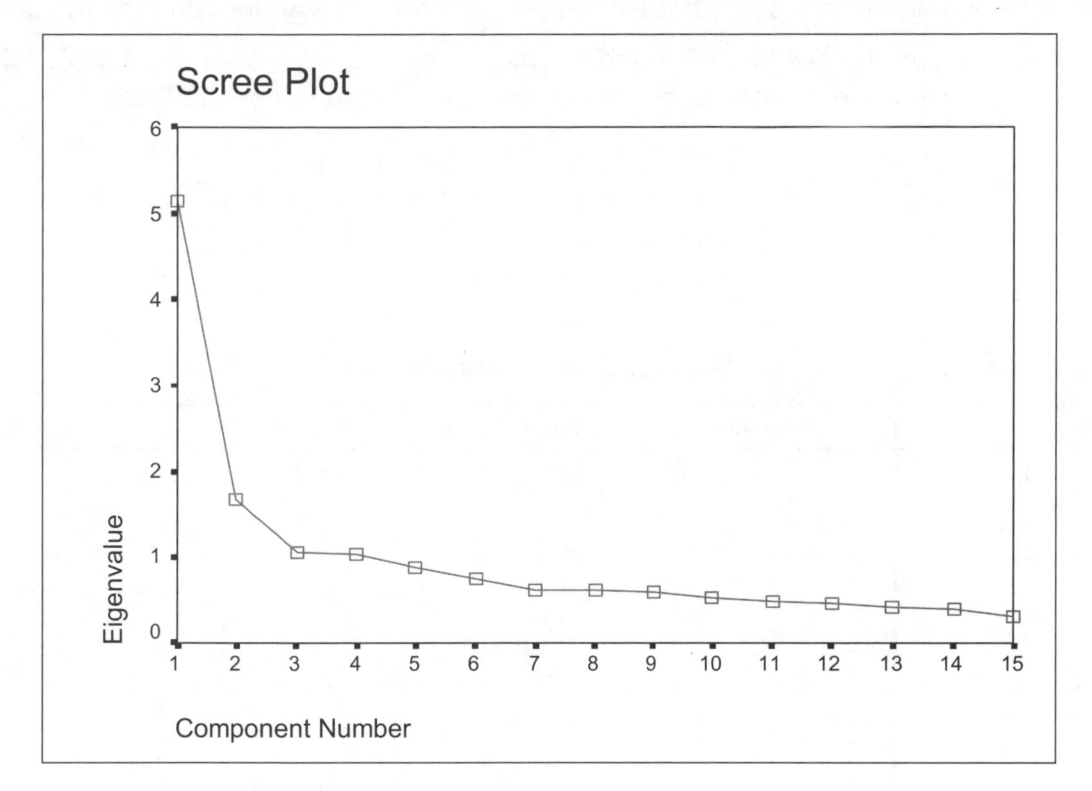

This is called a **Scree plot**. It plots the eigenvalues on a bicoordinate plane. It derives its name from the *scree* that is deposited at the base of a landslide. The scree plot is sometimes used to select how many factors to rotate to a final solution. The traditional construct for interpretation is that the scree should be ignored and that only factors on the steep portion of the graph should be selected and rotated. The SPSS default is to select and rotate any factor with an eigenvalue greater than 1.0. Since the default procedure is followed in this example, four (4) factors are selected for rotation; based on the scree, we might instead select two factors.

SPSS will next print out the *un*rotated component structure. This is rarely of interest to the researcher and, to save space, is not included here. What follows is the 4 × 4 factor transformation matrix. If you multiply the factor transformation matrix (below) by the original (unrotated) 4 × 15 factor matrix, the result would be the *rotated* factor matrix.

Component (Factor) Transformation Matrix

Component	1	2	3	4
1	.733	.405	.419	.352
2	-.441	.736	-.195	.420
3	-.113	-.538	.017	.835
4	-.506	.066	.859	-.043

The *rotated* factor structure is displayed next. Note that due to the selection of the **Sort by size** option, the factor loadings are sorted in two ways: (a) The highest factor loadings for each factor are selected and listed in separate blocks, and (b) within each block, the factor loadings are sorted from largest to smallest. The numbers in each column are the factor loadings for each factor, roughly the equivalent of the correlation between a particular item and the factor.

To aid in the interpretation of the rotated factor matrix, next to each efficacy variable (**effic1** to **effic15**) we listed the three categories of efficacy the questions were intended to test:

[Emot] Efficacy for emotional types of helping
[Inf] Efficacy for informational types of helping
[Instr] Efficacy for instrumental types of helping
[----] The open-ended question

Rotated Component Matrix

Variable	Component 1	Component 2	Component 3	Component 4
effic3 [Instr]	.702	.181	.022	.153
effic4 [Emot]	.644	.088	.302	-.044
effic7 [Emot]	.635	.345	-.090	.205
effic10 [Emot]	.626	.133	.472	-.186
effic13 [Emot]	.618	-.100	.506	.071
effic1 [Emot]	.617	.197	.162	.295
effic11 [Inf]	.573	.047	.274	.125
effic5 [Inf]	.059	.828	.081	.128
effic2 [Inf]	.167	.738	.024	.234
effic8 [Inf]	.313	.597	.226	-.057
effic14 [Emot]	.163	.045	.779	.254
effic15 [----]	.157	.251	.624	.113
effic12 [Instr]	.215	.064	.192	.738
effic9 [Instr]	.081	.418	.180	.631
effic6 [Instr]	.574	.089	-.009	.604

The initial reaction of the researcher who conducted this analysis was "pretty good factor structure!" The first factor is composed primarily of variables that measure efficacy for *emotional* types of helping. One question from instrumental helping ("To what extent did you have the ability to listen carefully or to appraise your friend's situation?") and one from informational helping ("Did you believe you were capable of reducing tension and helping your friend get his/her mind off the problem?") were included in the first factor. It is not difficult to see why these two items might load onto the same factor as efficacy for emotional helping.

Factor 2 is composed entirely of the remaining three measures of efficacy for *informational* types of giving. Factor 4 is composed entirely of the remaining three measures of efficacy for *instrumental* types of helping. Factor 3 is a rather strange factor and would likely not be used.

Included is the measure of efficacy for *other* types of helping—a measure that the majority of subjects did not answer. The other variable, **effic14**, is a somewhat strange measure that a number of subjects seemed confused about. It dealt with efficacy for helping the friend reduce self-blame. In many problem situations self-blame was not an issue.

This is the type of thinking a researcher does when attempting to interpret the results from a factor analysis. The present output seems to yield a fairly interpretable pattern of three types of efficacy: efficacy for emotional types of helping, efficacy for instrumental types of helping, and efficacy for informational types of helping. Factor 3, the strange one, would likely be dropped. Because the two variables in Factor 3 also have factor loadings on the other three factors, the researcher may omit these two variables and run the factor analysis again with just the 13 variables, to see if results differ.

21

CLUSTER ANALYSIS

DESCRIPTION OF any procedure is often most effectively accomplished by comparison to another procedure. To clarify an understanding of hierarchical cluster analysis, we will compare it to factor analysis. Cluster analysis is in some ways similar to factor analysis (Chapter 20), but it also differs in several important ways. If you are unfamiliar with factor analysis, please read through the introduction of Chapter 20 before attempting this chapter. The introductory paragraphs of this section will compare and contrast cluster analysis with factor analysis. Following this, the sequential steps of the cluster analysis procedure will be identified, along with a description of the data set used as an example to aid in the understanding of each step of the process.

CLUSTER ANALYSIS AND FACTOR ANALYSIS CONTRASTED

Cluster analysis and factor analysis are similar in that both procedures take a larger number of cases or variables and reduce them to a smaller number of factors (or clusters) based on some sort of similarity that members within a group share. However, the statistical procedure underlying each type of analysis and the ways that output is interpreted are often quite different:

1. Factor analysis is used to reduce a larger number of *variables* to a smaller number of factors that describe these variables. Cluster analysis is more typically used to combine cases into groups. More specifically, the SPSS procedure is primarily designed to use variables as criteria for grouping (agglomerating) *subjects* or *cases* (not variables) into groups, based on each subject's (or case's) scores on a given set of variables. For instance: With a data set of 500 subjects measured on 15 different types of helping behavior (the variables), factor analysis might be used to extract factors that describe 3 or 4 categories or types of helping based on the 15 help variables. Cluster analysis is more likely to be used with a data set that includes, for instance, 21 cases (such as brands of VCRs) with perhaps 15 variables that identify characteristics of each of the 21 VCRs. Rather than the *variables* being clustered (as in factor analysis), the cases (VCRs here) would be clustered into groups that shared similar features with each other. The result might be three or four clusters of brands of VCRs that share similarities such as price, quality of picture, extra features, reliability of the product, and so on.
2. Although cluster analysis is typically used for grouping cases, clustering of variables (rather than subjects or cases) is just as easy. Since cluster analysis is still more frequently used to cluster cases than variables, that procedure will be demonstrated in the present example. However, in the Step by Step section we will illustrate the sequence of steps for clustering variables.
3. The statistical procedures involved are radically different for cluster analysis and factor analysis. Factor analysis analyzes all variables at each factor extraction step to calculate the variance that each variable contributes to that factor. Cluster analysis calculates a *similarity* or a *distance* measure between each subject or case and every other subject or case and then it groups the two subjects/cases that have the greatest similarity or the least distance into a cluster of two. Then it computes the similarity or distance measures all over again and either combines the next two subjects/cases that are closest or (if the distance is smaller) combines the next case with the cluster of two already formed (yielding either two clusters of two cases each, or one cluster of three cases). This process

continues until all cases are grouped into one large cluster containing all cases. The researcher decides at which stage to stop the clustering process.

4. Reflecting the sentiments of an earlier paragraph, the **Hierarchical Cluster** command is more likely to be used in business, sociology, or political science (e.g., categories of the 25 best-selling brands of TVs; classification of 40 communities based on several demographic variables; groupings of the 30 most populous cities based on age, SES, and political affiliation) than for use in psychology. Psychologists are more often trying to find similarities between variables or underlying causes for phenomena than trying to divide their subjects into groups via some sort of mathematical construct; and when psychologists *are* trying to divide subjects into groups, discriminant analysis (Chapter 22) often accomplishes the process more efficiently than does cluster analysis. But these are only generalities, and there are many appropriate exceptions.
5. With the present ease of using cluster analysis to cluster variables into groups, it becomes immediately interesting to compare results of cluster analysis with factor analysis using the same data set. As with factor analysis, there are a number of variations to the way that the key components of cluster analysis (principally how distances/similarities are determined and how individual variables are clustered) may be accomplished. Since with both factor analysis and cluster analysis results can often be altered significantly by how one runs the procedure, researchers may shamelessly try them all and use whatever best supports their case. But: One should ideally use the process that best represents the nature of the data, considering carefully the rationale of whether a shared-variance construct (factor analysis) or a similarities-of-features procedure (cluster analysis) is better suited for a particular purpose.

PROCEDURES FOR CONDUCTING CLUSTER ANALYSIS

Cluster analysis goes through a sequence of procedures to come up with a final solution. To assist understanding, before we describe these procedures we will introduce the example constructed for this analysis and use it to clarify each step. Cluster analysis would not be appropriate for either the **grades.sav** file or any of the helping files. In each data set the focus is on the nature of the variables, and there is little rationale for wanting to cluster subjects. We turn instead to a data set that is better designed for cluster analysis. This is the only chapter in which these data will be used. It consists of a modified analysis from the magazine *Consumer Reports* on the quality of the top 21 brands of VCRs. The file associated with this data set is called **vcr.sav**. While this began as factual information, we have "doctored" the data to help create a clear cluster structure and thus serve as a convincing example. Because of this, the brand names in the data set are also fictional.

Step 1: Select variables to be used as criteria for cluster formation. In the **vcr.sav** file, cluster analysis will be conducted based on the following variables (listed in the order shown in the first Output display): Price, picture quality (5 measures), reception quality (3 measures), audio quality (3 measures), ease of programming (1 measure), number of events (1 measure), number of days for future programming (1 measure), remote control features (3 measures), and extras (3 measures).

Step 2: Select the procedure for measuring distance or similarity between each case or cluster (initially, each case is a cluster of one, the 21 brands of VCRs). The SPSS default for this measure is called the *squared Euclidean distance*, and it is simply the sum of the squared differences for each variable for each case. For instance, Brand A may have scores of 2, 3, 5 for the three audio ratings. Brand B may have ratings of 4, 3, 2, for the same three measures. Squared Euclidean distance between these two brands would be $(2 - 4)^2 + (3 - 3)^2 + (5 - 2)^2 = 13$. In the actual analysis, these squared differences would be summed for all 21 variables for each brand, yielding a numeric measure of the distance between each pair of brands. SPSS provides for the use of measures other than squared Euclidean distance to determine distance or similarity between clusters. These are described in the SPSS manuals.

A question that may come to mind is, will the cluster procedure be valid if the scales of measurement for the variables are different? In the **vcr.sav** file, most measures are rated on a 5-point scale from *much poorer than average*(1) to *much better than average*(5). But the listed prices fluctuate between $200 and $525, and events, days, and extras are simply the actual numbers associated with those variables. The solution suggested by SPSS is to standardize all variables—that is, change each variable to a z score (with a mean of 0 and a standard deviation of 1). There are other standardization options but the z score is so widely used that it has the advantage of familiarity. This will give each variable equal metrics, but will give them equal weight as well. If all variables are already in the same metric, or if you wish to retain the weighting of the original values, then it is not necessary to change to z scores.

Step 3: Form clusters. There are two basic methods for forming clusters, agglomerative and divisive. For agglomerative hierarchical clustering, SPSS groups cases into progressively larger clusters until all cases are in one large cluster. In divisive hierarchical clustering, the opposite procedure takes place. All cases are grouped into one large cluster, and clusters are split at each step until the desired number of clusters is achieved. The agglomerative procedure is the SPSS default and will be presented here.

Within the agglomerative framework, there are several options of mathematical procedures for combining clusters. The default procedure is called the *between-group linkage* or the *average linkage within groups*. SPSS computes the smallest average distance between all group pairs and combines the two groups that are closest. Note that in the initial phase (when all clusters are individual cases), the average distance is simply the computed distance between pairs of cases. Only when actual clusters are formed does the term *average* distance apply. The procedure begins with as many clusters as there are cases (21 in the **vcr.sav** file). At step one, the two cases with the smallest distance between them are clustered. Then SPSS computes distances once more and combines the two that are next closest. After the second step you will have either 18 individual cases and one cluster of 3 cases, or 17 individual cases and two clusters of two cases each. This process continues until all cases are grouped into one large cluster. Other methods of clustering cases are described in the SPSS Manuals.

Step 4: Interpreting results. Similar to factor analysis, interpretation and the number of clusters to accept as a final solution are largely a matter of the interpretation of the researcher. In the **vcr.sav** file a three-cluster solution seemed best. There appeared to be three primary qualities that differentiated among groups: The first group was highest in price (M = $511.67), highest in picture quality (M = 5.00), and had the greatest number of additional features (M = 10.8). The second group contained VCRs that were moderate in price, medium to low on picture quality, and had a smaller number of additional features ($400.00, 3.00, and 7.8, respectively). The third group contained the budget models ($262.22, 2.78, and 3.0). The other variables did not seem to contribute to the clustering in any systematic way.

The order followed for the description of cluster analysis in the Step by Step section will be similar to that followed in the chapters that present the more complex statistical procedures. Sequence steps 1-4 and 6-7 will be identical to other chapters. There will be three versions of step 5. The first (step 5) will be the simplest default procedure for conducting cluster analysis. The second (step 5a) will include a number of additional available options to tailor your program to fit your particular needs or to produce additional desired output. The last presentation (step 5b) will show how to conduct cluster analysis of variables. We will use the same variables used in the factor analysis chapter (Chapter 20), the 15 self-efficacy questions from the **helping2.sav** file. Although this procedure will be shown in the Step by Step section, space restraints prevent reproducing it in the Output section.

The best resource on cluster analysis known to the authors is by Brian S. Everitt (1993). Please see the reference section for additional information about this book.

STEP BY STEP

Hierarchical Cluster Analysis

To enter SPSS, a click on **Start** *in the taskbar (bottom of screen) activates the start menu:*

After clicking the SPSS program icon, Screen 1 (inside front cover) appears on the monitor.

Step 2

Create and name a data file or edit (if necessary) an already existing file (see Chapter 3).

Screens 1 and 2 (inside front cover) allow you to access the data file used in conducting the analysis of interest. The following sequence accesses the **vcr.sav** *file for further analyses:*

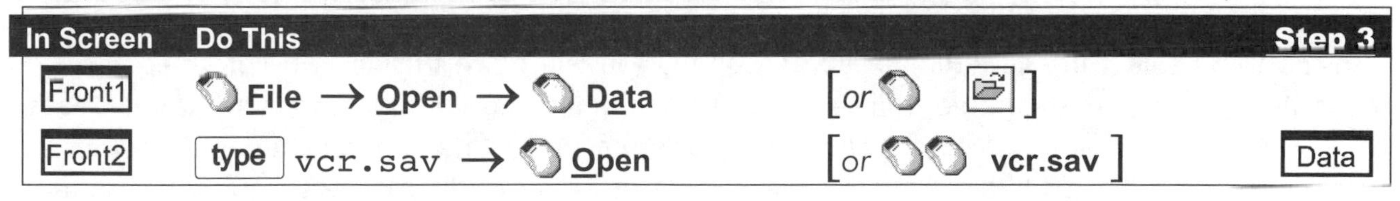

Whether first entering SPSS or returning from earlier operations the standard menu of commands across the top is required (shown below). As long as it is visible you may perform any analyses. It is not necessary for the data window to be visible.

This menu of commands disappears or modifies when using pivot tables or editing graphs. To uncover the standard menu of commands simply click on the or the icon.

After completion of Step 3 a screen with the desired menu bar appears. When you click a command (from the menu bar), a series of options will appear (usually) below the selected command. With each new set of options, click the desired item. The sequence to access cluster analysis begins at any screen with the menu of commands visible:

These three clicks open up the Hierarchical Cluster Analysis main dialog window (Screen 25.1, below). The use of this window will vary substantially depending on whether you choose to **Cluster Cases** or to **Cluster Variables**. If you cluster cases (as we do in our example in this chapter) then you will identify variables you wish to be considered in creating clusters for the cases (in our example, all variables except for **brand**). These will be pasted into the **Variables** box. Then you need to specify how you wish your cases to be identified. This will usually be an ID number or some identifying name. In the example, the **brand** variable (it lists the 21 brand names) will be pasted into the lower box to identify our 21 cases.

21.1

The Hierarchical Cluster Analysis Main Dialog Window

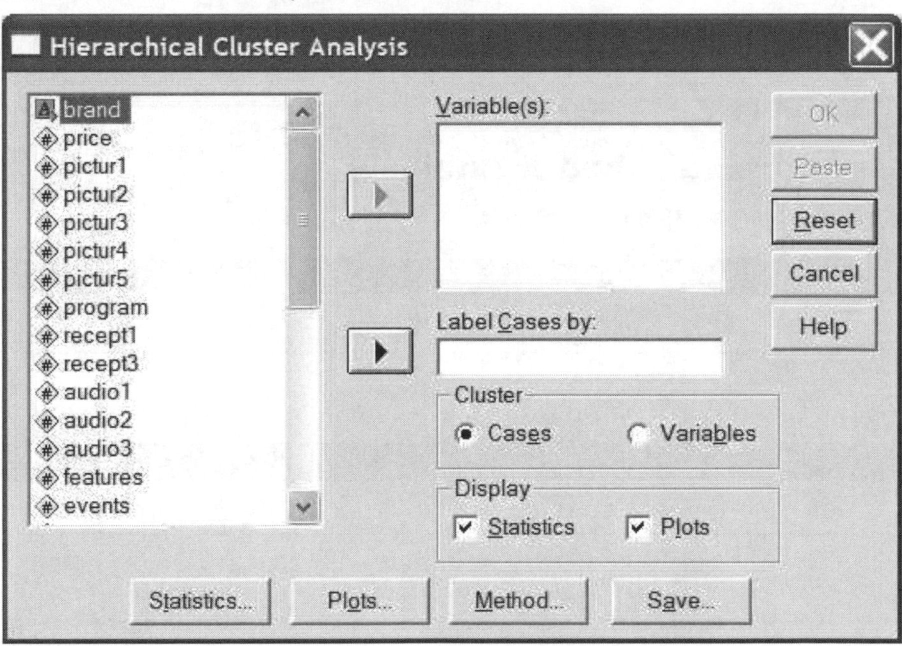

If instead you select the **Cluster Variables** option, you will paste the desired variables into the **Variables** box, but the variable *names* will serve to identify the variables and the **Label Cases by** box will remain empty. Under **Display**, **Statistics** and **Plots** are selected by default. In most cases you will wish to retain both these selections. Four additional sets of options are represented by the four pushbuttons along the bottom of the window. A description of frequently-used options for each of these buttons follows.

A click on **Statistics** opens a new dialog window (Screen 21.2, following page). The **Agglomeration schedule** is selected by default and represents a standard feature of cluster analysis (see the Output section for display and explanation). The distance matrix allows a look at distances between all cases and clusters. With a small file this might be useful, but with large files the number of comparisons grows geometrically, occupies many pages of output, and make this option impractical. In the **Cluster Membership** box, three options exist:

21.2

The Hierarchical Cluster Analysis: Statistics Window

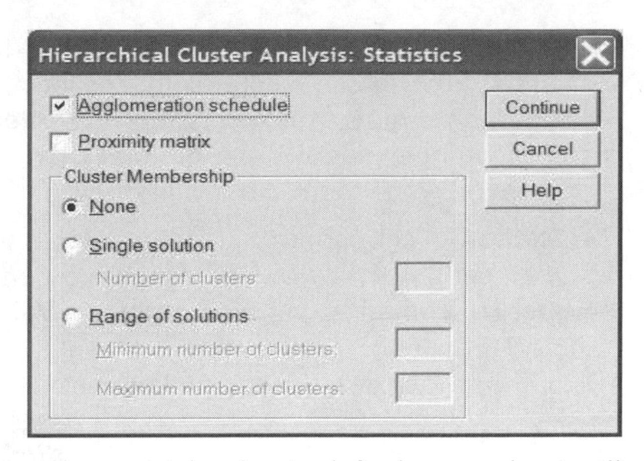

- **None**: This option actually lists *all* clusters since an icicle plot (a default procedure) will identify all possible solutions. What it does *not* do is identify cases included in each cluster for a *particular* solution.
- **Single solution**: Specify some number greater than 1 that indicates cluster membership for a specific number of clusters. For instance if you type 3 into the box indicating number of clusters, SPSS will print out a 3-cluster solution.
- **Range of solutions**: If you wish to see several possible solutions, type the value for the smallest number of clusters you wish to consider in the first box and the value for the largest number of clusters you wish to see in the second box. For instance if you typed in 3 and 5, SPSS would show case membership for a 3-cluster, a 4-cluster, and a 5-cluster solution.

The next available dialog box opens up options for inclusion or modifications of dendograms or icicle plots, plots that are particularly well suited to cluster analysis. Both **icicle plots** and **dendograms** are displayed and explained in the Output section. Please refer to those pages for visual reference if you wish. A click on the **Plots** pushbutton will open Screen 21.3, below.

21.3

The Hierarchical Cluster Analysis: Plots Window

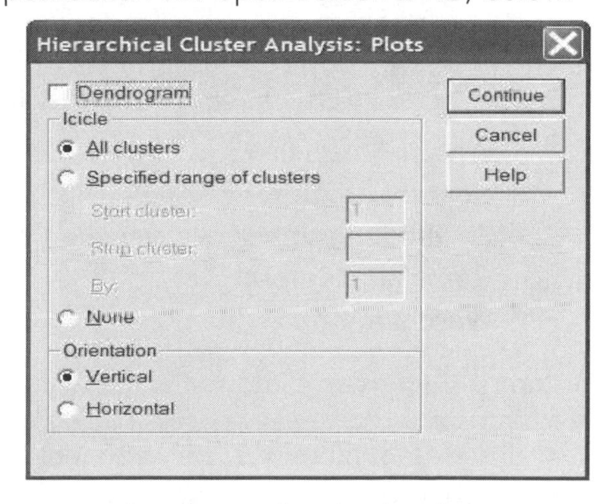

The **Dendogram** provides information similar to that covered in the icicle plot but features, in addition, a relative measure of the magnitude of differences between variables or clusters at each step of the process. An icicle plot of the entire clustering process is included by default. Since early steps in this process have little influence on the final solution, many researchers prefer to show only a relatively narrow range of clustering solutions. This may be accomplished by a click on the **Specific range of clusters** option, followed by typing in the smallest number of clusters you wish to consider, the largest number of clusters you wish to see in a solution, and the interval between those values. For instance if the numbers you entered were 3, 5, and 1, you would see 3-cluster, 4-cluster, and 5-cluster solutions. If the numbers you selected were 2, 10, and 2, SPSS would display solutions for 2, 4, 6, 8, and 10 clusters. A click on **None** eliminates any icicle plots from the output. The **Vertical** option is capable of displaying many more cases on a page (about 24) than the **Horizontal** option (about 14). You may select the horizontal option when there are too many variables or cases to fit across the top of a single page.

The **Method** pushbutton from Screen 21.1 opens the most involved of the cluster analysis windows, Screen 21.4, (below). There are a number of different options for the **Cluster Method**, the **Interval Measure**, and the **Transform Values: Standardize**. For each one we will describe the most frequently used option and simply list the others. The manual, *SPSS Base System Manual,* provides descriptions of the others.

21.4

The Hierarchical Cluster Analysis: Method Window

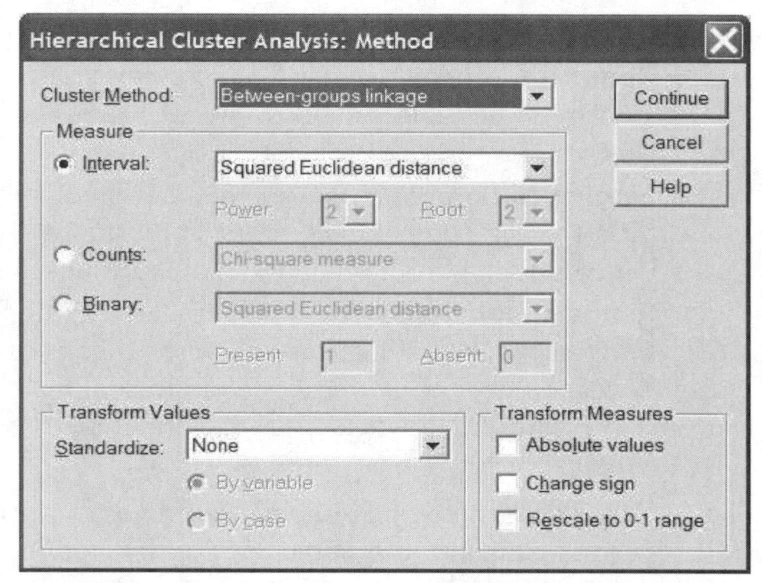

Between-groups linkage, (selected in Screen 21.4, above) also called the *average linkage within groups* simply joins the variables or clusters that have the least distance between them at each successive stage of the analysis. A more complete description of this method may be found on page 273. Other options (visible after a click on the ▾ button) include:

- **Within-group linkage** or *average linkage within groups*
- **Nearest neighbor** or *single linkage*
- **Furthest neighbor** or *complete linkage*
- **Centroid clustering**
- **Ward's method**

Squared Euclidean distance is the default method of determining distances or similarities between cases or clusters of cases. This is described in some detail on page 250. Briefly, it is the sum of squared differences between matching variables for each case. Other options include:

- **Cosine**: This is a similarity measure based on Cosines of vectors of values.
- **Pearson correlation**: This is a similarity measure determined by correlations of vectors of values.
- **Chebychev**: This is a distance measure, the maximum absolute difference between values for items.
- **Block**: This is a distance measure.
- **Minkowski**: This is a distance measure.
- **Customized**: This is a distance measure.

A procedure discussed in the introduction is accessed by the **Standardize** box. This option provides different methods for standardizing variables. The default for standardization is **None**, but when standardization does take place, **z scores** is the most frequently used option. Of alternatives to the z score, the first three standardization methods listed below will produce identical results. The two that follow, in which the mean or the standard deviation is allowed to vary, may produce different results. Selection of one or another is based upon that which is most appropriate to the data set or most convenient for the researcher. In addition to **z scores**, the other options include:

- **Range -1 to 1**: All variables are standardized to vary between –1 and 1.
- **Range 0 to 1**: All variables are standardized to vary between 0 and 1.
- **Maximum magnitude of 1**: All variables are standardized to have a maximum value of 1.
- **Mean of 1**: All variables are standardized to have a mean value of 1 (standard deviations may vary).
- **Standard deviation of 1**: All variables are standardized to have a standard deviation of 1 (mean values may vary).
- **Customized**: This is a researcher-determined distance measure.

The other **Transform Measures** (lower right corner of the window) simply takes the absolute value, reverses the sign of the original variable values, or rescales variable values on a 0 to 1 metric.

The final small dialog box available in the hierarchical cluster analysis procedure deals with saving new variables. The window is shown below (Screen 21.5) and allows three options. The default is **None**. If you choose to save new variables, they will appear in the form of a new variable in the last column of your data file and simply include a different coded number for each case. For instance, if you save as a new variable a 3-cluster solution, each case will be coded either 1, 2, or 3. This dialog box also allows you to save more than one solution as new variables. If you select the **Range of solutions** option and indicate **From** 3 **through** 5 **clusters**, three new variables will be saved: One variable that codes cases 1, 2, and 3; a second that codes cases 1, 2, 3, and 4; and a third that codes cases 1, 2, 3, 4, and 5.

21.5

The Hierarchical Cluster Analysis: Save New Variables Window

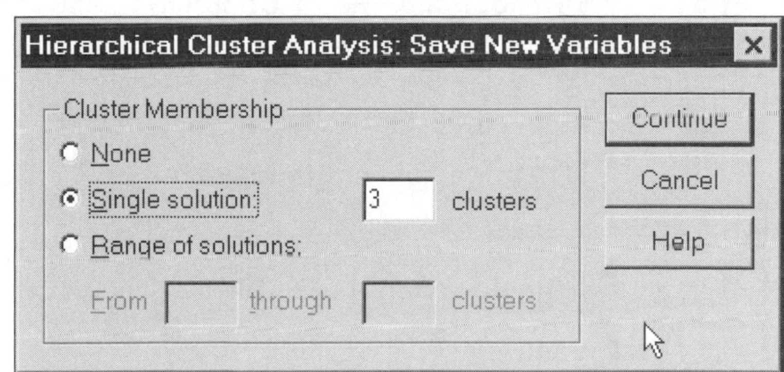

In the step-by-step sequences we will demonstrate three procedures: Step 5 accesses the simplest default cluster analysis, Step 5a provides a cluster analysis with a number of the options discussed earlier, and Step 5b a cluster analysis that clusters variables rather than cases. The italicized text before each box will specify the specifics of the analysis. Step 5a will be illustrated in the Output section.

To conduct the simplest default hierarchical cluster analysis (but standardizing the metrics of all variables) in which you attempt to cluster the 21 brands of VCRs based on similarities of features, perform the following sequence of steps. The starting point is Screen 21.1. Carry out the 4-step sequence (pages 265-266) to arrive at this screen.

In Screen	Do This	Step 5
21.1	**brand** → *lower* → *click and drag from* **price** *to* **exras3** *(highlighting all)* → *upper* → **Method**	
21.4	*to the right of* **Standardize** → **Z scores** → **Continue**	
21.1	**OK**	Back1

This procedure selects the brand name to identify different cases, selects all remaining variables as the basis on which clustering takes place, changes all variables to z scores to give each variable an equal metric and equal weight, determines distance between variables by the squared Euclidean distance, clusters variables based on between group linkage, and produces the agglomeration schedule and vertical icicle plot as output.

To conduct a hierarchical cluster analysis of the 21 VCR brands based on the 20 variables that describe each **brand***, to request the* **Agglomeration Schedule***, the* **Dendogram***, and the* **vertical icicle plot** *in the output, change all variables to* **z scores** *to yield equal metrics and equal weighting, select the* **Squared Euclidean distance** *as the method of determining distance between clusters and the* **Furthest neighbor** *method of clustering, and save a 3-cluster solution as a new variable, perform the following sequence of steps. The starting point is Screen 21.1(see page 266). Not demonstrated here, but strongly suggested: List all cases (page 45) with variables in the desired order for sake of reference during the cluster analysis procedure. The Output section includes this list of cases.*

In Screen	Do This	Step 5a
21.1	**brand** → *lower* → *click and drag from* **price** *to* **exras3** *(highlighting all),* → *upper* → **Plots**	
21.3	**Dendogram** → **Continue**	
21.1	**Method**	
21.4	*to the right of* **Cluster Method** → **Furthest neighbor**	
	to the right of **Standardize** → **Z scores** → **Continue**	
21.1	**Save**	
21.5	**Single solution** → type 3 *(In the box to the right)* → **Continue**	
21.1	**OK**	Back1

*To conduct a simplest default hierarchical cluster analysis in which variables (rather than cases) are clustered, for the 15 efficacy questions (***effic1** *to* **effic15***) from the* **helping3.sav** *file, using all standard defaults and without standardization (not needed since all variables have the same metric), perform the following sequence of steps. The starting point is Screen 21.1. Carry out the 4-step sequence (pages 265-266) to arrive at this screen.*

In Screen	Do This	Step 5b
21.1	**Variables** → *click and drag from* **effic1** *to* **effic15** *(highlighting all)*, → *upper* → **OK**	Back1

This simple procedure will conduct a cluster analysis of the 15 self-efficacy questions. What will be printed is a summary of variable information such as valid cases, missing values, etc., followed by an Agglomeration Schedule and a vertical icicle plot to indicate the nature of the clustering process. These results would be interesting to compare with the factor analysis (previous chapter) using the same variables.

Upon completion of any of steps 5, 5a, or 5b, the output screen will appear (Screen 1, inside back cover). All results from the just-completed analysis are included in the Output Navigator. Make use of the scroll bar arrows ( ) to view the results. Even when viewing output, the standard menu of commands is still listed across the top of the window. Further analyses may be conducted without returning to the data screen.

PRINTING RESULTS

Results of the analysis (or analyses) that have just been conducted requires a window that displays the standard commands (**File Edit Data Transform Analyze** . . .) across the top. A typical print procedure is shown below beginning with the standard output screen (Screen 1, inside back cover).

To print results, from the Output screen perform the following sequence of steps:

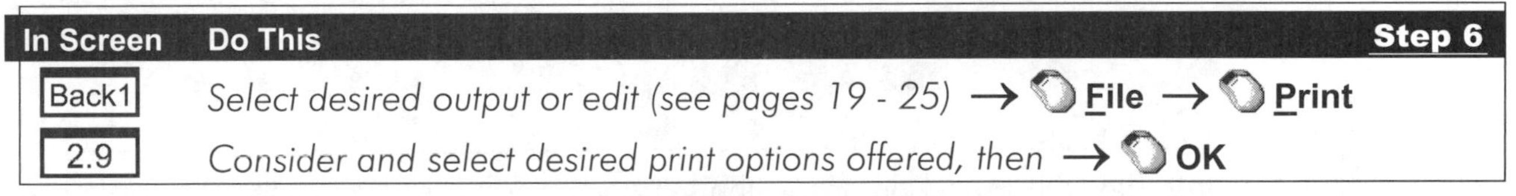

In Screen	Do This	Step 6
Back1	*Select desired output or edit (see pages 19 - 25)* → **File** → **Print**	
2.9	*Consider and select desired print options offered, then* → **OK**	

To exit you may begin from any screen that shows the **File** *command at the top.*

In Screen	Do This	Step 7
Menu	**File** → **Exit**	2.1

Note: After clicking **Exit**, there will frequently be small windows that appear asking if you wish to save or change anything. Simply click each appropriate response.

OUTPUT

Cluster Analysis

The following output is produced by sequence step 5a shown on page 270. What is reproduced here is a fairly close duplicate (both content and format) of what SPSS actually produces. We begin with a list of the variables (the 21 brands of VCRs) and the information in the data file about each one. This is followed by the agglomeration schedule, the vertical icicle plot, and the dendogram. Text between each section clarifies the meaning of the output.

Hierarchical Cluster Analysis: The Data File

	brand	price	pictur1	pictur2	pictur3	pictur4	pictur5	program	recept1	recept3	audio1	audio2	audio3	features	events	days	remote1	remote2	remote3	extras1	extras2	extras3
1	SONNY	520	5	5	5	5	5	4	3	3	5	3	3	5	8	31	4	4	4	13	13	13
2	ANGLER	535	5	5	5	5	5	3	4	3	5	4	4	4	6	365	3	4	4	12	12	12
3	MITTENSUB	515	5	5	5	5	5	4	3	3	5	3	3	5	8	28	4	4	4	13	13	13
4	SINGBO	470	5	5	5	5	5	3	3	3	3	3	4	4	6	365	2	3	3	7	7	7
5	WHACKACHY	525	5	5	5	5	5	2	4	3	4	3	4	4	6	365	4	3	3	11	11	11
6	SILVERMOON	370	4	4	4	4	4	3	4	3	3	5	4	5	6	365	4	3	3	8	8	8
7	EXPERTSEE	430	4	4	4	4	4	5	3	4	4	5	4	4	8	365	4	4	4	8	8	8
8	FROMSHEEBA	505	5	5	5	5	5	3	4	2	4	4	4	4	8	365	3	3	3	9	9	9
9	POTASONIC1	450	3	3	3	3	3	3	3	3	4	4	4	4	8	30	3	3	3	8	8	8
10	CLIMAX	365	4	4	4	4	4	2	3	3	4	4	4	3	8	365	4	2	2	4	4	4
11	POTASONIC2	435	3	3	3	3	3	3	3	3	4	3	4	4	4	30	4	4	4	9	9	9
12	MAGNESIA	265	3	3	3	3	3	3	4	2	4	4	3	3	8	365	4	4	4	3	3	3
13	DULL	200	3	3	3	3	3	3	4	2	4	4	4	4	8	365	3	3	3	3	3	3
14	BURSTINGSTAR	380	2	2	2	2	2	3	3	3	4	4	4	4	4	30	4	4	4	6	6	6
15	VCJ	420	3	3	3	3	3	2	4	3	4	4	4	4	8	365	2	3	2	6	6	6
16	ARC	335	2	2	2	2	2	5	4	3	5	3	4	5	8	365	3	2	3	8	8	8
17	MAGNETOX	205	3	3	3	3	3	3	4	1	3	4	3	3	8	365	4	4	4	3	3	3
18	RECALLAKING	200	3	3	3	3	3	3	4	2	3	4	4	4	8	365	3	3	3	4	4	4
19	PAULSANG	205	2	2	2	2	2	2	3	2	4	4	4	3	6	365	3	4	2	2	2	2
20	EG	275	2	2	2	2	2	3	4	2	4	4	4	3	8	365	3	3	3	2	2	2
21	RASES	225	2	2	2	2	2	3	4	2	4	4	4	3	8	365	3	4	2	0	0	0

This is simply the data file with brand names to the left and variable names along the top. The **Case Summaries** procedure (see Chapter 4) accesses this important output. Since the file with cluster analysis is usually fairly short, it is often a good idea to print it out because it makes an excellent reference as you attempt to interpret the clustering patterns. As mentioned in the introduction, this data file has been doctored (the brand names too) to create a more interpretable cluster pattern.

Following this listing, descriptive statistics for each variable will be printed out (mean, standard deviation, minimum, maximum, and *N*), followed by a listing of the new variable names after they have been changed to z scores (e.g., **price** becomes **zprice**, **pictur1** becomes **zpictur1**, **pictur2**, **zpictur2;** and so forth). Since these data are straightforward, we will save space by not reproducing them.

On the following page the agglomeration schedule is listed. The procedure followed by cluster analysis at Stage 1 is to cluster the two cases that have the smallest squared Euclidean distance between them. Then SPSS will recompute the distance measures between all single cases and clusters (there is only one cluster of two cases after the first step). Next, the 2 cases (or clusters) with the smallest distance will be combined, yielding either 2 clusters of 2 cases (with 17 cases unclustered) or one cluster of 3 (with 18 cases unclustered). This process continues until all cases are clustered into a single group. To clarify, we will explain Stages 1, 10, and 14 (following page).

Agglomeration Schedule

Squared Euclidean Distance with Complete Linkage

Stage	Cluster Combined		Coefficients	Stage Cluster First Appears		Next Stage
	Cluster 1	Cluster 2		Cluster 1	Cluster 2	
1	**1**	**3**	**.002**	**0**	**0**	**17**
2	13	18	2.708	0	0	9
3	12	17	4.979	0	0	14
4	20	21	5.014	0	0	7
5	11	14	8.509	0	0	10
6	5	8	11.725	0	0	8
7	19	20	11.871	0	4	14
8	2	5	13.174	0	6	13
9	13	15	14.317	2	0	12
10	**9**	**11**	**19.833**	**0**	**5**	**15**
11	6	7	22.901	0	0	15
12	10	13	23.880	0	9	16
13	2	4	28.378	8	0	17
14	**12**	**19**	**31.667**	**3**	**7**	**16**
15	6	9	40.470	11	10	18
16	10	12	44.624	12	14	19
17	1	2	47.720	1	13	20
18	6	16	49.963	15	0	19
19	6	10	64.785	18	16	20
20	1	6	115.781	17	19	0

At **Stage 1**, Case 1 is clustered with Case 3. The squared Euclidean distance between these two cases is .002. Neither variable has been previously clustered (the two zeros under Cluster 1 and Cluster 2), and the next stage (when the cluster containing Case 1 combines with another case) is Stage 17. (Note that at Stage 17, Case 2 joins the Case-1 cluster.)

At **Stage 10**, Case 9 joins the Case-11 cluster (Case 11 was previously clustered with Case 14 back in Stage 5, thus creating a cluster of 3 cases: Cases 9, 11, and 14). The squared Euclidean distance between Case 9 and the Case-11 cluster is 19.833. Case 9 has not been previously clustered (the zero under Cluster 1), and Case 11 was previously clustered at Stage 5. The next stage (when the cluster containing Case 9 clusters) is Stage 15 (when it combines with the Case-6 cluster).

At **Stage 14**, the clusters containing Cases 12 and 19 are joined. Case 12 had been previously clustered with Case 17, and Case 19 had been previously clustered with Cases 20 and 21, thus forming a cluster of 5 cases (Cases 12, 17, 19, 20, 21). The squared Euclidean distance between the two joined clusters is 31.667. Case 12 was previously joined at Stage 3 with Case 17. Case 19 was previously joined at Stage 7 with the Case-20 cluster. The next stage when the Case-12 cluster will combine with another case/cluster is Stage 16 (when it joins with the Case-10 cluster).

The icicle plot (that follows) displays the same information graphically. We have reverted to an equal width (Courier) font without borders because the table produced by SPSS will not fit on the page. The information, however, is identical to SPSS output.

Vertical Icicle Plot using Complete Linkage

```
(Down) Number of Clusters      (Across) Case Label and number

     -----------LOW-----------  -----MEDIUM-----  ------HIGH------
     R  E  P  M  M  V  R  D  C  A  B  P  P  E  S  S  F  W  A  M  S
     A  G  A  A  A  C  E  U  L  R  U  O  O  X  I  I  R  H  N  I  O
     S     U  G  G  J  C  L  I  C  R  T  T  P  L  N  O  A  G  T  N
     E     L  N  N     A  L  M     S  A  A  E  V  G  M  C  L  T  N
     S     S  E  E     L     A     T  S  S  R  E  B  S  K  E  E  Y
           A  T  S     L     X     I  O  O  T  R  O  H  A  R  N
           N  O  I     A           N  N  N  S  M     E  C     S
           G  X  A     K           G  I  I  E  O     E  H     U
                       I           S  C  C  E  O     B  Y     B
                       N           T  2        N     A

     2  2  1  1  1  1  1  1  1  1  1  1
     1  0  9  7  2  5  8  3  0  6  4  1  9  7  6  4  8  5  2  3  1
  1 |XXXXXXXXXXXXXXXXXXXXXXXXXXXXXXXXXXXXXXXXXXXXXXXXXXXXXXXXXXXXXX
  2 |XXXXXXXXXXXXXXXXXXXXXXXXXXXXXXXXXXXXXXXXXXX  XXXXXXXXXXXXXXXX
 3->|XXXXXXXXXXXXXXXXXXXXXXXXX  XXXXXXXXXXXXXXXX  XXXXXXXXXXXXXXXX<-
  4 |XXXXXXXXXXXXXXXXXXXXXXXXX  X  XXXXXXXXXXXXX  XXXXXXXXXXXXXXXX
  5 |XXXXXXXXXXXXXXXXXXXXXXXXX  X  XXXXXXXXXXXXX  XXXXXXXXXX  XXXX
  6 |XXXXXXXXXXXXX  XXXXXXXXXX  X  XXXXXXXXXXXXX  XXXXXXXXXX  XXXX
  7 |XXXXXXXXXXXXX  XXXXXXXXXX  X  XXXXXXX  XXXX  XXXXXXXXXX  XXXX
  8 |XXXXXXX  XXXX  XXXXXXXXXX  X  XXXXXXX  XXXX  XXXXXXXXXX  XXXX
  9 |XXXXXXX  XXXX  XXXXXXXXXX  X  XXXXXXX  XXXX  X  XXXXXXX  XXXX
 10 |XXXXXXX  XXXX  XXXXXXX  X  X  XXXXXXX  XXXX  X  XXXXXXX  XXXX
 11 |XXXXXXX  XXXX  XXXXXXX  X  X  XXXXXXX  X  X  X  XXXXXXX  XXXX
 12 |XXXXXXX  XXXX  XXXXXXX  X  X  XXXX  X  X  X  X  XXXXXXX  XXXX
 13 |XXXXXXX  XXXX  X  XXXX  X  X  XXXX  X  X  X  X  XXXXXXX  XXXX
 14 |XXXXXXX  XXXX  X  XXXX  X  X  XXXX  X  X  X  X  XXXX  X  XXXX
 15 |XXXX  X  XXXX  X  XXXX  X  X  XXXX  X  X  X  X  XXXX  X  XXXX
 16 |XXXX  X  XXXX  X  XXXX  X  X  XXXX  X  X  X  X  X  X  X  XXXX
 17 |XXXX  X  XXXX  X  XXXX  X  X  X  X  X  X  X  X  X  X  X  XXXX
 18 |X  X  X  XXXX  X  XXXX  X  X  X  X  X  X  X  X  X  X  X  XXXX
 19 |X  X  X  X  X  X  XXXX  X  X  X  X  X  X  X  X  X  X  X  XXXX
 20 |X  X  X  X  X  X  X  X  X  X  X  X  X  X  X  X  X  X  X  XXXX
```

This graph displays the same information as the agglomeration schedule table except that the value of the distance measure is not shown. The numbers to the left indicate the number of clusters at each level. For instance, the line opposite the "20" has 20 clusters, 19 individual cases, and brands 1 and 3 joined into a single cluster. Notice the bold face and the arrows to the right of the "3". At this stage there are 3 clusters, and the experimenters determined that 3 clusters was the most meaningful solution. As you may recall from the introduction, the cluster containing Brands 1, 2, 3, 4, 5, and 8 are VCRs characterized by high price, high picture quality, and many features. The second cluster, Brands 6, 7, 9, 11, 14, and 16, are VCRs characterized by medium-range price, lower picture quality, and fewer features. The final cluster, Brands 10, 12, 13, 15, 17, 18, 19, 20, and 21, are budget models with low price, medium picture quality, and few features.

The dendogram, a tree-type display of the clustering process, is the final output display. In this case, the output is identical to the SPSS output.

Dendogram using Complete Linkage

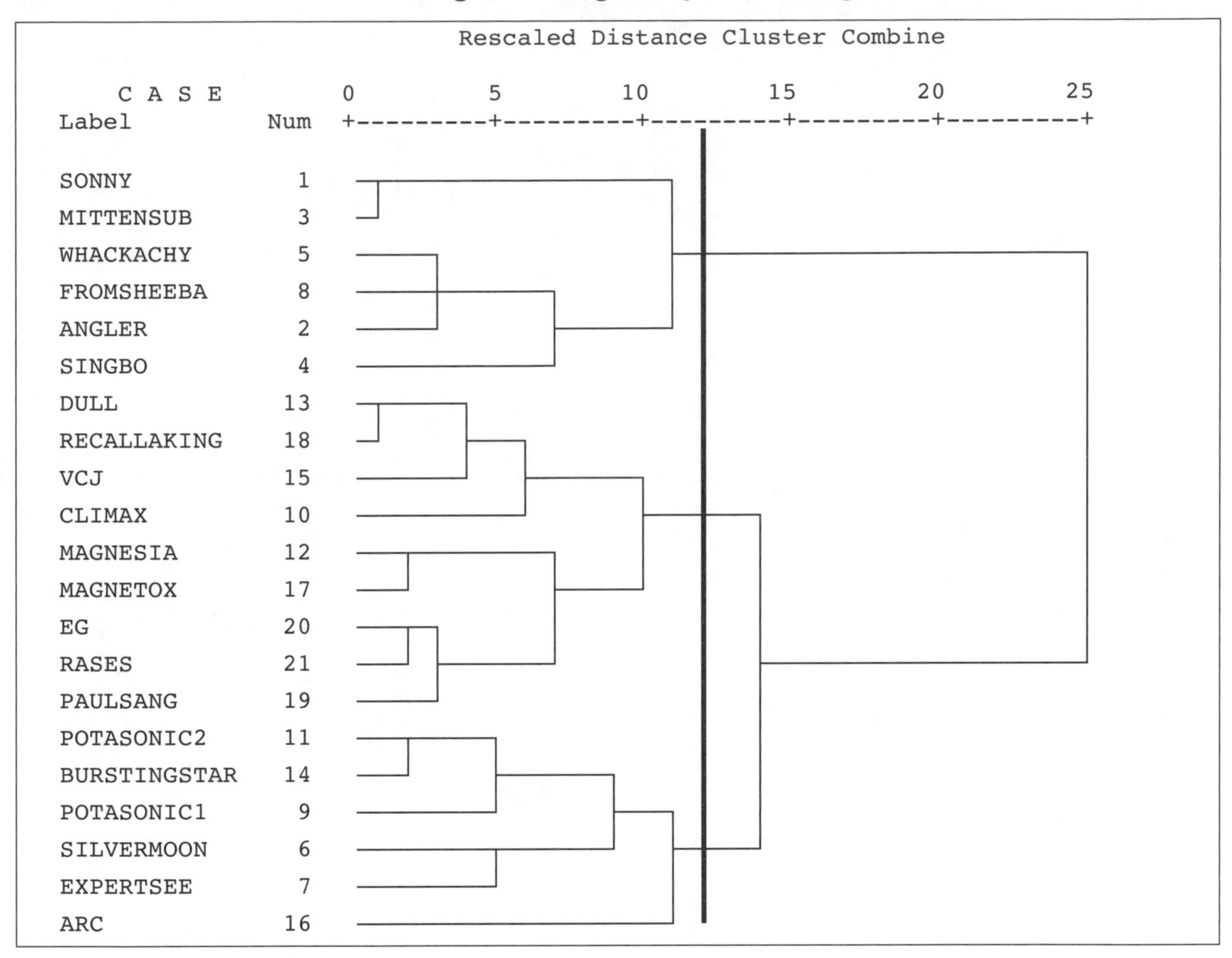

All three of the visual displays of the clustering procedure (the agglomeration table, the icicle plot, and the dendogram) provide slightly different information about the process. In addition to the branching-type nature of the dendogram, which allows the researcher to trace backward or forward to any individual case or clusters at any level, it adds an additional feature that the icicle plot does not. The 0 to 25 scale along the top of the chart gives an idea of how great the distance was between cases or groups that are clustered at a particular step. The distance measure has been rescaled from 0 to 25, with 0 representing no distance and 25 rating the greatest distance. While it is difficult to interpret distance in the early clustering phases (the extreme left of the graph), as you move to the right relative distances become more apparent.

Note the vertical line (provided by the authors) that designates the three-cluster solution. A similar vertical line placed in different positions on the graph will reveal the number of clusters at any stage by the number of horizontal lines that are crossed. To find the membership of a particular cluster, simply trace backwards down the branches (or is it roots?) to the name and case number.

Do you know that Kolmogorov-Smirnov is not a type of vodka and that LSD is not just a hallucinogenic compound? Are you constantly defending the meaning of "goodness of fit" to the multitudes of mall rats who just don't understand? If you recognize these symptoms, you may be suffering from SPSS—Statistical Psychosis of the Socially Skewed. There is help on the horizon. Through a dismembering and rejoining process called the "Guttman split half" you can regain control. Call 1-800-NOSTATS and regain 95% of your confidence."

Contributed by: Diane Anteau (Eastern Michigan University)

DISCRIMINANT ANALYSIS

DISCRIMINANT ANALYSIS is used primarily to predict membership in two or more mutually exclusive groups. The procedure for predicting membership is initially to analyze pertinent variables where the group membership is already known. For instance, a top research university wishes to predict whether applicants will complete a Ph.D. program successfully. Such a university will have many years of records of entry characteristics of applicants and additional information about whether they completed the Ph.D. and how long it took them. They thus have sufficient information to use discriminant analysis. They can identify two discrete groups (those who completed the Ph.D. and those who did not) and make use of entry information such as GPAs, GRE scores, letters of recommendation, and additional biographical information. From these variables it is possible to create a formula that maximally differentiates between the two groups. This formula (if it discriminates successfully) could be used to analyze the likelihood of success for future applicants. There are many circumstances where it is desirable to be able to predict with some degree of certainty outcomes based on measurable criteria: Is an applicant for a particular job likely to be successful or not? Can we identify whether mentally ill patients are suffering from schizophrenia, bipolar mood disorder, or psychosis? If a prisoner is paroled, is he or she likely to return to crime or become a productive citizen? What factors might influence whether a person is at risk to suffer a heart attack or not? The elements common to all five of these examples are that (1) group membership is known for many individuals, and (2) large volumes of information are available to create formulas that might predict future outcomes better than current instruments.

By way of comparison, discriminant analysis is similar to cluster analysis (Chapter 21) in that the researcher is attempting to divide individuals or cases (not *variables*) into discrete groups. It differs from cluster analysis in the procedure for creating the groups. Discriminant analysis creates a regression equation (Chapters 15 and 16) that makes use of a dependent (criterion) variable that is discrete rather than continuous. Based on preexisting data in which group membership is already known, a regression equation can be computed that maximally discriminates between the two groups. The regression equation can then be used to help predict group membership in future instances. Discriminant analysis is a complex procedure and the **Discriminant** command contains many options and features that extend beyond the scope of this book. We will therefore, throughout this chapter, refer you to *SPSS Base System Manual* for additional options when appropriate.

Like many of the more advanced procedures, **Discriminant** follows a set sequence of steps to achieve a final solution. Some of these steps are more experimenter-driven, and others are more data-driven. These steps will be presented here in the introduction. To assist in this explanation, we will introduce yet another example. Then material in the Step by Step and Output sections will provide additional detail.

THE EXAMPLE: ADMISSION INTO A GRADUATE PROGRAM

Neither the **grades.sav** nor the **helping.sav** files provide appropriate data sets to demonstrate the discriminant function. While it is possible to use discriminant analysis to predict, for instance, subject's gender or class standing, it is much easier, in the words of psychologist Gordon Allport, to "just ask them." What we have chosen is a topic where discriminant analysis might actually be used: Criteria for acceptance into a graduate program. Every year, selectors miss-guess and select students who are unsuccessful in their efforts to finish the degree. A wealth of information is collected about each applicant prior to acceptance, and department records indicate whether that student was successful in completing the course. Our example uses the information collected prior to acceptance to predict successful completion of a graduate program. The file is called **graduate.sav** and consists of 50 students admitted into the program between 7 and 11 years ago. The dependent variable is **category** (1 = finished the Ph.D., 2 = did not finish), and 17 predictor variables are utilized to predict category membership in one of these two groups:

- **gender**: 1 = female, 2 = male
- **age**: age in years at time of application
- **marital**: 1 = married, 2 = single
- **gpa**: overall undergraduate GPA
- **areagpa**: GPA in their area of specialty
- **grearea**: score on the major-area section of the GRE
- **grequant**: score on the quantitative section of the GRE
- **greverbal**: score on the verbal section of the GRE
- **letter1**: first of the three recommendation letters (rated 1 = weak through 9 = strong)
- **letter2**: second of the three recommendation letters (same scale)
- **letter3**: third of the three recommendation letters
- **motive**: applicant's level of motivation (1 = low through 9 = high)
- **stable**: applicant's emotional stability (same scale for this and all that follow)
- **resource**: financial resources and support system in place
- **interact**: applicant's ability to interact comfortably with peers and superiors
- **hostile**: applicant's level of inner hostility
- **impress**: impression of selectors who conducted an interview

Repeating once more before we move on, the dependent variable is called **category** and has two levels: 1 = those who finished the Ph.D., and 2 = those who did not finish the Ph.D. Finally, although the study has excellent face validity, the data are all fictional—created by the authors to demonstrate the discriminant function.

THE STEPS USED IN DISCRIMINANT ANALYSIS

Step 1: Selection of variables. Theoretical or conceptual concerns, prior knowledge of the researcher, and some bivariate analyses are frequently used in the initial stage. With the **graduate.sav** file, the number of predictor variables is small enough (17) that it would be acceptable to enter all of them into the regression equation. If, however, you had several hundred variables (e.g., number of questions on many personality tests), this would not be feasible due to conceptual reasons (e.g., collinearity of variables, loss of degrees of freedom) and due to practical reasons (e.g., limitations of available memory). A common first step is to compute a **correlation matrix** of predictor variables. This correlation matrix has a special meaning when applied to discriminant analysis: It is called the *pooled within-group correlation matrix* and involves the average correlation for the two or more correlation matrices for each variable pair. Also available are covariance matrices for separate groups, for pooled within-groups, and for the entire sample. Another popular option is to calculate a number of **t tests** between the two groups for each variable or to conduct **one-way ANOVA**s if you are discriminating into more than two groups. Because the goal of analysis is to find the best discriminant function possible, it is quite normal to try several procedures and several different sets of variables to achieve best results. In the present sample, in a series of *t*-tests significant differences were found between the two levels of **category** (those who finished the Ph.D. and those who did not) for **gpa**, **gre-quant**, **letter1**, **letter2**, **letter3**, **motive**, **age**, **interact**, and **hostile**. In the analyses described below, we will demonstrate first a forced-entry procedure that includes all variables that show significant bivariate relationships and then a stepwise procedure that accesses all 17 independent variables.

Step 2: Procedure. Two entry procedures will be presented in this chapter. The default procedure is designated **Enter independents together** and, in regression vernacular, is a *forced entry* process. The researcher indicates which variables will be entered, and *all* designated variables will be entered simultaneously. A second procedure, called **Wilks' lambda**, is a stepwise operation that is based on minimizing the Wilks' lambda (λ) after each new variable has been entered into the regression equation. As in stepwise multiple regression analysis, there is a criterion for entry into the regression equation ($F > 3.84$ is default) and also a criterion for removal from the equation once a variable has been entered if its contribution to the equation drops below a designated level ($F < 2.71$ is default). Wilks' λ is the ratio of within-groups sum of squares to the total sum of squares. It is designed (in this setting) to indicate whether a particular variable contributes significantly to explaining additional variance in the dependent (or criterion) variable. There is an *F* and *p*-value associated with Wilks' λ that indicates the level of significance. A more complete description is given in the Output section.

So, which of these two methods usually produces better results? As disturbing as it might seem, the computer (in a stepwise procedure with all variables entered) often does better than when the researcher preselects variables. However, there are frequently conceptual or practical reasons for not allowing the computer to make all of the decisions. With the present data set, when all variables were entered using the **Wilks' lambda** procedure (stepwise selection of variables), a regression equation was produced that misclassified only 3 of 50 cases. Using the **Wilks' lambda** procedure but entering only the variables that showed bivariate differences, 4 of 50 cases were misclassified. Using the **Enter independents together** procedure and entering all 17 variables, 4 of 50 were misclassified, and using the **Enter independents together** procedure with only the 9 variables that showed significant bivariate differences, 5 of 50 cases were misclassified. There are several other selection procedures offered by SPSS, and an extensive discussion of these selection methods is provided in *SPSS Base System Manual*.

Step 3: Interpretation and use. The rationale behind discriminant analysis is to make use of existing data pertaining to group membership and relevant predictor variables to create a formula that will accurately predict group membership, using the same variables with a new set of subjects. The formula created by the **Discriminant** command will generally not be as accurate in predicting new cases as it was on the original data. With the present data set, **Discriminant** classified 94% of the individuals correctly when predicting group membership for the same 50 subjects that were used to create the regression formula; but it is unlikely to do that well on subjects where membership is unknown. *SPSS Base System Manual* describes the *jackknife* procedure to identify a more likely percentage of correct classifications in applications when group membership is unknown. Over time, the discriminant formula can be improved and refined as additional data become available.

The best text known by the authors on discriminant analysis is by Geoffrey McLachlan (1992). Please see the reference section for additional information.

STEP BY STEP

Discriminant Analysis

To enter SPSS, a click on **Start** *in the taskbar (bottom of screen) activates the start menu:*

In Screen	Do This	Step 1
2.1	start → All Programs → SPSS for Windows → SPSS 13.0 for Windows	Front1

After clicking the SPSS program icon, Screen 1 (inside front cover) appears on the monitor.

	Step 2
Create and name a data file or edit (if necessary) an already existing file (see Chapter 3).	

Screens 1 and 2 (inside front cover) allow you to access the data file used in conducting the analysis of interest. The following sequence accesses the **graduate.sav** *file for further analyses:*

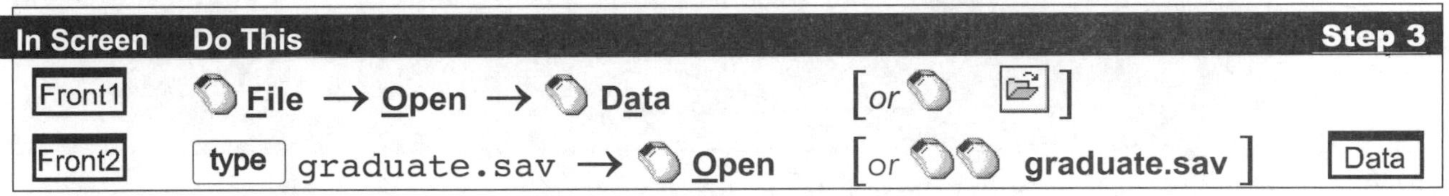

In Screen	Do This		Step 3
Front1	**File** → **Open** → **Data**	[*or*]	
Front2	type `graduate.sav` → **Open**	[or **graduate.sav**]	Data

Whether first entering SPSS or returning from earlier operations the standard menu of commands across the top is required (shown below). As long as it is visible you may perform any analyses. It is not necessary for the data window to be visible.

This menu of commands disappears or modifies when using pivot tables or editing graphs. To uncover the standard menu of commands simply click on the or the icon.

After completion of Step 3 a screen with the desired menu bar appears. When you click a command (from the menu bar), a series of options will appear (usually) below the selected command. With each new set of options, click the desired item. The sequence to access discriminant analysis begins at any screen with the menu of commands visible:

In Screen	Do This	Step 4
Menu	**Analyze** → **Classify** → **Discriminant**	22.1

The Step 4 sequence opens the main dialog window for discriminant analysis (Screen 22.1, following page). In addition to the traditional list of variables to the left and the five function pushbuttons to the right, this screen features:

1. a **Grouping Variable** box for entry of a single categorical dependent variable,
2. an **Independents** box where one may designate as many variables as desired to assist in predicting the levels of the dependent variable,

3. two options (**Enter independents together** and **Use stepwise method**) that identify the entry procedure for selecting variables, and
4. four pushbuttons along the bottom of the window that identify a number of different additional selections. For two of these alternatives (**Select** and **Save**) we do not display the associated dialog box due to their simplicity.

22.1

The Discriminant Analysis Main Dialog Window

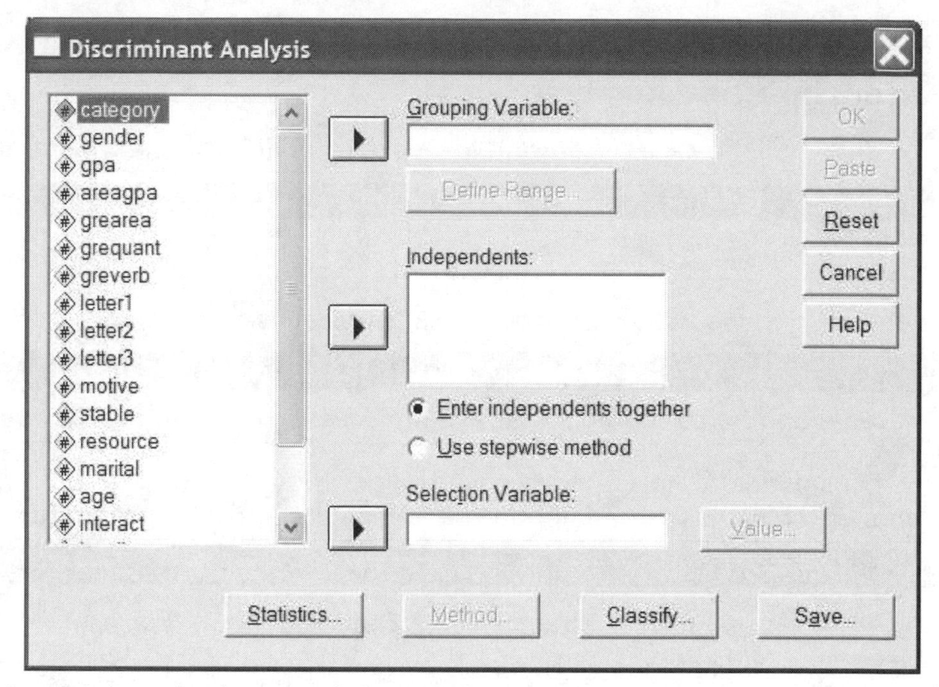

These two options will be briefly described in the following paragraph; then we will consider, with appropriate screen shots, the others.

The **Selection Variable** option operates like the **If** procedure described in Chapter 4. If, for instance, you paste **gender** into the Selection Variable window, the Value button would brighten up. A click on this button opens a tiny dialog window that allows you to select which level of the variable (in this case, 1 = women, 2 = men) you wish to select for the procedure. The analysis will then be conducted with the selected subjects or cases. This selection may be reversed later if you wish.

The **Save** option allows you to save as new variables:

- **Predicted group membership**
- **Discriminant scores**
- **Probabilities of group membership**

All three of these constructs will be discussed in some detail throughout this chapter.

The final item concerning the main dialog window for discriminant analysis is the coding of the dependent variable. Although discriminant analysis is used most frequently to categorize into one of two groups, it is also possible to create discriminant functions that group subjects into three or more groups. When you enter the **Grouping variable** (**category** in this case), the **Define range** pushbutton becomes bright and before you can conduct the analysis you must define the lowest and highest coded value for the dependent variable in the small dialog box that opens (Screen 22.2, following page). In this case, **category** has only two levels and you would enter 1 and 2 in the two boxes then click **Continue**. If you have a variable with more than two levels you may specify numbers that represent less than the full range of values.

22.2

The Discriminant Analysis: Define Range Window

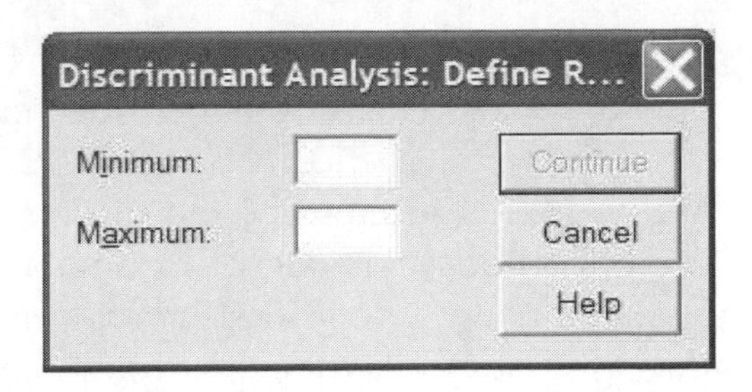

A click on the **Statistics** pushbutton opens a new window with a number of frequently desired options. This dialog box is shown as Screen 22.3 (below). Because all items in the **Statistics** box are frequently used, each will be described in the list that follows. Most of these items are used to analyze the bivariate nature of the variables prior to beginning the discriminant process.

22.3

Discriminant Analysis: Statistics Window

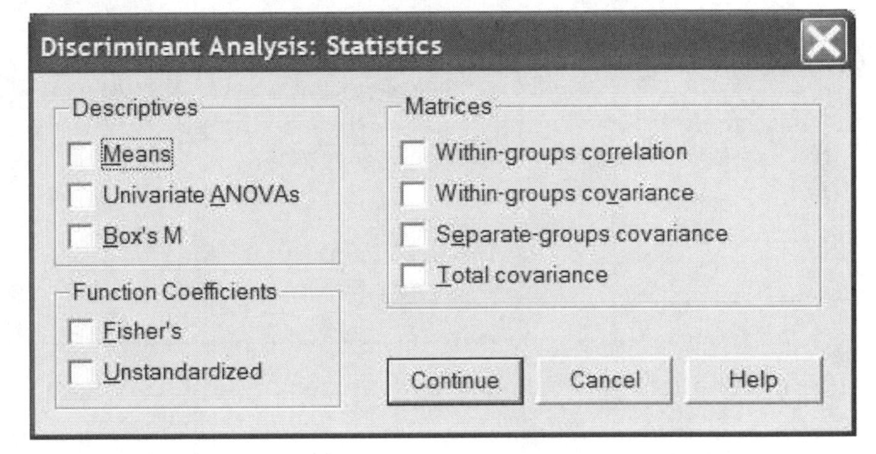

- **Means**: The means and standard deviations for each variable for each group (the two levels of **category** in this case) and for the entire sample.
- **Univariate ANOVAs**: If there are two levels of the dependent variable, these will be *t* tests (using an *F* rather than a *t* statistic). This compares the mean values for each group for each variable to see if there are significant univariate differences between means.
- **Box's M**: The test for equality of the covariances matrices for each level of the dependent variable. This is a test of the multivariate normality of your data.
- **Fisher's function coefficients**: These are the canonical discriminant function coefficients designed to discriminate maximally between levels of the dependent variable.
- **Unstandardized function coefficients**: The unstandardized coefficients of the discriminant equation based on the raw scores of discriminating variables.
- **Within-groups correlation matrix**: A matrix composed of the means of each corresponding value within the two (or more) matrices for each level of the dependent variable.
- **Within-groups covariance matrix**: Identical to the previous entry except a covariance matrix is produced instead of a correlation matrix.
- **Separate-groups covariance matrices**: A separate covariance matrix for members of each level of the dependent variable.
- **Total covariance matrix**: A covariance matrix for the entire sample.

A click on the **Method** button opens the **Stepwise Method** dialog box (Screen 22.4, below). This option is available only if you have previously selected the **Use stepwise methods** option from the main dialog box (Screen 22.1). This window provides opportunity to select the **Method** used in creating the discriminant equation, the **criteria** for entry of variables into the regression equation, and two **display** options. The **Wilks' lambda** is the only method we will describe here. The **Wilks' lambda** method is a stepwise procedure that operates by minimizing the Wilks' lambda value after each new variable has been entered into the regression equation. A more complete description of this may be found on page 291.

22.4

Discriminant Analysis: Stepwise Method Window

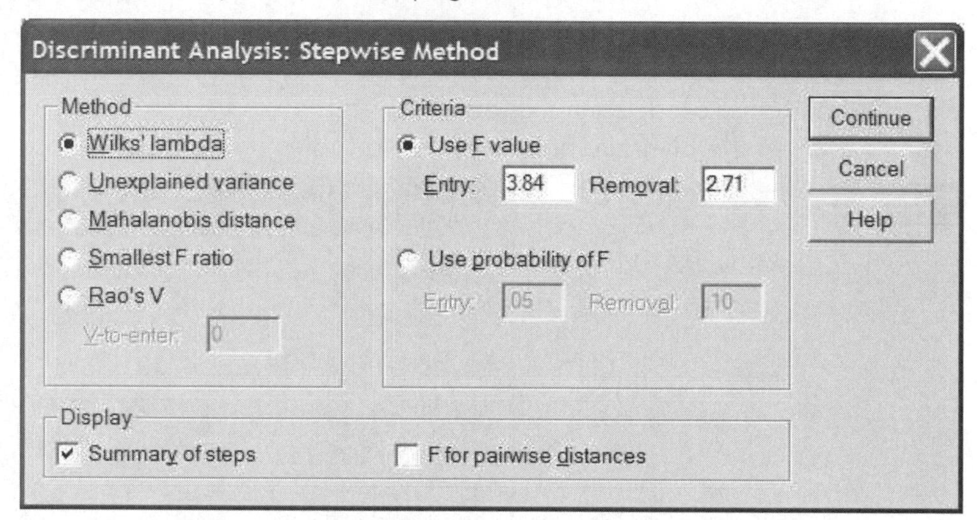

The **criteria** for entry sets default *F* values of 3.84 to enter a new variable and 2.71 to remove an already entered variable. These *F* values represent significance levels of approximately .05 and .10 respectively. In discriminant analysis, researchers often set more generous entry criteria and, selection of F values as low as 1.15 to enter and 1.0 to remove are not uncommon as the goal is to predict groups well with the data you have.

Under the **Display** section, the **Summary of steps** option is selected by default, is central to discriminant analysis output, and under normal circumstances is always retained. The **F for pairwise distances** relates only to the **Mahalanobis distance** method.

The final table considered here deals with the classification and display of discriminant results. A click of the **Classify** pushbutton opens the window shown as Screen 22.5. Most options in this box are of interest to the researcher and each will be described in the list that follows:

22.5

The Discriminant Analysis: Classification Window

Discriminant Analysis: Classification
Prior Probabilities
All groups equal
Compute from group sizes
Use Covariance Matrix
Within-groups
Separate-groups
Continue
Cancel
Help
Display
Casewise results
Limit cases to first:
Summary table
Leave-one-out classification
Plots
Combined-groups
Separate-groups
Territorial map
Replace missing values with mean

- **All groups equal**: Probability of membership into two or more groups created by the dependent variable are assumed to be equal.
- **Compute from group sizes**: If groups are of different sizes, the groups are weighted by the proportion of the number of cases in each group.
- **Combined-groups plot**: The histogram (if two groups) or scatterplot (if more than two groups) includes all groups on a single plot.
- **Separate-groups plots**: This option creates as many plots as there are groups and displays only one group on each plot.
- **Territorial map**: This chart shows centroids and boundaries in a graphic format; it is used only if there are three or more levels of the dependent variable.
- **Within-groups covariance matrix**: This is the default and will classify cases based on the pooled within-groups covariance matrix.
- **Separate-groups covariance matrices**: Classifies cases based on separate covariance matrices for each group.
- **Casewise results**: This is a useful output option if your file is not too large. It lists each case with information concerning the actual group, the group classification, probabilities for inclusion in the group, and the discriminant scores. This output is displayed in the Output section.
- **Summary table**: Another handy output device that sums up the number and percent of correct and incorrect classifications for each group. This is also displayed in the Output section.
- **Leave-one-out classification**: Each case in the analysis is classified by the functions derived from all cases other than that case. This is also known as the U method.
- **Replace Missing Values with mean**: As a high quality researcher, you have, of course, dealt with missing values long before this.

Is it logistic regression or factor analysis? Neither, it's discriminant analysis! Available at participating suppliers. Discriminant analysis comes complete with Wilks' lambdas (several types), Box's M's, Chi Squares, eigenvalues, group centroids, F's, significance levels, and a whole array of other "stuff". Discriminant analysis—for all your canonical discriminant functions. Get yours today!

Contributed by: Michael Smith (Eastern Michigan University)

In this chapter there will be a single Step 5 sequence. Yes, it is possible to create a simplest-default version of discriminant analysis, but anyone who has read this far would want to consider carefully a number of available options. The italicized text prior to the step-by-step box describes the included options. Then the Output section in narrative, definitions, and visual displays does an adequate job of describing the meaning of the results.

To conduct a discriminant analysis that predicts membership into two groups based on the dependent variable **category** *and creating the discriminant equation with inclusion of 17 independent variables (from* **age** *to* **stable***) selected by a* **stepwise** *procedure based on the minimization of* **Wilks' lambda** *at each step with an F-to-enter of 1.15 and an F-to-remove of 1.00; further, selecting* **Means**, **Box's M**, *and* **Univariate ANOVAs**, *to gain a fuller understanding of the univariate nature of independent variables; requesting the* **unstandardized** *discriminant functions coefficients, and selecting for output the* **Combined-groups plot**, *the* **Results for each case**, *and the* **Summary table**, *perform the following sequence of steps. The starting point is Screen 22.1. Carry out the 4-step sequence (page 281) to arrive at this screen.*

In Screen	Do This	Step 5
22.1	category → upper ▸ → **Define Range**	
22.2	type 1→ press TAB → type 2 → **Continue**	
22.1	*click and drag from* **gender** *to* **impress** *(highlight all 17),* → *middle* ▸ → **Use stepwise method** → **Statistics**	
22.3	**Means** → **Univariate ANOVAs** → **Box's M** → **Unstandardized** → **Continue**	
22.1	**Method**	
22.4	press **TAB** twice → type `1.15` → press **TAB** once → type `1.00` → → **Continue**	
22.1	**Classify**	
22.5	**Combined groups** → **Casewise results** → **Summary table** → **Continue**	
22.1	**OK**	Back1

OUTPUT

Discriminant Analysis

The following output is produced by sequence step 5 (above). As with factor analysis, SPSS's efforts to create a coherent and concise output format provide considerable opportunity for future growth. Their output occupies 18 bewildering pages. We present, below, results that we feel are most relevant to the discriminant process. We apologize that you may have to do a substantial amount of hunting to find the corresponding information in the SPSS output.

Number of Cases by Group

Category	Label	Unweighted cases
1	Finished Ph.D.	25
2	Did not finish Ph.D.	25
Total		50

At the top of the output, beneath *Number of Cases by Group*, is information concerning the number of cases (weighted and unweighted) for each level of the dependent variable (**category**) and the labels for each level. The table on the following page (actually four tables in the SPSS output, which have been combined) identifies basic preliminary *uni*variate information (the means for the two levels and the overall mean for each variable); Wilks' lambda, *F*, and significance values contribute *bi*variate information about the differences between means for each variable. The *F* and *Signif* values identify for which variables the two groups differ significantly. This is the type of information that the researcher considers before running a discriminant analysis. Definitions of terms listed across the top of the table follow.

Group means, Wilks' Lambda (U-statistic) and Univariate F-ratio

Variable	Category 1 Mean	Category 2 Mean	Total Mean	Wilks' Lambda	F	Signif.
gender	1.240	1.480	1.360	.938	3.200	.080
gpa	3.630	3.390	3.510	.795	12.360	.001
areagpa	3.822	3.734	3.778	.951	2.943	.121
grearea	655.600	648.800	652.200	.998	.113	.738
grequant	724.000	646.800	685.400	.628	28.390	.000
greverb	643.200	620.000	631.200	.974	1.283	.263
lettter1	7.720	6.160	6.940	.650	25.890	.000
letter2	7.640	6.360	7.000	.756	15.480	.000
letter3	7.960	6.160	7.060	.534	41.970	.000
motive	8.360	7.280	7.820	.679	22.720	.000
stable	6.400	6.360	6.380	1.000	.007	.934
resource	5.920	5.640	5.780	.993	.370	.564
marital	1.640	1.560	1.600	.993	.322	.573
age	29.960	25.120	27.540	.768	14.530	.000
interact	7.000	6.160	6.580	.904	5.079	.029
hostile	2.120	3.080	2.600	.787	13.020	.001
impress	7.280	6.880	7.080	.972	1.378	.246

Term	Definition/Description
VARIABLE	Names of the independent variables.
CATEGORY 1 MEAN	Mean value for each variable for the first level of **category**, those who finished the PhD.
CATEGORY 2 MEAN	Mean value for each variable for the second level of **category**, those who did not finish.
TOTAL MEAN	The mean value for each variable for all subjects. Since there are equal numbers of subjects in each level of **category**, this number is simply the average of the other two means.
WILKS' LAMBDA	The ratio of the within-groups sum of squares to the total sum of squares. This is the proportion of the total variance in the discriminant scores *not* explained by differences among groups. A lambda of 1.00 occurs when observed group means are equal (all the variance is explained by factors *other than* difference between these means), while a small lambda occurs when within-groups variability is small compared to the total variability. A small lambda indicates that group means appear to differ. The associated significance values indicate whether the difference is significant.
F	*F* values are the same as those calculated in a one-way analysis of variance. This is also the square of the *t* value calculated from an independent samples *t* test.
SIGNIFICANCE	The *p*-value: Likelihood that the observed *F*-value could occur by chance.

Next is specification information concerning the stepwise discriminant procedure about to take place.

Stepwise Variable Selection: Selection rule: Minimize Wilks' Lambda

Max. number of steps	Min. Tolerance Level	Min. partial F to enter	Max. partial F to remove
34	.001	1.150	1.000

Prior probability for each group is .5000

Term	Definition/Description
STEPWISE VARIABLE SELECTION	This procedure enters variables into the discriminant equation, one at a time, based on a designated criterion for inclusion ($F \geq 3.84$ is default); but will drop variables from the equation if the inclusion requirement drops below the designated level when other variables have been entered.
SELECTION RULE	The procedure selected here is to minimize Wilks' Lambda at each step.
MAXIMUM NUMBER OF STEPS	2 times the number of variables designated in the **ANALYSIS** subcommand line ($2 \times 17 = 34$). This number reflects that it is possible to enter a maximum of 17 variables and to remove a maximum of 17 variables.
MINIMUM TOLERANCE LEVEL	The tolerance level is a measure of linear dependency between one variable and the others. If a tolerance is less than .001, this indicates a high level of linear dependency, and SPSS will not enter that variable into the equation.
MAXIMUM F TO ENTER, MINIMUM F TO REMOVE	This indicates that any variable with an *F* ratio greater than 1.15 (indicating some influence on the dependent variable) will enter the equation (in the order of magnitude of *F*) and any variable that has an *F* ratio drop below 1.00 after entry into the equation will be removed from the equation.
PRIOR PROBABILITY FOR EACH GROUP	The .5000 value indicates that groups are weighted equally.

What follows is the output that deals with the regression analysis. This output displays the variables that were included in the discriminant equation with associated statistical data; the variables that did not meet the requirements for entry with associated statistical data; and the order in which variables were entered (or removed), along with Wilks' λ, significance, and variable labels.

Variables in the analysis after step 11

Variable	Tolerance	F to remove	Wilks' Lambda	Label
gender	.736	5.816	.330	women = 1, men = 2
grearea	.005	4.632	.321	score on area GRE
greverb	.005	5.035	.324	score on verbal section of the GRE
letter2	.358	1.955	.302	letter-of-recommendation number 2
letter3	.704	3.813	.315	letter-of-recommendation number 3
motive	.762	4.026	.317	rating of student motivation
resource	.845	3.428	.305	personal and financial resources
age	.745	5.268	.326	age at time of entry into program
impress	.878	2.851	.308	selector's overall impression of candidate

Variables not in the analysis after step 11

Variable	Tolerance	Min Tolerance	F to enter	Wilks' λ	Label
gpa	.747	.004	.104	.287	overall undergraduate GPA
areagpa	.747	.004	.027	.288	GPA in major field
grequant	.548	.004	.000	.288	score on quantitative GRE
letter1	.692	.004	.058	.287	First letter of recommendation
stable	.817	.004	.088	.287	emotional stability
marital	.914	.006	.046	.287	marital status
interact	.694	.005	.746	.282	ability at interaction with others
hostile	.765	.005	.001	.288	level of hostility

Entry and Removal of Variables: Summary Table

Step	Action		Number of variables in	Wilks' Lambda	Signif.	Label
	Entered	Removed				
1	letter3		1	.534	.000	third letter of recommendation
2	motive		2	.451	.000	students motivation
3	letter1		3	.415	.000	first letter of recommendation
4	age		4	.391	.000	age in years at entry
5	gender		5	.363	.000	1 = women 2 = men
6	impress		6	.332	.000	rating or selectors impression
7	resource		7	.319	.000	financial/personal resources
8	greverb		8	.310	.000	GRE score on verbal
9	grearea		9	.398	.000	GRE score on major area
10		letter1	8	.302	.000	first letter of recommendation
11	letter2		9	.288	.000	second letter of recommendation

These three charts indicate which variables were included in the final discriminant function and which ones were not. Notice that all variables in the analyses after 11 steps have higher than the acceptable tolerance level (.001) and have *F* values greater than 1.15. Variables not in the equation all have acceptable tolerance levels, but the *F*-to-enter value is less than 1.00 for each of them. The third table gives a summary of the step-by-step procedure. Notice that in a total of 11 steps 9 variables were entered, but at step 10 one variable (**letter1**) was dropped due to an *F*-value falling below 1.00. Then at step 11 the final variable was added (**letter2**).

This result raises an important experimental concern. When the values for the three letters were originally entered, no attention was given to which was stronger than which. Thus there is no particular rationale to drop **letter1** in favor of **letter2**, as suggested by these results. The appropriate response for the researcher would be to go back into the raw data file and reorder letters from strongest to weakest for each subject (e.g., **letter1** is strongest, **letter2**, next strongest and **letter3** weakest) and update the data file. Then, when a discriminant analysis is conducted, if one of the letters has greater influence than another, it may have meaning. It might indicate, for instance, that **letter1** (the strongest letter) has a significant influence but that **letter3** (the weakest) does not contribute significantly to the discriminating process. While this is representative of the thinking the researcher might do, remember that these data are fictional and thus do not reflect an objective reality.

Most of the terms used here are defined on the previous pages. Definitions of terms unique to this section follow:

Term	Definition/Description
NUMBER OF VARIABLES IN	Indicates the numbers of variables in the discriminant equation at each step.
SIGNIF	This is a measure of multivariate significance, not the significance of each new variable's unique contribution to explaining the variance.

Next is the test for multivariate normality of the data.

Test of equality of group covariance matrices using Box's M

Group	Label	Rank	Log Determinant	Box's M	Approx F	df1	df2	signif
1	Finished PhD	9	11.340					
2	Not finished PhD	9	12.638					
Pooled within-groups		9	13.705	82.381	1.461	45	7569.06	.024

Term	Definition/Description
RANK	Rank or size of the covariance matrix. The **9** indicates that this is a 9 × 9 matrix, the number of variables in the discriminant equation.
LOG DETERMINANT	Natural log of the determinant of each of the two (the two levels of the dependent variable, **CATEGORY**) covariance matrices.
POOLED WITHIN-GROUPS COVARIANCE MATRIX	A matrix composed of the means of each corresponding value within the two 9 × 9 matrices of the two levels of **category**.
BOX'S M	Based on the similarities of determinants of the covariance matrices for the two groups. It is a measure of multivariate normality.
APPROXIMATE F	A transformation that tests whether the determinants from the two levels of the dependent variable differ significantly from each other. It is conceptually similar to the *F* ratio in ANOVA, in which the between-groups variability is compared to the within-groups variability.
SIGNIFICANCE	A significance value of .024 suggests that data do differ significantly from multivariate normal. However, a value less than .05 does not automatically disqualify the data from the analysis. It has been found that even when multivariate normality is violated, the discriminant function can still often perform surprisingly well. Since this value is low, it would be well to look at the univariate normality of some of the included variables. For instance, we know that the **gender** variable is not normally distributed, but inclusion of **gender** improves the discriminating function of the equation.

What follows is information concerning the canonical discriminant function, correlations between each of the discriminating variables and the canonical discriminant function, the unstandardized discriminant function coefficients (these are the coefficients for the discriminant equation), and the group centroids.

Structure Matrix Pooled-within-groups correlations (ordered by size of correlation)	
letter3	.59437
grequant	.48949
motive	.43734
letter1	.42874
letter2	.36093
age	.34968
hostile	.32595
gpa	.26962
areagpa	.17158
gender	.16412
impress	.10769
greverb	.10391
interact	.07031
stable	.05865
resource	.05326
grearea	.03084
marital	.01318

Unstandardized Canonical Discriminant Function Coefficients	
gender	-1.038
grearea	-.079
greverb	.080
letter2	-.372
letter3	.424
motive	.512
resource	.181
age	.104
impress	.271
(constant)	-8.106

Canonical Discriminant Functions evaluated at Group Means (Group Centroids)	
group	Function 1
1	1.54143
2	-1.54143

Eignevalues and Wilks' Lambda

Function	Eigenvalue	% of Variance	Cum %	Canonical Correlation	Test of function	Wilks' Lambda	χ^2	df	Sig
1	2.475	100.00	100.00	.844	1	.288	54.18	9	.00

The definition of terms will provide the entire discussion of the tables presented in the previous page.

Term	Definition/Description
UNSTANDARDIZED CANONICAL DISCRIMINANT FUNCTION COEFFICIENTS	This is the list of coefficients (and the constant) of the discriminant equation. Each subject's discriminant score would be computed by entering his or her variable values for each of the 9 variables in the equation. The discriminant equation follows: D = –8.11 + –1.03(**gender**) + –.79(**grearea**) + .80(**greverb**) + –.37(**letter2**) + .42(**letter3**) + .51(**motive**) + .18(**resource**) + .10(**age**) + .27(**impress**)
FUNCTION TEST OF FUNCTION	The one (1) designates information about the *only* discriminant function created with two levels of the criterion variable. If there were three levels of the criterion variable, there would be information listed about two discriminant functions.
WILKS' LAMBDA	Ratio of the within-groups sum of squares to total sum of squares (more complete definition earlier in this chapter).
CHI-SQUARE (χ^2)	A measure of whether the two levels of the function significantly differ from each other based on the discriminant function. A high chi-square value indicates that the function discriminates well.
DF	Degrees of freedom is equal to the number of variables used in the discriminant function (9).
SIG.	p-value associated with the chi-square function.
EIGENVALUE	Between-groups sums of squares divided by within-groups sums of squares. A large eigenvalue is associated with a strong function.
% OF VARIANCE, CUM %	The function always accounts for 100% of the variance.
CANONICAL CORRELATION	To those who have heard the term for years and never quite known what it was, your moment has come: The canonical correlation is a correlation between the discriminant scores and the levels of the dependent variable. Please refer to the chart on the following page to further clarify. Note the scores in the extreme right column. Those are discriminant scores. They are determined by substituting into the discriminant equation the relevant variable measures for each subject. There are 50 subjects; thus there will be 50 discriminant scores. There are also 50 codings of the dependent variable (**category**) that show 25 subjects coded **1** (finished the Ph.D.), and 25 subjects coded **2** (didn't finish). The canonical correlation is the correlation between those two sets of numbers. A high correlation indicates a function that discriminates well. The present correlation of .844 is extremely high (1.00 is perfect).
POOLED-WITHIN-GROUP-CORRELATIONS	"Pooled within group" differs from "values for the entire (total) group" in that the pooled values are the average (mean) of the group correlations. If the *N*s are equal (as they are here), then this would be the same as the value for the entire group. The list of 17 values is the correlations between each variable of interest and the discriminant scores. For instance, the first correlation listed (**letter3** .59437) is the correlation between the 50 ratings for **letter3** and the 50 discriminant scores (extreme right column on the chart that follows).
GROUP CENTROIDS	The average discriminant score for subjects in the two groups. More specifically, the discriminant score for each group when the variable means (rather than individual values for each subject) are entered into the discriminant equation. Note that the two scores are equal in absolute value but have opposite signs. The dividing line between group membership is zero (0).

The following chart is titled "Casewise Statistics" and gives straightforward information about group membership for each subject, probability of group membership, and discriminant scores. Only 16 of the 50 cases are shown to conserve space. On the first table, the asterisks identify cases that were misclassified.

Casewise Statistics

Case #	Actual Group	Highest Group: Predicted Group	Highest Group: P(D>d | G = g)	Highest Group: P(G = g | D = d)	Second-Highest Group: 2nd highest Group	Second-Highest Group: P(G = g | D = d)	Discriminant Score
1	1	1	.9878	.9918	2	.0082	1.5567
2	1	1	.4790	.9990	2	.0010	2.2494
3	1	1	.6217	.9981	2	.0019	2.0348
4	1	1	.7209	.9747	2	.0253	1.1842
5	2	1**	.6242	.9264	2	.0376	1.0516
6	1	1	.4719	.9991	2	.0009	2.2609
--	--	--	--	--	--	--	--
--	--	--	--	--	--	--	--
37	2	2	.6807	.9702	1	.0298	1.1299
38	2	2	.7121	.9738	1	.0262	1.1724
39	1	2**	.3141	.8287	1	.1613	.5347
40	2	2	.6393	.9980	1	.0020	2.0101
--	--	--	--	--	--	--	--
--	--	--	--	--	--	--	--
45	2	2	.4204	.9062	1	.0938	.7357
46	2	2	.3297	.9996	1	.0004	2.5162
47	1	2**	.9757	.9922	1	.0078	1.5719
48	2	2	.2761	.9997	1	.0003	2.6305
49	2	2	.8198	.9957	1	.0043	1.7692
50	2	2	.3528	.9995	1	.0005	2.4705

Classification Results

Actual Group	Number of Cases	Predicted Group Membership: Category 1	Predicted Group Membership: Category 2
Category 1 --- Finished PhD	25	23 (92%)	2 (08%)
Category 2 --- Not Finished PhD	25	1 (04%)	24 (96%)

Once again, the definitions of terms represents the entire commentary on the two tables shown above.

Term	Definition/Description
ACTUAL GROUP	Indicates the actual group membership of that subject or case.
PREDICTED GROUP	Indicates the group the discriminant function assigned this subject or case to. Asterisks (**) indicate a misclassification.
P(D > d | G = g)	Given the discriminant value for that case (*D*), what is the probability of belonging to that group (*G*)?
P(G = g | D = d)	Given that this case belongs to a given group (*G*), how likely is the observed discriminant score (*D*)?
2ND GROUP	What is the second most likely assignment for a particular case? Since there are only two levels in the present data set, the second most likely group will always be the "other one."
DISCRIMINANT SCORES	Actual discriminant scores for each subject, based on substitution of variable values into the discriminant formula (presented earlier).
CLASSIFICATION RESULTS	Simple summary of number and percent of subjects classified correctly and incorrectly.

23

General Linear Models: MANOVA and MANCOVA

THIS IS the first chapter describing a procedure that uses several dependent variables concurrently within the same analysis: Multivariate Analysis of Variance (MANOVA) and Covariance (MANCOVA). The **General Linear Model** procedure is used to perform MANOVA in SPSS and is one of the more complex commands in SPSS. It can be used, in fact, to compute multivariate linear regression, as well as MANOVA and MANCOVA. This chapter describes how to perform MANOVA and MANCOVA, and Chapter 24 goes on to illustrate multivariate analysis of variance using within-subjects designs and repeated-measures designs. Because the procedures are so complex, we will restrict our discussion here to the most frequently used options. The procedures described in this chapter are an extension of ANOVA; if you are unfamiliar with ANOVA, you should ground yourself firmly in those operations (see Chapters 12-14) before attempting to conduct a MANOVA.

As described in earlier chapters, an independent-samples *t* test indicates whether there is a difference between two separate groups on a particular dependent variable. This simple case (one independent variable with two levels, and one continuous dependent variable) was then extended to examine the effects of

- an independent variable with more than two levels (one-way ANOVA),
- multiple independent variables (two- and three-way ANOVA), and
- including the effects of covariates (ANCOVA).

Throughout this progression from very simple to quite complex independent variable(s), the dependent variable has remained a single continuous variable. The good news is that (at least until the next chapter), the independent variables and covariates involved in MANOVA and MANCOVA procedures will not get more complicated.

There are times, however, when it may be important to determine the effects of one or more independent variables on *several* dependent variables simultaneously.

For example, it might be interesting to examine the differences between men and women (**gender**, an independent variable with two levels) on both previous GPAs (**gpa**) and scores on the final exam (**final**). One popular approach to examining these gender differences is to do two separate *t* tests or a one-way ANOVA (remember, because $t^2 = F$, the tests are equivalent). This approach has the advantage of conceptual clarity and ease of interpretation; however, it does have disadvantages. In particular, when several separate tests are performed (in this case, two: one for each dependent variable), the experimentwise (or family wise) error increases —that is, the chance that one or more of your findings may be due to chance increases. Furthermore, when dependent variables are correlated (and previous GPA and score on a final exam usually are) with each other, doing separate tests may not give the most accurate picture of the data.

In response to these and other problems associated with doing multiple *t* tests or multiple ANOVAs, Hotelling's T^2 was developed to replace the *t* test, and MANOVA—Multivariate Analysis of Variance—was developed to replace ANOVA. SPSS performs these tests using the **General Linear Models – Multivariate** procedure. This procedure can also analyze covariates, allowing the computation of Multivariate Analysis of Covariance (MANCOVA). These tests examine whether there are differences among the dependent variables *simultaneously*; one test does it all. Further analyses allow you to examine the *pattern* of changes in the dependent variables, either by conducting a series of univariate *F* tests or by using other post-hoc comparisons.

Please note that whenever you are using multiple dependent variables, it is important to be certain that the dependent variables do not exhibit linear dependency on each other. For example, it would be incorrect to analyze class percent (**percent**) and total points received (**total**) together because the percent for the class depends on the total points received.

Just as in univariate analysis of variance (ANOVA), MANOVA produces an *F* statistic to determine whether there are significant differences among the groups. MANOVA is designed to test for interactions as well as for main effects. Since more than one dependent variable is present, however, multivariate *F* statistics involve matrix algebra and examine the differences between all of the dependent variables simultaneously.

For the example in this chapter, the effects of students' section (**section**) as well as whether they are upper or lower division (**lowup**) are examined in terms of their influence on students' scores on the five quizzes (**quiz1** to **quiz5**). In addition to this example, further analysis will be performed using previous GPA (**gpa**) as a covariate; that is, the effects of **gpa** on the dependent variables are removed before the MANOVA itself is performed. Although this is a moderately complex example (with two independent variables, one covariate, and five dependent variables), this procedure provides an example by which, in your own analyses, it is easy to increase or decrease the number of variables. So, if you have more or less independent or dependent variables than our example, you may simply use as many variables as you need. At the minimum, you need two dependent variables (if you have fewer dependent variables, it's not MANOVA, right?) and one independent variable with two levels. There is no theoretical maximum number of independent and dependent variables in your analysis, but in reality the sample size will limit you to a few variables at a time. Covariates are optional.

STEP BY STEP

General Linear Models: MANOVA and MANCOVA

To enter SPSS, a click on **Start** *in the taskbar (bottom of screen) activates the start menu:*

After clicking the SPSS program icon, Screen 1 (inside front cover) appears on the monitor.

Step 2
Create and name a data file or edit (if necessary) an already existing file (see Chapter 3).

Screens 1 and 2 (inside front cover) allow you to access the data file used in conducting the analysis of interest. The following sequence accesses the **grades.sav** *file for further analyses:*

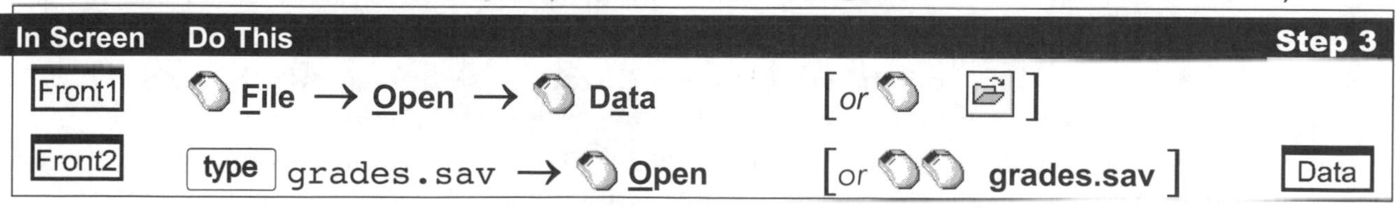

Whether first entering SPSS or returning from earlier operations the standard menu of commands across the top is required (shown below). As long as it is visible you may perform any analyses. It is not necessary for the data window to be visible.

This menu of commands disappears or modifies when using pivot tables or editing graphs. To uncover the standard menu of commands simply click on the or the icon.

After completion of Step 3 a screen with the desired menu bar appears. When you click a command (from the menu bar), a series of options will appear (usually) below the selected command. With each new set of options, click the desired item. The sequence to access General Linear Models begins at any screen with the menu of commands visible:

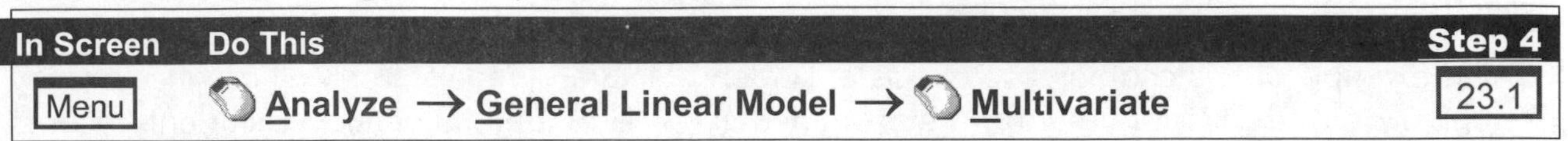

After this sequence step is performed, a window opens (Screen 23.1, below) in which you specify the dependent variables and independent variables (called fixed factors in this screen). The box to the left of the window contains all variables in your data file. Any variables that you wish to treat as dependent variables in the analysis should be moved into the **Dependent Variables** box by clicking on the top ▸. In our example, we will examine two independent variables and five dependent variables in the **grades.sav** file. As dependent variables, we will examine the five quizzes. For the independent variables, or fixed factors, we will examine the three different class sections (**section**) and whether the student is upper- or lower-divisions (**lowup**).

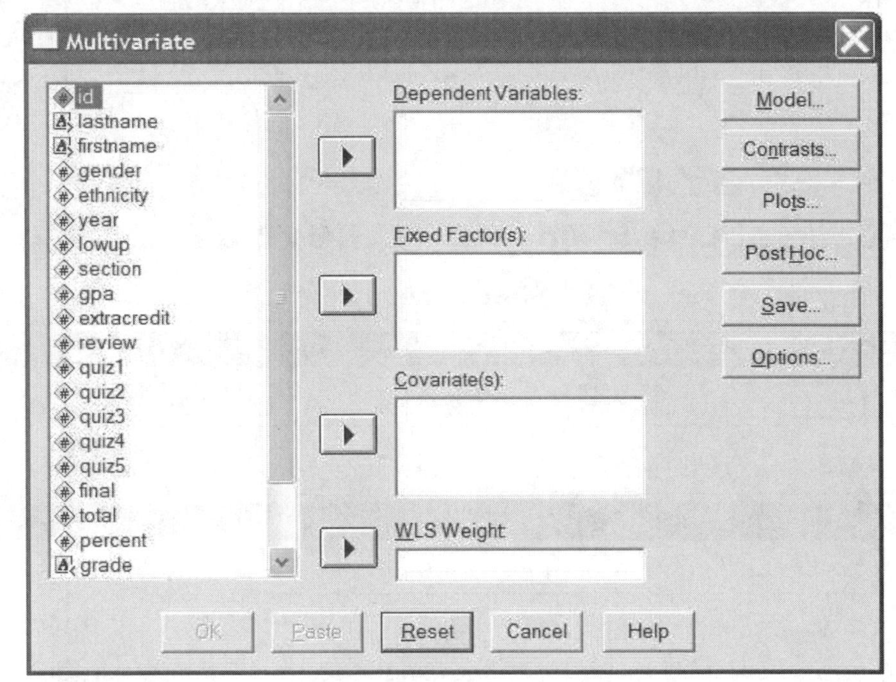

23.1

The General Linear Models – Multivariate Window (for MANOVA and MANCOVA)

Covariates may also be entered in Screen 23.1, if desired. If you wish to perform a MANOVA, without any covariates, then you should leave the **Covariate(s)** box empty. If, however, you wish to factor out the effect of one or more variables from the MANOVA, then you should enter these variables in the **Covariate(s)** box. Do this by selecting each variable from the box on the left of the window, and clicking on the ▸ to the left of the **Covariate(s)** box.

After the covariates (if desired) are specified, there are four more buttons in the main dialog box that may need to be selected: **Model**, **Plots**, **Post Hoc**, and **Options**. (The **Contrasts** and **Save** buttons are more advanced than this introductory chapter will consider.) Each of these important selections will be described.

In most cases, a full factorial model is desired: This produces a model with all main effects and all interactions included, and is the default. At times, you may wish to test only certain main effects and interactions. To accomplish this, use the **Model** button (from Screen 23.1), which calls up Screen 23.2 (below). To select which particular main effects and interactions you wish to test, select the **Custom** button, and move any main effects and interactions desired in the model from the **Factor & Covariates** box to the left of the window, to the **Model** box to the right of the window, by clicking the ▸ button.

23.2

Model Selection Dialog Window for Multivariate General Linear Model

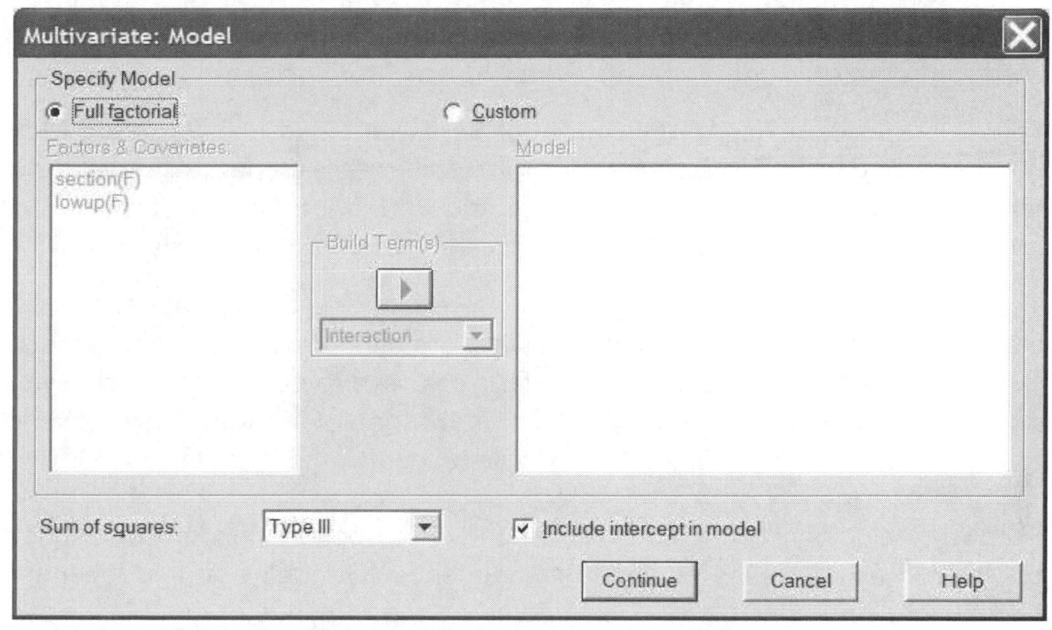

The Model dialog box also allows you to select the type of **Sum of Squares**. You may select Types I, II, III, or IV: Type III is the default, and is appropriate for most situations. Type IV is usually more appropriate if you have missing cells in your design (that is, some cells in your MANOVA model that do not have any participants).

When you are interpreting MANOVA or MANCOVA results, it is often useful to see plots of the means of the dependent variables as determined by the different levels of the factors. To do this, select the **Plots** button from Screen 23.1; this calls up another dialog box, shown in Screen 23.3 (below). This dialog allows you to specify how to plot each factor. Separate plots will be produced for each dependent variable in your analysis. Note that if your analysis includes one or more covariates, the plots produced will not include the *actual* means of your data, but instead will include the *estimate* of the means *adjusted* for the covariate(s). In that case, it is often useful to produce a set of profile **Plots** without the covariate.

23.3

Profile Plot Selection Dialog Box for Multivariate General Linear Models

On the left side of Screen 23.3 are listed the **Factors** in your model. In the center are three boxes; into each of these boxes you can move one of the **Factors** by clicking on the ▸ button. These center boxes let you specify which factor you want to be drawn with separate categories across the **Horizontal Axis**, which factor you want to have drawn with **Separate Lines** (as many separate lines as there are levels of your factor), and which factor (if any) you want to be used to produce several **Separate Plots** (one plot for each level of that factor).

You must specify a **Horizontal Axis**, but **Separate Lines** and **Separate Plots** are optional. If you want **Separate Plots**, however, you must specify **Separate Lines** first. Once you have specified a plot (or series of plots), click **Add** to add it to the list of plots to be produced at the bottom of the dialog. Click **Continue** once you are finished specifying which plots you wish.

In addition to using plots to help you interpret any significant main effects and interactions you may find, **Post Hoc Multiple Comparisons** are often used to determine the specific meaning of main effects or interactions. *Post hoc* tests are used to determine which levels of a variable are significantly different from other levels of that variable; these tests are done for each factor that you specify, and are produced for each dependent variable that is in your analysis.

Because including covariates goes beyond using only dependent and independent variables, but instead examines the effect of the independent variables on the dependent variables above and beyond the effects of the covariates on the dependent variables, *post hoc* tests are not available for analyses that include a covariate.

To select *post hoc* tests, click on the **Post Hoc** button (on Screen 23.1). SPSS produces a truly dizzying array of *post hoc* tests (Screen 23.4) from which you may select whichever of the 18 tests you want. Before selecting tests, you need to specify which **Factor(s)** you want to perform **Post Hoc Tests for**, by selecting one or more variables in the **Factor(s)** box and moving them to the **Post Hoc Tests for** box by clicking on the ▸ button.

23.4

Post Hoc Multiple Comparisons for Observed Means Dialog Box for Multivariate General Linear Models

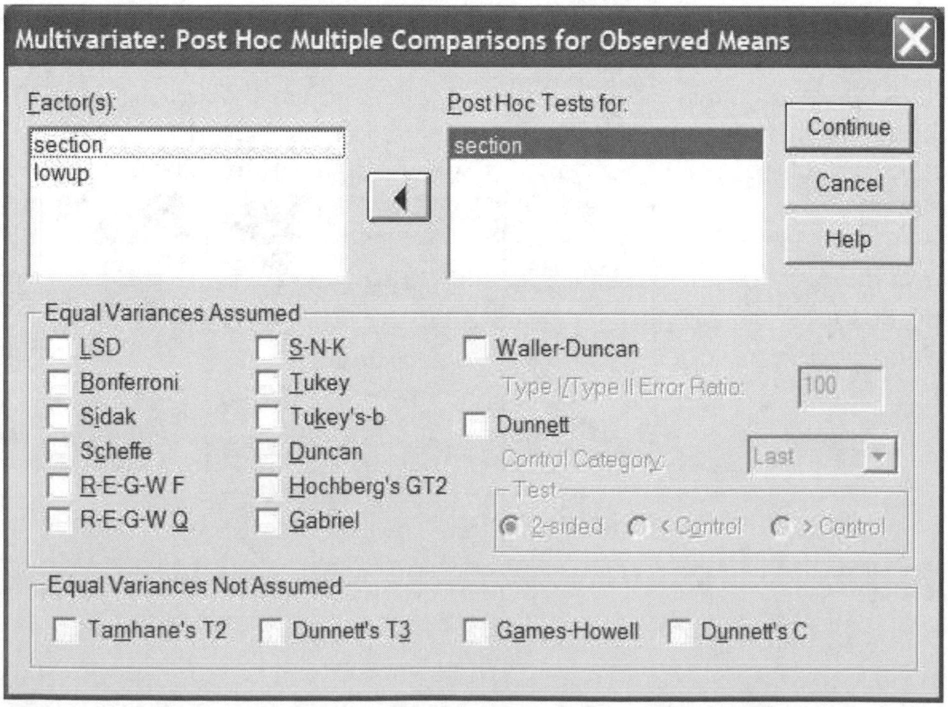

Multivariate: Post Hoc Multiple Comparisons for Observed Means

Factor(s): section, lowup

Post Hoc Tests for: section

Continue | Cancel | Help

Equal Variances Assumed

LSD, Bonferroni, Sidak, Scheffe, R-E-G-W F, R-E-G-W Q, S-N-K, Tukey, Tukey's-b, Duncan, Hochberg's GT2, Gabriel, Waller-Duncan, Type I/Type II Error Ratio: 100, Dunnett, Control Category: Last, Test: 2-sided, < Control, > Control

Equal Variances Not Assumed

Tamhane's T2, Dunnett's T3, Games-Howell, Dunnett's C

Once you have selected one or more factors, you may select which of the *post hoc* tests you wish to perform. We don't recommend that you run all 18; in fact, many of these tests are very rarely used. Our discussion will focus on several of the most frequently used *post hoc* tests:

- **LSD**: LSD stands for "least significant difference," and all this test does is to perform a series of *t* tests on all possible combinations of the independent variable on each dependent variable. This is one of the more "liberal" tests that you may choose: Because it does not correct for experimentwise error, you may find significant differences due to chance alone. For example, if you are examining the effects of an independent variable with 5 levels, the LSD test will perform 10 separate *t* tests (one for each possible combination of the levels of the factor). With an alpha of .05, each individual test has a 5% chance of being significant purely due to chance; because 10 tests are done, however, the chance is quite high that at least 1 of these tests will be significant due to chance. This problem is known as experimentwise or family wise error.
- **Bonferroni**: This test is similar to the **LSD** test, but the Bonferroni test adjusts for experimentwise error by dividing the alpha value by the total number of tests performed. It is therefore a more "conservative" test than the least significant difference test.
- **Scheffé**: The Scheffé test is still more conservative than the Bonferroni test, and uses *F* tests (rather than the *t* tests in the least significant difference and Bonferroni tests). This is a fairly popular test.
- **Tukey**: This test uses yet a different statistic, the Studentized range statistic, to test for significant differences between groups. This test is often appropriate when the factor you are examining has many levels.

It should be noted that the four *post hoc* tests described here, as well as the majority of the *post hoc* tests available, assume that the variances of your cells are all equal. The four tests listed at the bottom of Screen 23.4 do not make this assumption, and should be considered if your cells do not have equal variances. To determine whether or not your cells have equal variances, use the **Homogeneity tests** (described in the **Options** section, following Screen 23.5).

23.5

Options Dialog Box for Multivariate General Linear Models

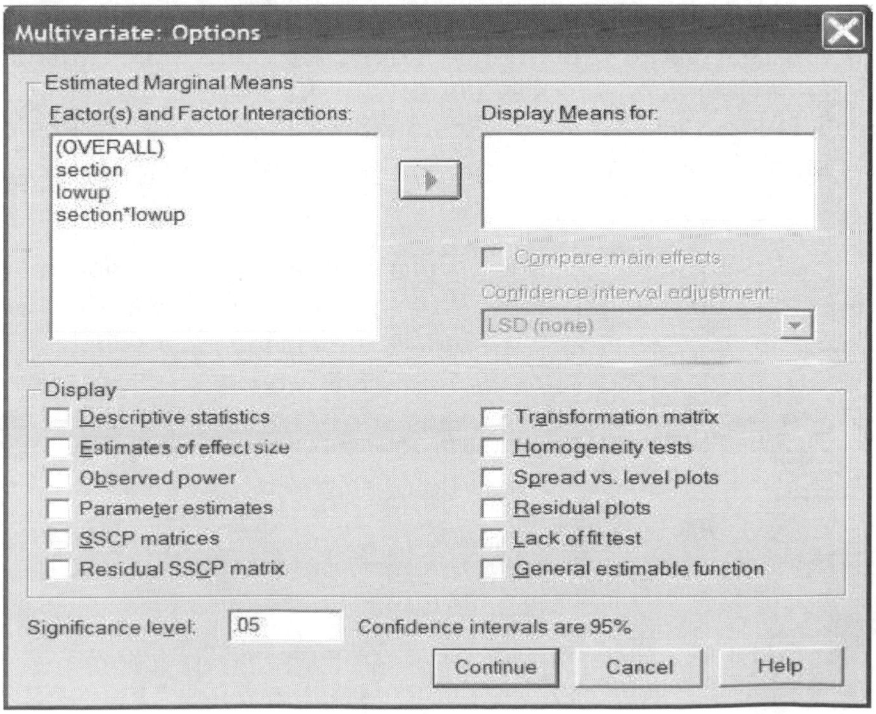

Multivariate: Options

Estimated Marginal Means

Factor(s) and Factor Interactions:

(OVERALL)
section
lowup
section*lowup

Display Means for:

Compare main effects

Confidence interval adjustment:

LSD (none)

Display

Descriptive statistics
Estimates of effect size
Observed power
Parameter estimates
SSCP matrices
Residual SSCP matrix
Transformation matrix
Homogeneity tests
Spread vs. level plots
Residual plots
Lack of fit test
General estimable function

Significance level: .05 Confidence intervals are 95%

Continue Cancel Help

In the **Options** dialog box, the default has nothing selected (no means, descriptive statistics, estimates of effect size, or homogeneity tests). If this is what you wish, then you do not need to click **Options**. If you do wish to choose these or other options, then click on **Options** and Screen 23.5 appears. Many selections are included here that will not be described due to space limitations; the most critical and frequently used ones follow:

- **Estimated Marginal Means**: By selecting each factor in the left box, and clicking on the ▸ button, means will be produced for each dependent variable at each level of that factor. If you do not have a covariate in your analysis, then the means produced will be the actual means of your data; if you have one or more covariates, then means will be adjusted for the effect of you covariate(s). In the **Factor(s) and Factor Interactions** box, **(Overall)** refers to the grand mean across all cells. Checking the **Compare main effects** box produces a series of *post hoc* tests examining the differences between cells in each factor. You may choose an LSD or Bonferroni comparison (described above), as well as a Sidak test (more conservative than the Bonferroni test). If you want other tests, use the **Post Hoc** option (described above) when you are not using covariates.
- **Descriptive Statistics**: Produces means and standard deviations for each dependent variable in each cell.
- **Estimates of effect size:** Produces eta squared (η^2), the effect-size measure. This indicates how much of the total variance is explained by the independent variables.
- **Observed power:** Allows you to see the probability of finding a significant effect, given your sample size and effect size.
- **Parameter estimates**: Produces parameter estimates (and significance values) for each factor and covariate in the model. This is particularly useful when your model has one or more covariates.
- **Homogeneity tests:** Examines whether the variance-covariance matrices are the same in all cells.
- **Significance level and Confidence interval**: Lets you specify the alpha value that you want to use for the analysis. You specify the significance level (the default is .05), and SPSS calculates the confidence intervals based on that significance level.

As you can see in Screen 23.5, many other options are available in SPSS. The SPSS manuals describe them fully, but they are beyond the scope of this book. Also, some background information useful in interpreting **Residual SSCP matrix**es and **Residual plots** is found in Chapter 28 of this book.

*The following steps compute a multivariate analysis of variance (MANOVA) with the five quizzes (**quiz1** to **quiz5**) as the dependent variables, and with the section number (**section**) and status (lower- or upper-division; **lowup**) as the independent variables. Please perform whichever of steps 1-4 (pages 295-296) are necessary to arrive at Screen 23.1.*

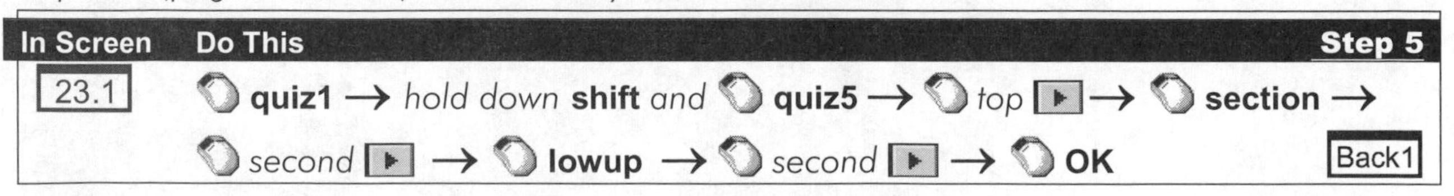

*If you want to include one or more covariates in the analysis, they may be specified in the **Covariate(s)** box. In the following variation of Step 5, we include **gpa** as a covariate.*

In Screen	Do This	Step 5a
23.1	quiz1 → *hold down* **shift** *and* **quiz5** → *top* ▶ → **section** → *second* ▶ → **lowup** → *second* ▶ → **gpa** → *third* ▶ → **OK**	Back1

The following variation of Step 5 includes a variety of useful supplementary statistics and graphs. Please note that, because post hoc tests are included, a covariate is not included. If you wish to include a covariate, you may do so, but you will not be able to select post hoc tests, and any graphs requested will produce means adjusted for the covariate, rather than the actual means of the data. Also note that post hoc tests are only performed on **section**, *because post hoc tests are not very useful when applied to a factor with only two levels (such as* **lowup**).

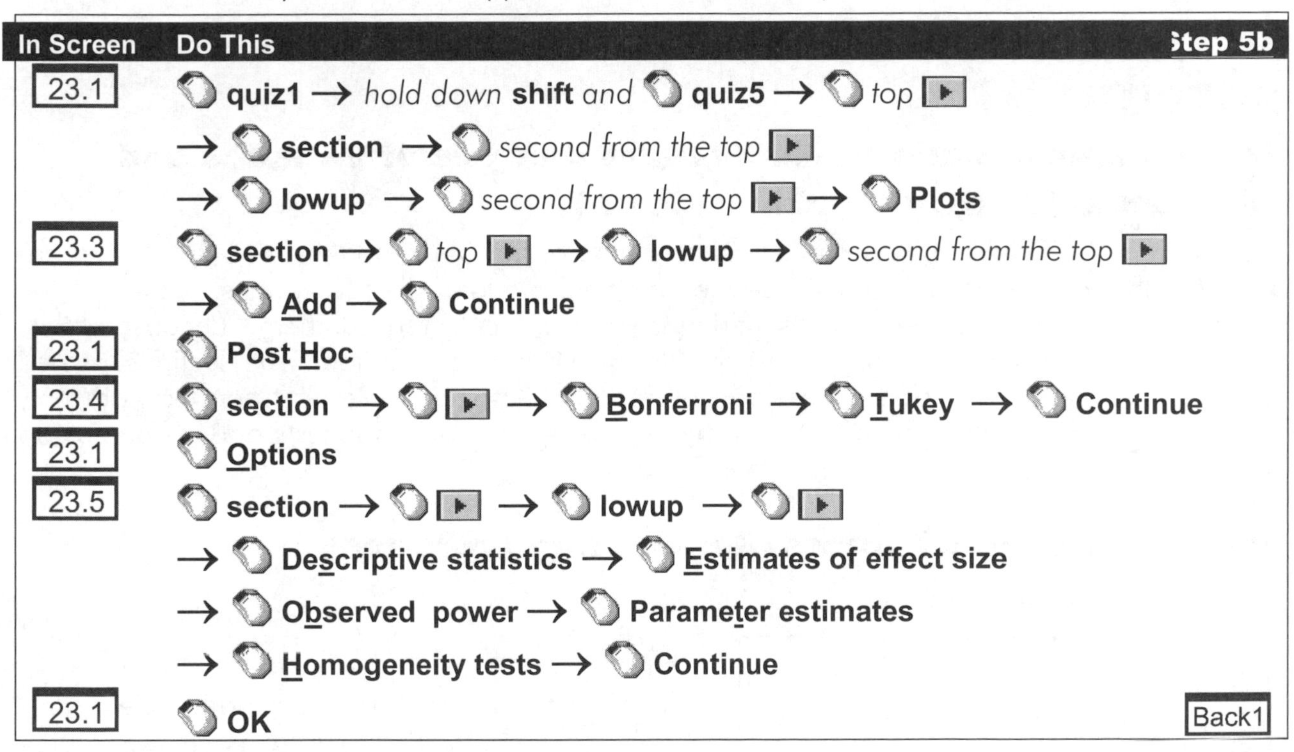

In Screen	Do This	Step 5b
23.1	**quiz1** → *hold down* **shift** *and* **quiz5** → *top* ▶ → **section** → *second from the top* ▶ → **lowup** → *second from the top* ▶ → **Plots**	
23.3	**section** → *top* ▶ → **lowup** → *second from the top* ▶ → **Add** → **Continue**	
23.1	**Post Hoc**	
23.4	**section** → ▶ → **Bonferroni** → **Tukey** → **Continue**	
23.1	**Options**	
23.5	**section** → ▶ → **lowup** → ▶ → **Descriptive statistics** → **Estimates of effect size** → **Observed power** → **Parameter estimates** → **Homogeneity tests** → **Continue**	
23.1	**OK**	Back1

Upon completion of steps 5, 5a, or 5b, the output screen will appear (Screen 1, inside back cover). All results from the just-completed analysis are included in the Output Navigator. Make use of the scroll bar arrows (▲▼▶◀) to view the results. Even when viewing output, the standard menu of commands is still listed across the top of the window. Further analyses may be conducted without returning to the data screen.

PRINTING RESULTS

Results of the analysis (or analyses) that have just been conducted requires a window that displays the standard commands (**File Edit Data Transform Analyze** . . .) across the top. A typical print procedure is shown below beginning with the standard output screen (Screen 1, inside back cover).

To print results, from the Output screen perform the following sequence of steps:

In Screen	Do This	Step 6
Back1	*Select desired output or edit (see pages 19 - 25)* → **File** → **Print**	
2.9	*Consider and select desired print options offered, then* → **OK**	

To exit you may begin from any screen that shows the **File** *command at the top.*

In Screen	Do This	Step 7
Menu	**File** → **Exit**	2.1

Note: After clicking **Exit**, there will frequently be small windows that appear asking if you wish to save or change anything. Simply click each appropriate response.

OUTPUT

General Linear Models: Multivariate Analysis of Variance and Covariance (MANOVA and MANCOVA)

The following is partial output from sequence steps 5a and 5b on (pages 300-301). We abbreviated the following output, because the complete printout takes 23 pages. Some of the sections of the report may not be present unless you specifically request them. Because of this, we list each section of the report separately, along with its interpretation. The sections labeled "General Interpretation" will be present in all reports. Also, because *post hoc* tests are only available if no covariates are included, sections dealing with *post hoc* tests and covariates do not both appear within a single output.

Between-Subjects Factors (General Interpretation)

Between-Subjects Factors

		Value Label	N
section	1		33
	2		39
	3		33
lowup	1	Lower	22
	2	Upper	83

This output simply lists each independent variable, along with the levels of each (with value labels) and the sample size (*N*).

Descriptive Statistics

Descriptive Statistics

	section	lowup	Mean	Std. Deviation	N
quiz1	1	Lower	9.43	.976	7
		Upper	7.85	2.493	26
		Total	8.18	2.338	33
	2	Lower	5.45	2.115	11

(output continues)

For each cell in the model, means, standard deviations, and sample size (*N*) is displayed. Output is given for each dependent variable in the model.

Box's Test of Equality of Covariance Matrices

Box's M	56.108
F	1.089
df1	45
df2	6109.950
Sig.	.317

This statistic tests whether the covariance matrices for the dependent variables are significantly different. In our example, they are not ($p > .05$); if they did differ significantly, then we might believe that the covariance matrices are different. (That would be bad, because the equality of covariance matrices is an assumption of MANOVA.) This test is very sensitive, so just because it detects differences between the variance-covariance matrices does not necessarily mean that the *F* values are invalid.

Multivariate Tests (General Interpretation)

Effect		Value	F	Hypothesis df	Error df	Sig.	Partial Eta Squared	Observed Power
Intercept	Pillai's Trace	0.948	347.556	5.000	95.000	.000	.948	1.000
⋮	⋮	⋮	⋮	⋮	⋮	⋮	⋮	⋮
section	Pillai's Trace	0.127	1.299	10.000	192.000	.233	.063	.657
	Wilks' Lambda	0.876	1.302	10.000	190.000	.232	.064	.658
	Hotelling's Trace	0.139	1.305	10.000	188.000	.231	.065	.659
	Roy's Largest Root	0.112	2.155	5.000	96.000	.065	.101	.686
lowup	Pillai's Trace	0.031	0.598	5.000	95.000	.702	.031	.210
⋮	⋮	⋮	⋮	⋮	⋮	⋮	⋮	⋮
section * lowup	Pillai's Trace	0.234	2.540	10.000	192.000	.007	.117	.949
⋮	⋮	⋮	⋮	⋮	⋮	⋮	⋮	⋮

Information presented in this section includes tests of each main effect and interaction possible in your design. The Intercept refers to the remaining variance (usually the error variance). In this example, we can see that there is a significant interaction between **section** and **lowup** ($F(10,192)=2.54$, $p < .01$). Note that if a covariate is included, it will be included in this table as well; in our example, our covariate **gpa** is not significantly related to the quiz scores.

Term	Definition/Description
VALUE	Test names and values indicate several methods of testing for differences between the dependent variables due to the independent variables. Pillai's method is considered to be one of the more robust tests.
F	Estimate of the *F* value.
HYPOTHESIS DF	(Number of DV's – 1) x (Levels of IV_1 – 1) x (Levels of IV_2 – 1) x ... For example, the degrees of freedom for the **section** x **lowup** effect is: (5 DV's – 1) x (3 levels of **section** – 1) x (2 levels of **lowup** – 1) = 4 x 2 x 1 = 10
ERROR DF	Calculated differently for different tests.
SIG.	*p* value (level of significance) for the *F*.
PARTIAL ETA SQUARED	The effect-size measure. This indicates how much of the total variance is explained by each main effect or interaction. In this case, for example, the **section** by **lowup** interaction accounts for 11.7% of the variance in the five quizzes.
OBSERVED POWER	The probability that a result will be significant in a sample drawn from a population with an effect size equal to the effect size of your sample, and a sample size equal to the sample size of your sample. For example, if the effect size of **section** for the population of all classes was equal to .063 (the effect size in this sample), there is a 65.7% chance of finding that effect to be significant in a sample of 105 students (the sample size analyzed here).

Levine's Test of Equality of Error Variances

Levene's Test of Equality of Error Variances

	F	df1	df2	Sig.
quiz1	2.560	5	99	.032
quiz2	1.101	5	99	.365
quiz3	1.780	5	99	.124
quiz4	2.287	5	99	.052
quiz5	.912	5	99	.477

This test examines the assumption that the variance of each dependent variable is the same as the variance of all other dependent variables. Levene's test does this by doing an ANOVA on the *differences* between each case and the mean for that variable, rather than for the *value* of that variable itself. In our example, **quiz1** is significant, so we should interpret our results with caution (but the *F* isn't large, so we don't need to panic just yet).

Tests of Between-Subjects Effects (General Interpretation)

In addition to the multivariate tests that the general linear model procedure performs, it also does simple *univariate F* tests on each of the dependent variables. Although this procedure does not have the main advantage of MANOVA—examining all of the dependent variables simultaneously—it is often helpful in interpreting results from MANOVA.

Source	Dependent Variable	Type III Sum of Squares	df	Mean Square	F	Sig.	Partial Eta Squared	Observed Power
⋮	⋮	⋮	⋮	⋮	⋮	⋮	⋮	⋮
lowup	quiz1	0.071	1	0.071	0.013	.909	.000	.051
	quiz2	0.611	1	0.611	0.239	.626	.002	.077
	quiz3	0.504	1	0.504	0.106	.746	.001	.062
	quiz4	2.653	1	2.653	0.541	.464	.005	.113
	quiz5	0.314	1	0.314	0.113	.737	.001	.063
section * lowup	quiz1	73.236	2	36.618	6.796	.002	.121	.912
	quiz2	10.356	2	5.178	2.027	.137	.039	.409
	quiz3	45.117	2	22.559	4.744	.011	.087	.780
	quiz4	37.712	2	18.856	3.842	.025	.072	.685
	quiz5	39.870	2	19.935	7.213	.001	.127	.928
Error	quiz1	533.462	99	5.389				
	quiz2	252.945	99	2.555				
	quiz3	470.733	99	4.755				
	quiz4	485.846	99	4.908				
	quiz5	273.593	99	2.764				
⋮	⋮	⋮	⋮	⋮	⋮	⋮	⋮	⋮

In this abbreviated output, we list ANOVA results for **lowup** and **section** x **lowup**; we can see that there is a significant univariate interaction of **section** by **lowup** on **quiz1**, **quiz3**, **quiz4**, and **quiz5**. Note that it is possible to have one or more significant univariate tests on an effect without the multivariate effect being significant, or for the multivariate test to be significant on an effect without any of the univariate tests reaching significance. If your analysis includes any covariates, they will be listed here as well, and ANCOVA will be performed on each dependent variable.

Term	Definition/Description
SUM OF SQUARES	For each main effect and interaction, the between-groups sum of squares; the sum of squared deviations between the grand mean and each group mean, weighted (multiplied) by the number of subjects in each group. For the error term, the within-groups sum of squares; the sum of squared deviations between the mean for each group and the observed values of each subject within that group.
DF	Degrees of freedom. For main effects and interactions, DF = (Levels of $IV_1 - 1$) x (Levels of $IV_2 - 1$) x ...; for the error term, DF = N – DF for main effects and interactions – 1.
MEAN SQUARE	Sum of squares for that main effect or interaction (or for the error term) divided by degrees of freedom.
F	Hypothesis mean square divided by the error mean square.
SIG.	*p*-value (level of significance) for the *F*.
PARTIAL ETA SQUARED	The univariate effect-size measure reported for each dependent variable. This indicates how much of the total variance is explained by the independent variable.
OBSERVED POWER	The probability that a result will be significant in a sample drawn from a population with an effect size equal to the effect size of your sample, and a sample size equal to the sample size of your sample.

Parameter Estimates

Parameter Estimates

Dependent Variable	Parameter	B	Std. Error	t	Sig.	95% Confidence Interval Lower Bound	95% Confidence Interval Upper Bound	Partial Eta Squared
quiz1	Intercept	4.647	.883	5.262	.000	2.895	6.400	.220
	gpa	.796	.303	2.627	.010	.195	1.398	.066
	[section=1]	.807	.623	1.295	.198	-.430	2.044	.017

(table continues)

Although we have only shown a portion of the total listing of parameter estimates, many more parameters will be listed to fully describe the underlying General Linear Model. Because the independent variables are typically interpreted through examination of the main effects and interactions, rather than the parameter estimates, we will limit our description to the covariate (**gpa**). The table above shows only the effects of **gpa** on **quiz1**; the full table includes all dependent variables.

Term	Definition/Description
B	The coefficient for the covariate in the model.
STANDARD ERROR	A measure of the stability of the *B* values. It is the standard deviation of the *B* value given a large number of samples drawn from the same population.
t	*B* divided by the standard error of *B*.
SIG.	Significance of *t*; the probability that these *t* values could occur by chance; the probability that *B* is not significantly different from zero. Because this *B* is significant, we know that **gpa** did have a significant effect on **quiz1**.
LOWER AND UPPER 95% CONFIDENCE LIMITS	Based on the *B* and the standard error, these values indicate that there is (in this example) a 95% chance that *B* is between .195 and 1.398.
PARTIAL ETA SQUARED	The effect-size measure. This indicates how much of the total variance is explained by the independent variable.

Estimated Marginal Means

2. Lower or upper division

Dependent Variable	Lower or upper division	Mean	Std. Error	95% Confidence Interval Lower Bound	95% Confidence Interval Upper Bound
quiz1	lower	7.628	.538	6.560	8.696
	upper	7.560	.255	7.053	8.066
quiz2	lower				

(table continues for the other four quizzes)

For each of the dependent variables, marginal means and standard errors are given for each level of the independent variables. Standard error is the standard deviation divided by the square root of *N*.

Post Hoc Tests

Multiple Comparisons

Dependent Variable		(I) section	(J) section	Mean Difference (I-J)	Std. Error	Sig.	95% Confidence Interval Lower Bound	95% Confidence Interval Upper Bound
quiz1	Tukey HSD	1	2	.80	.549	.319	-.51	2.10
			3	1.33	.571	.056	-.03	2.69
		2	1	-.80	.549	.319	-2.10	.51
			3	.54	.549	.593	-.77	1.84
		3	1	-1.33	.571	.056	-2.69	.03
			2	-.54	.549	.593	-1.84	.77
	Bonferroni	1	2	.80	.549	.449	-.54	2.13
			3	1.33	.571	.065	-.06	2.73
		2	1	-.80	.549	.449	-2.13	.54
			3	.54	.549	.994	-.80	1.87
		3	1	-1.33	.571	.065	-2.73	.06
			2	-.54	.549	.994	-1.87	.80

For each dependent variable, *post hoc* tests are computed. There are 18 *post hoc* tests you may select from; we show here only two of the most popular tests, the Tukey's HSD and the Bonferroni. Pairwise comparisons are computed for all combinations of levels of the independent variable (in this example, section 1 and section 2, section 1 and section 3, section 2 and section 3, etc.). SPSS displays the difference between the two means, the standard error of that difference, as well as whether that difference is significant and the 95% confidence interval of the difference. In this example, there are no significant differences between the different sections, but both the Tukey test and the Bonferroni suggest that sections 1 and 3 nearly reach significance at the .05 level.

EXERCISES

Answers to selected exercises are available for download at **www.ablongman.com/george6e**.

1. Using the **grade.sav** file, compute and interpret a MANOVA examining the effect of whether or not students completed the extra credit project on the total points for the class and the previous GPA.
2. Using the **grades.sav** file, compute and interpret a MANOVA examining the effects of **section** and **lowup** on **total** and **GPA**.
3. Why would it be a bad idea to compute a MANOVA examining the effects of **section** and **lowup** on **total** and **percent**?
4. A researcher wishes to examine the effects of high- or low-stress situations on a test of cognitive performance and self-esteem levels. Participants are also divided into those with high- or low-coping skills. The data is shown after question 5 (ignore the last column for now). Perform and interpret a MANOVA examining the effects of stress level and coping skills on both cognitive performance and self-esteem level.
5. Coping skills may be correlated with immune response. Include immune response levels (listed below) in the MANOVA performed for Question 4. What do these results mean? In what way are they different than the results in Question 4? Why?

Stress Level	Coping Skills	Cognitive Performance	Self-Esteem	Immune Response
High	High	6	19	21
Low	High	5	18	21
High	High	5	14	22
High	Low	3	8	15
Low	High	7	20	22
High	Low	4	8	17
High	High	6	15	28
High	Low	5	7	19
Low	Low	5	20	16
Low	Low	5	17	18

General Linear Models: Repeated-Measures MANOVA

THE PREVIOUS chapter discussed designs with more than one dependent variable, and multiple independent variables with two or more levels each (MANOVA and MANCOVA). These analyses have all involved between-subjects designs, in which each subject is tested in only one level of the independent variable(s). There may be times, however, when a within-subjects design is more appropriate. Each subject is tested for more than one level of the independent variable or variables in a within-subjects or repeated-measures design.

This chapter will describe three different within-subjects procedures that the **General Linear Model – Repeated Measures** procedure may compute. The first is a completely within-subjects design, in which each subject experiences every experimental condition and produces values for each cell in the design. Mixed-design analyses are then described, in which one or more of the independent variables are within subjects, and one or more are between subjects. This example also includes a covariate. Finally, doubly-multivariate designs are discussed. Doubly-multivariate designs are similar to standard within-subjects designs, except that there are multiple dependent variables tested within subjects.

The **grades.sav** file is again the example. In this chapter, however—in order to demonstrate a within-subjects design—the meaning of **quiz1** through **quiz4** will be (perhaps somewhat capriciously) redefined. In particular, instead of being the scores on four different quizzes, **quiz1** through **quiz4** will refer to scores on the same quiz taken under four different conditions (or, alternatively, equivalent tests taken under four different conditions). These four conditions are based on a 2 (colors of paper) by 2 (colors of ink) within-subjects design. Students are given the same quiz (or equivalent quizzes) on either blue or red paper, printed in either green or black ink. This design produces the following combinations of paper colors and ink colors, with **quiz1** through **quiz4** assigned to each of the cells as noted:

	Ink Color	
Paper Color	**Green**	**Black**
Blue	Blue paper, green ink **quiz1**	Blue paper, black ink **quiz2**
Red	Red paper, green ink **quiz3**	Red paper, black ink **quiz4**

Because this chapter is an extension of the previous chapter, some of the advanced options and output refer you to the previous chapter. Also, if you don't have a basic understanding of between-subjects MANOVA or MANCOVA, you should read the previous chapter before trying to interpret the output from this chapter.

STEP BY STEP

General Linear Models: Within-Subjects and Repeated-Measures MANOVA

To enter SPSS, a click on **Start** *in the taskbar (bottom of screen) activates the start menu:*

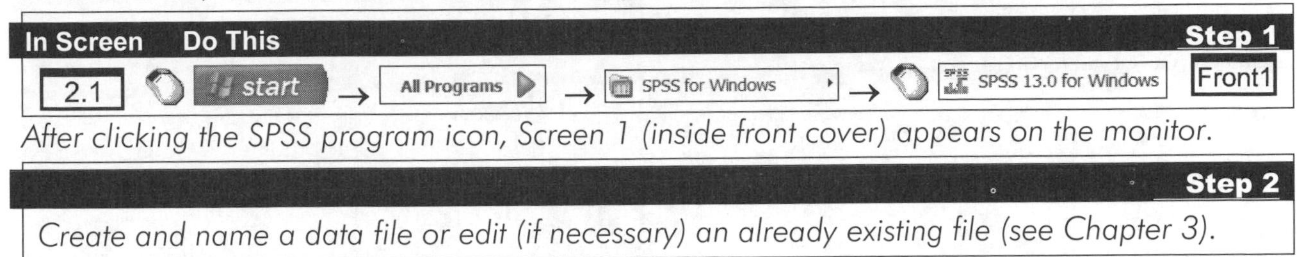

In Screen	Do This	Step 1
2.1	start → All Programs → SPSS for Windows → SPSS 13.0 for Windows	Front1

After clicking the SPSS program icon, Screen 1 (inside front cover) appears on the monitor.

Step 2
Create and name a data file or edit (if necessary) an already existing file (see Chapter 3).

Screens 1 and 2 (inside front cover) allow you to access the data file used in conducting the analysis of interest. The following sequence accesses the **grades.sav** *file for further analyses:*

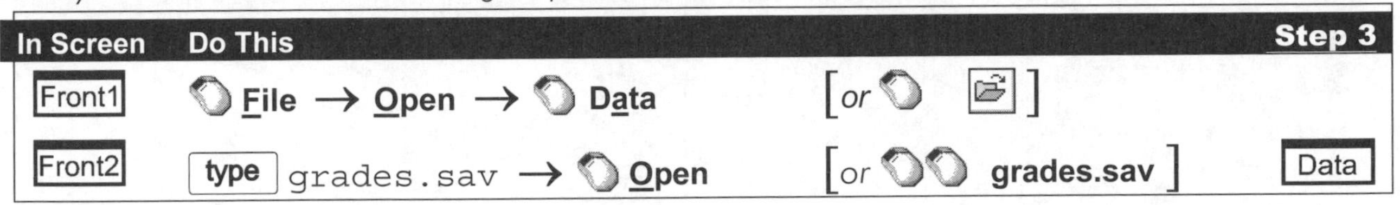

In Screen	Do This		Step 3
Front1	**File** → **Open** → **Data**	[or]	
Front2	type `grades.sav` → **Open**	[or **grades.sav**]	Data

Whether first entering SPSS or returning from earlier operations the standard menu of commands across the top is required (shown below). As long as it is visible you may perform any analyses. It is not necessary for the data window to be visible.

This menu of commands disappears or modifies when using pivot tables or editing graphs. To uncover the standard menu of commands simply click on the or the icon.

After completion of Step 3 a screen with the desired menu bar appears. When you click a command (from the menu bar), a series of options will appear (usually) below the selected command. With each new set of options, click the desired item. The sequence to access general linear models begins at any screen with the menu of commands visible:

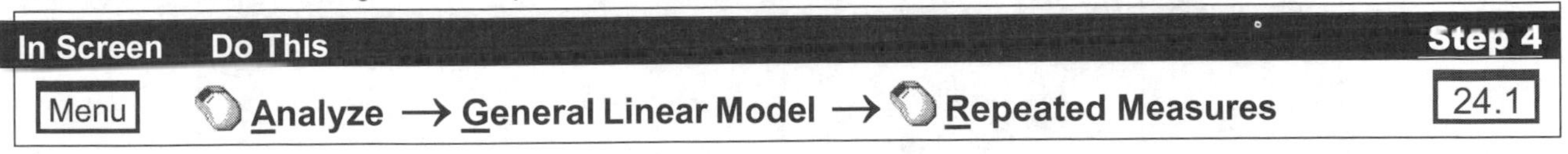

In Screen	Do This	Step 4
Menu	**Analyze** → **General Linear Model** → **Repeated Measures**	24.1

After this sequence step is performed, a window opens (Screen 24.1, below) in which you specify the names of the within-subjects independent variables. The names that you type in the **Within-Subject Factor Name** box are *not* names of variables that currently exist in your data file; they are *new* variable names that will only be used by the **Repeated Measures** command.

In our example, we define two independent variables as **papercol** (the color of the paper on which the quiz was printed) and **inkcolor** (the color of the ink with which the quiz was printed). To specify each within-subjects variable name, first type the name of the variable in the **Within-Subject Factor Name** box; the name can be up to eight characters long. Then, type the number of levels of that independent variable in the **Number of Levels** box, and click the **Add** button. If you wish to do doubly-multivariate ANOVA, in which there are multiple dependent variables for each level of the within-subjects design, then you will use the lower portion of the dialog box. In the **Measure Name** box, you can then type the name of each dependent variable used in the analysis and click on the **Add** button when finished.

24.1

General Linear Model – Repeated Measures Define Factor(s) Dialog Window

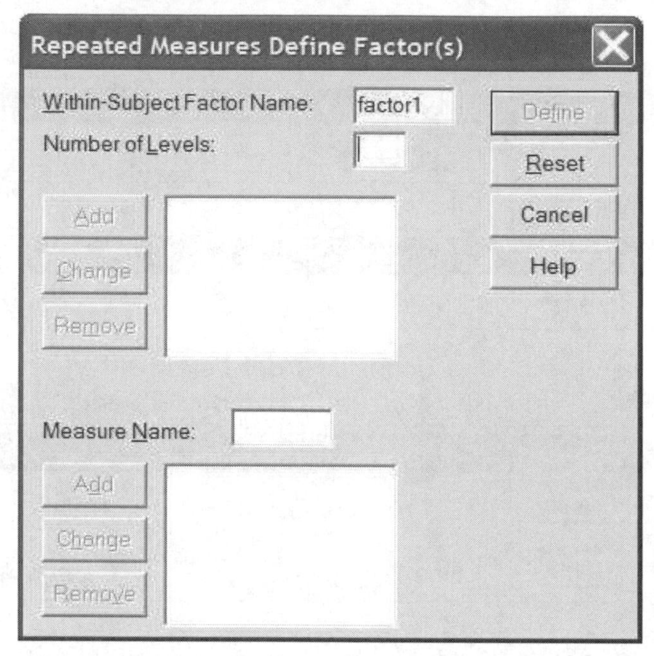

After you have specified the **Within-Subject Factor** name or names, and (optionally) entered the dependent variable names for a doubly-multivariate design, click on the **Define** button to specify the rest of your analysis.

Screen 24.2 now appears (following page). The box to the left of the window contains all variables in your data file. The **Within-Subjects Variables** box, in the top center of the window, lets you specify what variable in your data file represents each level of each within-subjects variable in the analysis. In this case, we have defined **papercol** and **inkcolor** (paper color, and ink color), with two levels each, so we need to specify four variables from the box on the left, and match them up with the cells specified in the **Within-Subjects Variables** box. The numbers in parenthesis in that box refer to the cells in the design. By way of explanation, observe the chart that follows:

Quiz number	Cell	Paper color	Ink color
1	(1,1)	blue = 1	green = 1
2	(1,2)	blue = 1	black = 2
3	(2,1)	red = 2	green = 1
4	(2,2)	red = 2	black = 2

Note that the ink color changes more rapidly than paper color as we move from **quiz1** to **quiz4**. Here we demonstrate with two variables with two levels each, but the same rationale applies with a greater number of variables that have more than two levels. If your design is doubly-multivariate, with more than one dependent measure for each cell in the within-subjects design, then those multiple dependent measures will be listed in the **Within-Subjects Variables** box.

24.2

General Linear Model – Repeated Measures Dialog Window

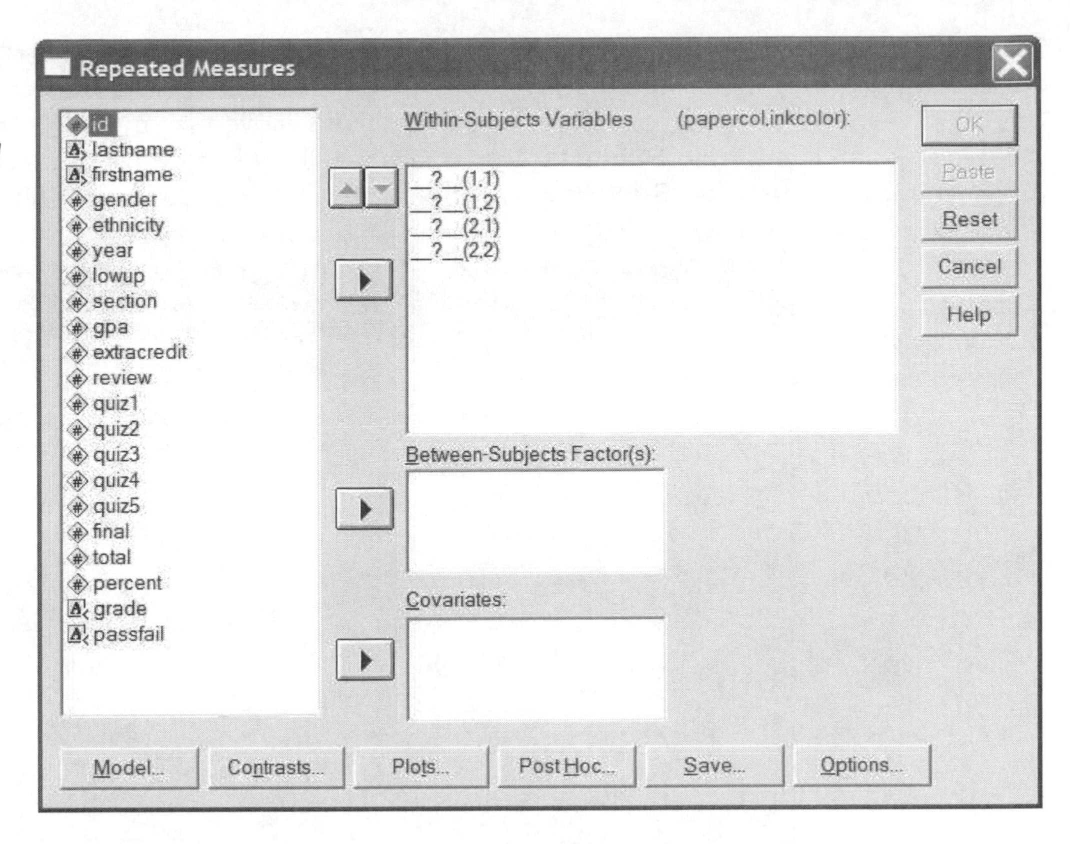

For each variable in the within-subjects analysis, select the variable in the box to the left of the window, and click on the top ▶. By default, SPSS assumes that the first variable you select will go in the top cell in the **Within-Subjects Variables** box [cell (1,1) in our example], the second variable you select will go in the second cell in the **Within-Subjects Variables** box [(1,2) in our example], and so forth. If a variable does not go in the proper cell in the **Within-subject Variables** box, then click on the up and down arrow buttons (▲▼) to move the variable up and down the cell list.

Once the within-subjects dependent variables have been specified, define any between-subjects variables or covariates in your model. Select any variables in the variable list box to the left of Screen 24.2 that you wish to treat as between-subjects variables or covariates, and click on the appropriate ▶ button to move it into the **Between-Subjects Factor(s)** box in the middle of the window or the **Covariates** box near the bottom of the window.

Six buttons at the bottom of Screen 24.2 may be selected: **Model**, **Contrasts**, **Plots**, **Post Hoc**, **Save**, and **Options**. The **Options** button produces a large array of selections, and is identical to Screen 23.5 in the previous chapter; see that description (page 299) for those choices. The **Plots** button functions identically to the **Plots** button in the previous chapter (see page 297), but now you can plot both within-subjects and between-subjects factors. The **Post Hoc** button also works similarly to the way it worked in the previous chapter (page 298), but in this case it is important to remember that you can only select *post hoc* tests for between-subjects factors. The **Contrasts** and **Save** buttons are beyond the scope of this book, and will not be discussed.

The **Model** button calls up Screen 24.4 (following page). In most cases, a **Full Factorial** model is desired: This produces a model with all main effects and all interactions included. At times, you may wish to test only certain main effects and interactions. In this case, select the **Custom** button, and move any main effects (chosen by selecting one variable) and interactions (chosen by selecting two or more variables) desired in the model from the left boxes to the right boxes, by clicking the ▶ button. Variables and interactions moved from the **Within-Subjects** box at the left go to the **Within-Subjects Model** box on the right, and variables and interactions moved from the **Between-Subjects** box on the left go to the **Between-Subjects Model** on the right; you cannot specify interactions between within-subjects and between-subjects variables.

24.3

General Linear Model – Repeated Measures Model Specification Dialog Window

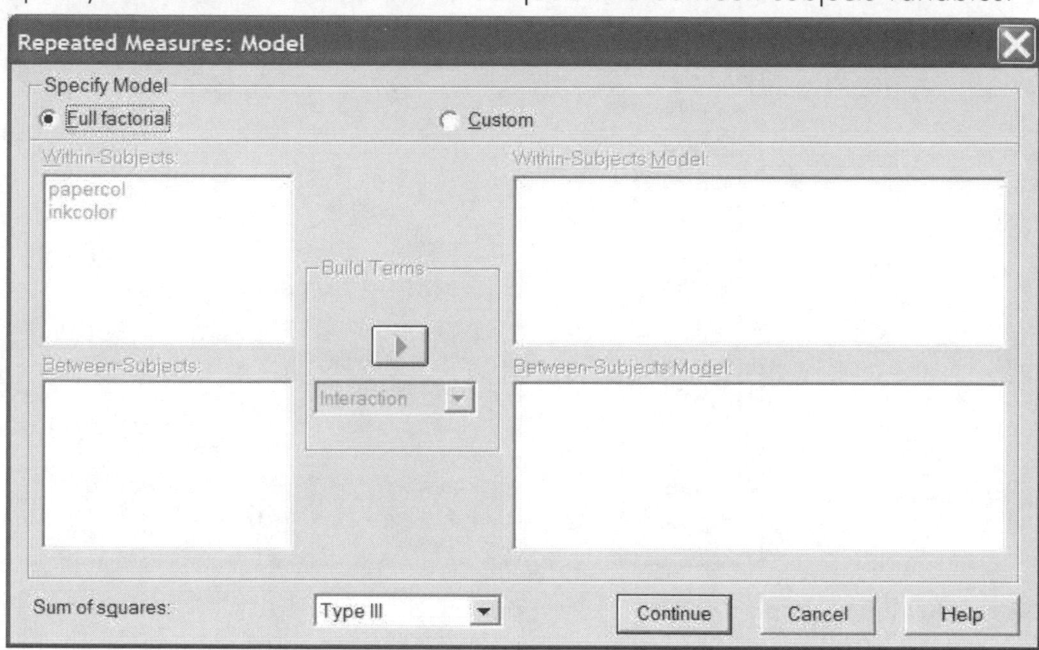

Once you have specified the model, click **Continue**, then click the **OK** to calculate the within-subjects or repeated-measures MANOVA or MANCOVA.

Three different versions of sequence step 5 will be included in this chapter; the first demonstrates within-subjects MANOVA, the second includes both a within- and a between-subjects variable along with a covariate, and the third a doubly-multivariate design.

This step instructs SPSS to perform a 2 (color of paper) x 2 (color of ink) within-subjects analysis of variance, with **quiz1** *through* **quiz4** *referring to the cells of the model as presented in the table in the introductory section of this chapter.*

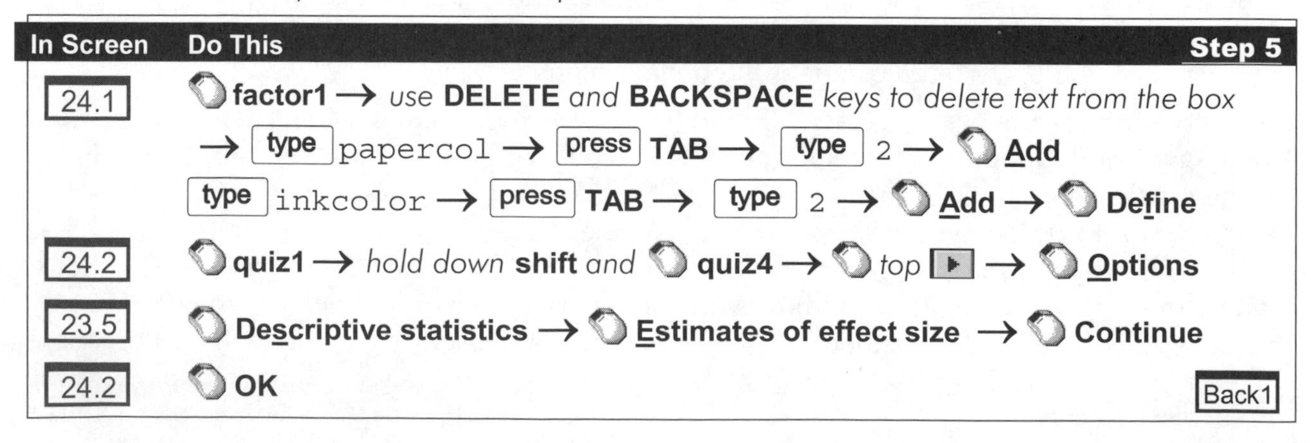

In Screen	Do This	Step 5
24.1	**factor1** → *use* **DELETE** *and* **BACKSPACE** *keys to delete text from the box* → type papercol → press **TAB** → type 2 → **Add** type inkcolor → press **TAB** → type 2 → **Add** → **Define**	
24.2	**quiz1** → *hold down* **shift** *and* **quiz4** → *top* ▶ → **Options**	
23.5	**Descriptive statistics** → **Estimates of effect size** → **Continue**	
24.2	**OK**	Back1

The following step illustrates a mixed design, in which some of the factors are within-subjects and some are between-subjects. In this case a 2 (ink color) x 2 (gender of subject) MANOVA is performed, using **gpa** *as a covariate.*

In Screen	Do This	Step 5a
24.1	factor1 → *use* **DELETE** *and* **BACKSPACE** *keys to delete text from the box* → type inkcolor → press **TAB** → type 2 → **Add** → **Define**	
24.2	**quiz1** → *hold down* **shift** *and* **quiz2** → *top* ▶ **gender** → *middle* ▶ → **gpa** → *bottom* ▶ → **Options**	
23.5	**Descriptive statistics** → **Estimates of effect size** → **Continue**	
24.2	**OK**	Back1

The following step illustrates a doubly-multivariate design, in which there are two or more dependent variables measured in different levels of one or more within-subjects factors. In this example, **quiz1** *and* **quiz2** *are short-answer quizzes, and* **quiz3** *and* **quiz4** *are multiple-choice quizzes.* **quiz1** *and* **quiz3** *were taken in the green-ink color condition, and* **quiz2** *and* **quiz4** *were taken in the black-ink color condition.*

In Screen	Do This	Step 5b
24.1	**factor1** → *use* **DELETE** *and* **BACKSPACE** *keys to delete text from the box* → type inkcolor → press **TAB** → type 2 → **Add** → *in the* **Measure Name** *box* type essay → **Add** → type mc → **Add** → **Define**	
24.2	**quiz1** → *hold down* **shift** *and* **quiz4** → *top* ▶ → **Options**	
23.5	**Descriptive statistics** → **Estimates of effect size** → **Continue**	
24.2	**OK**	Back1

Upon completion of step 5, 5a, or 5b, the output screen will appear (Screen 1, inside back cover). All results from the just-completed analysis are included in the Output Navigator. Make use of the scroll bar arrows (▲▼▶◀) to view the results. Even when viewing output, the standard menu of commands is still listed across the top of the window. Further analyses may be conducted without returning to the data screen.

PRINTING RESULTS

Results of the analysis (or analyses) that have just been conducted requires a window that displays the standard commands (**File Edit Data Transform Analyze** . . .) across the top. A typical print procedure is shown below beginning with the standard output screen (Screen 1, inside back cover).

To print results, from the Output screen perform the following sequence of steps:

In Screen	Do This	Step 6
Back1	*Select desired output or edit (see pages 19 - 25)* → **File** → **Print**	
2.9	*Consider and select desired print options offered, then* → **OK**	

To exit you may begin from any screen that shows the **File** *command at the top.*

In Screen	Do This	Step 7
Menu	File → Exit	2.1

Note: After clicking **Exit**, there will frequently be small windows that appear asking if you wish to save or change anything. Simply click each appropriate response.

OUTPUT

General Linear Models – Within-Subjects and Repeated-Measures MANOVA

Most of the output from the General Linear Models procedure using within-subjects designs is similar to the output for between-subjects designs (Chapter 23), so in order to conserve space, we present only representative portions of the complete output here. We have included all of the *essential* output and focus on material that is unique to within-subjects designs. Also, because there are three example analyses done in this chapter, each subheading will be followed by a note indicating which of the three types of analyses are applicable to that output section, as well as which step (step 5, 5a, or 5b) is used as the sample output.

Multivariate Tests

- Applies to: Step 5 (Within-Subjects Designs), Step 5a (Mixed Designs), Step 5b (Doubly-multivariate Designs).
- Example from: Step 5 (Within-Subjects Designs).

Multivariate Tests

Effect		Value	F	Hypothesis df	Error df	Sig.	Partial Eta Squared
papercol	Pillai's Trace	0.027	2.866	1	104	.093	.027
	Wilks' Lambda	0.973	2.866	1	104	.093	.027
	Hotelling's Trace	0.028	2.866	1	104	.093	.027
	Roy's Largest Root	0.028	2.866	1	104	.093	.027
inkcolor	Pillai's Trace	0.020	2.124	1	104	.148	.020
⋮	⋮	⋮	⋮	⋮	⋮	⋮	⋮
papercol * inkcolor	Pillai's Trace	0.081	9.160	1	104	.003	.081
	⋮	⋮	⋮	⋮	⋮	⋮	⋮

SPSS will present information similar to that given here for each main effect and interaction possible in your design. In our example, the interaction effect of paper color by ink color is significant ($p = .003$), so the means of the cells must be examined to interpret the interaction effect, either by examining the means themselves or by performing additional t tests. In this case the means demonstrate an interaction shown in the figure below: Black ink produces higher quiz scores than green ink, when the quiz is printed on blue paper; however, green ink produces higher quiz scores than black ink when the quiz is printed on red paper.

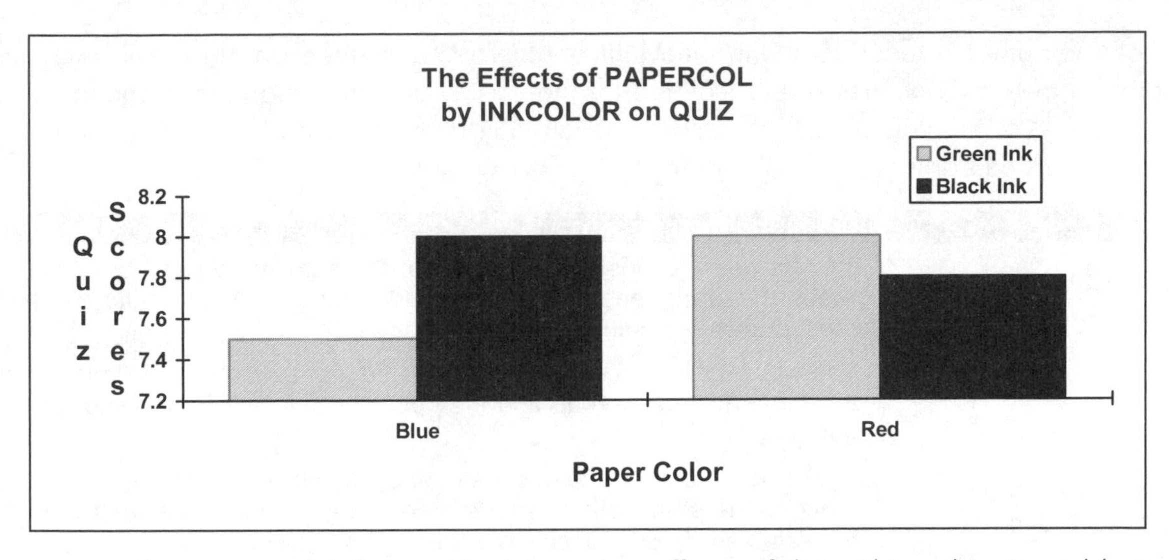

If you are analyzing a mixed design, the interaction effects of the within-subjects and between-subjects variables will also be produced. If you have covariates in the design, the interaction effects between your covariates and the within-subjects variables will also be given. Interpreting these interactions is tricky: You have to look for differences in the regression coefficients of your covariates for different levels of your within-subject variables.

Term	Definition/Description
VALUE	Test names and values indicate several methods of testing for differences between the dependent variables due to the independent variables. Pillai's method is considered to be one of the more robust tests.
F	Estimate of the F-value.
HYPOTHESIS DF	(Levels of $IV_1 - 1$) x (Levels of $IV_2 - 1$) x ... For doubly-multivariate designs, this is multiplied by (Number of DV's – 1).
ERROR DF	Calculated differently for different tests.
SIG.	*p*-value (level of significance) for the *F*.
PARTIAL ETA SQUARED	The effect-size measure. This indicates how much of the total variance is explained by each main effect or interaction. In this case, for example, the papercol by inkcolor interaction accounts for 8.1% of the variance in the four quizzes.

Tests of Within-Subjects Contrasts

- Applies to: Step 5 (Within-Subjects Designs), Step 5a (Mixed Designs).
- Example from: Step 5 (Within-Subjects Designs).

Tests of Within-Subjects Contrasts

Source	Type III Sum of Squares	df	Mean Square	F	Sig.	Partial Eta Squared
papercol	2.917	1	2.917	2.866	.093	.027
Error(papercol)	105.833	104	1.018			
inkcolor	2.917	1	2.917	2.124	.148	.020
Error(inkcolor)	142.833	104	1.373			
papercol*inkcolor	12.688	1	12.688	9.160	.003	.081
Error(papercol*inkcolor)	144.062	104	1.385			

Most of this output is redundant with the Multivariate Tests described on the previous page. In this table, however, *F* values are calculated through using sum of squares and mean squares. Because most of the values in this output have already been described, only those values that are not presented earlier are defined here.

Term	Definition/Description
SUM OF SQUARES FOR MAIN EFFECTS AND INTERACTIONS	The between-groups sum of squares; the sum of squared deviations between the grand mean and each group mean, weighted (multiplied) by the number of subjects in each group.
SUM OF SQUARES FOR ERROR	The within-groups sum of squares; the sum of squared deviations between the mean for each group and the observed values of each subject within that group.
MEAN SQUARE FOR MAIN EFFECTS AND INTERACTIONS	Hypothesis sum of squares divided by hypothesis degrees of freedom. Since there is only one degree of freedom for the hypothesis, this value is the same as hypothesis sum of squares.
MEAN SQUARE FOR ERROR	Error sum of squares divided by error degrees of freedom.

Univariate Tests

- Applies to: Step 5b (Doubly-multivariate Designs).
- Example from: Step 5b (Doubly-multivariate Designs).

Source	Measure		Type III Sum of Squares	df	Mean Square	F	Sig.	Partial Eta Squared
inkcolor	essay	Sphericity Assumed	13.886	1	13.886	8.247	.005	.073
		Greenhouse-Geisser	13.886	1.000	13.886	8.247	.005	.073
		Huynh-Feldt	13.886	1.000	13.886	8.247	.005	.073
		Lower-bound	13.886	1.000	13.886	8.247	.005	.073
	mc	Sphericity Assumed	1.719	1	1.719	1.599	.209	.015
		Greenhouse-Geisser	1.719	1.000	1.719	1.599	.209	.015
		Huynh-Feldt	1.719	1.000	1.719	1.599	.209	.015
		Lower-bound	1.719	1.000	1.719	1.599	.209	.015
Error (inkcolor)	essay	Sphericity Assumed	175.114	104	1.684			
		Greenhouse-Geisser	175.114	104	1.684			
		Huynh-Feldt	175.114	104	1.684			
		Lower-bound	175.114	104	1.684			
	mc	Sphericity Assumed	111.781	104	1.075			
		Greenhouse-Geisser	111.781	104	1.075			
		Huynh-Feldt	111.781	104	1.075			
		Lower-bound	111.781	104	1.075			

For any doubly-multivariate designs (with more than one dependent variable), univariate *F* tests are performed for each main effect and interaction. In this example, there is a significant effect of ink color for essay exams, but not for multiple choice exams. Several different methods of calculating each test are used (Sphericity Assumed, Greenhouse-Geisser, etc.), but it is very rare for them to produce different results.

Term	Definition/Description
SUM OF SQUARES	For each main effect and interaction, the between-groups sum of squares; for the error term, the within-groups sum of squares.
DF	Degrees of freedom. For main effects and interactions, DF = (Levels of $IV_1 - 1$) x (Levels of $IV_2 - 1$) x ...; for the error term, DF = N – DF for main effects and interactions – 1.
MEAN SQUARE	Sum of squares for that main effect or interaction (or for the error term) divided by degrees of freedom.
F	Hypothesis mean square divided by the error mean square.
SIG.	p-value (level of significance) for the F.

Tests of Between-Subjects Effects

- Applies to: Step 5a (Mixed Designs).
- Example from: Step 5a (Mixed Designs).

Source	Type III Sum of Squares	df	Mean Square	F	Sig.	Partial Eta Squared
Intercept	489.802	1	489.802	73.411	.000	.419
gpa	53.083	1	53.083	7.956	.006	.072
gender	.736	1	.736	.110	.740	.001
Error	680.548	102	6.672			

The between-subjects effects are listed together, along with the effects of any covariates. Most of this output is redundant with the Multivariate Tests, described above. In this table, however, as in the Tests of Within-Subjects Effects table, *F* values are calculated through using sum of squares and mean squares. All of the statistics reported in this table have been defined above in either the Multivariate Tests or the Tests of Within-Subjects Effects.

EXERCISES

Answers to selected exercises are available for download at **www.ablongman.com/george6e**.

1. Imagine that in the **grades.sav** file, the five quiz scores are actually the same quiz taken under different circumstances. Perform repeated-measures ANOVA on the five quiz scores. What do these results mean?
2. To the analysis in exercise 1, add whether or not students completed the extra credit project (**extrcred**) as a between-subjects variable. What do these results mean?
3. A researcher puts participants in a highly stressful situation (say, performing repeated-measures MANCOVA) and measures their cognitive performance. He then puts them in a low-stress situation (say, lying on the beach on a pleasant day). Participant scores on the test of cognitive performance are reported below. Perform and interpret a within-subjects ANOVA on these data.

Case Number:	1	2	3	4	5	5	6	7	8	10
High Stress:	76	89	86	85	62	63	85	115	87	85
Low Stress:	91	92	127	92	75	56	82	150	118	114

4. The researcher also collects data from the same participants on their coping ability. They scored (in case number order) 25, 9, 59, 16, 23, 10, 6, 43, 44, and 34. Perform and interpret a within-subjects ANCOVA on these data.

5. The researcher just discovered some more data...in this case, physical dexterity performance in the high-stress and low-stress situations (listed below, in the same case number order as in the previous two exercises). Perform and interpret a 2 (stress level: high, low) by 2 (kind of performance: cognitive, dexterity) ANCOVA on these data.

Case Number:	1	2	3	4	5	5	6	7	8	10
High Stress:	91	109	94	99	73	76	94	136	109	94
Low Stress:	79	68	135	103	79	46	77	173	111	109

25

LOGISTIC REGRESSION

LOGISTIC REGRESSION is an extension of multiple regression (Chapter 16) in which the dependent variable is not a continuous variable. In logistic regression, the dependent variable may have only two values. Usually these values refer to membership-nonmembership, inclusion-noninclusion, or yes-no.

Because logistic regression is an extension of multiple regression, we will assume that you are familiar with the fundamentals of multiple regression analysis: namely, that several variables are regressed onto another variable, using forward, backward, or other selection processes and criteria. These basic concepts are the same for logistic regression as for multiple regression. However, the meaning of the regression equation is somewhat different in logistic regression. In a standard regression equation, a number of weights are used with the predictor variables to predict a value of the criterion or dependent variable. In logistic regression the value that is being predicted represents a probability, and it varies between 0 and 1. In addition to this, it is possible to use a categorical predictor variable, using an indicator-variable coding scheme. This is described in the Step by Step and Output sections in more detail, but it essentially breaks up a single categorical predictor variable into a series of variables, each coded as 1 or 0 indicating whether or not the subjects are in a particular category.

At this point, the basic mathematics of logistic regression will be summarized, using the example that will be applied in this chapter. The example utilizes another helping file named **helping3.sav**. This is a file of real data ($N = 537$) that deals with issues similar to the **helping1.sav** file presented earlier, but variable names are different in several instances. It is described in greater detail in the *Data Disk* section of the Appendix. In the **helping1.sav** file used in Chapter 16, the amount of help given to a friend was predicted by feelings of: (a) sympathy aroused in the helper in response to the friend's need (**sympathy**); (b) feelings of anger or irritation aroused in the helper by the friend's need (**anger**); and (c) self-efficacy of the helper in relation to the friend's need (**efficacy**). In this case, however, instead of predicting the *amount* of help given to a friend, our model will predict a different dependent variable: whether the friend thought that the help given was useful or not. This is coded as a yes-or-no (dichotomous) variable. It should be noted that, although the rest of the data in the **helping3.sav** file are real data, this categorical dependent was created for this example. Specifically, anyone who gave more than the average amount of help was coded "helpful" and anyone who gave less than the average amount of help was coded "unhelpful".

Mathematics of Logistic Regression

If you want to understand logistic regression, then probabilities, odds, and the logarithm of the odds must be understood. Probabilities are simply the likelihood that something will happen; a probability of .20 of rain means that there is a 20% chance of rain. In the technical sense used here, odds are the ratio of the probability that an event will occur divided by the probability that an event will not occur. If there is a 20% chance of rain, then there is an 80% chance of no rain; the odds, then, are:

$$\text{Odds} = \frac{\text{prob(rain)}}{\text{prob(no rain)}} = \frac{.20}{.80} = \frac{1}{4} = .25$$

Although probabilities vary between 0 and 1, odds may be greater than 1: For instance, an 80% chance of rain has odds of .80/.20 = 4.0. A 50% chance of rain has odds of 1.

A key concept in logistic regression analysis is a construct known as a *logit*. A logit is the natural logarithm (*ln*) of the odds. If there is a 20% chance of rain, then there is a logit of:

$$ln(.25) = -1.386...$$

In the example used in Chapter 16, the regression equation looked something like this:

$$\text{HELP} = B_0 + B_1 \times \text{SYMPATHY} + B_2 \times \text{ANGER} + B_3 \times \text{EFFICACY}$$

The amount of helping was a function of a constant, plus coefficients times the amount of sympathy, anger, and efficacy. In the example used in this chapter, in which we examine whether or not the subjects' help was useful instead of how much they helped, the regression equation might look something like this. Please note that we have now switched to the variable names used in the **helping3.sav** file: Sympathy (**sympatht**), anger (**angert**), and efficacy (**effict**).

$$ln\left[\frac{\text{prob(helping)}}{\text{prob(not helping)}}\right] = B_0 + B_1(\textbf{sympatht}) + B_2(\textbf{angert}) + B_3(\textbf{effict})$$

In this equation, the log of the odds of helping is a function of a constant, plus a series of weighted averages of sympathy, anger, and efficacy. If you wish to think in terms of the odds-of-helping or the probability-of-helping instead of the log-odds-of-helping, this equation may be converted to the following:

$$\frac{\text{prob(helping)}}{\text{prob(not helping)}} = e^{B_0} \times e^{B_1(\textbf{sympatht})} \times e^{B_2(\textbf{angert})} \times e^{B_3(\textbf{effict})}$$

or

$$\text{prob(helping)} = \frac{1}{1 + e^{-B_0} \times e^{-B_1(\textbf{sympatht})} \times e^{-B_2(\textbf{angert})} \times e^{-B_3(\textbf{effict})}}$$

This equation is probably not very intuitive to most people (several of my students would certainly agree with this!); it takes a lot of experience before interpreting logistic regression equations becomes intuitive. Because of this as well as other problems in selecting an appropriate model (since model selection involves both mathematical and theoretical considerations), you should use extreme caution in interpreting logistic regression models.

Due to this complexity we refer you to three sources if you wish to gain a greater understanding of logistic regression: The *SPSS 13.0 Manuals* cover logistic regression in much greater detail than we do here, and then there are textbooks by McLachlan (1992) and Wickens (1989), both of which are quite good.

STEP BY STEP

Logistic Regression

To enter SPSS, a click on **Start** *in the taskbar (bottom of screen) activates the start menu:*

After clicking the SPSS program icon, Screen 1 (inside front cover) appears on the monitor.

Step 2

Create and name a data file or edit (if necessary) an already existing file (see Chapter 3).

Screens 1 and 2 (inside front cover) allow you to access the data file used in conducting the analysis of interest. The following sequence accesses the **helping3.sav** *file for further analyses:*

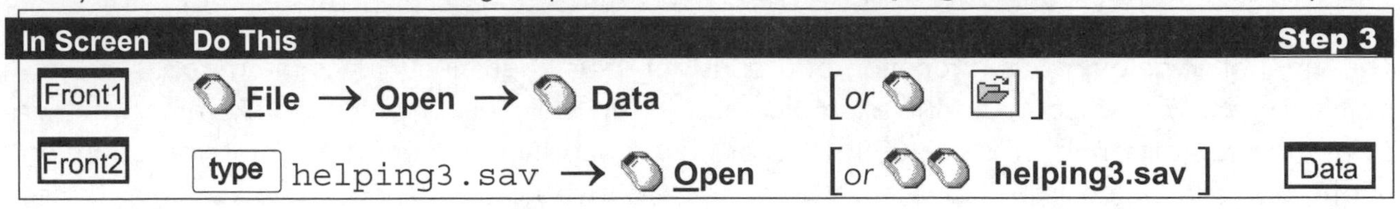

In Screen	Do This	Step 3
Front1	File → Open → Data [or]	
Front2	type helping3.sav → Open [or helping3.sav]	Data

Whether first entering SPSS or returning from earlier operations the standard menu of commands across the top is required (shown below). As long as it is visible you may perform any analyses. It is not necessary for the data window to be visible.

This menu of commands disappears or modifies when using pivot tables or editing graphs. To uncover the standard menu of commands simply click on the or the icon.

After completion of Step 3 a screen with the desired menu bar appears. When you click a command (from the menu bar), a series of options will appear (usually) below the selected command. With each new set of options, click the desired item. The sequence to access logistic regression begins at any screen with the menu of commands visible:

In Screen	Do This	Step 4
Menu	Analyze → Regression → Binary Logistic	25.1

After this sequence step is performed, a window opens (Screen 25.1, following page) in which you select the predicted variable, predictor variables, and other specifications that allow you to do a logistic regression analysis. On the left side of the window is the list of all of the variables in the file; in the top center of the window is the **Dependent** box. In this box, you will place the categorical variable you wish to predict based on your other variables. In our example, we will use the variable **cathelp**, which equals 1 if help was *not* rated as useful, and 2 if it *was* rated useful. Because logistic regression uses 0's and 1's (instead of 1's and 2's) to code the dependent variable, SPSS will recode the variable from 1 and 2, to 0 and 1. It does this automatically; you don't have to worry about it.

In the center of the window is the **Covariates** box. In this box, you will enter the predictor variables by clicking on each variable that you want to include in the analysis, and then clicking on the ▶ button. It is possible to enter interaction terms into the regression equation, by selecting all of the variables that you want in the interaction term and then clicking on the >a*b> button. It is often difficult to interpret interaction terms, so you probably shouldn't use interactions unless you are familiar with them, and you have strong theoretical rationale for including them.

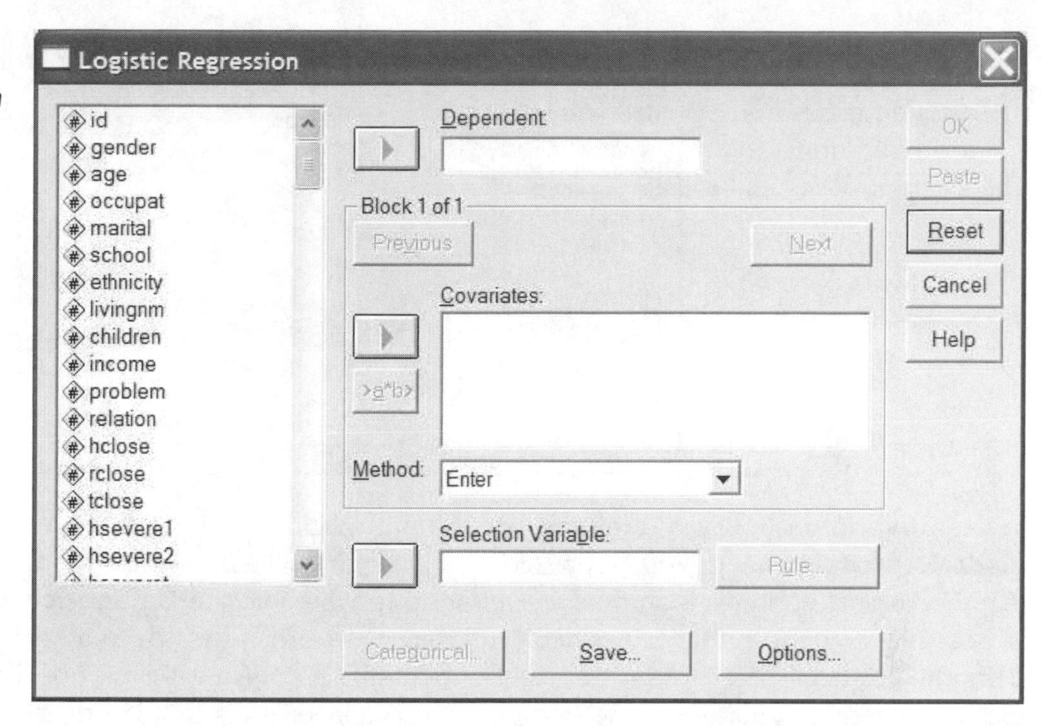

25.1

Logistic Regression Dialog

The **Method** specifies the way that SPSS will build the regression equation. **Enter**, shown in Screen 25.1, tells SPSS to build the equation by entering all of the variables at once, whether or not they significantly relate to the dependent variable. In **Forward: LR**, SPSS builds the equation by entering variables one at a time, using likelihood ratio estimates to determine which variable will add the most to the regression equation. In the **Backward: LR** method, SPSS builds the equation starting with all variables and then removes them one by one if they do not contribute enough to the regression equation. These are the most commonly used methods.

If you wish to use a different method for different variables, then you will need to use the **Previous** and **Next** buttons to set up different blocks of variables. For each block you set up, you can enter several variables and the method that you want SPSS to use to enter those variables (SPSS will analyze one block at a time, and then move on to the next block). Then, click the **Next** button to set up the next block. Different blocks can have different methods. In our example, we will only use a single block, and use the **Forward: LR** method.

The **Selection Variable** option operates like the **If** procedure described in Chapter 4. If, for instance, you paste **gender** into the Selection Variable window, the Value button would brighten up. A click on this button opens a tiny dialog window that allows you to select which level of the variable (in this case, 1 = women, 2 = men) you wish to select for the procedure. The analysis will then be conducted with the selected subjects or cases. This selection may be reversed later if you wish.

Once you have set up your Covariates (predictor variables), you need to select the **Categorical** button if a categorical predictor variable is included. In our examples, individuals may be White, Black, Hispanic, Asian, or other. Because there is no particular order to these ethnicities, the **Categorical** button is clicked and Screen 25.2 appears (following page).

25.2

Logistic Regression Categorical Variables Dialog Window

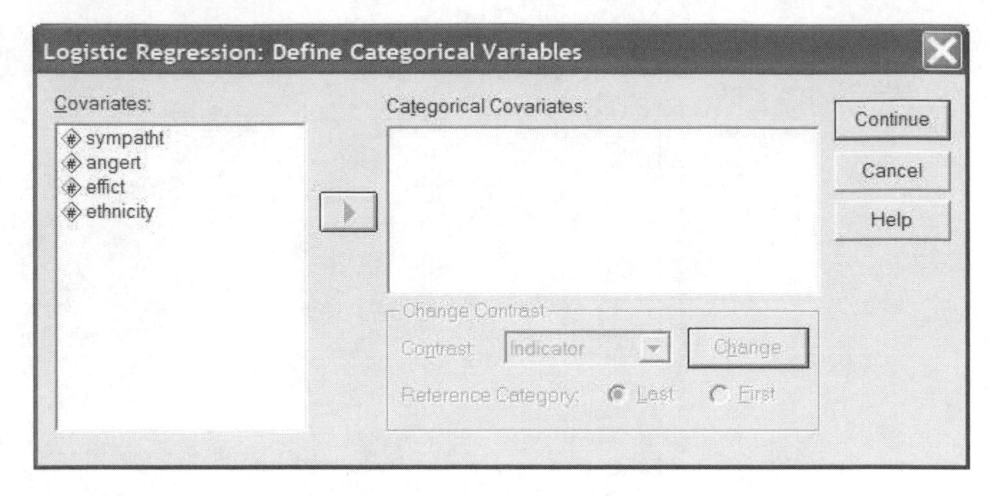

In the **Covariates** box to the left are all of the predictor variables that were entered in the **Logistic Regression** window (Screen 25.1). For each of these variables that are categorical, select the variable and click on the ▶ button to move it into the **Categorical Covariates** box. It is possible to handle the categorical variable in several ways, by setting the **Contrast** in the **Change Contrast** box in the lower right of Screen 25.2. We will limit our discussion to the **Deviation** contrast (**Indicator** is the default), which converts this simple categorical variable into a series of variables (each of them a contrast between one of the first variables and the final variable) that may be entered into the logistic regression equation. Note that the **Change Contrast** box allows you to change the current contrast to a different contrast, required, of course in the shift from **Indicator** to **Deviation**.

The **Options** button (Screen 25.1) allows you to select a number of additional options that are often useful. Upon clicking this button, Screen 25.3 (below) appears that allows you to select which additional statistics and plots you wish, as well as the probability for entry and removal of variables when performing stepwise regression.

25.3

Logistic Regression: Options Window

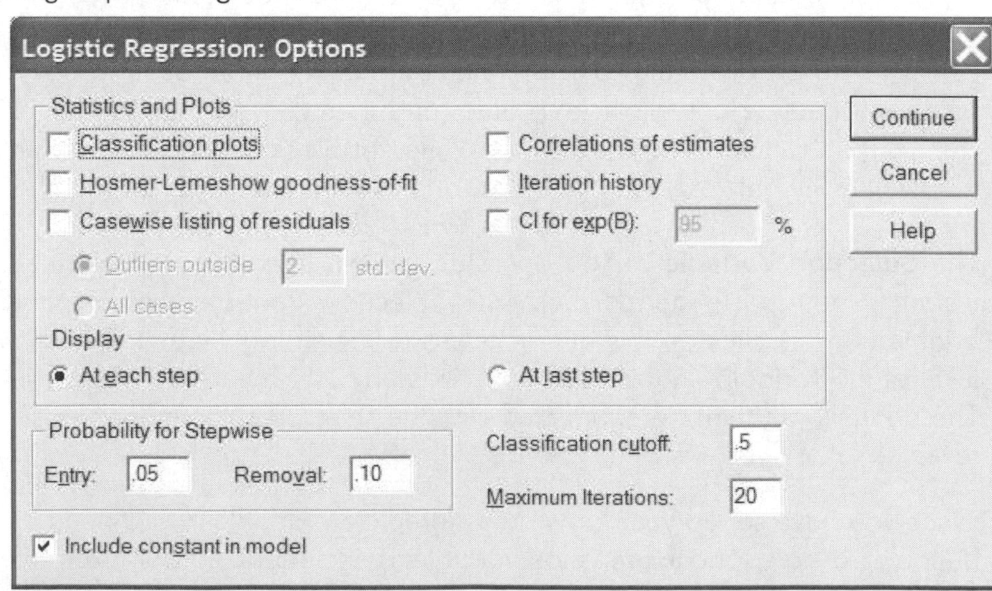

Clicking on the **Classification plots** box will produce a graph in which each case is plotted along its predicted probability from the regression equation produced, indicating its classification based on the predicted variable in the original data. This provides a graphical representation of how well the regression equation is working. **Correlations of estimates** will produce correlations between all variables that have been entered into the regression equation. Values for the **Probability for Stepwise Entry** and **Removal** are important when stepwise regression (the Forward and Backward methods) is used. The **Probability for Stepwise Entry** specifies the probability value used to add a variable to the regression equation. Similarly, the **Probability for Stepwise Removal** specifies the values used to drop a variable from the regression equation. As default, SPSS uses a probability for **Entry** of .05, and a probability for **Removal** of .10. These values may be changed if the researcher desires.

To perform a logistic regression analysis with **cathelp** *as the dependent variable, and* **sympatht**, **angert**, **effict**, *and* **ethnic** *as predictor variables, perform the following sequence of steps. The starting point is Screen 25.1. Please perform whichever of steps 1-4 (pages 323-324) are necessary to arrive at this screen.*

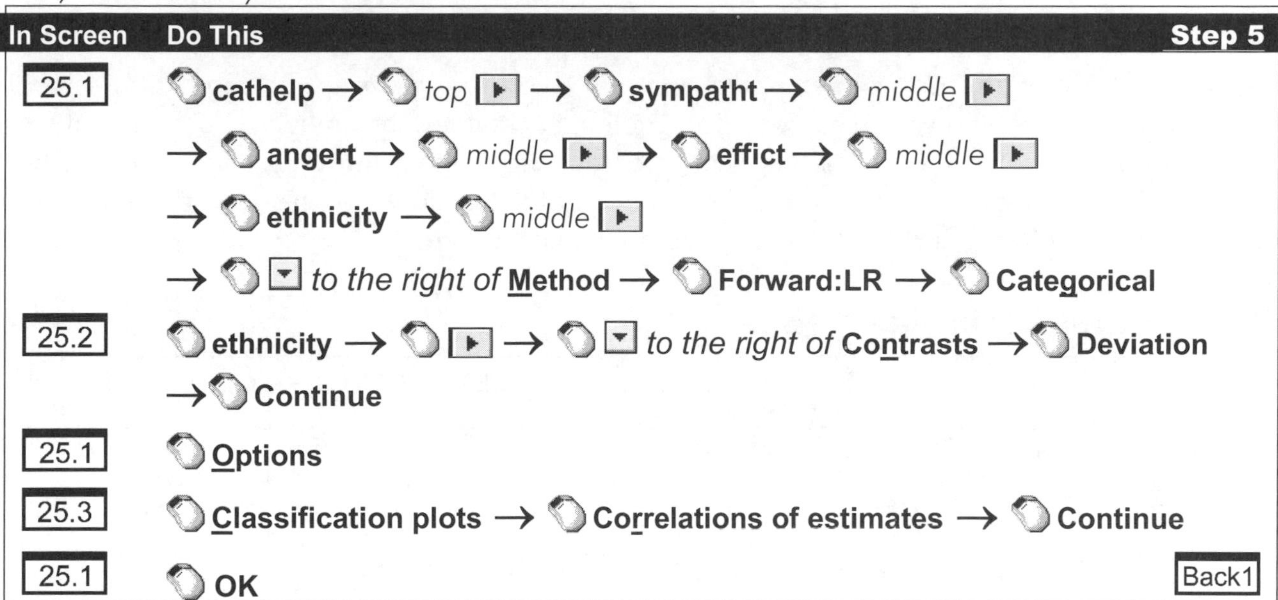

In Screen	Do This	Step 5
25.1	**cathelp** → *top* ▶ → **sympatht** → *middle* ▶ → **angert** → *middle* ▶ → **effict** → *middle* ▶ → **ethnicity** → *middle* ▶ → ▼ *to the right of* **Method** → **Forward:LR** → **Categorical**	
25.2	**ethnicity** → ▶ → ▼ *to the right of* **Contrasts** → **Deviation** → **Continue**	
25.1	**Options**	
25.3	**Classification plots** → **Correlations of estimates** → **Continue**	
25.1	**OK**	Back1

Upon completion of step 5, the output screen will appear (Screen 1, inside back cover). All results from the just-completed analysis are included in the Output Navigator. Make use of the scroll bar arrows (▲ ▼ ▶ ◀) to view the results. Even when viewing output, the standard menu of commands is still listed across the top of the window. Further analyses may be conducted without returning to the data screen.

PRINTING RESULTS

Results of the analysis (or analyses) that have just been conducted requires a window that displays the standard commands (**File Edit Data Transform Analyze** . . .) across the top. A typical print procedure is shown below beginning with the standard output screen (Screen 1, inside back cover).

To print results, from the Output screen perform the following sequence of steps:

In Screen	Do This	Step 6
Back1	*Select desired output or edit (see pages 19-25)* → **File** → **Print**	
2.9	*Consider and select desired print options offered, then* → **OK**	

To exit you may begin from any screen that shows the **File** *command at the top.*

In Screen	Do This	Step 7
Menu	File → Exit	2.1

Note: After clicking **Exit**, there will frequently be small windows that appear asking if you wish to save or change anything. Simply click each appropriate response.

OUTPUT

Logistic Regression Analysis

What follows is partial output from sequence step 5 on page 327.

Because forward- and backward-stepping analyses involve reanalyzing the regression several times (adding or deleting variables from the equation each time), they may produce several sets of output, as illustrated here. The final set of output shows the final regression equation, but output is produced for each step in the development of the equation to see how the regression equation was formed. Only the final set of output is shown here.

Categorical Variables Codings

			Parameter coding			
		Frequency	(1)	(2)	(3)	(4)
ethnicity	WHITE	293	1.000	.000	.000	.000
	BLACK	50	.000	1.000	.000	.000
	HISPANIC	80	.000	.000	1.000	.000
	ASIAN	70	.000	.000	.000	1.000
	OTHER/DTS	44	.000	.000	.000	.000

If **Contrasts** are used in the analysis, a table will be produced at the beginning of the output that shows how the computer has converted from the various values of your variable (the rows in the table) to coding values of several different variables within the computer (the columns). In this case, **ethnic** had five levels in the original data and has been broken down into four new variables, labeled **ethnic(1)** through **ethnic(4)**. These variables are a series of contrasts between the various ethnicities.

Omnibus Tests of Model Coefficients

Omnibus Tests of Model Coefficients

		Chi-square	df	Sig.
Step 1	Step	114.843	1	.000
	Block	114.843	1	.000
	Model	114.843	1	.000
Step 2	Step	29.792	1	.000
	Block	144.635	2	.000
	Model	144.635	2	.000

Model Summary

Step	-2 Log likelihood	Cox & Snell R Square	Nagelkerke R Square
1	629.506	.193	.257
2	599.713	.236	.315

Term	Definition/Description
STEPS 1 AND 2	Using the Forward: LR method of entering variables into the model, it took two steps for SPSS to enter all variables that significantly improved the model. We will find **which** variables are entered later (in the "Variables in Equation" section), but for now we will have to be content knowing that there are two variables in the model.
STEP, BLOCK, AND MODEL χ^2	These values test whether or not all of the variables entered in the equation (for model), all of the variables entered in the current block (for block), or the current increase in the model fit (for step) have a significant effect. χ^2 values are provided for each step of the model. In this case, a high χ^2 value for the model and step for Step 1 (as Step 1 is the first step, it is the entire model; note that we are only working with one block, so χ^2 values for the block are the same as for the model throughout) indicates that the first variable added to the model significantly impact the dependent variable. The step χ^2 for Step 2 indicates that adding a second variable significantly improves the model, and the model χ^2 for Step 2 indicates that the model including two variables is significant. Note that the χ^2 for the model in Step 2 is equal to the sum of the model in Step 1 and the step χ^2 in Step 2.
-2 LOG LIKELIHOOD	This and the other model summary measures are used to indicate how well the model fits the data. Smaller -2 log likelihood values mean that the model fits the data better; a perfect model has a -2 log likelihood value of zero.
COX & SNELL AND NAGELKERKE R SQUARE	Estimates of the R^2 value, indicating what percentage of the dependent variable may be accounted for by all included predictor variables.

Classification Table

			Predicted		
			CATHELP		Percentage Correct
	Observed		NOT HELPFUL	HELPFUL	
Step 1	CATHELP	NOT HELPFUL	176	89	66.4
		HELPFUL	79	193	71.0
	Overall Percentage				68.7
Step 2	CATHELP	NOT HELPFUL	181	84	68.3
		HELPFUL	75	197	72.4
	Overall Percentage				70.4

The classification table compares the predicted values for the dependent variable, based on the regression model, with the actual observed values in the data. When computing predicted values, SPSS simply computes the probability for a particular case (based on the current regression equation) and classifies it into the two categories possible for the dependent variable based on that probability. If the probability is less than .50, then SPSS classifies it as the first value for the dependent variable (*Not Helpful* in this example); and if the probability is greater than .50, then SPSS classifies that case as the second value for the dependent variable (*Helpful*). This table compares these predicted values with the values observed in the data. In this case, the Model-2 variables can predict which value of **cathelp** is observed in the data 70% of the time.

Variables in the Equation

		B	S.E.	Wald	df	Sig.	Exp(B)
Step 1[a]	EFFICT	1.129	.122	85.346	1	.000	3.094
	Constant	-5.302	.585	82.019	1	.000	.005
Step 2[b]	SYMPATHT	.467	.089	27.459	1	.000	1.596
	EFFICT	1.114	.128	76.197	1	.000	3.046
	Constant	-7.471	.775	93.006	1	.000	.001

a. Variable(s) entered on step 1: EFFICT.

b. Variable(s) entered on step 2: SYMPATHT.

At last, we learn which variables have been included in our equation. For each step in building the equation, SPSS displays a summary of the effects of the variables that are currently in the regression equation. The constant variable indicates the constant B_0 term in the equation.

Term	Definition/Description
B	The weighting value of *B* used in the equation; the magnitude of *B*, along with the scale of the variable that *B* is used to weight, indicates the effect of the predictor variable on the predicted variable. In this case, for example, sympathy and efficacy both have a positive effect on helping.
S.E.	Standard error; a measure of the dispersion of *B*.
WALD	A measure of the significance of *B* for the given variable; higher values, in combination with the degrees of freedom, indicate significance.
SIG	The significance of the **WALD** test.
EXP(B)	e^B, used to help in interpreting the meaning of the regression coefficients (as you may remember from the introduction to this chapter, the regression equation may be interpreted in terms of *B* or e^B).

Correlation Matrix

		Constant	EFFICT	SYMPATHT
Step 1	Constant	1.000	-.986	
	EFFICT	-.986	1.000	
Step 2	Constant	1.000	-.825	-.619
	SYMPATHT	-.619	.085	1.000
	EFFICT	-.825	1.000	.085

This is the correlation matrix for all variables in the regression equation, and it is only printed if you request **Correlations of estimates.** It is useful because if some variables are highly correlated, then the regression may have multicollinearity and be unstable.

Model if Term Removed

Variable		Model Log Likelihood	Change in -2 Log Likelihood	df	Sig. of the Change
Step 1	EFFICT	-372.174	114.843	1	.000
Step 2	SYMPATHT	-314.753	29.792	1	.000
	EFFICT	-350.086	100.458	1	.000

All variables in the model are tested here to see if they should be removed from the model. If the significance of the change in the –2-Log-Likelihood-for-the-model-if-a-variable-is-dropped is larger than the value entered in the **Probability for Stepwise Removal** in the Options window (Screen 25.3), then that variable will be dropped. In this example, since all of the Significance of the Change values are less than .10 (the default), no variables are removed.

Variables not in the Equation

			Score	df	Sig.
Step 2	Variables	ANGERT	.640	1	.424
		ETHNIC	4.892	4	.299
		ETHNIC(1)	.464	1	.496
		ETHNIC(2)	2.304	1	.129
		ETHNIC(3)	.357	1	.550
		ETHNIC(4)	2.169	1	.141
	Overall Statistics		5.404	5	.369

All variables that are not entered into the equation that could possibly be entered are listed here. The **Sig** indicates for each variable whether it has a significant impact on the predicted variable, independently from the other predictor variables. Here, **angert** does not have an impact on **cathelp**. Note here that **ethnic** is divided into four variables, indicating the four different contrasts that SPSS is performing. Because no variables can be deleted or added, the logistic regression equation is now complete. SPSS will not try to add or delete more variables from the equation.

Observed Groups and Predicted Probabilities

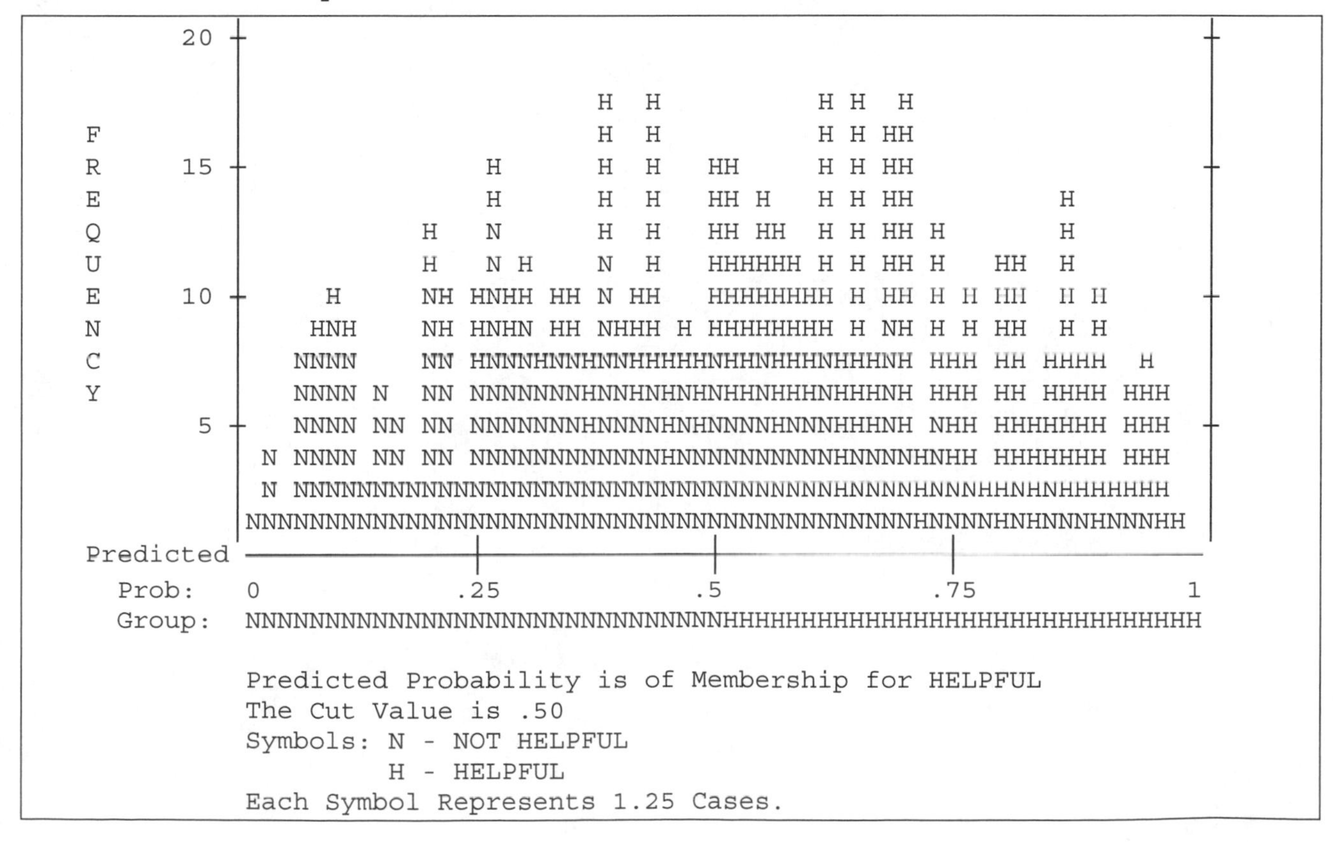

This graph is produced if you request **Classification plots** on Screen 25.3. First note that this graph will simply use the first letter of the value label for coding. With labels of *helpful* and *not helpful*, the graph is composed of Ns and Hs. These codes are plotted along the *x*-axis based on its predicted probability from the regression equation. If the logistic regression equation were perfect, all of the N's would be to the left of all of the H's.

Hierarchical LOGLINEAR MODELS

BOTH THIS chapter and the chapter that follows focus on log-linear models: the present chapter on *hierarchical* models, and the next (Chapter 27) on *non*hierarchical models. The first part of this introduction will discuss log-linear models in general and is applicable (and, indeed, necessary) for both this chapter and the next.

Log-Linear Models

Log-linear models allow the analysis of chi-square-type data using regression-like models, and in many ways they appear to be similar to ANOVAs. As in chi-square analysis, log-linear models deal with frequency tables in which several categorical variables categorize the data. If only two categorical variables are in the data for analysis, then chi-square analyses are indeed the simplest analysis available. For example, in Chapter 8 on chi-square analysis, the relationship between gender and ethnicity was examined in the **grades.sav** file. If, however, you wish to analyze several different categorical variables together, then it quickly becomes difficult or impossible to interpret chi-square tables visually. For example, a chi-square table and analysis of **gender** by **ethnicity** by **year** (in school) by **review** (whether or not the student attended the review session) is virtually impossible to interpret. It is for this purpose that log-linear models were developed.

In this chapter, the example will be drawn from the same **helping3.sav** file ($N = 537$) used in the previous chapter. In particular, hierarchical log-linear models will be tested for **gender** × **ethnicity** × **income** level × **cathelp** (whether or not the person receiving help thought the help was useful or not). Note that in this example, **income** is a categorical variable, indicating whether subjects earned less than $15,000/year, between $15,000 and $25,000, between $25,000 and $50,000, or greater than $50,000/year.

Log-linear models are essentially multiple linear regression models in which the classification variables (and their interaction terms) are the independent (predictor) variables, and the dependent variable is the natural logarithm of the frequency of cases in a cell of the frequency table. Using the natural log (*ln*) of the frequencies produces a linear model. A log-linear model for the effects of gender, ethnicity, income level, and their interactions on a particular cell in the crosstabulation table, might be represented by the following equation:

$$ln(\text{FREQUENCY}) = \mu + \lambda^{G} + \lambda^{E} + \lambda^{I} + \lambda^{G\times E} + \lambda^{G\times I} + \lambda^{E\times I} + \lambda^{G\times E\times I}$$

There will be different values for each of these variables for each of the cells in the model. In this equation, *Frequency* represents the frequency present within a particular cell of the data. μ represents the overall grand mean of the effect; it is equivalent to the constant in multiple regression analysis. Each of the λs (lambdas) represents the effect of one or more independent variables. λ^{G} represents the main effect (here's where log-linear models start sounding similar to ANOVAs) of **gender,** λ^{E} represents the main effect (also known as first-order effect) of **ethnicity**, and λ^{I} represents the main effect of **income** level on the frequency. λ values with multiple superscripts are interaction terms; for example, $\lambda^{G\times E}$ represents the two-way (second-order) interactive effect of **gender** and **ethnicity** on frequency; and $\lambda^{G\times E\times I}$ represents the three-way (third-order) interactive effect of **gender, ethnicity**, and **income** on the frequency. The model presented here is a *saturated* model because it contains *all* possible main effects and interaction terms. Because it is a saturated model, it can perfectly reproduce the data; however, it is not parsimonious and usually not the most desirable model.

The purpose of SPSS's **Model selection** option of the **Loglinear** procedures is to assist you in choosing an unsaturated log-linear model that will fit your data, as well as to calculate parameters of the log-linear model (the μ's and λ's).

The Model Selection Log-Linear Procedure

Although any of the terms may be deleted from the saturated model in order to produce a simpler, more parsimonious model, many researchers explore *hierarchical* log-linear models. In hierarchical models, in order for an effect of a certain order to be present, all effects of a lower order must be present. In other words, in a hierarchical model, in order for a two-way interaction (second-order effect) of **gender** and **ethnicity** to be present, there must also be main effects (first-order effects) of both **gender** and **ethnicity**. Likewise, in order for a third-order interactive effect of **gender**, **ethnicity**, and **income** level to be present, second-order interactive effects of **gender** by **ethnicity**, **gender** by **income** level, and **ethnicity** by **income** level must be present.

There are three primary techniques that SPSS provides in assisting with model selection. All three techniques are useful and will usually yield similar or identical results. Ultimately, however, the choice of which model or models you will use has to rely on both the statistical results provided by SPSS *and* your understanding of the data and what the data mean. The three techniques are summarized here, with a more detailed example provided in the Output section:

- **Examine parameter estimates**: One technique used in developing a model is to calculate parameter estimates for the saturated model. SPSS provides, along with these parameter estimates, *standardized* parameter estimates. If these standardized parameter estimates are small, then they probably do not contribute very much to the model and might be considered for removal.
- **Partitioning the chi-square statistic**: SPSS can, in addition to providing parameter estimates for the model, calculate a chi-square value that indicates how well the model fits the data. This chi-square value may also be subdivided and may be useful in selecting a model. SPSS can test whether all one-way and higher effects are nonsignificant, whether all two-way and higher effects are nonsignificant, whether all three-way and higher effects are nonsignificant, and so on. The program can also test that all one-way effects are zero, all two-way effects are zero, and so on. These tests examine a combination of all first-order effects, second-order effects, and so forth. However, just because the second-order effects *overall* may not be significant doesn't mean that *none* of the individual second-order effects are significant. Similarly, just because the second-order effects are significant overall doesn't mean that *all* of the second-order effects are significant. Because of this, SPSS can also examine partial chi-square values for individual main effects and interactive effects.
- **Backward elimination**: Another way to select a model is to use backward elimination; this is very similar to backwards elimination in multiple regression analysis. In backward elimination, SPSS starts with a saturated model and removes effects that are not contributing to the model significantly. This model-building technique is subject to the constraints of hierarchical log-linear modeling; third-order effects are not examined as can-

didates for exclusion if fourth-order effects are present, since if they were removed the assumptions of hierarchical models would be violated. The model is considered to fit best when all remaining effects contribute significantly to the model's fit.

We acknowledge that hierarchical log-linear models are complex and direct you to three other sources for a more complete picture than has been presented here: The SPSS manuals do a fairly thorough job of description and have the advantage of explaining the procedure within the context of SPSS documentation. Two other textbooks are quite good: Agresti (1990) and Wickens (1989); please see the reference section for a more complete description of these sources.

STEP BY STEP

Model Selection (Hierarchical) Log-Linear Models

To enter SPSS, a click on **Start** *in the taskbar (bottom of screen) activates the start menu:*

After clicking the SPSS program icon, Screen 1 (inside front cover) appears on the monitor.

	Step 2
Create and name a data file or edit (if necessary) an already existing file (see Chapter 3).	

Screens 1 and 2 (inside front cover) allow you to access the data file used in conducting the analysis of interest. The following sequence accesses the **helping3.sav** *file for further analyses:*

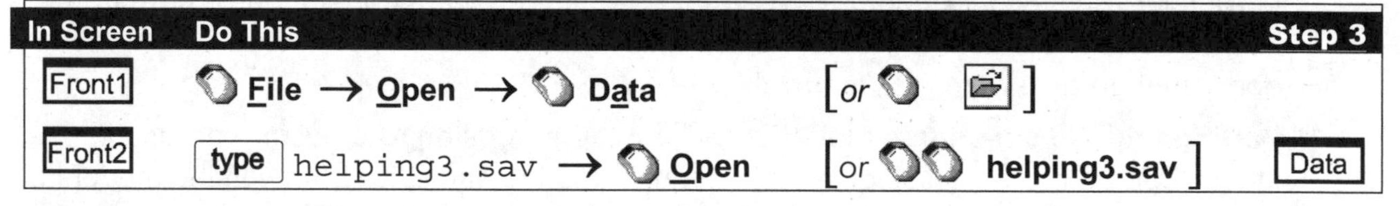

Whether first entering SPSS or returning from earlier operations the standard menu of commands across the top is required (shown below). As long as it is visible you may perform any analyses. It is not necessary for the data window to be visible.

This menu of commands disappears or modifies when using pivot tables or editing graphs. To uncover the standard menu of commands simply click on the or the icon.

After completion of Step 3 a screen with the desired menu bar appears. When you click a command (from the menu bar), a series of options will appear (usually) below the selected command. With each new set of options, click the desired item. The sequence to access hierarchical log-linear models begins at any screen with the menu of commands visible:

After this sequence step, a window opens (Screen 26.1, below) in which you specify the factors to include in the analysis and select the model building procedure. In the box at the left of Screen 26.1 is the list of variables in your data file. Any variables that will be included in the analysis should be moved to the **Factor(s)** box to the right of the variable list, by clicking on the upper ▶. Because all variables used in the analysis will be categorical, it is important to click on **Define Range** for each variable. SPSS will prompt you (Screen 26.2, below) to enter the **Minimum** and **Maximum** values of the variable. For example, in most cases gender variables will have a range of 1 through 2.

26.1

Model Selection (Hierarchical) Log-linear Analysis Dialog Window

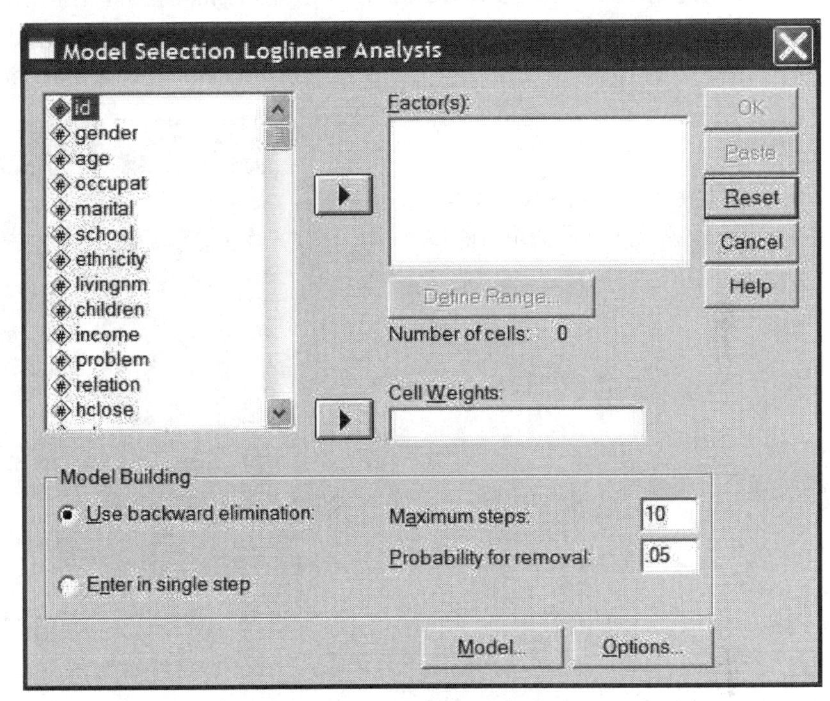

Cell Weights are used if your model contains structural zeros. Structural zeros are beyond the scope of this book; please see the SPSS manuals for details.

26.2

Log-linear Analysis Define Range Dialog Window

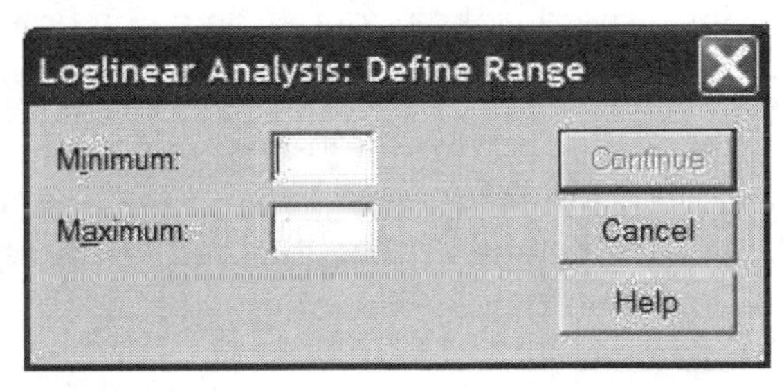

The **Use backward elimination** selection in the **Model Building** box at the bottom of Screen 26.1 instructs SPSS to start with a saturated model and remove terms that do not significantly contribute to the model, through a process of backward elimination. Use the **Enter in single step** option if you do not wish to eliminate terms from the model, but test the model containing all of the terms. Assuming you do want to use backward elimination, the **Probability for removal** provides a *p* value that SPSS will use as a criterion for dropping effects from the model. SPSS drops effects that are not significant at the level of the designated *p* value. The default is $p=.05$, so you do not need to enter a value here unless you wish to specify a different *p* value. Each time that SPSS removes an effect from the model, this is called a *step*. **Maximum steps**

allows you to set the maximum number of steps that SPSS will perform, to something other than the default of 10. You may need to increase the maximum number of steps if you are testing a very large model.

Once the factors for analysis are defined, and the model building procedure is identified, the **Options** and **Model** may be specified. When the **Options** button is clicked, Screen 26.3 appears. This window allows you to select which output you desire, as well as specifying some technical aspects of the way model selection will proceed (these technical aspects are not covered in this book). **Frequencies** produces a listing of the number of subjects in each cell of the model. **Residuals** displays the difference between the actual number of subjects within each cell, and the number of subjects predicted to be in that cell based on the model.

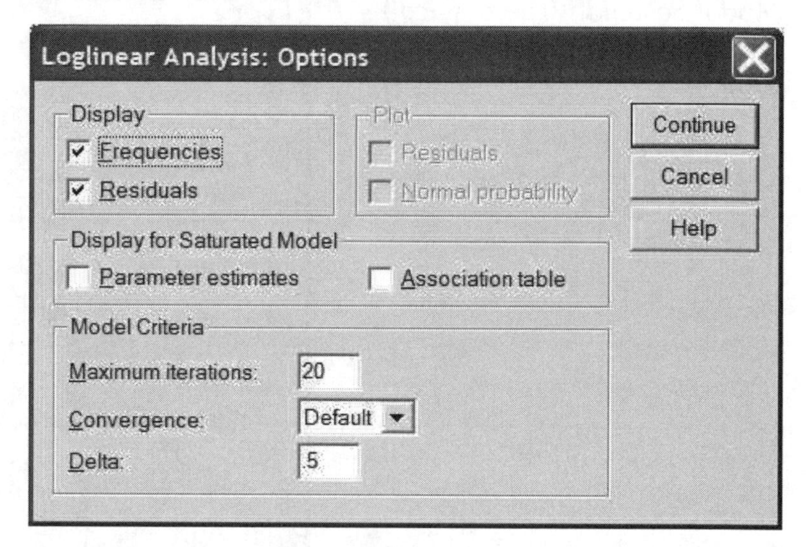

26.3

Model Selection (Heirarchical) Log-linear Analysis Options Dialog Window

Other frequently used options include **Parameter estimates** and an **Association table**. For a saturated model, these options will produce parameter estimates (which may be useful in model selection; see the introduction to this chapter) and an association table that partitions the chi-square values for each of the main and interactive effects (again useful in model selection). These options are strongly recommended when model selection with a saturated model is desired.

After any options desired are selected (Screen 26.3), one more button on Screen 26.1 may be needed: **Model**. If you wish to test a saturated model (in which all effects and interaction effects are included), then this step is not necessary. The **Model** button should only be used when you wish to use a non-saturated model, using only some of the first-order effects and interactive effects.

The **Model** button calls up Screen 26.4 (shown on the following page with a custom model already specified). After **Custom** is selected, **Factors** may be selected by clicking on them, in the list at the left of the window. Once they have been selected, the type of effects (both main effects and interactions) may be chosen from the list underneath the ▶. **Factors** may be selected as **Main effects**, **Interactions**, or **All 2-way**, **All 3-way**, **All 4-way**, or **All 5-way**. In Screen 26.4, we have selected **All 3-way** interactions. Because this is a hierarchical model, all 3-way interactions also assume that all 2-way interactions and main effects are also present. **Custom** models are generally not used unless the researcher has strong theoretical reasons for choosing to use a non-saturated model, and/or the saturated model would contain higher-order interactions that would be too difficult to interpret.

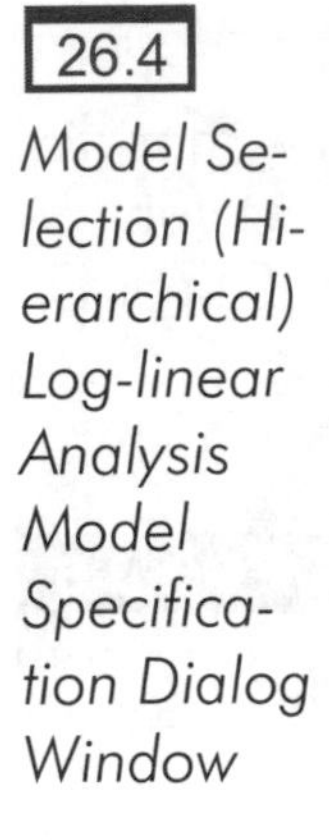

26.4

Model Selection (Hierarchical) Log-linear Analysis Model Specification Dialog Window

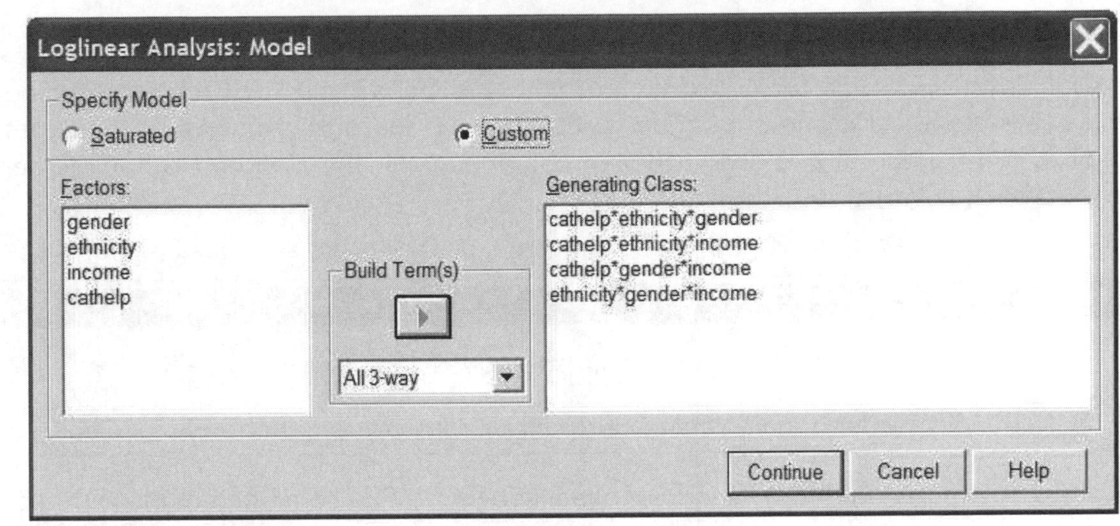

The following sequence instructs SPSS to perform a hierarchical log-linear analysis of **gender** *(2 levels),* **ethnicity** *(5 levels),* **income** *(4 levels), and* **cathelp** *(2 levels, whether or not the individuals helped felt that they had benefited). The saturated model is tested, and* **Parameter estimates** *will be produced, along with partitioning the chi-square into individual effects. Finally,* **backward elimination** *will be used to build a model containing only effects that significantly contribute to the model. The step begins with the initial dialog box, Screen 26.1*

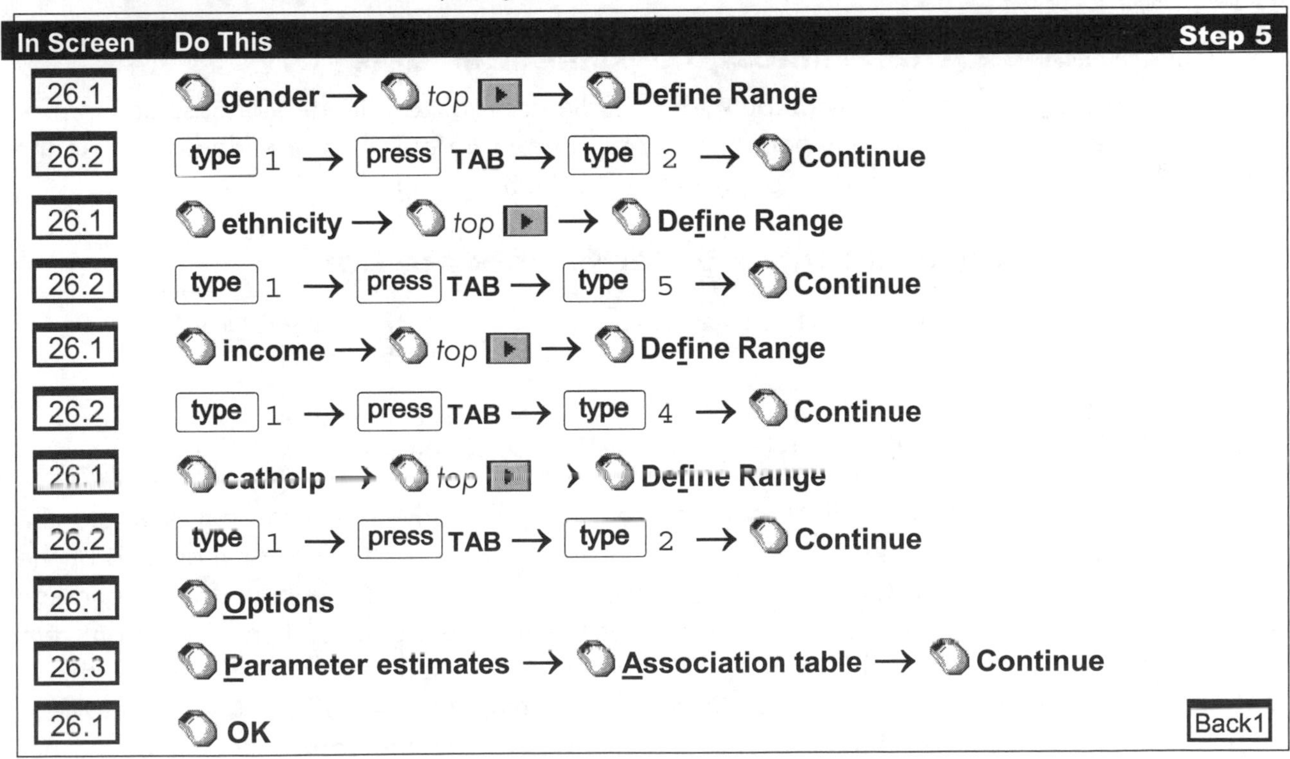

In Screen	Do This	Step 5
26.1	**gender** → *top* ▶ → **Define Range**	
26.2	type 1 → press **TAB** → type 2 → **Continue**	
26.1	**ethnicity** → *top* ▶ → **Define Range**	
26.2	type 1 → press **TAB** → type 5 → **Continue**	
26.1	**income** → *top* ▶ → **Define Range**	
26.2	type 1 → press **TAB** → type 4 → **Continue**	
26.1	**cathelp** → *top* ▶ → **Define Range**	
26.2	type 1 → press **TAB** → type 2 → **Continue**	
26.1	**Options**	
26.3	**Parameter estimates** → **Association table** → **Continue**	
26.1	**OK**	Back1

Upon completion of step 5 the output screen will appear (Screen 1, inside back cover). All results from the just-completed analysis are included in the Output Navigator. Make use of the scroll bar arrows (▲ ▼ ▶ ◀) to view the results. Even when viewing output, the standard menu of commands is still listed across the top of the window. Further analyses may be conducted without returning to the data screen.

PRINTING RESULTS

Results of the analysis (or analyses) that have just been conducted requires a window that displays the standard commands (**File Edit Data Transform Analyze** . . .) across the top. A typical print procedure is shown below beginning with the standard output screen (Screen 1, inside back cover).

To print results, from the Output screen perform the following sequence of steps:

In Screen	Do This	Step 6
Back1	*Select desired output or edit (see pages 19 - 25)* → **File** → **Print**	
2.9	*Consider and select desired print options offered, then* → **OK**	

To exit you may begin from any screen that shows the **File** *command at the top.*

In Screen	Do This	Step 7
Menu	**File** → **Exit**	2.1

Note: After clicking **Exit**, there will frequently be small windows that appear asking if you wish to save or change anything. Simply click each appropriate response.

OUTPUT

Model Selection (Hierarchical) Log-Linear Models

Below is an explanation of the key output from the Hierarchical Log-linear Model selection procedure (sequence step 5, previous page). Only representative output is provided, because the actual printout from this procedure is approximately 20 pages long.

Tests That K-Way and Higher Order Effects are Zero

K	DF	L.R. Chisq	Prob	Pearson Chisq	Prob	Iteration
4	12	13.408	.3401	12.330	.4195	4
3	43	62.361	.0283	59.044	.0524	3
2	70	112.633	.0009	102.374	.0070	2
1	79	428.287	.0000	618.095	.0000	0

These tests examine whether all effects at a certain order and above are zero. For example, the first line in this table tests whether all fourth-order effects are equal to zero. As indicated by the small chi-square values and the large *p*-values, there is no fourth-order effect. The second line ($K = 3$) indicates that the third- and fourth-order effects may be significantly different from zero (note *p* values of .028 and .052), and the third line ($K = 2$) indicates that the second-, third-, and fourth-order effects *are* significantly different from zero (note $p = .007$). The last line ($K = 1$) suggests that, in the first- through fourth-order effects, there are some effects that are not equal to zero.

Term	Definition/Description
K	The order of effects for each row of the table (4 = fourth-order and higher effects, 3 = third-order and higher effects, etc.).
DF	The degrees of freedom for *K*th and higher-order effects.

Term	Definition/Description
L.R. CHISQ	The likelihood-ratio chi-square value testing that *K*th and higher-order effects are zero.
PEARSON CHISQ	The Pearson chi-square value testing that *K*th and higher-order effects are zero.
PROB	The probability that *K*-order and higher effects are equal to zero; small *p*-values suggest that one or more of the effects of order *K* and higher are not equal to zero.
ITER	The number of iterations that SPSS took to estimate the chi-square values.

Tests That K-Way Effects are Zero

K	DF	L.R. Chisq	Prob	Pearson Chisq	Prob	Iteration
1	9	315.653	.0000	515.721	.0000	0
2	27	50.272	.0042	43.331	.0242	0
3	31	48.953	.0213	46.714	.0348	0
4	12	13.408	.3401	12.330	.4195	0

Here SPSS examines whether or not effects of each *particular* order in the design are equal to zero. Significant *p*-values suggest that the effects of a particular level are not equal to zero; in this case, the one-, two-, and three-way effects are not equal to zero.

Term	Definition/Description
K	The order of effects for each row of the table (1 = first-order effects, 2 = second-order effects, etc.).
DF	The degrees of freedom for *K*th-order effects.
L.R. CHISQ	The likelihood-ratio chi-square value testing that *K*th-order effects are zero.
PEARSON CHISQ	The Pearson chi-square value testing that *K*th-order effects are zero.
PROB	The probability that *K*-order effects are equal to zero; small *p*-values suggest that one or more of the effects of order *K* are not equal to zero.
ITER	For this table, this column will always be zero.

Tests of Partial Associations

Effect Name	DF	Partial Chisq	Prob	Iter
gender*ethnicity*income	12	17.673	.1260	4
gender*ethnicity*cathelp	4	7.939	.0938	3
gender*income*cathelp	3	12.470	.0059	4
ethnicity*income*cathelp	12	14.680	.2594	4
gender*ethnicity	4	3.214	.5227	3
gender*income	3	1.605	.6583	3
ethnicity*income	12	32.394	.0012	3
gender*cathelp	1	4.169	.0412	3
ethnicity*cathelp	4	5.399	.2488	3
income*cathelp	3	4.918	.1779	3
gender	1	10.248	.0014	2
ethnicity	4	236.046	.0000	2
income	3	66.886	.0000	2
cathelp	1	2.473	.1158	2

This table (previous page) examines partial chi-square values for each effect in the saturated model. Each partial chi-square examines the unique contribution of that effect to the model; those with low *p* values contribute to the model significantly. In this case, there are main effects of **gender**, **ethnicity**, and **income**, two-way interactions between **gender** by **cathelp** and **ethnicity** by **income**, as well as a three-way interaction of **gender** by **income** by **cathelp**. Because these partial chi-squares are not necessarily independent, their partial associations (displayed here) as a portion of the total chi-square for the saturated model may not be equivalent to the chi-square in nonsaturated models.

Term	Definition/Description
DF	These degrees of freedom refer to the particular effects listed in each line.
PARTIAL CHI-SQUARE	The partial chi-square for each effect.
PROB	The probability that the effect is equal to zero; small probabilities indicate that the given effect has a large contribution to the model.
ITER	The number of iterations that the computer took to calculate the partial association for each effect. This procedure takes quite a while; the number of iterations is there to remind you how long it took to compute each partial association.

Estimates for Parameters

```
                .
                .
ethnicity*income
Parameter         Coeff.     Std. Err.      Z-Value  Lower 95 CI  Upper 95 CI
        1  -.1094269287        .19057      -.57421      -.48294       .26409
        2  -.1701748589        .22173      -.76749      -.60476       .26441
        3  -.1126158596        .15926      -.70710      -.42478       .19954
        4  -.6824807624        .36984     -1.84535     -1.40736       .04240
        5   .7257495472        .26866      2.70140       .19918      1.25232
        6  -.0233976743        .25276      -.09257      -.51881       .47201
        7  -.0966040450        .24469      -.39480      -.57619       .38299
        8   .2256749092        .24800       .90998      -.26040       .71175
        9   .1096424852        .20073       .54621      -.28379       .50308
       10   .7330105611        .23621      3.10327       .27005      1.19597
       11  -.4831889921        .37115     -1.30189     -1.21063       .24426
       12  -.4276172574        .26355     -1.62252      -.94418       .08894
                .
                .
 income
Parameter         Coeff.     Std. Err.      Z-Value  Lower 95 CI  Upper 95 CI
        1  -.1832940387        .14238     -1.28734      -.46236       .09577
        2  -.4908628679        .15797     -3.10726      -.80049      -.18124
        3   .2435742286        .11668      2.08756       .01488       .47226
```

Parameter estimates are provided for each main effect and interaction. These are the λ's from the log-linear equation presented in the introduction of this chapter. In this example, only one interaction (**ethnicity × income**) and one main effect (**income**) are presented. Because the parameters are constrained to sum to zero, only some of the parameters appear here. In this example, the **ethnicity** × **income** parameters may be interpreted as shown in the following table, where parameters in **bold** print are calculated by SPSS and those in *italics* are calculated by summing each row and column in the table to zero.

Income	Ethnicity					SUM
	Caucasian	Black	Hispanic	Asian	Other	
<15,000	**-.109**	**-.682**	**-.097**	**.733**	*.155*	0
<25,000	**-.170**	**.726**	**.226**	**-.483**	*-.299*	0
<50,000	**-.112**	**-.023**	**.110**	**-.428**	*.453*	0
>50,000	*.391*	*-.021*	*-.239*	*.178*	*-.309*	0
SUM	0	0	0	0	0	0

These parameters suggest that Whites with higher incomes are far more frequent in this sample than those with lower incomes, and that Blacks, Hispanics, and Asians are more common with lower incomes. Additional detail may be observed: Notice that the Asians, although well represented in the lowest income level are also quite prominent in the highest income level. The parameters for **income** may be interpreted in the same way as the **ethnicity** × **income** interaction:

Income

<15,000	<25,000	<50,000	>50,000	SUM
-.183	-.491	.244	*.430*	0

These parameters suggest that there are few individuals with less than $25,000/year income, and more with greater than $50,000/year income. In this case, the fact that the *z* values for most of these parameters are large (greater than 1.96 or less than −1.96) imply that the parameters are significant (at $p < .05$) and support this interpretation of the data.

Term	Definition/Description
PARAMETER	The number of the parameter. Parameters go from low codes to high codes of the independent variables. See the above **gender** × **ethnicity** table for an example of the way interactions are coded.
COEFFICIENT	The λ value in the log-linear model equation.
STANDARD ERROR	A measure of the dispersion of the coefficient.
Z-VALUE	A standardized measure of the parameter coefficient; large *z* values (those whose absolute value is greater than 1.96) are significant ($\alpha = .05$).
LOWER AND UPPER 95% CONFIDENCE INTERVAL	There is a 95% chance that the actual (rather than estimated) coefficient is between the lower and upper confidence interval.

Backward Elimination

In the output on the following page, the chi-square values, probability values, and interactions involved in backward elimination were already explained and are not different here from the previous printouts in which the chi-square values were partitioned. Backward elimination begins with, in this case, the saturated model. SPSS calculates a likelihood ratio chi-square for the model including the fourth-order and all first-, second-, and third-order effects; this chi-square is 0, because it is a saturated model. SPSS calculates that if the 4-way effect was deleted, the chi-square change would not be significant. This suggests that the 4-way interactive effect does not contribute significantly to the model, and it is eliminated.

```
Backward Elimination (p = .050) for DESIGN 1 with generating class

gender*ethnicity*income*cathelp
.
.
.

If Deleted Simple Effect is              DF   L.R. Chisq Change    Prob  Iter
 gender*ethnicity*income*cathelp         12        13.408         .3401     4

Step 1
  The best model has generating class
      gender*ethnicity*income
      gender*ethnicity*cathelp
      gender*income*cathelp
      ethnicity*income*cathelp

  Likelihood ratio chi square =    13.40805    DF = 12   P =  .340
.
.
.

Step 6
  The best model has generating class
      gender*income*cathelp
      ethnicity*income
                  (output continues)
  Likelihood ratio chi square =    59.37980    DF = 48   P =  .126

If Deleted Simple Effect is           DF   L.R. Chisq Change    Prob  Iter
 gender*income*cathelp                 3        11.809         .0081     3
 ethnicity*income                     12        31.567         .0016     2

Step 7
  The best model has generating class
      gender*income*cathelp
      ethnicity*income

  Likelihood ratio chi square =    59.37980    DF = 48   P =  .126

The final model has generating class
    gender*income*cathelp
    ethnicity*income
```

In Step 1, the model with all 3-way interactions is tested, and all effects are tested to see whether or not they significantly contribute to the model. This process continues (through 7 steps in this example); lower-order effects are evaluated and considered for elimination only if they may be removed from the model without violating the hierarchical assumptions. In this case, the final model is composed of **gender × income × cathelp** and **ethnicity × income** interactions. The model has a chi-square of 59.38 with 48 degrees of freedom; because this is not significant, the model does appear to fit the data well.

Observed, Expected Frequencies and Residuals

For output on the following page, observed and expected frequencies are presented for each cell in the design. Note that in order to avoid cells with zero frequencies, SPSS adds .5 to each cell frequency when saturated models are examined. Here, expected frequencies for the final model are chosen through backward elimination. Large residuals indicate possible problems with the model. Because the chi-square values are not large and the *p*-values are not small, the model does seem to fit the data well.

```
    Factor        Code        OBS count  EXP count  Residual   Std Resid
gender        FEMALE
   ethnicity        CAUCASIAN
    income       <15,000
     cathelp       NOT HELPFUL       4.0        5.8      -1.75         -.73
     cathelp       HELPFUL          14.0       13.9        .10          .03
    income       <25,000
     cathelp       NOT HELPFUL       7.0        5.4       1.61          .69
     cathelp       HELPFUL           9.0        7.8       1.16          .41
    income       <50,000
     cathelp       NOT HELPFUL      11.0       13.7      -2.68         -.72
     cathelp       HELPFUL          16.0       17.9      -1.92         -.45
    income       >50,000
     cathelp       NOT HELPFUL      20.0       26.7      -6.68        -1.29
     cathelp       HELPFUL          34.0       31.1       2.87          .52
           .
           .
           .
   ethnicity        BLACK
    income       <15,000
     cathelp       NOT HELPFUL       1.0         .5        .51          .72
     cathelp       HELPFUL           2.0        1.2        .81          .74
           .
           .
           .
 Goodness-of-fit test statistics:
    Likelihood ratio chi square =       59.37980       DF = 0   P = .126
             Pearson chi square =       54.59835       DF = 0   P = .238
```

Term	Definition/Description
OBSERVED COUNT	The observed cell count derives from the data and indicates the number of cases in each cell.
EXPECTED COUNT	The expected cell counts indicate the expected frequencies for the cells, based on the model being tested.
RESIDUAL AND STANDARDIZED RESIDUAL: OBSERVED-EXPECTED	Cells with high values here are those that are not well predicted by the model.
GOODNESS-OF-FIT TEST STATISTICS	The likelihood ratio chi-square and the Pearson chi-square statistics examine the fit of the model. Large chi-square values and small *p*-values indicate that the model does not fit the data well; in this case, since $p > .05$, the data do fit the model adequately.

27

Nonhierarchical LOGLINEAR MODELS

LOG-LINEAR modeling, using the SPSS General Loglinear procedure, is very flexible. Because of its great flexibility, it is difficult to provide a simple step-by-step approach to these operations. Using the General Loglinear procedure often involves executing the procedure multiple times, testing various models in an attempt to discover the best fit. Because of these subtleties in using log-linear models, this chapter will attempt to (a) provide a general introduction to log-linear modeling for those who need a review of the basic ideas involved, and (b) provide step-by-step computer instructions for running the most common types of models. For more complex models, beyond the scope of this book, we refer you to the *SPSS* manuals for details.

If you are not familiar with the basic concepts of log-linear models, and if you have not read Chapter 26, then be sure to read the first portion of that chapter before continuing with this chapter. The general introduction there applies to creating both hierarchical log-linear models (Chapter 26) and general, nonhierarchical log-linear models (this chapter).

Nonhierarchical versus Hierarchical Log-Linear Models

In hierarchical log-linear models (using procedures described in the previous chapter), if an effect of a given order is present, then all effects of lesser order must also be present. This constraint is unique to *hierarchical* log-linear models. In the General Loglinear procedure of SPSS, this is not a necessary constraint: You may include any main or interactive effect, or omit an effect if you don't want the effect included. Furthermore, whereas the **Loglinear → Model Selection** procedure can calculate only parameter estimates for saturated models, the general loglinear procedure can calculate parameter estimates whether or not the model is saturated.

As in the previous two chapters, we use the **helping3.sav** file ($N = 537$) in this chapter. A model is tested including the main effects of **gender**, **ethnicity**, and **cathelp** (whether or not the help given was useful or not), along with the two-way interactive effects of **gender** by **ethnicity** and **gender** by **cathelp**. One thing to remember: Testing a single model, as is done in the example in this chapter, is usually only one step among many in the quest for the best possible model.

Using Covariates with Log-Linear Models

It is possible to use one or more variables as covariates in the model. Because these covariates are not categorical variables, it is possible to use covariates to test for particular types of trends within categorical variables. For example, in the **helping3.sav** file, **income** seems to be related to the *number of people* in each income category: There are 73 people with less than \$15,000/year income, 51 people between \$15,000 and \$25,000, 106 people between \$25,000 and \$50,000, and 159 people with greater than \$50,000 per year. In the Step by Step section of this chapter, we discuss a model that examines **income** and **income**^{3} (**income** cubed) to see if a model that includes these two factors predicts well the number of subjects from each income group (that is, produces a model that fits the data well).

Logit Models

SPSS has the ability to work with logit models, through the **Loglinear → Logit** procedure. Conceptually, this modeling procedure is very similar to general log-linear modeling, with several exceptions. First of all, logit models allow dichotomous variables to be treated as dependent variables and one or more categorical variables to be treated as independent variables. Second, instead of predicting the frequency within a particular cell, a dichotomous dependent variable is designated and membership into one of two distinct categories (the logit—see Chapter 25 for a description) is predicted for each cell.

The execution of logit models in SPSS is very similar to the execution of general loglinear models. One difference is important: A dependent variable, always dichotomous, is specified. In the step-by-step section of this chapter, we discuss a model in which **gender** and a **gender** by **ethnicity** interaction (the two categorical independent variables) predict **cathelp** (the dichotomous dependent variable).

A Few Words about Model Selection

The procedures described in this chapter assume that you know what model you want to test. In practice, this is not an easy task. Model selection usually depends on a tight interplay of theory and testing of multiple models. If the different models tested have different degrees of freedom, then they may be compared using chi-square differences (using the tables in the back of your statistics textbooks you thought you'd never need again) in order to determine whether or not one model is significantly better than another model. The goal of model selection is to find a model with the best fit possible *and* as parsimonious as possible. Obviously, this process is too complex and involves too much artistry to fully describe in a step by step format. However, the model selection process is likely to *use* (if not consist of) the procedures described in the Step by Step section.

Types of Models Beyond the Scope of This Chapter

In addition to the complexities of model selection, this chapter does not describe several types of models used with the general log-linear procedure because their complexity makes it difficult to describe in fewer words than are used in the SPSS manuals. Procedures and techniques discussed in the SPSS manuals, but not in this chapter, include:

- equiprobability models,
- linear-by-linear association models,
- row- and column-effects models,
- cell weights specification for models,
- tables with structural zeros,
- linear combinations of cells, and
- using contrasts.

For additional information about log-linear models that extends beyond the SPSS manuals, we refer you to Agresti (1990) and Wickens (1989).

STEP BY STEP

General (Nonhierarchical) Log-Linear Models

To enter SPSS, a click on **Start** *in the taskbar (bottom of screen) activates the start menu:*

After clicking the SPSS program icon, Screen 1 (inside front cover) appears on the monitor.

Step 2
Create and name a data file or edit (if necessary) an already existing file (see Chapter 3).

Screens 1 and 2 (inside front cover) allow you to access the data file used in conducting the analysis of interest. The following sequence accesses the **helping3.sav** *file for further analyses:*

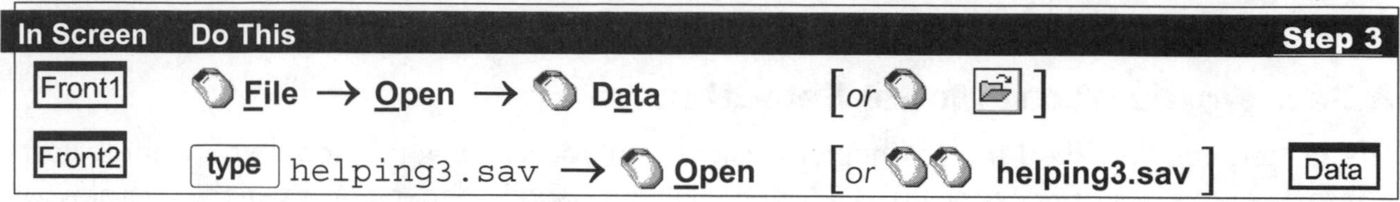

Whether first entering SPSS or returning from earlier operations the standard menu of commands across the top is required (shown below). As long as it is visible you may perform any analyses. It is not necessary for the data window to be visible.

This menu of commands disappears or modifies when using pivot tables or editing graphs. To uncover the standard menu of commands simply click on the ▬ or the ⧉ icon.

After completion of Step 3 a screen with the desired menu bar appears. When you click a command (from the menu bar), a series of options will appear (usually) below the selected command. With each new set of options, click the desired item. The sequence to access general linear models begins at any screen with the menu of commands visible:

After this sequence step, a window opens (Screen 27.1, next page) in which you specify the factors and covariates to include in the analysis. In the box at the left of Screen 27.1 is the list of variables in your data file. Any categorical variables that will be included in the analysis should be moved to the **Factor(s)** box to the right of the variable list, by clicking on the upper ▶. In the example in sequence step 5 (page 353), we will examine **gender**, **ethnicity**, and **cathelp**. Because these are all categorical variables, they will all be moved into the **Factor(s)** box.

If you wish to examine the effects of any non-categorical variables on the categorical factor(s), then these variables should be moved into the **Cell Covariate(s)** box by selecting the variable and clicking on the second ▶ from the top. For example, if you wished to test for the influence of linear and cubed effects of **income**, the procedure would be:

- Click **Transform**, click **Compute**, then compute two new variables, **income1** (equal to **income**) and **income3** (equal to **income**^{3}).
- Access Screen 27.1, and enter **income** in the **Factor(s)** box. Then, enter **income1** and **income3** in the **Cell Covariate(s)** box.
- Be sure that when you select a model (Screen 27.3, on page 352) you include the covariates in the model.

27.1

General Log-Linear Analysis Dialog Window

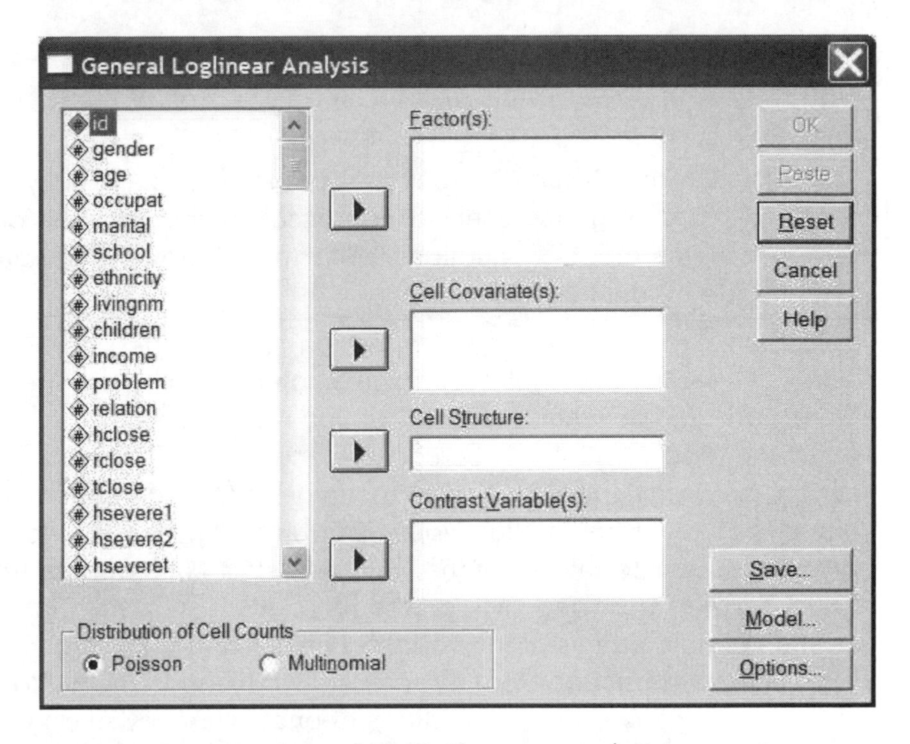

The **Cell Structure**, **Contrast Variables**, **Distribution of Cell Counts**, and **Save** options on Screen 27.1 are not frequently used and extend beyond the scope of this book. The **Model** and **Options** buttons will, however, be used quite often. We will begin our discussion with the **Options** button, used to specify desired output and criteria used in the analysis.

When the **Options** button is selected, Screen 27.2 (below) appears. The **Plot** options produce charts; these are useful if you are familiar with residuals analysis, or if you wish to examine the normal probabilities involved in the analysis. They will not be described in this chapter; for a discussion of residuals in general, see Chapter 28.

27.2

General Log-Linear Analysis Options Dialog Window

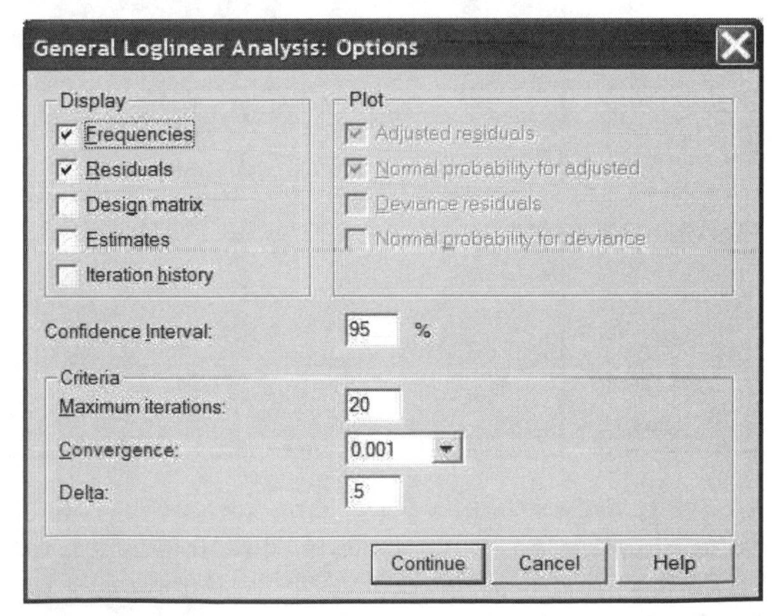

The **Display** options are frequently used. **Frequencies** outputs tables of observed and estimated frequencies for all cells in the model. **Residuals** produces actual and standardized residuals from these cells, and **Estimates** produces parameter estimates for the model. The **Design matrix** option is rarely desired; it produces information about the internal design matrix used by SPSS in producing the log-linear model.

Occasionally, the **Criteria** options may be needed, but usually these settings do not need to be changed. **Convergence** specifies the accuracy that must be attained in order to accept convergence of the model's equations. **Maximum iterations** specifies how long SPSS should keep iterating the model to achieve a fit.

By default, SPSS produces a saturated log-linear model, including all main effects and all interactions. This is often not desired; the saturated model perfectly predicts the data, but is no more parsimonious than the data. If you wish to specify a custom, non-saturated model, then select the **Model** button (on Screen 27.1) to activate Screen 27.3. When you click the **Custom** button, you may specify which terms (main effects and interactions) you want included in the model. For each main effect you wish to include, click on the variable name in the **Factors & Covariates** box to the left of the window, select **Main effects** from the drop-down menu in the **Build Term(s)** box, and click on the ▶ button. For each interaction you wish included, click on the variable names of the variables involved in the interaction (in the **Factors & Covariates** box), select **Interaction** from the drop-down menu in the **Build Term(s)** box, and click the ▶ button. To enter several interactions at once, you may select all variable you want in the interactions, and select **All 2-way**, **All 3-way**, **All 4-way**, or **All 5-way** interactions from the drop-down menu in the **Build Term(s)** box. Screen 27.3 shows the following custom model already specified: Main effects for **gender**, **ethnicity**, and **cathelp**; plus **gender × ethnicity** and **gender × cathelp** interactions.

27.3

General Log-Linear Model Specification Dialog Window

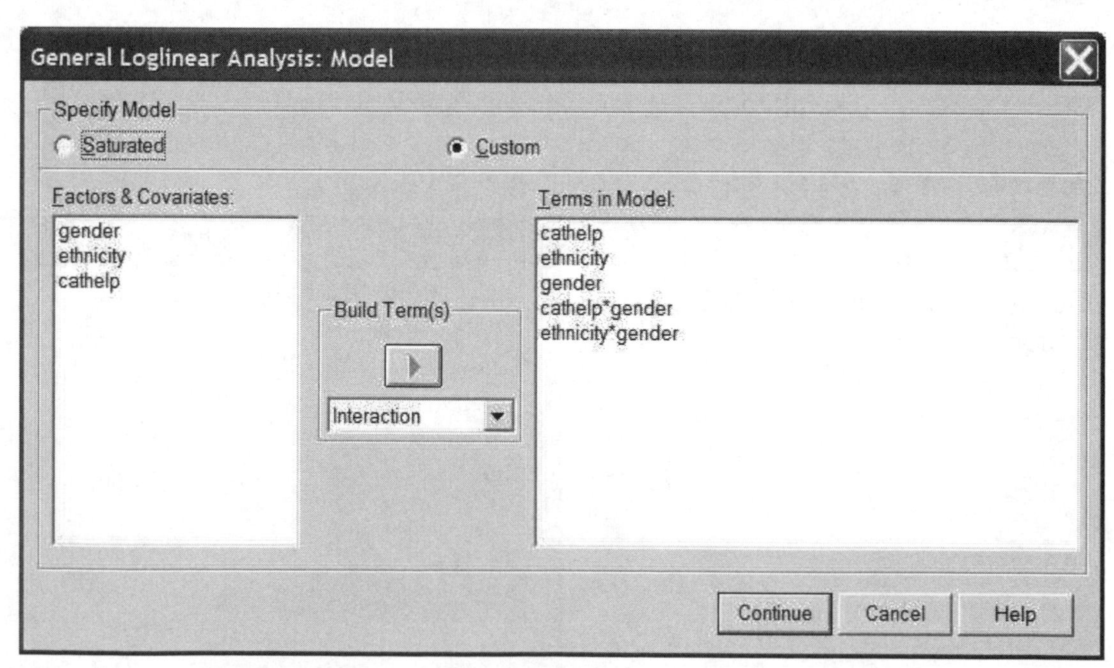

SPSS has the ability to test logit models, in which one or more dichotomous dependent variables are included in the analysis. This procedure is virtually identical to the General Loglinear procedure, and is selected through the **Analyze → Loglinear → Logit** sequence, instead of the **Analyze → Loglinear → General** sequence.

This sequence calls up Screen 27.4. This procedure is nearly identical to the General Loglinear procedure, and Screen 27.4 is nearly identical to Screen 27.1. The primary difference is that this window has an additional box, **Dependent**, in which one or more categorical dependent variables may be placed.

27.4

Logit Log-Linear Analysis Window

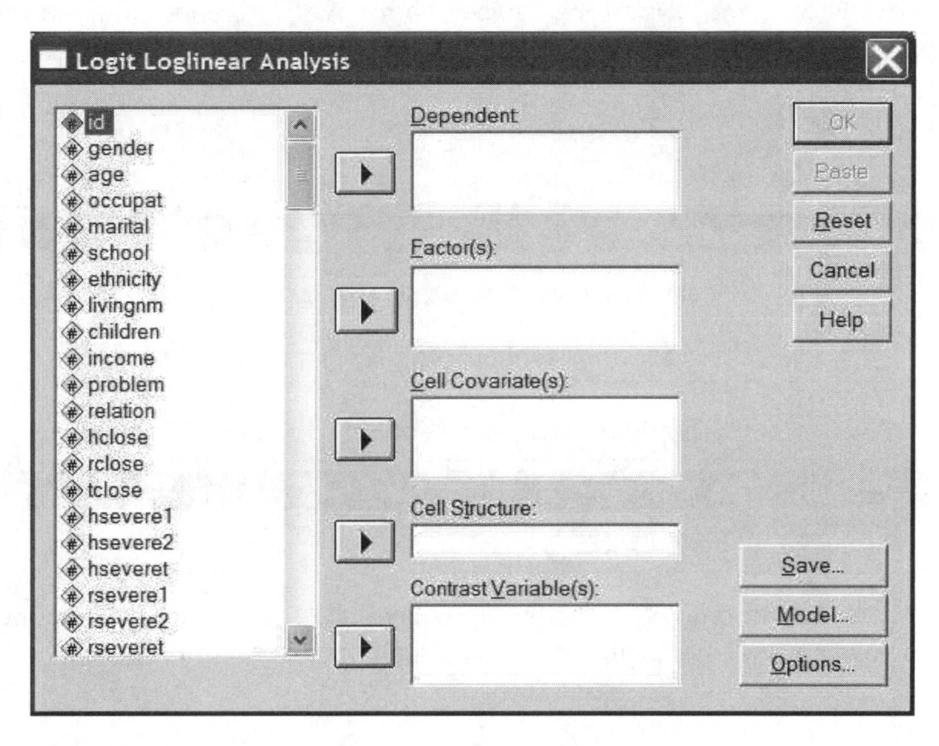

The following procedure instructs SPSS to test a log-linear model including the main effects of **gender**, **ethnicity**, *and* **cathelp**, *as well as interactive effects of* **gender** *by* **ethnicity** *and* **gender** *by* **cathelp**.

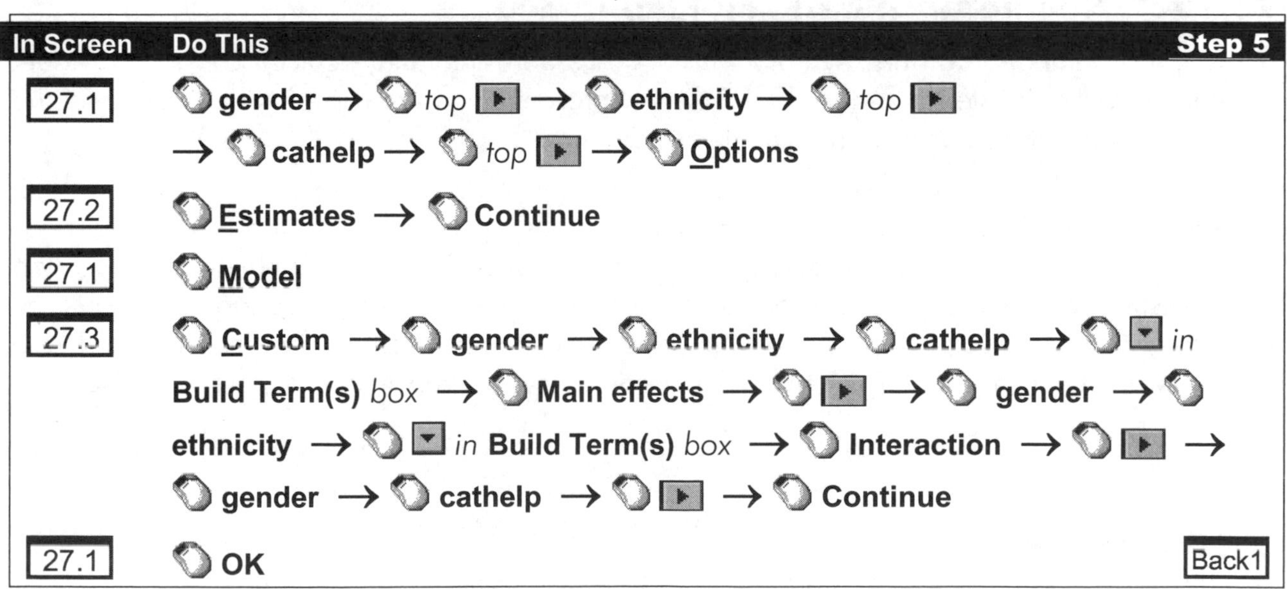

In Screen	Do This	Step 5
27.1	**gender** → *top* ▶ → **ethnicity** → *top* ▶ → **cathelp** → *top* ▶ → **Options**	
27.2	**Estimates** → **Continue**	
27.1	**Model**	
27.3	**Custom** → **gender** → **ethnicity** → **cathelp** → ▼ *in* **Build Term(s)** *box* → **Main effects** → ▶ → **gender** → **ethnicity** → ▼ *in* **Build Term(s)** *box* → **Interaction** → ▶ → **gender** → **cathelp** → ▶ → **Continue**	
27.1	**OK**	Back1

Upon completion of step 5 the output screen will appear (Screen 1, inside back cover). All results from the just-completed analysis are included in the Output Navigator. Make use of the scroll bar arrows (▲ ▼ ▶ ◀) to view the results. Even when viewing output, the standard menu of commands is still listed across the top of the window. Further analyses may be conducted without returning to the data screen.

PRINTING RESULTS

Results of the analysis (or analyses) that have just been conducted requires a window that displays the standard commands (**File Edit Data Transform Analyze** . . .) across the top. A typical print procedure is shown below beginning with the standard output screen (Screen 1, inside back cover).

To print results, from the Output screen perform the following sequence of steps:

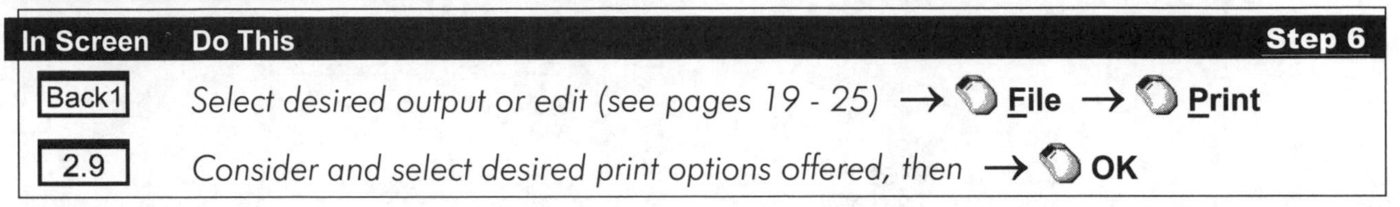

In Screen	Do This	Step 6
Back1	*Select desired output or edit (see pages 19 - 25)* → **File** → **Print**	
2.9	*Consider and select desired print options offered, then* → **OK**	

To exit you may begin from any screen that shows the **File** *command at the top.*

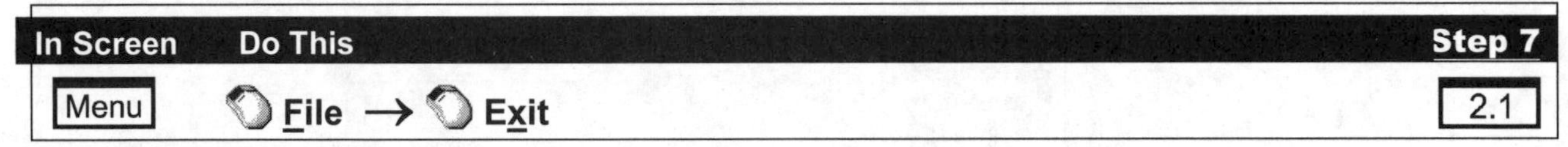

In Screen	Do This	Step 7
Menu	**File** → **Exit**	2.1

Note: After clicking **Exit**, there will frequently be small windows that appear asking if you wish to save or change anything. Simply click each appropriate response.

OUTPUT

General (Nonhierarchical) Log-Linear Models

Below is an explanation of the key output from the General (nonhierarchical) Log-linear Model procedure (sequence step 5, page 353). Only representative output is provided, because the actual printout from this procedure is many pages long.

Note: Because the interpretation of log-linear models with covariates (or designs that are logit models) is virtually identical to log-linear models without covariates (or that are not logit models), we include only one sample output. Logit models also produce an analysis of dispersion and measures of association. These measures indicate the dependent variable's dispersion, and how much of the total dispersion of the dependent variable comes from the model. These measures are difficult to interpret without considerable experience and so are not discussed here.

Goodness-of-Fit Tests

	Value	df	Sig.
Likelihood Ratio	8.281	8	.406
Pearson Chi-Square	8.298	8	.405

The likelihood-ratio chi-square and the Pearson chi-square examine the fit of the model. Large chi-square values and small *p*-values indicate that the model does *not* fit the data well. Note that this is the opposite of thinking in interpretation of most types of analyses. Usually one looks for a large test statistic and a small *p*-value to indicate a significant effect. In this case, a large chi-square and a small *p*-value would indicate that your data differs significantly (or does not fit well) the model you have created. The degrees of freedom refers to the number of non-zero cells, minus the number of parameters in the model.

Predicted and Actual Cell Counts and Residuals

gender	ethnicity	cathelp	Observed		Expected		Residual	Standardized Residual	Adjusted Residual	Deviance
			Count	%	Count	%				
FEMALE	WHITE	NOT HELPFUL	70	13.0%	74.123	13.8%	-4.123	-.479	-.920	-.483
		HELPFUL	95	17.7%	90.877	16.9%	4.123	.433	.920	.429
	BLACK	NOT HELPFUL	13	2.4%	15.274	2.8%	-2.274	-.582	-.829	-.597
		HELPFUL	21	3.9%	18.726	3.5%	2.274	.525	.829	.515
	HISPANIC	NOT HELPFUL	22	4.1%	22.462	4.2%	-.462	-.097	-.143	-.098
		HELPFUL	28	5.2%	27.538	5.1%	.462	.088	.143	.088
	ASIAN	NOT HELPFUL	28	5.2%	20.215	3.8%	7.785	1.731	2.513	1.635
		HELPFUL	17	3.2%	24.785	4.6%	-7.785	-1.564	-2.513	-1.659
	OTHER/DTS	NOT HELPFUL	13	2.4%	13.926	2.6%	-.926	-.248	-.352	-.251
		HELPFUL	18	3.4%	17.074	3.2%	.926	.224	.352	.222
MALE	WHITE	NOT HELPFUL	75	14.0%	71.849	13.4%	3.151	.372	.892	.369
		HELPFUL	53	9.9%	56.151	10.5%	-3.151	-.420	-.892	-.425
	BLACK	NOT HELPFUL	8	1.5%	8.981	1.7%	-.981	-.327	-.514	-.334
		HELPFUL	8	1.5%	7.019	1.3%	.981	.370	.514	.362
	HISPANIC	NOT HELPFUL	15	2.8%	16.840	3.1%	-1.840	-.448	-.731	-.457
		HELPFUL	15	2.8%	13.160	2.5%	1.840	.507	.731	.496
	ASIAN	NOT HELPFUL	15	2.8%	14.033	2.6%	.967	.258	.415	.255
		HELPFUL	10	1.9%	10.967	2.0%	-.967	-.292	-.415	-.296
	OTHER/DTS	NOT HELPFUL	6	1.1%	7.297	1.4%	-1.297	-.480	-.748	-.496
		HELPFUL	7	1.3%	5.703	1.1%	1.297	.543	.748	.524

Term	Definition/Description
OBSERVED COUNT AND %	The observed cell count and percentage derives from the data and indicates the number of cases in each cell.
EXPECTED COUNT AND %	The expected cell counts and percentages indicate the expected frequencies for the cells, based on the model being tested.
RESIDUAL, STANDARDIZED, AND ADJUSTED RESIDUAL	The residuals are the observed counts minus the expected counts. Standardized residuals are adjusted for the standard error; adjusted residuals are standardized residuals adjusted for *their* standard error. High residuals indicate that the model is not adequate.
DEVIANCE	The sign (positive or negative) of the deviance is always the same as the residual; the magnitude of the deviance (or, how far away it is from zero) is an indication of how much of the chi-square statistic comes from each cell in the model.

Parameter Estimates

Parameter	Estimate	Std. Error	Z	Sig.	95% Confidence Interval	
					Lower Bound	Upper Bound
Constant	1.741	.288	6.044	.000	1.176	2.305
[cathelp = 1.00]	.247	.138	1.781	.075	-.025	.518
[cathelp = 2.00]	0[a]	.	.	.	.	.
[ethnicity = 1]	2.287	.291	7.857	.000	1.717	2.858
[ethnicity = 2]	.208	.373	.556	.578	-.524	.939
[ethnicity = 3]	.836	.332	2.518	.012	.185	1.487
[ethnicity = 4]	.654	.342	1.912	.056	-.016	1.324
[ethnicity = 5]	0[a]	.	.	.	.	.
[gender = 1]	1.097	.343	3.196	.001	.424	1.769
[gender = 2]		.	.	.	.	.
[gender = 1] * [cathelp = 1.00]	-.450	.178	-2.533	.011	-.799	-.102
[gender = 1] * [cathelp = 2.00]	0[a]	.	.	.	.	.
[gender = 2] * [cathelp = 1.00]	0[a]	.	.	.	.	.
[gender = 2] * [cathelp = 2.00]	0[a]	.	.	.	.	.
[gender = 1] * [ethnicity = 1]	-.615	.351	-1.754	.080	-1.303	.072
[gender = 1] * [ethnicity = 2]	-.115	.448	-.257	.797	-.994	.764
[gender = 1] * [ethnicity = 3]	-.358	.403	-.889	.374	-1.148	.432
[gender = 1] * [ethnicity = 4]	-.281	.414	-.679	.497	-1.093	.530
[gender = 1] * [ethnicity = 5]	0[a]	.	.	.	.	.
[gender = 2] * [ethnicity = 1]	0[a]	.	.	.	.	.
[gender = 2] * [ethnicity = 2]	0[a]	.	.	.	.	.
[gender = 2] * [ethnicity = 3]	0[a]	.	.	.	.	.
[gender = 2] * [ethnicity = 4]	0[a]	.	.	.	.	.
[gender = 2] * [ethnicity = 5]	0[a]	.	.	.	.	.

a This parameter is set to zero because it is redundant.

Parameter estimates are provided for each effect. Because some of the parameters are not necessary to predict the data, many parameters are zero (that is, they really aren't part of the model). For example, if you know the parameter for **cathelp** = 1, you don't need to know the parameter for **cathelp** = 2, because people who are not in the **cathelp** = 1 condition will be in the **cathelp** = 2 condition.

Term	Definition/Description
ESTIMATE	The λ value in the log-linear model equation.
STANDARD ERROR	A measure of the dispersion of the coefficient.
Z-VALUE	A standardized measure of the parameter coefficient: Large z values (those whose absolute value is greater than 1.96) are significant (α = .05).
LOWER AND UPPER 95% CONF INTERVAL	There is a 95% chance that the coefficient is between the lower and upper confidence interval.

RESIDUALS: Analyzing Left-Over Variance

THIS IS a somewhat unusual chapter. It does not contain step-by-step instructions for performing a particular analysis in SPSS; indeed, it doesn't even deal with a single SPSS procedure. Instead, it deals with residuals, a statistical concept that is present in many SPSS procedures. The crosstabs, ANOVA and MANOVA models, regression, and loglinear models procedures can all analyze residuals. This chapter takes a didactic format, rather than step-by-step format, because the analysis of residuals takes quite a bit of statistical finesse and experience. Because of this it is useful to examine the way residuals work separately, so that the common aspects of residuals in all applicable SPSS procedures can be examined at once. If residuals were discussed in each procedure in which they are present, it would add several pages to each chapter, rather than several pages for the entire book. Also, for those with little experience in analyzing residuals (or little desire to gain experience), placing the discussion of residuals throughout the book would likely confuse rather than clarify.

With that apology for the unusual chapter format, we begin. This chapter will commence with a brief description of residuals, and their primary usefulness. The bulk of the chapter will be taken up with two case studies, in which a linear regression and a general log-linear model are examined. Each case study will illustrate a different use of residuals. The chapter will conclude with a summary of the various methods of analyzing residuals within SPSS, along with the ways of accessing the appropriate output from SPSS.

RESIDUALS

Most of the sophisticated statistical procedures commonly done are designed to test a theoretical model. For example:

- In 2 x 2 Chi-square analyses, expected frequencies are computed; these expected frequencies are based on a model that the effects of the two categorical variables are independent: **Expected frequencies = effects of the categorical variables + residuals**.
- In log-linear modeling, the relationship between categorical variables is examined, and a model is developed containing main effects and interactions between the variable: **Expected frequencies main effects + interactions + residuals**.
- In regression analyses, one or more predicted variables predict a criterion or dependent variable. The regression analysis procedure examines whether or not each predictor variable significantly relates to the dependent variable. Thus, a model of predictor variables is developed: **Dependent variable = effects of predictor variables + residual**.
- In an ANOVA, some main effects and interactions are examined. These main effects and interactions may be considered a model of independent variable(s) that influence the dependent variable(s). It is with this understanding that SPSS groups most of the ANOVA procedures under the heading "General Linear Model":
  Dependent variable(s) = main effects + interactions + covariates + residuals.

In all of these techniques, models are developed to predict the data that have been collected. The goal is, of course, to develop a model that predicts the data perfectly, and parsimoniously. In other words, the best model will predict the data collected, and yet also be very simple. As a result of this desire for parsimony, there is generally some discrepancy between the actual, observed data, and the expected data predicted by the model. This discrepancy is the residual, also known as error. The analysis of residuals can be a useful, for instance, for determining how good the model is that you are testing. It can also be useful in examining unusual cases in your data that do not fit the model will, or of finding patterns in the data that you hadn't noticed or predicted.

Our first case study will test a simple linear regression analysis, to determine the adequacy of the model developed by SPSS.

LINEAR REGRESSION: A CASE STUDY

A researcher hypothesizes that students with low anxiety do poorly on an exam and those with high anxiety do better. This is the same example used in Chapter 15. The researcher expects a linear relationship. The description of the model developed is described fully in Chapter 15, pp. 178-182, and the SPSS procedure for developing this model is described in Steps 4 and 5 in that chapter. In summary, the model is that:

$$\textbf{exam}_{(\text{true})} = \text{some } \textbf{constant} + \text{a coefficient} \times \textbf{anxiety} + \textbf{residual}$$

or

$$\textbf{exam}_{(\text{true})} = 64.247 + 2.818(\textbf{anxiety}) + \textbf{residual}$$

In order to examine how well this model of anxiety predicts exam scores, we need to look at the residual values for all of the cases to see if there is any pattern to the residuals. In this case, we can request a Histogram, Normal probability plot, as well as customized graphs relating a number of different forms (e.g., standardized and studentized) of the predicted values, residuals, and dependant measure. Graphs are requested from the plots window, selected from the Linear Regression window in Screen 15.3.

The Histogram (at right) displays the standardized residuals (the residuals divided by the estimate of the standard error) across the horizontal axis, and the number of subjects within each range of standardized residuals along the vertical axis. The standard deviations will always be 1 (or practically 1, due to rounding error) and the mean will always be 0. This is because the residuals are standardized (therefore the standard deviation is 1) and the model is based on the data (therefore the mean is 0).

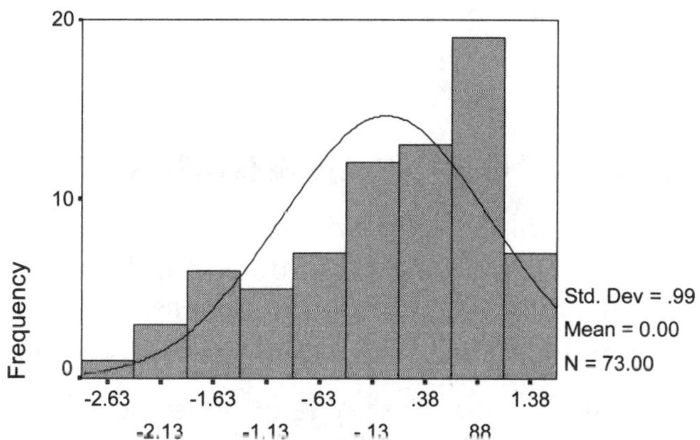

A line is drawn on the histogram, which shows where the normal curve predicts the residual to fall. What is important in the analysis of residuals is any pattern in the relationship between the histogram bars and the normal probability curve. In this case, the bars in the left side of the graph tend to be lower than the normal curve, and the bars on the right side of the graph tend to be higher than the normal curve. This indicates that there may be a curvilinear relationship between the anxiety and exam scores.

The normal probability plot takes a bit more explanation, but in this case makes the curvilinear relationship between the anxiety and exam scores even clearer. The plot places the observed cumulative probability along the horizontal axis, and the expected cumulative probability along the vertical axis. Cumulative probabilities are calculated by ranking all of the subjects, and then comparing the value of that variable with the value expected from a normal distribution. If the residuals are normally distributed (that's what we want with a good model) then there should be a linear relationship between the expected and observed cumulative probabilities. Any other pattern suggests a problem or weakness in the model.

In our example, we can see that values near the middle (probabilities around .5) tend to have the higher expected values than observed values, and for values near the ends (probabilities near 0 or 1) tend to have lower expected values than observed values. This indicates the possibility of a non-linear relationship between **anxiety** and **exam**. An analysis of the curvilinear relationship between **anxiety** and **exam** (in Chapter 15) indeed reveals that the inclusion of the anxiety^2 variable improves the fit of the model.

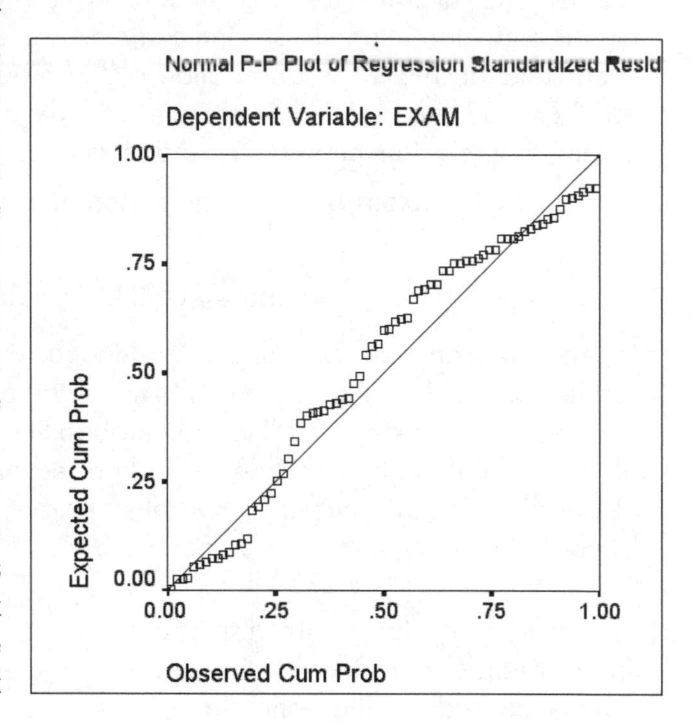

Other non-linear patterns on the normal probability plot will indicate different problems in the regression equation. Nearly any pattern is possible and it is beyond the scope of this book to describe them all, but the histogram and the normal probability plot are good places to look for these patterns.

GENERAL LOG-LINEAR MODELS: A CASE STUDY

In addition to looking for curvilinear relationships present in data but not in the model being examined, the examination of residuals can be useful in searching for additional variables that should be included in a model. This is particularly true in log-linear modeling, in which the development of the model is a fairly complex procedure.

In this case study we will examine a model based on the **helping3.sav** file. This model will include main effects of **gender**, **ethnicity**, and **cathelp** (referring to whether or not help was helpful, a dichotomous variable), and an interactive effect of **gender** by **cathelp**. Residuals-related output may be requested by selecting **Display Residuals** and all four of the **Plot** options in the **Options** window (Screen 27.2). Note that plots are not available if a saturated model is selected (i.e., a model containing all main effects and interactions), because a saturated model predicts the data perfectly, and thus all residuals are equal to zero.

The residuals output includes a list of residuals, adjusted residuals (like the standardized residuals), and the deviance residuals (based on the probability for being in a particular group). Although there are no cells in this table with very high adjusted residuals (the highest is 2.58), there are quite a number of cells with somewhat high adjusted residuals. It appears that there may be an interaction between **gender** and **ethnicity**, because the pattern of residuals is quite different across ethnicities for males and females.

Gender	Ethnicity	Cathelp	Residual	Adj. Residuals	Dev. Residuals
MALE	White	Helpful	-9.66	-1.88	-1.11
		Not Helpful	-2.67	-.49	-.27
	Black	Helpful	-.59	-.20	-.16
		Not Helpful	4.33	1.37	1.02
	Hispanic	Helpful	.25	.07	.05
		Not Helpful	1.33	.34	.26
	Asian	Helpful	8.97	2.58	1.92
		Not Helpful	-6.33	-1.72	-1.38
	Other/DTS	Helpful	1.04	.37	.30
		Not Helpful	3.33	1.11	.84
FEMALE	White	Helpful	10.07	2.10	1.22
		Not Helpful	2.26	.52	.31
	Black	Helpful	-3.08	-1.10	-.97
		Not Helpful	-.66	-.26	-.23
	Hispanic	Helpful	-2.73	-.80	-.67
		Not Helpful	1.15	.37	.30
	Asian	Helpful	-.51	-.16	-.13
		Not Helpful	-2.12	-.72	-.63
	Other/DTS	Helpful	-3.75	-1.42	-1.29
		Not Helpful	-.62	-.26	-.23

The adjusted residuals plot (at right) is, at first glance, quite confusing. Upon closer examination, however, it is manageable. It is essentially a correlation table with scatter plots instead of correlation values within each cell. As in a correlation table, cells below the diagonal are reflected in cells above the diagonal. The diagonal cells identify each row and column. So, the left center cell is a scatter plot of observed and expected cell counts, the bottom left cell is a scatterplot of observed cell counts with adjusted residual, and the middle cell at the bottom shows expected counts with adjusted residuals. The other scatter plots are mirror images of these bottom left three scatterplots

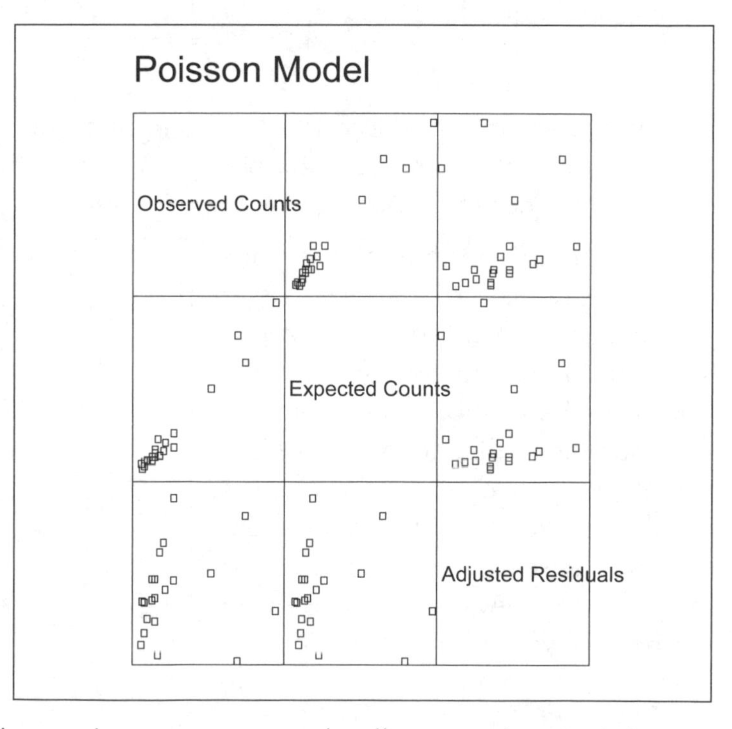

Two particularly important plots are the observed- versus expected-cell-count plot, and the expected-cell-count versus adjusted-residual plot. A good model will produce a linear relationship between the observed and the expected cell counts—that is, the observed data will match the data predicted by the model. A good model wouldn't show a relationship between the expected cell count and the adjusted residual, either. If there is a relationship between these variables, then another variable may need to be added to the model, or else some assumption of the model my be violated (like homogeneity of variance).

In this example, there is a generally linear relationship between the observed and the expected values (though not perfectly linear). There does, however, appear to be a pattern in the scatter plot between expected cell counts and the adjusted residuals.

The normal plot of adjusted residuals, to the right, displays adjusted residuals and expected normal values of those residuals. A good model should produce a linear relationship between expected normalized residuals and the actual residuals. Note that this graph is similar to the bottom center cell of the figure (previous page), except that the expected cell counts have been standardized. In this case, because there is little non-linear trend present, no problems are present. If there was a non-linear trend present, that pattern in the residuals could suggest problems or inadequacies in the model.

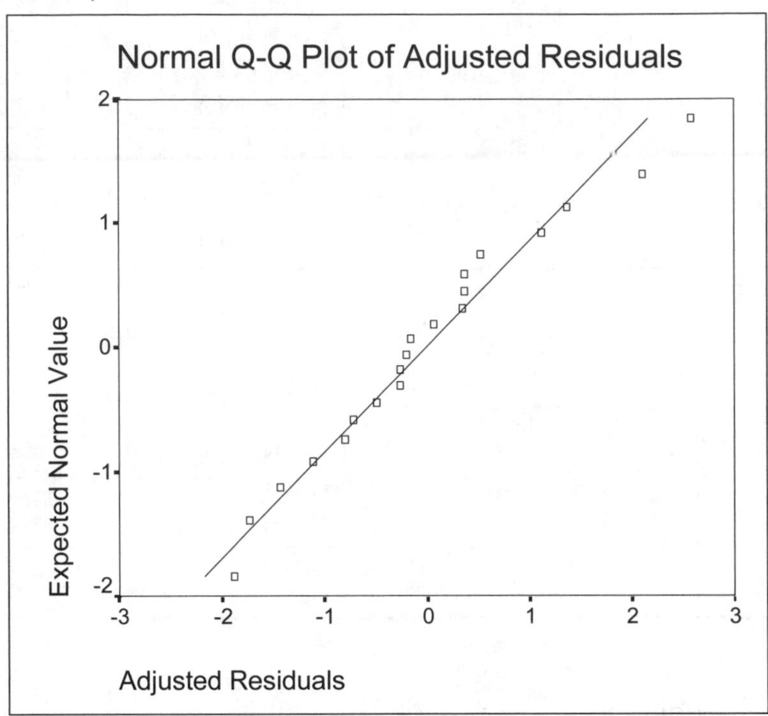

It does appear that the points to the far right of the graph have higher residual than expected, indicating that there may be something unusual going on with those cells. However, the cells are probably not far enough away from the trend line to be a problem.

The detrended normal plot of adjusted residuals, to the right, is a variation on the normal plot of adjusted residuals, above. In this case, however, the diagonal trend line in the chart above is moved to a horizontal line, and the vertical axis graphs how far away the cell is from the normal trend line. This serves to magnify any deviations away from the normal value, and is particularly useful in examining small, subtle patterns that may be present in a normal plot of residuals in which no cells are very far from the normal trend line (as in our example).

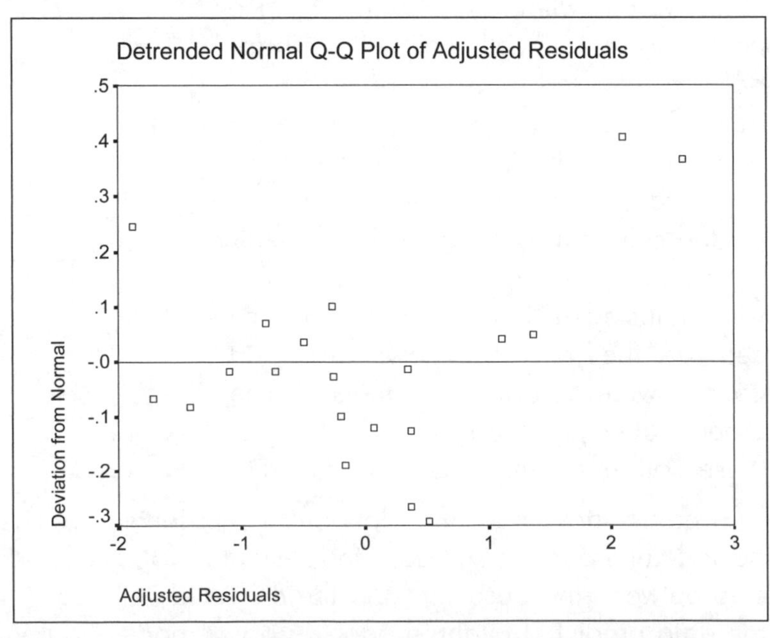

Note that the two "outlying" cells clearly stand out, on the top right of the chart. However, an examination of the scale on the vertical axis points out that even these cells are actually quite close to the expected normal values.

Based on an examination of the pattern of adjusted residuals, another log-linear model was tested, adding an interaction of **gender x ethnicity**. This model produced smaller adjusted residuals overall, although there was still one cell with a relatively high adjust residual (2.51). As can be seen in the new adjusted residuals plot (at right), the plot of observed versus expected cell counts is even more linear than before, and the plot of expected cell count versus adjusted residuals (as well as observed counts versus adjusted residual) does not show as strong a pattern as before. Adding the **gender x ethnicity** interaction did, indeed, improve the model.

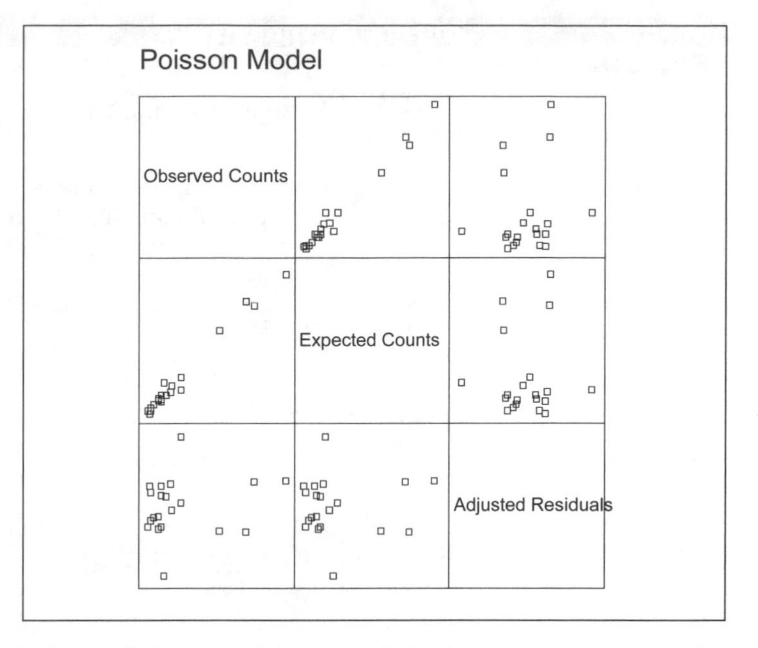

It should be noted that the residual list and plots of this model are a little bit too symmetrical and perfect. This is because the authors created the cathelp variable to provide a good fit for the model tested. So if you examine these plots, don't read too much meaningful significance into the patterns present.

ACCESSING RESIDUALS IN SPSS

We have only begun to explain the analysis of residuals. Many SPSS procedures allow you to save residuals in your data file, using new variable names; this allows you to perform any statistical analysis on those residuals that you desire.

We conclude our brief foray into residuals with a table summarizing the various SPSS options to analyze residuals. For each entry in the table, we present the chapter that describes the SPSS procedure, the window and options within that window needed to access residuals, and a brief description of the output produced.

Chapter	Window	Options	Descriptions
8 **Crosstabs**	Cells 8.2	**Unstandardized** **Standardized** **Adj. Standardized Residuals**	Produces residuals for each cell in the crosstabs table. Note that standardized residuals are transformed so that there is a mean of 0 and a standard deviation of 1, and the adjusted standardized residual is transformed to indicate the number of standard deviations above or below the mean.
15, 16 **Regression**	Save 16.3	**Unstandardized** **Standardized** **Studentized** **Deleted** **Studentized deleted**	Saves various kinds of residuals in the data file, with new variable names. SPSS will produce the new variable names, and the output will define them. Studentized residuals are standardized on a case by case basis, depending on the distance of the case from the mean. Deleted residuals are what the residual for a case would be if the case were excluded from the analysis.

Chapter	Window	Options	Descriptions
Regression (contd.)	Plots 16.5	**Histogram** **Normal probability plot** And customized plots of the dependent variable with various kinds of residuals	Produces histograms and normal probability plots, as well as letting you produce scatterplots for any two of the following: the dependent variable, the standardized predicted values and residuals, the deleted residuals, the adjusted predicted residuals, studentized residuals, and deleted residuals.
23, 24 **General Linear Models (MANOVA)**	Plots 23.1 24.2	**Residual plots** **Residual SSCP Matrix**	Residual plots produces for each dependent variable scatterplots for the observed predicted, and standardized residuals. The residual sum of squares and the cross-product matrix allows you to examine if there are any correlations between the dependent variables' residuals.
25 **Logistic Regression**	Options 25.3	**Casewise listing of residuals**	Lists residuals for each case, for either all cases or outliers more than a specified number of standard deviations from the mean (the default is 2).
	Save	**Unstandardized** **Logit** **Studentized** **Standardized** **Deviance** **Residuals**	Saves various kinds of residuals in the data file, with new variable names. SPSS will choose the new variable names. Studentized residuals are standardized on a case-by-case basis, depending on the distance of the case from the mean. Deviance residuals are based on the probability of a case being in a group. Logit residuals are residuals predicted on logit scales.
26 **Hierarchical Loglinear Models**	Options 26.3	**Residual Display** **Residuals Plots** or **Normal probability Plots**	Displays residuals (including standardized residuals), produces scatterplots of expected and observed values with residuals, and normal probability plots (see Glossary for a description of normal probability plots).
27 **General Loglinear Models**	Options 27.2	**Display Residuals** **Plot Adjusted residuals** **Normal probability: adjusted residuals** **Deviance residuals** **Normal probability: deviance residuals**	Displays residuals for each case, and produces plots of adjusted residuals, and normal and detrended normal probability plots for adjusted residuals and deviance residuals. Note: Plots are only available when a non-saturated model is tested.
	Save	**Residuals** **Standardized residuals** **Adjusted residuals** **Deviance residuals** **Predicted values**	Saves various kinds of residuals in the data file with new variable names. SPSS will select the new variable names.

Note: If you want to analyze residuals for an ANOVA, you must use the GLM command.

DATA FILES

Available for download from **www.ablongman.com/george6e** are 11 different data files that have been used to demonstrate procedures in this book. There is also a 12th file not utilized to demonstrate procedures in the book but included on the data disk (and in the Instructor's Manual) for additional exercises. There are also data files for all of the exercises, with the exception of the exercises for Chapter 3 (as the whole point of those exercises is to practice entering data!). You can tell from the name of the data file which exercise it goes with: for example, **ex14-1.sav** is the dataset for Chapter 14, Exercise 1. The **grades.sav** file is the most thoroughly documented and demonstrates procedures in 16 of the 28 chapters. This file is described in detail in Chapters 1 and 3. For analyses described in the other 12 chapters, it was necessary to employ different types of data to illustrate. On the web site, *all* files utilized in this book are included. What follows are brief narrative introductions to each file, and when appropriate, critical variables are listed and described. Before presenting this, we comment briefly on how to read a data file from a disk (as opposed to reading it from the hard drive of your computer, as is illustrated in all chapters of this book).

In every one of Chapters 6 through 27, there is a sequence step 3 that gives instructions on how to access a particular file. The instructions assume that you are reading a file saved in your hard drive. Depending on your computer set-up, you may have downloaded the files to your hard drive or to a floppy drive. To read from a floppy disk instead of a hard drive, refer to sequence step 3 and Screen 2.4 on page 16, and then perform either one of the following steps:

1. Click on the down arrow (▾) next to the **Drives** box, then click on the "a" drive or the "b" drive, depending on which represents your disk drive. At this point, all file names will appear in the **File Name** box. Double click on the desired file.
2. Type "a:\filename.sav" under **File Name**, then click the **Open** button.

GRADES.SAV: The complete data file is reproduced on pages 39-41. This is a fictional file (N = 105) created by the authors to demonstrate a number of statistical procedures. This file is used to demonstrate procedures in Chapters 3-14, 17, and 23-24. In addition to the data, all variable names and descriptions are also included on page 39. Be aware that in addition to the key variable names listed there, additional variables are included in the file:

- **total:** Sum of the five quizzes and the final
- **percent:** The percent of possible points in the class
- **grade:** The grade received in the class (A, B, C, D, or F)
- **passfail:** Whether or not the student passed the course (P or F)

GRADEROW.SAV: This file includes 10 additional subjects with the same variables as those used in the grades.sav file. It is used to demonstrate merging files in Chapter 4.

GRADECOL.SAV: This file includes the same 105 subjects as the **grades.sav** file but includes an additional variable (IQ) and is used in Chapter 4 to demonstrate merging files.

ANXIETY.SAV: A fictional data file (*N* = 73) that lists values to show the relationship between pre-exam anxiety and exam performance. It is used to demonstrate simple linear and curvilinear regression (Chapter 15). It contains two variables:

- **exam:** The score on a 100-point exam
- **anxiety:** A measure of pre-exam anxiety measured on a *low*(1) the *high*(10) scale

HELPING1.SAV: A file of real data (*N* = 81) created to demonstrate the relationship between several variables and the amount of time spent helping a friend in need. It is used to demonstrate multiple regression analysis (**Chapter 16**). Although there are other variables in the file, the ones used to demonstrate regression procedures include:

- **zhelp:** *Z*-scores of the amount of time spent helping a friend on a –3 to +3 scale
- **sympathy:** Sympathy felt by helper in response to friend's need on a *little*(1) to *much*(7) scale
- **anger:** Anger felt by helper in response to friend's need; same 7-point scale
- **efficacy:** Self-efficacy of helper in relation to friend's need; same scale
- **severity:** Helper's rating of the severity of the friend's problem; same scale
- **empatend:** Empathic tendency of the helper as measured by a personality test

HELPING2.SAV: A file of real data (*N* = 517) dealing with issues similar to those in the **helping1.sav** file. Although the file is large (both in number of subjects and number of variables), only the 15 measures of self-efficacy and the 14 empathic tendency questions are used to demonstrate procedures: reliability analysis (Chapter 18) and factor analysis (Chapter 20). (The full dataset is included in **helping2a.sav**; the file has been reduced to work with the student version of SPSS.) Variable names utilized in analyses include:

- **effic1** to **effic15:** The 15 self-efficacy questions used to demonstrate factor analysis
- **empathy1** to **empath14:** The 14 empathic tendency questions used to demonstrate reliability analysis

HELPING3.SAV: A file of real data (*N* = 537) dealing with issues similar to the previous two files. This is the same file as **HELPING2.SAV** except it has been expanded by 20 subjects and all missing values in the former file have been replaced by predicted values. Although the N is 537, the file represents over 1000 subjects because both the helpers and help recipients responded to questionnaires. In the book these data are used to demonstrate logistic regression (Chapter 25) and log-linear models (Chapters 26 and 27). We describe it here in greater detail because it is a rich data set that is able to illustrate every procedure in the book. Many exercises from this data set are included at the end of chapters and in the Instructor's Manual. Key variables include:

- **thelplnz**: Time spent helping (z-score scale, -3 to +3)
- **tqualitz**: Quality of the help given (z-score scale, -3 to +3)
- **tothelp**: A help measure that weights time and quality equally (z-score scale, -3 to +3)

- **empahelp:** Amount of time spent in empathic helping (10-point scale, 0 = little, 10 = much)
- **insthelp:** Amount of time spent in instrumental (doing things) helping (0 = little, 10 = much)
- **infhelp:** Time spent in informational (e.g. giving advice) helping (0 = little, 10 = much)
- **gender**: 1 = female, 2 = male
- **age**: Range from 17 to 89
- **school**: 7-point scale; from 1(lowest level education, <12 yr) to 7 (the highest, >19 yr)
- **angert**: Amount of anger felt by the helper toward the needy friend (7-
- **effict**: Helper's feeling of self-efficacy (competence) in relation to the friend's problem
- **empathyt**: Helper's empathic tendency as rated by a personality test
- **hclose**: Helper's rating of how close the relationship was
- **hcontrot**: Helper's rating of how controllable the cause of the problem was
- **hcopet**: Helper's rating of how well the friend was coping with his or her problem
- **hseveret**: Helper's rating of the severity of the problem
- **obligat**: The feeling of obligation the helper felt toward the friend in need
- **sympathi**: The extent to which the helper felt sympathy toward the friend
- **worry**: Amount the helper worried about the friend in need

[Note: all variables from **angert** to **worry** are scored on 7-point scales ranging from *low*(1) to *high*(7).]

GRADUATE.SAV: A fictitious data file (*N* = 50) that attempts to predict success in graduate school based on 17 classifying variables. This file is utilized to demonstrate discriminant analysis (Chapter 22). All variables and their metrics are described in detail in that chapter.

GRADES-MDS.SAV: A fictitious data file (*N* = 20) is used in the multidimensional scaling chapter (Chapter 19). For variables **springer** through **shearer**, the rows and columns of the data matrix represent a 20 x 20 matrix of disliking ratings, in which the rows (cases) are the raters, and the variables are the ratees. For variable quiz1 through quiz5, these are quiz scores for each student. Note that these names and quiz scores are derived from the **grades.sav** file.

GRADES-MDS2.SAV: A fictitious data file used to demonstrate individual differences multidimensional scaling (Chapter 19). This file includes four students' ratings of the similarity between four television shows. The contents and format of this data file are described on page 222.

VCR.SAV: A fictitious data file (*N* = 21) that compares 21 different brands of VCRs on 21 classifying features. The fictitious brand names and all variables are described in detail in the cluster analysis chapter (Chapter 21).

DIVORCE.SAV: This is a file of 229 divorced individuals recruited from communities in central Alberta. The objective of researchers was to identify cognitive or interpersonal factors that assisted in recovery from divorce. No procedures in this book utilize this file, but there are a number of exercises at the end of chapters and in the Instructor's Manual that do. Key variables include:

- **lsatisfy**: A measure of life satisfaction based on weighted averages of satisfaction in 12 different areas of life functioning. This is scored on a 1 (low satisfaction) to 7 (high satisfaction) scale.
- **trauma**: A measure of the trauma experienced during the divorce recovery phase based on the mean of 16 different potentially traumatic events, scored on a 1 (low trauma) to 7 (high trauma) scale.
- **sex**: Gender [women(1), men(2)]
- **age**: Range from 23 to 76
- **sep**: Years separated accurate to one decimal
- **mar**: Years married prior to separation, accurate to one decimal
- **status**: Present marital status [married(1), separated(2), divorced(3), cohabiting(4)]
- **eth**: Ethnicity [White(1), Hispanic(2), Black(3), Asian(4), other or DTS(5)]
- **school**: [1-11yr(1), 12yr(2), 13yr(3), 14yr(4), 15yr(5), 16yr(6), 17yr(7), 18yr(8), 19+(9)]
- **childneg**: Number of children negotiated in divorce proceedings
- **childcst**: Number of children presently in custody
- **income**: [DTS(0), <10,000(1), 10-20(2), 20-30(3), 30-40(4), 40-50(5), 50+(6)]
- **cogcope**: Amount of cognitive coping during recovery [little(1) to much(7)]
- **behcope**: Amount of behavioral coping during recovery [little(1) to much(7)]
- **avoicop**: Amount of avoidant coping during recovery [little(1) to much(7)]
- **iq**: Intelligence or ability at abstract thinking [low(1) to high(12)]
- **close**: Amount of physical (non-sexual) closeness experienced [little(1) to much(7)]
- **locus**: Locus of control [external locus(1) to internal locus(10)]
- **asq**: Attributional style questionnaire [pessimistic style(≈-7) to optimistic style(≈+9)]
- **socsupp**: Amount of social support experienced [little(1) to much(7)]
- **spirituality**: Level of personal spirituality [low(1) to high(7)]
- **d**-variables: There are 16 variables (all beginning with the letter "d") that specify 16 different areas that may have been destructive in their efforts at recovery. Read labels in the data file to find what each one is. Great for bar chart, *t* test, or reliability demonstrations.
- **a**-variables: There are 13 variables (all beginning with the letter "a") that specify 13 different areas that may have assisted in their efforts toward recovery. Read labels in the data file to find what each one is. Great for bar chart, *t* test, or reliability demonstrations.

GLOSSARY

ADJUSTED R SQUARE: In multiple regression analysis, R^2 is an accurate value for the sample drawn but is considered an optimistic estimate for the *population* value. The Adjusted R^2 is considered a better population estimate and is useful when comparing the R^2 values between models with different numbers of independent variables.

AGGLOMERATIVE HIERARCHICAL CLUSTERING: A procedure used in cluster analysis by which cases are clustered one at a time until all cases are clustered into one large cluster. See Chapter 25 for a more complete description.

ALPHA: Also, *coefficient alpha* or α is a measure of internal consistency based on the formula $\alpha = rk/[1 + (k - 1)r]$, where k is the number of variables in the analysis and r is the mean of the inter-item correlations. The alpha value is inflated by a larger number of variables, so there is no set interpretation as to what is an *acceptable* alpha value. A rule of thumb that applies to most situations is: $\alpha > .9$—excellent, $\alpha > .8$—good, $\alpha > .7$—acceptable, $\alpha > .6$—questionable, $\alpha > .5$—poor, $\alpha < .5$—unacceptable.

ALPHA IF ITEM DELETED: In reliability analysis, the resulting alpha if the variable to the left is deleted.

ANALYSIS OF VARIANCE (ANOVA): A statistical test that identifies whether there are any significant differences between three or more sample means. See Chapters 12-14 for a more complete description.

ASYMPTOTIC VALUES: Determination of parameter estimates based on asymptotic values (the value a function is never expected to exceed). This process is used in nonlinear regression and other procedures where an actual value is not possible to calculate.

B: In regression output, the *B* values are the regression coefficients and the constant for the regression equation. The *B* may be thought of as a weighted constant that describes the magnitude of influence a particular independent variable has on the dependent variable. A positive value for *B* indicates a corresponding increase in the value of the dependent variable, whereas a negative value for *B* decreases the value of the dependent variable.

BAR GRAPH: A graphical representation of the frequency of categorical data. A similar display for continuous data is called a *histogram*.

BARTLETT TEST OF SPHERICITY: This is a measure of the multivariate normality of a set of distributions. A significance value < .05 suggests that the data do not differ significantly from multivariate normal. See page 256 for more detail.

BETA (β): In regression procedures, the standardized regression coefficients. This is the *B*-value for standardized scores (*z*-scores) of the variables. These values will vary strictly between plus-and-minus 1.0 and may be compared directly with beta values in other analyses.

BETA IN: In multiple regression analysis, the beta values for the *excluded* variables if these variables were actually in the regression equation.

BETWEEN-GROUPS SUM OF SQUARES: The sum of squared deviations between the grand mean and each group mean weighted (multiplied) by the number of subjects in each group.

BINOMIAL TEST: A nonparametric test that measures whether a distribution of values is binomially distributed (each outcome equally likely). For instance, if you tossed a coin 100 times, you would expect a binomial distribution (approximately 50 heads and 50 tails).

BONFERRONI TEST: A *post hoc* test that adjusts for experimentwise error by dividing the alpha value by the total number of tests performed.

BOX'S M: A measure of multivariate normality based on the similarities of determinants of the covariance matrices for two or more groups.

CANONICAL CORRELATION: In discriminant analysis, the canonical correlation is a correlation between the discriminant scores for each subject and the levels of the dependent variable for each subject. See Chapter 26 to see how a canonical correlation is calculated.

CANONICAL DISCRIMINANT FUNCTIONS: The linear discriminant equation(s) calculated to maximally discriminate between levels of the dependent (or criterion) variable. This is described in detail in Chapter 26.

CHI-SQUARE ANALYSIS: A nonparametric test that makes comparisons (usually of crosstabulated data) between two or more samples on the *observed frequency* of values with the *expected frequency* of values. Also used as a test of the goodness-of-fit of log-linear and structural models. For the latter, the question being asked is: Does the actual data differ significantly from results predicted from the model that has been created? The formula for the Pearson chi-square is:

$$\chi^2 = \Sigma[(f_o - f_e)^2/f_e]$$

CLUSTER ANALYSIS: A procedure by which subjects, cases, or variables are clustered into groups based on similar characteristics of each.

COCHRAN'S C and BARTLETT-BOX F: Measure whether the variances of two or more groups differ significantly from each other (heteroschedasticity). A high probability value (for example, $p > .05$) indicates that the variances of the groups do *not* differ significantly.

COLUMN PERCENT: A term used with crosstabulated data. It is the result of dividing the frequency of values in a particular cell by the frequency of values in the entire column. Column percents sum to 100% in each column.

COLUMN TOTAL: A term used with crosstabulated data. It is the total number of subjects in each column.

COMMUNALITY: Used in factor analysis, a measure designed to show the proportion of variance that factors contribute to explaining a particular variable. In the SPSS default procedure, communalities are initially assigned a value of 1.00.

CONFIDENCE INTERVAL: The range of values within which a particular statistic is likely to fall. For instance, a 95% confidence interval for the mean indicates that there is a 95% chance that the true population mean falls within the range of values listed.

CONVERGE: To converge means that after some number of iterations, the value of a particular statistic does not change more than a prespecified amount and parameter estimates are said to have "converged" to a final estimate.

CORRECTED ITEM-TOTAL CORRELATION: In reliability analysis, correlation of the designated variable with the sum of all other variables in the analysis.

CORRELATION: A measure of the strength and direction of association between two variables. See Chapter 10 for a more complete description.

CORRELATION BETWEEN FORMS: In split-half reliability analysis, an estimate of the reliability of the measure if each half had an equal number of items.

CORRELATION COEFFICIENT: A value that measures the strength of association between two variables. This value varies between ±1.0 and is usually designated by the lowercase letter *r*.

COUNT: In crosstabulated data, the top number in each of the cells indicating the actual number of subjects or cases in each category.

COVARIATE: A variable that has substantial correlation with the dependent variable and is included in an experiment as an adjustment of the results for differences existing among subjects prior to the experiment.

CRAMER'S *V*: A measure of the strength of association between two categorical variables. Cramer's *V* produces a value between 0 and 1 and (except for the absence of a negative relation) may be interpreted in a manner similar to a correlation. Often used within the context of chi-square analyses. The equation follows. (Note: *k* is the smaller of the number of rows and columns.)

$$V = \sqrt{\chi^2 / [N(k-1)]}$$

CROSSTABULATION: Usually a table of frequencies of two or more categorical variables taken together. However, crosstabulation may also be used for different *ranges* of values for continuous data. See Chapter 8.

CUMULATIVE FREQUENCY: The total number of subjects or cases having a given score or any score lower than the given score.

CUMULATIVE PERCENT: The total percent of subjects or cases having a given score or any score lower than the given score.

DEGREES OF FREEDOM (DF): The number of values that are free to vary, given one or more statistical restrictions on the entire set of values. Also, a statistical compensation for the failure of a range of values to be normally distributed.

DENDOGRAM: A branching-type graph used to demonstrate the clustering procedure in cluster analysis. See Chapter 21 for an example.

DETERMINANT OF THE VARIANCE-COVARIANCE MATRICES: The determinant provides an indication of how strong a relationship there is among the variables in a correlation matrix. The smaller the number, the more closely the variables are related to each other. This is used primarily by the computer to compute the Box's *M* test. The determinant of the *pooled* variance-covariance matrix refers to all the variance-covariance matrices present in the analysis.

DEVIATION: The distance and direction (positive or negative) of any raw score from the mean.

(DIFFERENCE) MEAN: In a *t*-test, the difference between the two means.

DISCRIMINANT ANALYSIS: A procedure that creates a regression formula to maximally discriminate between levels of a categorical dependent variable. See Chapter 22.

DISCRIMINANT SCORES: Scores for each subject, based on substitution of values for the corresponding variables into the discriminant formula.

EIGENVALUE: In factor analysis, the proportion of variance explained by each factor. In discriminant analysis, the between-groups sums of squares divided by within-groups sums of squares. A large eigenvalue is associated with a strong discriminant function.

EQUAL-LENGTH SPEARMAN-BROWN: Used in split-half reliability analysis when there is an unequal number of items in each portion of the analysis. It produces a correlation value that is inflated to reflect what the correlation would be if each part had an equal number of items.

ETA: A measure of correlation between two variables when one of the variables is discrete.

ETA SQUARED: The proportion of the variance in the dependent variable accounted for by an independent variable. For instance, an eta squared of .044 would indicate that 4.4% of the variance in the dependent variable is due to the influence of the independent variable.

EXP(B): In logistic regression analysis, e^B is used to help in interpreting the meaning of the regression coefficients. (Remember that the regression equation may be interpreted in terms of B or e^B.)

EXPECTED VALUE: In the crosstabulation table of a chi-square analysis, the number that would appear if the two variables were perfectly independent of each other. In regression analysis, it is the same as a *predicted value*, that is, the value obtained by substituting data from a particular subject into the regression equation.

FACTOR: In factor analysis, a factor (also called a component) is a combination of variables whose shared correlations explain a certain amount of the total variance. After rotation, factors are designed to demonstrate underlying similarities between groups of variables.

FACTOR ANALYSIS: A statistical procedure designed to take a larger number of constructs (measures of some sort) and reduce them to a smaller number of *factors* that describe these measures with greater parsimony. See Chapter 20.

FACTOR TRANSFORMATION MATRIX: If the original *unrotated* factor matrix is multiplied by the *factor transformation matrix*, the result will be the *rotated* factor matrix.

F-CHANGE: In multiple regression analysis, the *F*-change value is associated with the additional variance explained by a new variable.

F-RATIO: In an analysis of variance, an *F*-ratio is the between-groups mean square divided by the within-groups mean square. This value is designed to compare the between-groups variation to the within-groups variation. If the between-groups variation is substantially larger than the within-groups variation, then significant differences between groups will be demonstrated. In multiple regression analysis, the *F*-ratio is the mean square (regression) divided by the mean square (residual). It is designed to demonstrate the strength of association between variables.

FREQUENCIES: A listing of the number of times certain events take place.

FRIDMAN 2-WAY ANOVA: A nonparametric procedure that tests whether three or more groups differ significantly from each other, based on average rank of groups rather than comparison of means from normally distributed data.

GOODNESS-OF-FIT TEST STATISTICS: The likelihood-ratio chi-square and the Pearson chi-square statistics examine the fit of log-linear models. Large chi-square values and small *p*-values indicate that the model does *not* fit the data well. Be aware that this is the opposite of thinking in interpretation of most types of analyses. Usually one looks for a large test statistic and a small *p*-value to indicate a significant effect. In this case, a large chi-square and a small *p*-value would indicate that your data differs significantly from (or does not fit well) the model you have created.

GROUP CENTROIDS: In discriminant analysis, the average discriminant score for subjects in the two (or more) groups. More specifically, the discriminant score for each group is determined when the variable means (rather than individual values for each subject) are entered into the discriminant equation. If you are discriminating between exactly two outcomes, the two scores will be equal in absolute value but have opposite signs. The dividing line between group membership in that case will be zero (0).

GUTTMAN SPLIT-HALF: In split-half reliability, a measure of the reliability of the overall test, based on a lower-bounds procedure.

HYPOTHESIS SS: The between-groups sum of squares; the sum of squared deviations between the grand mean and each group mean, weighted (multiplied) by the number of subjects in each group.

ICICLE PLOT: A graphical display of the step-by-step clustering procedure in cluster analysis. See Chapter 21 for an example.

INTERACTION: The idiosyncratic effect of two or more independent variables on a dependent variable over and above the independent variables' separate (main) effects.

INTERCEPT: In regression analysis, the point where the regression line crosses the Y-axis. The intercept is the predicted value of the vertical-axis variable when the horizontal-axis variable value is zero.

INTER-ITEM CORRELATIONS: In reliability analysis, this is descriptive information about the correlation of each variable with the sum of all the others.

ITEM MEANS: In reliability analysis (using coefficient alpha), this is descriptive information about all subjects' means for all the variables. On page 230, an example clarifies this.

ITEM VARIANCES: A construct similar to that used in *item means* (the previous entry). The first number in the SPSS output is the mean of all the variances, the second is the lowest of all the variances, and so forth. Page 230 clarifies this with an example.

ITERATION: The process of solving an equation based on preselected values, then replacing the original values with the computer-generated values and solving the equation again. This process continues until some criterion (in terms of amount of change from one iteration to the next) is achieved.

K: In hierarchical log-linear models, the order of effects for each row of the table (1 = first-order effects, 2 = second-order effects, and so on).

KAISER-MAYER-OLKIN: A measure of whether the distribution of values is adequate for conducting factor analysis. A measure > .9 is generally thought of as excellent, > .8 as good, > .7 as acceptable, > .6 as marginal, > .5 as poor, and < .5 as unacceptable.

KOLMOGOROV-SMIRNOV ONE-SAMPLE TEST: A nonparametric test that determines whether the distribution of the members of a single group differ significantly from a *normal* (or *uniform*, *poisson,* or *exponential*) distribution.

K-SAMPLE MEDIAN TEST: A nonparametric test that determines whether two or more groups differ on the number of instances (within each group) greater than the grand median value or less than the grand median value.

KURTOSIS: A measure of deviation from normality that measures the peakedness or flatness of a distribution of values. See Chapter 7 for a more complete description.

LEVINE'S TEST: A test that examines the assumption that the variance of each dependent variable is the same as the variance of all other dependent variables.

LOG DETERMINANT: In discriminant analysis, the natural log of the determinant of each of the two (or more) covariance matrices. This is used to test the equality of group covariance matrices using Box's *M*.

LSD: "Least Significant Difference" post hoc test; performs a series of t tests on all possible combinations of the independent variable on each dependent variable. A liberal test.

MAIN EFFECTS: The influence of a single independent variable on a dependent variable. See Chapters 13 and 14 for examples.

MANCOVA: A MANOVA that includes one or more covariates in the analysis.

MANN-WHITNEY AND WILCOXON RANK-SUM TEST: A nonparametric alternative to the *t*-test that measures whether two groups differ from each other based on ranked scores.

MANOVA: Multivariate analysis of variance. A complex procedure similar to ANOVA except that it allows for more than one dependent variable in the analysis.

MANOVA REPEATED MEASURES: A multivariate analysis of variance in which the same set of subjects experiences several measurements on the variables of interest over time. Computationally, it is the same as a within-subjects MANOVA.

MANOVA WITHIN-SUBJECTS: A multivariate analysis of variance in which the same set of subjects experience all levels of the dependent variable.

MANTEL-HAENSZEL TEST FOR LINEAR ASSOCIATION: Within a chi-square analysis, this procedure tests whether the two variables correlate with each other. This measure is often meaningless unless there is some logical or numeric relation to the order of the levels of the variables.

MAUCHLY'S SPHERICITY TEST: A test of multivariate normality. SPSS computes a χ^2 approximation for this test, along with its significance level. If the significance level associated with Mauchly's sphericity test is small (i.e., $p < .05$), then the data may not be spherically distributed.

MAXIMUM: Largest observed value for a distribution.

MEAN: A measure of central tendency; the sum of a set of scores divided by the total number of scores in the set.

MEAN SQUARE: Sum of squares divided by the degrees of freedom. In ANOVA, the most frequently observed mean squares are the within-groups sum of squares divided by the corresponding degrees of freedom and the between-groups sum of squares divided by the associated degrees of freedom. In regression analysis, it is the regression sum of squares and the residual sum of squares divided by the corresponding degrees of freedom. For both ANOVA and regression, these numbers are used to determine the *F*-ratio.

MEDIAN: A measure of central tendency; the middle point in a distribution of values.

MEDIAN TEST: See K-Sample Median Test.

MINIMUM: Lowest observed value for a distribution.

MINIMUM EXPECTED FREQUENCY: A chi-square analysis identifies the value of the cell with the minimum expected frequency.

–2 LOG LIKELIHOOD: This is used to indicate how well a log-linear model fits the data. Smaller –2 log likelihood values mean that the model fits the data better; a perfect model has a –2 log likelihood value of zero. Significant χ^2 values indicate that the model differs significantly from the theoretically "perfect" model.

MODE: A measure of central tendency; it is the most frequently occurring value.

MODEL χ^2: In logistic regression analysis, this value tests whether or not all the variables entered in the equation have a significant effect on the dependent variable. A high χ^2 value indicates that the variables in the equation significantly impact the dependent variable. This test is functionally equivalent to the overall *F*-test in multiple regression.

MULTIPLE REGRESSION ANALYSIS: A statistical technique designed to predict values of a dependent (or criterion) variable from knowledge of the values of two or more independent (or predictor) variables. See Chapter 16 for a more complete description.

MULTIVARIATE TEST FOR HOMOGENEITY OF DISPERSION MATRICES: Box's *M* test examines whether the variance-covariance matrices are the same in all cells. To evaluate this test, SPSS calculates an *F* or χ^2 approximation for the *M*. These values, along with their associated *p*-values, appear in the SPSS output. Significant *p*-values indicate *differences* between the variance-covariance matrices for the two groups.

MULTIVARIATE TESTS OF SIGNIFICANCE: In MANOVA, there are several methods of testing for differences between the dependent variables due to the independent variables. Pillai's method is considered the best test by many, in terms of statistical power and robustness.

95% CONFIDENCE INTERVAL: See *confidence interval*.

NONLINEAR REGRESSION: A procedure that estimates parameter values for intrinsically nonlinear equations.

NONPARAMETRIC TESTS: A series of tests that make no assumptions about the distribution of values (usually meaning the distribution is not normally distributed) and performs statistical analyses based upon rank order of values, comparisons of paired values, or other techniques that do not require normally distributed data.

NORMAL DISTRIBUTION: A distribution of values that, when graphed, produces a smooth, symmetrical, bell-shaped distribution that has skewness and kurtosis values equal to zero.

OBLIQUE ROTATIONS: A procedure of factor analysis in which rotations are allowed to deviate from orthogonal (or from perpendicular) in an effort to achieve a better simple structure.

OBSERVED VALUE or COUNT: In a chi-square analysis, the frequency results that are actually obtained when conducting an analysis.

ONE-SAMPLE CHI-SQUARE TEST: A nonparametric test that measures whether observed scores differ significantly from expected scores for levels of a single variable.

ONE-TAILED TEST: A test in which significance of the result is based on deviation from the null hypothesis in only *one* direction.

OVERLAY PLOT: A type of scatter plot that graphs two or more variables along the horizontal axis against a single variable on the vertical axis.

PARAMETER: A numerical quantity that summarizes some characteristic of a population.

PARAMETRIC TEST: A statistical test that requires that the characteristics of the data being studied be normally distributed in the population.

PARTIAL: A term frequently used in multiple regression analysis. A partial effect is the unique contribution of a new variable after variation from other variables has already been accounted for.

PARTIAL CHI-SQUARE: The chi-square value associated with the unique additional contribution of a new variable on the dependent variable.

P(D/G): In discriminant analysis, given the discriminant value for that case (*D*), what is the probability of belonging to that group (*G*)?

PEARSON PRODUCT-MOMENT CORRELATION: A measure of correlation ideally suited for determining the relationship between two continuous variables.

PERCENTILE: A single number that indicates the percent of cases in a distribution falling below that single value. See Chapter 6 for an example.

P(G/D): In discriminant analysis, given that this case belongs to a given group (*G*), how likely is the observed discriminant score (*D*)?

PHI COEFFICIENT: A measure of the strength of association between two categorical variables, usually in a chi-square analysis. Phi is computed from the formula:

$$\phi = \sqrt{\chi^2 / N}$$

POOLED WITHIN-GROUP CORRELATIONS: *Pooled within group* differs from *values for the entire (total) group* in that the pooled values are the average (mean) of the group correlations. If the *N*s are equal, then this would be the same as the value for the entire group.

POOLED WITHIN-GROUPS COVARIANCE MATRIX: In discriminant analysis, a matrix composed of the means of each corresponding value within the two (or more) matrices for each level of the dependent variable.

POPULATION: A set of individuals or cases who share some characteristic of interest. Statistical inference is based on drawing samples from populations to gain a fuller understanding of characteristics of that population.

POWER: Statistical power refers to the ability of a statistical test to produce a significant result. Power varies as a function of the type of test (parametric tests are usually more powerful than nonparametric tests) and the size of the sample (greater statistical power is usually observed with large samples than with small samples).

PRINCIPAL-COMPONENTS ANALYSIS: The default method of factor extraction used by SPSS.

PRIOR PROBABILITY FOR EACH GROUP: The .5000 value usually observed indicates that groups are weighted equally.

PROBABILITY: Also called **SIGNIFICANCE**. A measure of the rarity of a particular statistical outcome given that there is actually no effect. A significance of $p < .05$, is the most widely accepted value by which researchers accept a certain result as statistically significant. It means that there is less than a 5% chance that the given outcome could have occurred by chance.

QUARTILES: Percentile ranks that divide a distribution into the 25th, 50th, and 75th percentiles.

R: The multiple correlation between a dependent variable and two or more independent (or predictor) variables. It varies between 0 and 1.0 and is interpreted in a manner similar to a bivariate correlation.

R^2: Also called the multiple coefficient of determination. The proportion of variance in the dependent (or criterion) variable that is explained by the combined influence of two or more independent (or predictor) variables.

R^2 CHANGE: This represents the unique contribution of a new variable added to the regression equation. It is calculated by simply subtracting the R^2 value for the given line from the R^2 value of the previous line.

RANGE: A measure of variability; the difference between the largest and smallest scores in a distribution.

RANK: Rank or size of a covariance matrix.

REGRESSION: In multiple regression analysis, this term is often used to indicate the amount of *explained* variation and is contrasted with **RESIDUAL**, which is *unexplained* variation.

REGRESSION ANALYSIS: A statistical technique designed to predict values of a dependent (or criterion) variable from knowledge of the values of one or more independent (or predictor) variable(s). See Chapters 15 and 16 for greater detail.

REGRESSION COEFFICIENTS: The *B* values. These are the coefficients of the variables within the regression equation plus the constant.

REGRESSION LINE: Also called the *line of best fit*. A straight line drawn through a scatter plot that represents the best possible fit for making predictions from one variable to the other.

REGRESSION PLOT: A scatter plot that includes the intercepts for the regression line in the vertical axes.

RESIDUAL: Statistics relating to the *un*explained portion of the variance. See Chapter 27.

RESIDUALS AND STANDARDIZED RESIDUALS: In log-linear models, the residuals are the observed counts minus the expected counts. High residuals indicate that the model is not adequate. SPSS calculates the adjusted residuals using an estimate of the standard error. The distribution of adjusted residuals is a standard normal distribution, and numbers greater than 1.96 or less than –1.96 are *not* likely to have occurred by chance ($\alpha = .05$).

ROTATION: A procedure used in factor analysis in which axes are rotated in order to yield a better simple structure and a more interpretable pattern of values.

ROW PERCENT: A term used with crosstabulated data. It is the result of dividing the frequency of values in a particular cell by the frequency of values in the entire row. Row percents sum to 100% in each row.

ROW TOTAL: The total number of subjects in a particular row.

RUNS TEST: A nonparametric test that determines whether the elements of a single dichotomous group differ from a random distribution.

SAMPLE: A set of individuals or cases taken from some population for the purpose of making inferences about characteristics of the population.

SAMPLING ERROR: The anticipated difference between a random sample and the population from which it is drawn based on chance alone.

SCALE MEAN IF ITEM DELETED: In reliability analysis, for each subject all the variables (excluding the variable to the left) are summed. The values shown are the means for all variables across all subjects.

SCALE VARIANCE IF ITEM DELETED: In reliability analysis, the variance of summed variables when the variable to the left is deleted.

SCATTER PLOT: A plot showing the relationship between two variables by marking all possible pairs of values on a bicoordinate plane. See Chapter 10 for greater detail.

SCHEFFÉ PROCEDURE: The Scheffé test allows the researcher to make pair-wise comparisons of means after a significant *F*-value has been observed in an ANOVA.

SCREE PLOT: A plot of the eigenvalues in a factor analysis that is often used to determine how many factors to retain for rotation.

SIGNIFICANCE: Frequently called **PROBABILITY**. A measure of the rarity of a particular statistical outcome given that there is actually no effect. A significance of $p < .05$ is the most widely accepted value by which researchers accept a certain result as statistically significant. It means that there is less than a 5% chance that the given outcome could have occurred by chance.

SIGN TEST: A nonparametric test that determines whether two distributions differ based on a comparison of paired scores.

SINGULAR VARIANCE-COVARIANCE MATRICES: Cells with only one observation or with singular variance-covariance matrices indicate that there may not be enough data to accurately compute MANOVA statistics or that there may be other problems present in the data, such as linear dependencies (where one variable is dependent on one or more of the other variables). Results from any analysis with only one observation or with singular variance-covariance matrices for some cells should be interpreted with caution.

SIZE: The number associated with an article of clothing in which the magnitude of the number typically correlates positively with the size of the associated body part.

SKEWNESS: In a distribution of values, this is a measure of deviation from symmetry. Negative skewness describes a distribution with a greater number of values above the mean; positive skewness describes a distribution with a greater number of values below the mean. See Chapter 7 for a more complete description.

SLOPE: The angle of a line in a bicoordinate plane based on the amount of change in the *Y* variable per unit change in the *X* variable. This is a term most frequently used in regression analysis and can be thought of as a weighted constant indicating the influence of the independent variable(s) on a designated dependent variable.

SPLIT-HALF RELIABILITY: A measure of reliability in which an instrument is divided into two equivalent sections (or different forms of the same test or the same test given at different times) and then intercorrelations between these two halves are calculated as a measure of internal consistency.

SQUARED EUCLIDEAN DISTANCE: The most common method (and the SPSS default) used in cluster analysis to determine how cases or clusters differ from each other. It is the sum of squared differences between values on corresponding variables.

SQUARED MULTIPLE CORRELATION: In reliability analysis, these values are determined by creating a multiple regression equation to generate the *predicted correlation* based on the correlations for all other variables.

SSCON = 1.000E–08: This is the default value at which iteration ceases in nonlinear regression. 1.000E–08 is the computer's version of 1.000×10^{-8}, scientific notation for .00000001 (one hundred-millionth). This criterion utilizes the residual sum of squares as the value to determine when iteration ceases.

STANDARD DEVIATION: The standard measure of variability around the mean of a distribution. The standard deviation is the square root of the variance (the sum of squared deviations from the mean divided by $N - 1$).

STANDARD ERROR: This term is most frequently applied to the *mean* of a distribution but may apply to other measures as well. It is the standard deviation of the statistic-of-interest given a large number of samples drawn from the same population. It is typically used as a measure of the stability or of the sampling error of the distribution and is based on the standard deviation of a single random sample.

STANDARDIZED ITEM ALPHA: In reliability analysis, this is the alpha produced if the included items are changed to *z*-scores before computing the alpha.

STATISTICS FOR SUMMED VARIABLES: In reliability analysis, there are always a number of variables being considered. This line lists descriptive information about the *sum* of all variables for the entire sample of subjects.

STEPWISE VARIABLE SELECTION: This procedure enters variables into the discriminant equation, one at a time, based on a designated criterion for inclusion ($F \geq 1.00$ is default) but will drop variables from the equation if the inclusion requirement drops below the designated level when other variables have been entered.

SUM OF SQUARES: A standard measure of variability. It is the sum of the square of each value subtracted from the mean.

***t*-TEST**: A procedure used for comparing exactly two sample means to see if there is sufficient evidence to infer that the means of the corresponding population distributions also differ.

***t*-TEST—INDEPENDENT SAMPLES**: A *t*-test that compares the means of two distributions of some variable in which there is no overlap of membership of the two groups being measured.

***t*-TEST—ONE SAMPLE**: A *t*-test in which the mean of a distribution of values is compared to a single fixed value.

***t*-TEST—PAIRED SAMPLES**: A *t*-test in which the same subjects experience both levels of the variable of interest.

***t*-TESTS IN REGRESSION ANALYSIS**: A test to determine the likelihood that a particular correlation is statistically significant. In the regression output, it is *B* divided by the standard error of *B*.

TOLERANCE LEVEL: The tolerance level is a measure of linear dependency between one variable and the others. In discriminant analysis, if a tolerance is less than .001, this indicates a high level of linear dependency, and SPSS will not enter that variable into the equation.

TOTAL SUM OF SQUARES: The sum of squared deviations of every raw score from the overall mean of the distribution.

TUKEY'S HSD: (Honestly Significant Difference). A value that allows the researcher to make pair-wise comparisons of means after a significant *F*-value has been observed in an ANOVA.

TWO-TAILED TEST: A test in which significance of the result is based on deviation from the null hypothesis in *either* direction (larger or smaller).

UNEQUAL-LENGTH SPEARMAN-BROWN: In split-half reliability, the reliability calculated when the two "halves" are not equal in size.

UNIVARIATE F-TESTS: An *F*-ratio showing the influence of exactly one independent variable on a dependent variable.

UNSTANDARDIZED CANONICAL DISCRIMINANT FUNCTION COEFFICIENTS: This is the list of coefficients (and the constant) of the discriminant equation.

VALID PERCENT: Percent of each value excluding missing values.

VALUE: The number associated with each level of a variable.

VALUE LABEL: Names or number codes for levels of different variables.

VARIABILITY: The way in which scores are scattered around the center of a distribution. Also known as variance, dispersion, or spread.

VARIABLE LABELS: These are labels entered when formatting the raw data file. They allow up to 40 characters for a more complete description of the variable than is possible in the 8-character name.

VARIABLES IN THE EQUATION: In regression analysis, after each step in building the equation, SPSS displays a summary of the effects of the variables that are currently in the regression equation.

VARIANCE: A measure of variability about the mean, the square of the standard deviation, used largely for computational purposes. The variance is the sum of squared deviations divided by $N - 1$.

WALD: In log-linear models, a measure of the significance of *B* for the given variable. Higher values, in combination with the degrees of freedom, indicate significance.

WILCOXON MATCHED-PAIRS SIGNED-RANKS TEST: A nonparametric test that is similar to the sign test except the positive and negative signs are weighted by the mean rank of positive versus negative comparisons.

WILKS' LAMBDA: The ratio of the within-groups sum of squares to the total sum of squares. This is the proportion of the total variance in the discriminant scores *not* explained by differences among groups. A lambda of 1.00 occurs when observed group means are equal (all the variance is explained by factors *other than* difference between these means), whereas a small lambda occurs when within-groups variability is small compared to the total variability. A small lambda indicates that group means appear to differ. The associated significance values indicate whether the difference is significant.

WITHIN-GROUPS SUM OF SQUARES: The sum of squared deviations between the mean for each group and the observed values of each subject within that group.

Z-SCORE: Also called *standard score*. A distribution of values that standardizes raw data to a mean of zero (0) and a standard deviation of one (1.0). A z-score is able to indicate the direction and degree that any raw score deviates from the mean of a distribution. *Z*-scores are also used to indicate the significant deviation from the mean of a distribution. A z-score with a magnitude greater than ±1.96 indicates a significant difference at $p < .05$ level.

REFERENCES

Four SPSS manuals cover (in great detail) all procedures that are included in the present book:

SPSS 13.0 Base User's Guide (2005). Chicago, IL: SPSS, Inc.

SPSS 13.0 Brief Guide (2005). Chicago, IL: SPSS, Inc.

SPSS 13.0 Regression Guide (2005). Chicago, IL: SPSS, Inc.

SPSS 13.0 Advanced Models (2003). Chicago, IL: SPSS, Inc.

Good introductory statistics texts that cover material through Chapter 13 (one-way ANOVA), and Chapter 18 (reliability):

Fox, James; Levin, Jack; & Harkins, Stephen. (1993). *Elementary Statistics in Behavioral Research*. New York: Harper Collins College Publishers.

Hopkins, Kenneth; Glass, Gene; & Hopkins, B.R. (1987). *Basic Statistics for the Behavioral Sciences*. Boston: Allyn and Bacon.

Moore, David; & McCabe, George. (1993). *Introduction to the Practice of Statistics, Second Edition*. New York: W.H. Freeman and Company.

Welkowitz, Joan; Ewen, Robert; & Cohen, Jacob. (1991). *Introductory Statistics for the Behavioral Sciences, Fourth Edition*. New York: Harcourt Brace Jovanovich.

Witte, Robert S. (1985). *Statistics, Second Edition*. New York: Holt, Rinehart and Winston.

Comprehensive coverage of Analysis of Variance:

Keppel, Geoffrey. (1973). *Design and Analysis: A Researcher's Handbook*. Englewood Cliffs, NJ: Prentice Hall.

Lindman, Harold R. (1992). *Analysis of Variance in Experimental Design*. New York: Springer-Verlag.

Schulman, Robert S. (1998). *Statistics in Plain English with Computer Applications*. New York: Van Nostrand Reinhold.

Comprehensive coverage of MANOVA and MANCOVA:

Lindman, Harold R. (1992). *Analysis of Variance in Experimental Design*. New York: Springer-Verlag.

Comprehensive coverage of simple and multiple regression analysis:

Chatterjee, Samprit; & Price, Bertram. (1997). *Regression Analysis by Example, Second Edition*. New York: John Wiley & Sons.

Gonick, Larry; & Smith, Woolcott. (1993). *The Cartoon Guide to Statistics*. New York: Harper Perennial.

Kerlinger, Fred N.; & Pedhazur, Elazar J. (1973). *Multiple Regression in Behavioral Research*. New York: Holt, Rinehart and Winston.

Schulman, Robert S. (1998). *Statistics in Plain English with Computer Applications*. New York: Van Nostrand Reinhold.

Sen, Ashish; & Srivastava, Muni. (1999). *Regression Analysis: Theory, Methods, and Applications*. New York: Springer-Verlag.

Weisberg, Sanford. (1985). *Applied Linear Regression, Second Edition*. New York: John Wiley & Sons.

Comprehensive coverage of factor analysis:

Comrey, Andrew L.; & Lee, Howard B. (1991). *A First Course in Factor Analysis*. Hillsdale, NJ: Lawrence Erlbaum Associates.

Gorsuch, R. L. (1983). *Factor Analysis*. Hillsdale, NJ: Lawrence Erlbaum Associates.

Comprehensive coverage of cluster analysis:

Everitt, Brian S. (1993). *Cluster Analysis, Third Edition*. London: Edward Arnold.

Comprehensive coverage of discriminant analysis:

McLachlan, Geoffrey J. (1992). *Discriminant Analysis and Statistical Pattern Recognition*. New York: John Wiley & Sons.

Comprehensive coverage of nonlinear regression:

Seber, G .A . F.; & Wild, C. J. (1989). *Nonlinear Regression*. New York: John Wiley & Sons.

Comprehensive coverage of logistic regression analysis and loglinear models:

Agresti, Alan. (1990). *Categorical Data Analysis*. New York: Wiley & Sons.

McLachlan, Geoffrey J. (1992). *Discriminant Analysis and Statistical Pattern Recognition*. New York: John Wiley & Sons.

Wickens, Thomas D. (1989). *Mulltiway Contingency Tables Analysis for the Social Sciences*. Hillsdale, NJ: Lawrence Erlbaum Associates.

Comprehensive coverage of nonparametric tests:

Siegel, Sidney; & Castellan, N. John, Jr. (1988). *Nonparametric Statistics for the Behavioral Sciences, Second Edition*. New York: McGraw-Hill.

Comprehensive coverage of multidimensional scaling:

Davison, M.L. (1983). *Multidimensional scaling*. New York: Wiley.

Young, F.W. & Hamer, R.M. (1987). *Multidimensional scaling: History, theory, and applications*. Hillsdale, Nj: L. Erlbaum Associates.

INDEX

H

I

J

K

L

M

S

T

U

V

W

Z

NOTES

NOTES

NOTES

NOTES